# RECENT TRENDS IN LIFE SCIENCES

# RECENT TRENDS IN LIFE SCIENCES

*Editors*

**M.H. Fulekar**

*Professor & Dean*

School of Environment and Sustainable Development
Central University of Gujarat
Gandhinagar, Gujarat

**R.K. Kale**

*Vice Chancellor*

Central University of Gujarat
Gandhinagar, Gujarat

**I.K. International Publishing House Pvt. Ltd.**

NEW DELHI • BANGALORE

*Published by*
I.K. International Publishing House Pvt. Ltd.
S-25, Green Park Extension
Uphaar Cinema Market
New Delhi–110 016 (India)
E-mail: info@ikinternational.com
Website: www.ikbooks.com

ISBN 978-93-82332-25-1



10 9 8 7 6 5 4 3 2 1



Published by Krishan Makhijani for I.K. International Publishing House Pvt. Ltd., S-25, Green Park Extension, Uphaar Cinema Market, New Delhi–110 016 and Printed by Rekha Printers Pvt. Ltd., Okhla Industrial Area, Phase II, New Delhi–110 020.

# PREFACE

*"The life sciences comprise the field of science that involves the scientific study of living organisms, like plants, animals and human beings. While biology remains the centerpiece of the life sciences, technological advances in molecular biology and biotechnology has led to a burgeoning of specializations and new, often interdisciplinary fields".*

Life sciences, research plays an important role in the R&D system, for appliances in the multidisciplinary areas, related to living creatures. It generates new knowledge and understanding that provide the foundation for applied research and development. The basic research provides reliable information on multidisciplinary area of the biology for application in the life system, more intensive knowledge creation through basic research could be seen as a way to enhance innovative activities. However, basic research has come under increasing pressure or the past decade, because of the short-term needs and economic priorities. Hence, there has been tendency to increase the share of applied research and development in the R&D.

Environmental science and forestry have evolved new methods for managing sustainable resources that will help our expanding population to pass on, more of its natural wealth to future generations. Medical science has provided fundamental understanding of the molecular basis of numerous diseases which has led to elimination of some and the containment of many. Advances in molecular biology not only have spawned the economically important biotechnology industry but have contributed fundamental knowledge about the structure of genes and the behaviour of biological macromolecules. These advances have yielded new insights into the relationships among organisms and into the continuum of structure and function that connects living and nonliving things. The long-range implications of all the rapidly evolving knowledge are hard to predict, but many additional benefits are now on horizon.

The life sciences advances go beyond molecular base science. Improvement also puts ahead behavioural sciences and ecology. The greatest advances of the recent past requirement combination of technology which include ecosystem and biodiversity study, biotechnology aspect, proteomic and genomic studies leading to metagenomic and the health care using the diagnosis, prevention of disorder and disease. The combination of multidisciplinary research in plants, animals, microorganisms their interaction and the molecular biology, genetic engineering approaches and their advances in cell biology research has takes the life sciences, to the greatest extent for the wide horizons of life. The present book deals with life sciences which includes biodiversity and its conservation, sustainable development, bioprospecting, carbon dioxide sequestration, microbial interactions, bioinformatics, bioelectromagnetics, phytoremediation, reuse and recycling of industrial waste, smog pollution, metagenomics, cancer research, immune system, sleep benefits and climate change.

This book deals with the recent trends in life sciences and will be beneficial for the postgraduates and research scholars of life sciences, environmental sciences, molecular biologist and biotechnologist.

**Prof. M.H. Fulekar**
**Prof. R.K. Kale**

# CONTRIBUTORS

**R.K. Kale**
Vice Chancellor
Central University of Gujarat, Gandhinagar, India

**M.H. Fulekar**
Professor and Dean
School of Environment and Sustainable Development
Central University of Gujarat, Gandhinagar, India

**Dheeren Panwar**
BNF & Stress Biology Laboratory
Department of Botany, J.N. Vyas University, Jodhpur-342001, Rajasthan, India

**Nisha Tak**
BNF & Stress Biology Laboratory
Department of Botany, J.N. Vyas University, Jodhpur-342001, Rajasthan, India

**H.S. Gehlot**
BNF & Stress Biology Laboratory
Department of Botany, J.N. Vyas University, Jodhpur-342001, Rajasthan, India

**Narpat Singh Shekhawat**
BNF & Stress Biology Laboratory
Department of Botany, J.N. Vyas University, Jodhpur-342001, Rajasthan, India

**Raghunath T. Mahajan**
Department of Biotechnology
KCE Society's, Moolji Jaitha College, Jalgaon-425001, Maharashtra, India.

**S.C. Garkoti**
School of Environmental Sciences
Jawaharlal Nehru University, New Delhi-110067
E-mail: sgarkoti@yahoo.com

**Shamkant B. Badgujar**
Department of Biotechnology,
KCE Society's, Moolji Jaitha College, Jalgaon-425001, Maharashtra, India.
E-mail: sham83badgujar@gmail.com

**Sibi Balachandran**
SIES Indian Institute of Environment Management
Sri. Chandrasekarendra Saraswati Vidyapuram, Plot No. 1E, Sector-V, Nerul, Navi Mumbai-400706

**Seema Mishra**
SIES Indian Institute of Environment Management
Sri. Chandrasekarendra Saraswati Vidyapuram, Plot No. 1E, Sector-V, Nerul, Navi Mumbai-400706
E-mail: seema.mishra03@gmail.com

**N. Kumar**
Department of Information Technology
School of Information Science and Technology, Babasaheb Bhimrao Ambedkar University, Lucknow (Central University), Raebareli Road, Lucknow-226025, UP, India

**M.Y. Khan**
Department of Information Technology
School of Information Science and Technology, Babasaheb Bhimrao Ambedkar University, Lucknow (Central University), Raebareli Road, Lucknow-226025, UP, India
Email: neerajmtech@gmail.com

**R.A. Khan**
Department of Information Technology
School of Information Science and Technology, Babasaheb Bhimrao Ambedkar University, Lucknow (Central University), Raebareli Road, Lucknow-226025, UP, India

**Mohammad Miransari**
Professor of Soil Science
Department of Soil Science, College of Agricultural Sciences, Shahed University, Tehran, Iran
Tel: 982151212469 E-mail: miransari1@gmail.com Miransari@Shahed.Ac.Ir

**Anwesha M. Bhaduri**
Research Scholar
Environmental Biotechnology Laboratory
Department of Life Sciences, University of Mumbai, Santacruz (East). Mumbai-400098

**A. Swabna**
Dept. of Environmental Biotechnology
School of Environmental Sciences, Bharathidasan University, Tiruchirapalli-620024

**Jyoti Fulekar**
M.Phil. Student
School of Environment and Sustainable Development, Central University of Gujarat
Sector–30, Gandhinagar

**Karthiga**
Dept. of Environmental Biotechnology
School of Environmental Sciences, Bharathidasan University, Tiruchirapalli-620024

**Leena Appavoo**
Department of Chemical and Environmental Engineering
University of Mauritius, Réduit, Mauritius
Email: ackmezchem@yahoo.co.uk,

**Ackmez Mudhoo**
Department of Chemical and Environmental Engineering
University of Mauritius, Réduit, Mauritius

**Sanyogita**
Jawaharlal Nehru University, New Delhi-110067

**Amit Prakash**
Jawaharlal Nehru University, New Delhi-110067

**V.K. Jain**
School of Environment Science
Jawaharlal Nehru University, New Delhi-110067
E-mail: vkj0400@mail.jnu.ac.in

**Rana P. Singh**
Cancer Biology Laboratory
School of Life Sciences, Jawaharlal Nehru University, New Delhi-110067, India
E-mail: ranaps@hotmail.com; ranasingh@cug.ac.in

**Mahendra Phulwaria**
Biotechnology Unit
Department of Botany, Jai Narain Vyas University, Jodhpur-342001, India

**Harish**
Biotechnology Unit
Department of Botany, Jai Narain Vyas University, Jodhpur-342001, India

**Amit Kumar Gupta**
Biotechnology Unit
Department of Botany, Jai Narain Vyas University, Jodhpur-342001, India

**Kheta Ram**
Biotechnology Unit
Department of Botany, Jai Narain Vyas University, Jodhpur-342001, India

**Smita**
Biotechnology Unit
Department of Botany, Jai Narain Vyas University, Jodhpur-342001, India

**Bhawana Pathak**
School of Environment and Sustainable Development
Central University of Gujarat, Gandhinagar, India
Email: bhawanasp@hotmail.com

**Vibha JB**
Biotechnology Unit
Department of Botany, Jai Narain Vyas University, Jodhpur-342001, India

**Manoj Kumar Rai**
Biotechnology Unit
Department of Botany, Jai Narain Vyas University, Jodhpur-342001, India

**Vinod Kataria**
Biotechnology Unit
Department of Botany, Jai Narain Vyas University, Jodhpur-342001, India

**Gurdeep Kaur**
Biotechnology Unit
Department of Botany, Jai Narain Vyas University, Jodhpur-342001, India

**Harchand Ram Dagla**
Biotechnology Unit
Department of Botany, Jai Narain Vyas University, Jodhpur-342001, India

**Vibha Madan**
School of Life Sciences
Jawaharlal Nehru University, New Delhi-110067, India

**Sushil K. Jha**
School of Life Sciences
Jawaharlal Nehru University, New Delhi-110067, India

**Sukrat Sinha**
Postdoctoral Fellow
Centre for Biotechnology, Nehru Science Complex, University of Allahabad, Allahabad-211002, U.P, India

**Kankal, N.C.**
National Environmental Engineering Research Institute (NEERI), Nehru Marg, Nagpur-440020, India
Email: nc_kankal@neeri.res.in

**Sharda Dhadse**
National Environmental Engineering Research Institute (NEERI), Nehru Marg, Nagpur-440020, India
Email: nc_kankal@neeri.res.in

**Anitha Kartha**
School of Life Sciences
Central University of Gujarat, Gandhinagar, India

**Jaya Sharma**
Research Scholar
Environmental Biotechnology Laboratory
Department of Life Sciences
University of Mumbai, Santacruz (East). Mumbai-400098

**M. Vasanthy**
Dept. of Environmental Biotechnology
School of Environmental Sciences, Bharathidasan University, Tiruchirapalli-620024.
Email: drvasanthy02@gmail.com

**Thamarai, C.**
Dept. of Biotechnology, Mother Teresa University for Women, Kodaikanal

**Ujjwal Kumar**
KNMI, Royal Netherlands Metrological Institute, Nederlands

**Pradip Bauri**
School of Environment and Sustainable Development
Central University of Gujarat, Sector–30, Gandhinagar, Gujarat

**Krishan Kumar**
Jawaharlal Nehru University, New Delhi-110067

**Arvinda Kumar Ragen**
Department of Chemical and Environment Engineering
University of Mauritius, Reduits, Mauritius

# CONTENTS

# 1

# Biodiversity and Its Conservation

**Bhawana Pathak and M.H. Fulekar**
*School of Environment and Sustainable Development*
*Central University of Gujarat*
*Gandhinagar-382030, Gujarat*

**ABSTRACT**

Biodiversity is the sum of genetic, species and ecosystem richness on the planet. Biological diversity has direct consumptive value in food, agriculture, medicine and industry. Biodiversity maintains the ecological balance and is responsible for continuous evolutionary processes. Conservation of biological diversity and sustainable use of natural resources are of vital importance for sustainable development. The ecological significance, economic important reasons for loss of biodiversity, conservation of biodiversity including Biological Diversity Act, biodiversity conservation strategy, both in situ and ex situ, have been discussed.

**Keywords:** Biodiversity, Conservation of Biodiversity, Conservation, Ecological, Economic Importance.

## INTRODUCTION

Biodiversity means: biological and diversity. Biodiversity refers to the variability among living organism from all sources including inter alias, terrestrial, marine and other aquatic ecosystems and ecological complexes of which they are part; this includes diversity within the species, between the species and of ecosystems. Biodiversity includes diversity of species, genetic material and ecosystem. Species diversity refers to the variety of different species such as plants, animals, and microorganisms. Species are the fundamental unit of biodiversity because without them the other part would not exist. Habitats are made up of groups of species, and they are lost or drastically changed when their components species become extinct. Genetic diversity corresponds to the variety of genes contained in plants, animals, fungi and microorganisms. It occurs within the species and between the species. The genetic variation also occurs in the amount of DNA per cell and in the number of chromosome within an individual. Both these factors also contribute to genetic diversity of organism. The fusion of genetic material from two separate individuals of a species during reproduction means that new combinations are formed that generate diversity. The real value of biodiversity lies in the information that is encoded in the genes and molecules. The greatest significance of genetic diversity is that it allows the process of evolution to take place. Biodiversity also includes the variety of ecosystems which make up the habitats or communities in which the organisms live. Ecosystem refers to all the different habitats or places that exist, like tropical or temperate forest, hot and cold deserts, wetlands, rivers, mountains, coral reefs, etc. Each ecosystem corresponds to a series of complex relationship between the biotic (living) components such as plants and animals, and abiotic (non- living) components, which include sunlight, air, water, minerals and nutrients.

Biodiversity is the sum of genetic, species and ecosystem richness on the planet. It is the fundamental basis for evolution and continuing adaption to changing circumstances. The present diversity of life on earth is a response to the planetary range of conditions, what we used to call ecological niches available for organisms. The diversity of nature is itself a spur to more diversity, and diversification is a fundamental attribute of life. Tomorrow's diversity will be the result of processes that build on what we have today, replacing some species by others and producing quite new forms. Diversity is thus the foundation of future, and is critically important to both integrity and productivity.

Conservation of biological diversity and the sustainable use of plant and animal species was highlighted in the first Global Comprehensive Agreement, namely the Convention on Biological Diversity signed in 1992 at the UN Earth Summit in Rio de Janeiro. Biodiversity conservation involves all the ecological processes that have maintained the area's biodiversity such as predation, pollution, parasitism, seed dispersal and complex interactions between several species of plants and animals. Protection requires specialized knowledge and/or solutions and an understanding of ecological processes. In order to bring about sustainable resource conservation and management, it is essential to adopt several different approaches for managing the biodiversity.

## IMPORTANCE OF BIODIVERSITY

Biological diversity has direct consumptive value in food, agriculture, medicine and industry. It also has aesthetic and recreational value. Biodiversity maintains the ecological balance and continuous evolutionary processes. The indirect ecosystem services provided through biodiversity are photosynthesis, pollination, transportation, chemical cycling, nutrient cycling, soil maintenance, climate regulation, air and water system management, waste treatment and pest control.

**Table 1** Ecological importance

| *Ecological Parameters* | *Ecological Implications* |
|---|---|
| Habitat | Trees provide habitat and food for birds, insects, other plants and animals, fungi and microorganisms. |
| Pollination | Insects, bats, birds and other animals serve as pollinators. |
| Natural population control | Parasites and predators act as natural population control. |
| Soil fertility | Various organisms such as earthworms and bacteria are responsible for recycling organic materials and maintaining the productivity of soils. |
| Respiration | Green plants take in carbon dioxide from the atmosphere and replenish it with oxygen. |
| Carbon sinks | Forests are particularly important "sinks" for the absorption of carbon dioxide and are the key factors in reducing global climate change. |
| Flood resistance — interface | Wetlands serve as a sponge to reduce the impact of floods and to cleanse streams by filtering sediments, nutrients and contaminants from inflowing water. |
| Restoration ecology | Bacteria are instrumental in the function of virtually all communities and ecosystems, yet they remain poorly understood. The restoration or rehabilitation of brown field or industrial waste sites typically involves deep mixing of the soil and planning vegetation. Natural processes then slowly result in some degree of a return towards a more fully functioning community and ecosystem. |

## Ecological Significance

The interaction of all these natural processes forms a complex web of life. If any part of this web suffers or breaks down, the future of other parts is threatened. Humans are, in many cases, degrading and destroying the ability of biological diversity to perform the services mentioned above. Diversity is not simply a passive response to environmental factors and species interaction; diversity itself can have significant effects on other ecological processes and properties. Some of the recent work explores the extent to which diversity controls community structure and ecosystem functioning.

## Economic Importance

Biodiversity is widely used as food pantry, genetic store house for biotechnology; it also provides critical indirect benefit to humans. This benefit encompasses ecosystem services. Natural ecosystems also help absorb the wastes we create and render them non-toxic. It is well known that biodiversity has provided us many benefits, some of which come in the form of goods that can be directly valued.

**Table 2** Economic importance of resources

| *Economic Parameter* | *Importance* |
|---|---|
| Food | Species are hunted and cultivated for agriculture. |
| Fuels | Timber, coal, oil and natural gases are natural resources used to produce energy. |
| Shelter and warmth | Timber and other forest products are used as building materials and for shelter. Fibres such as wool and cotton are used to make clothes. |
| Medicines | Processed drugs are obtained from biodiversity, e.g., penicillin is obtained from pencillium; codeine is obtained from poppies, digitalis from fox glove and quinine from the bark of cinchona. |
| Other goods | Minerals, oils and metals are obtained from raw materials provided by the earth's diversity. |

## INDIA'S BIODIVERSITY

India is one of the 12 mega biodiversity countries of the world. In about 70% of the total geographical area surveyed so far, 46,000 plant species and 81,000 animal species have been described. India with 2.4% of the world's area has 8.1% of the world's total biodiversity with a species count of about 0.130 million. Some salient features of India's biodiversity are mentioned in Table 3.

**Table 3** Present status of India's biodiversity

| *Biodiversity Criteria* | *Biodiversity Status* |
|---|---|
| Biodiversity hot spot | India has a rich variety of biodiversity. Two of the world's 18 hot spots of biodiversity are in India. |
| Species | India has 26 recognized endemic centres. Of the estimated 5-50 million species of the world's biota, only 1.4 million have been described and distribution is highly uneven. |
| Biodiversity centres/sites | India is one of the 12 centres of cultivated plants. It has 5 world heritage sites, 14 biosphere reserves and 6 Ransar wetlands. |
| Protected areas | The total protected areas is about 3.16 million $km^2$ (4.4% of the geographical area). |
| | The country has extensive ecosystems with a marine habitat extending over 7500 km of its coastline. |

**Table 3** Contd.

| *Biodiversity Criteria* | *Biodiversity Status* |
|---|---|
| Plant species | Of the 49,219 plant species, 5150 are endemic and distributed into 141 genera under 47 families corresponding to about 30% of the world's recorded flora, i.e., 30% of the world's recorded floras are endemic to India. |
| Animal species | With 81,25 animal species, India accounts for 6.67% of the world's total animal species |
| Agro biodiversity | India's record in agro biodiversity is equally impressive; there are 166 crop species and 320 wild relatives. India is seventh in terms of contribution to world agriculture. |

## LOSS OF BIODIVERSITY

Biodiversity is a new name for species richness (plants, animals and microorganisms) occurring as an interacting system in a given habitat. All forms of life are so clearly interlinked that any disturbance in one gives rise to imbalance in the other. Basically, the problem of biodiversity is one of conflict resolution between man on one side, and most living organisms occurring on land, in freshwater bodies, and the marine environment on the other side. Biodiversity is an irreplaceable resource. Its extinction is forever; at present there is no way to recreate extinct plants and animals. Our own species is subjected to evolutionary processes as any other life in the struggle for survival. Losses of biodiversity have generally been improved, viz., natural and man-made sources. The natural causes include floods, earthquakes, landslides, lack of pollination and diseases. The man-made threat to biological diversity is due to destruction of the ecosystem, especially in the tropics, and the disappearance of habitats in the wake of developmental activities like industrialization, urbanization, overgrazing, overexploitation, fire, etc. Habitat losses due to human intervention for various reasons are the prime concern for species extinction. The main factors that threaten biodiversity are given in Table 4.

**Table 4** Loss of biodiversity

| *Reasons for Loss of Biodiversity* | *Impact on Biodiversity* |
|---|---|
| **Habitat loss and destruction** | The population growth has increased the human activities usually responsible for habitat loss and destruction of species, for example, cutting of forests, digging of mines, construction of buildings, and construction of roads destroy habitats resulting in loss of species, population and ecosystem. Species living in the habitats compete with each other and humans for limited resources. Some species population is very small; they have fewer mates, with whom to have offspring, diminishing the genetic diversion of their populations. Small species have greater difficulties migrating from one habitat to another habitat hence there is destruction in the ecosystem. |
| **Introduction of alien (exotic or non-native) species** | The numbers of species introduced to areas of which they are not native are expected to increase as the scale of international trade, transportation and tourism continues to grow. This can disrupt the |

**Table 4** Contd.

| *Reasons for Loss of Biodiversity* | *Impact on Biodiversity* |
|---|---|
| | entire ecosystem and have a major impact on the populations of native plants and animals. Invades can affect native species by eating them, infecting them, competing with them or mating with them. Invasion can happen through many different ways: marine species are transported around the world through ballast water from ships and insects can be introduced through the international trade of food or timber. |
| **Pollution and contaminants** | Industrialization, urbanization, energy generation, developmental activities generate waste (solid waste, nuclear waste causing depletion and destruction of environment causing pollution of soil, water leading to damaging the species habitat. The increased agricultural activities involving use of pesticides and chemical fertilizers causes chemical pollution that affects all levels of biodiversity. |
| **Population Growth** | Increasing population on the earth demands for more utilization of natural resources to meet their requirements such as food, water, medicines, clothes, shelters and fuels. This causes impact on species and habitat of the species. |
| **Overexploitation** | Biodiversity is important both for the health of the biosphere and for agriculture, animal husbandry, fisheries, forestry and the pharmaceutical industry, which is overexploited. Overexploitation of natural resources can lead to depletion and disappearance of the species. Many of the world's natural resources are being used by humans faster than they can replace themselves. |
| **Climate change** | The use of fossil fuels, such as coal, oil, natural gas used for energy generation releases greenhouse gases causing the global warming. Methane and nitrous oxide released in the atmosphere by the human activities also contributes to the global warming. Carbon dioxide ($CO_2$) acts like a greenhouse gas letting in the sunlight and heat into the lower atmosphere but not letting the heat out, resulting into heat trapping on the earth surface leading to green house effect. As a result the average temperature on the earth is increasing. Changes in the climate modify the ecosystem in which living things exist. Some species and populations become extinct if they are unable to adapt to the changed weather conditions or relocate to suitable habitats, while other species such as certain microbes and pests may flourish and expand their newer varieties. Climate change will alter the environmental conditions, which cause the impact on the ecosystem resulting into loss of biodiversity. |
| **Species—Ecological theory** | Conserving biological diversity equals conserving ecosystems. Species do not exist alone; in addition to being genetically diverse populations, they are members of a community and have a distinct niche and role within it. Each species is dependent on others within the community in some way. A species is gained by migration from outside or from production of new species within the region. Species loss could be due to: extinction, unpredictable drastic changes in physical conditions, stochastic events, and biological factors including competition, predation, pathogens and disposal agents. Species diversity is a product of the history of species accumulation and disappearance. |

## CONVENTION OF BIODIVERSITY

The Convention on Biological Diversity (CBD) initiated at the Rio Earth Summit in June 1992 was certainly a great event in the history of global biodiversity conservation, particularly for developing countries. This convention has become almost an international law as far as biodiversity and its global utilization is concerned. The three objectives of the convention are:

- conservation of biological diversity.
- sustainable use of components of biological diversity; and
- fair and equitable sharing of benefits arising out of utilization of genetic resources.

**Table 5** Relevant articles of the Convention on Biological Diversity (CBD).

| *Article No.* | *Main Relevant CBD Articles* |
|---|---|
| *Article 1.* | *Objectives* |
| *Article 2.* | *Use of Terms* |
| *Article 3.* | *Principle* |
| *Article 4.* | *Jurisdictional Scope* |
| *Article 5.* | *Cooperation* |
| *Article 6.* | *General Measures for Conservation and Sustainable Use* |
| *Article 7.* | *Identification and Monitoring* |
| *Article 8.* | *In situ Conservation* |
| *Article 9.* | *Ex situ Conservation* |
| *Article 10.* | *Sustainable Use of Components of Biological Diversity* |
| *Article 11.* | *Incentive Measures* |
| *Article 12.* | *Research and Training* |
| *Article 13.* | *Public Education and Awareness* |
| *Article 14.* | *Impact Assessment and Minimizing Adverse Impacts* |
| *Article 15.* | *Access to Genetic Resources* |
| *Article 16.* | *Access to and Transfer of Technology* |
| *Article 17* | *Exchange of Information* |
| *Article 18.* | *Technical and Scientific Cooperation* |
| *Article 19.* | *Handling of Biotechnology and Distribution of its Benefits* |
| *Article 20.* | *Financial Resources* |
| *Article 21.* | *Financial Mechanism* |
| *Article 22.* | *Relationship with Other International Conventions* |
| *Article 23.* | *Conference of the Parties* |
| *Article 24.* | *Secretariat* |
| *Article 25.* | *Subsidiary Body on Scientific, Technical and Technological Advice* |
| *Article 26.* | *Reports* |
| *Article 27.* | *Settlement of Disputes* |
| *Article 28.* | *Adoption of Protocols* |
| *Article 29.* | *Amendment of the Convention or Protocol* |
| *Article 30.* | *Adoption and Amendment of Annexes* |
| *Article 31.* | *Right to Vote* |
| *Article 32.* | *Relationship between this Convention and Its Protocols* |
| *Article 33.* | *Signature* |
| *Article 34.* | *Ratification, Acceptance or Approval* |
| *Article 35.* | *Accession* |

**Table 5** Contd.

| *Article No.* | *Main Relevant CBD Articles* |
|---|---|
| *Article 36.* | *Entry Into Force* |
| *Article 37.* | *Reservations* |
| *Article 38.* | *Withdrawals* |
| *Article 39.* | *Financial Interim Arrangements* |
| *Article 40.* | *Secretariat Interim Arrangements* |
| *Article 41.* | *Depositary* |
| *Article 42.* | *Authentic texts* |

## BIOLOGICAL DIVERSITY ACT

India is a Party to the Convention on Biological Diversity (1992), which recognizes the sovereign rights of the States to use their own biological resources; the convention facilitates access to genetic resources. The Convention on Biological Diversity (Article 8j) recognizes contributions of local and indigenous communities to the conservation and sustainable utilization of biological resources through traditional knowledge, practices and innovations, and provides equitable sharing of benefits with such people.

Biodiversity is a multidisciplinary field with diverse activities and actions. The stakeholders in biological diversity include the Central Governments, institutions of local self-governmental organizations, industry, etc. India has one of the major challenges in adopting objectives of equitable sharing of benefits enshrined in the Convention on Biological Diversity. The Central Government has brought Biological Diversity Act, 2002, with the following salient features:

i. To regulate access to biological resources of the country with the purpose of securing equitable share in benefits arising out of the use of biological resources, and associated knowledge relating to biological resources.
ii. To conserve and sustainably use biological diversity.
iii. To respect and protect the knowledge of local communities related to biodiversity.
iv. To secure sharing of benefits with local people as conservers of biological resources and holders of knowledge and information relating to the use of biological resources.
v. Conservation and development of areas of importance from the standpoint of biological diversity by declaring them as biological diversity heritage sites.
vi. Protection and rehabilitation of threatened species.
vii. Involvement of institutions of state governments in the broad scheme of implementation of the Biological Diversity Act through constitution of committees.

The Act provides for conservation of biological diversity, sustainable use of its components, and fair and equitable sharing of the benefits arising out of utilization of genetic resources and to give effect to the Convention, knowledge and for matters connected therewith. The Act is called the Biological Diversity Act (2002) and extends to the whole of India.

India is rich in biological diversity and associated traditional and contemporary knowledge system, and is a party to the United Nations Convention on Biological Diversity signed at Rio de Janeiro on 5th day of June, 1992.

The Act states that ***"Biological Diversity" means the variability among living organisms from all sources and the ecological complexes of which they are part and includes diversity within the species or between the species and of ecosystems.***

The regulations of access of biological diversity states that:

- To undertake biodiversity related activities with the approval of National Biodiversity Authority.
- The result of research permitted to transfer only with the approval of Biodiversity Authority.
- Provision permit for collaborative research projects involving biological research between the institutions.
- Application for intellectual property rights to be made with the approval of National Biodiversity Authority.
- Prior intimation to State Biodiversity Board for obtaining biological resources for certain purposes.

Established National Biodiversity Authority regulates the function under the Biodiversity Act and Rules.

The function and power of the national Biodiversity Authority includes the following:

- Regulate access to biological resources and for equitable benefit sharing.
- The National Biodiversity Authority may approve for undertaking any activity related to biodiversity.
- Advise to Central Government on matters related to the conservation of biodiversity, sustainable use of its components, and equitable sharing arising out of the utilization of biological resources.
- Advise the State Governments in the selection of areas of biodiversity important as heritage sites, and measures for the management of such heritage sites.
- To perform some other functions related to biodiversity for carrying out the provisions of the Act.
- The National Biodiversity Authority on behalf of the Central Government takes necessary measures for the grant of IPR on biological resources in any country outside India.

## BIODIVERSITY CONSERVATION

Conservation and sustainable use of biological resources is essential to avoid or minimize adverse impacts on biological diversity. It requires protecting and encouraging the use of biological resources in accordance with traditional cultural practices that are compatible with conservation or sustainable use requirements. The knowledge of local population should be considered to develop and implement remedial action in degraded areas where biological diversity has been reduced. For sustainable use of biological resources, there is a need to encourage governmental authorities and private sector for developing measures for conservation of biological resources.

The specific goals of biodiversity conservation include the following:

- Measures for conservation of all natural ecosystems such as forests, grasslands and wetlands and their species (wild and domestic).
- Measures to conserve the natural and cultural diversity for maintenance and/ or recovery of viable populations of species in their natural habitat.
- Measures for conservation of floral and faunal genetic resources.
- Measures for sustainable use of natural resources to conserve biodiversity and ecological processes so as to meet the rational needs of people of the present and future generation.
- Measures to develop incorporation of biological diversity conservation with developmental planning and implementation.
- Measures for biodiversity conservation and management related to traditional knowledge, practices and use of resources of indigenous communities for their protection and promotion.
- Measures for conservation of biodiversity through ecosystems and landscape approach so as to reconcile with human activity in a way that is sustainable through time and space.
- Measures for integration of biodiversity conservation and natural resources management with cross-sectoral planning.
- Measures for forest resource management with emphasizing the primary role of forests in maintenance of ecological processes and hydrological systems.
- Measures the productivity enhancement of plantation, afforestation in degraded lands, and efficiency of farm forestry and the use of renewable materials.
- Measures for empowerment of public and women for sustainable management of biodiversity.
- The dissemination of knowledge for awareness among all the sections of society to understand the importance of biodiversity.

## Conservation Methods

The earth is a home to a rich and diverse array of living organisms, whose genetic diversity and relationship with each other constitute planet's biodiversity. It includes species as well as varieties found within the same species, as also the enormous ecosystems and habitats—forests, grasslands, oceans and its depths, peats and peat bogs. One-fourth of the earth's total biological diversity, amounting to about 1.6 million species, which might be useful to mankind in one way or other, is in the serious risk of existence, and if we ignore the biodiversity crisis, we would lose these species in the next 25 years. The challenge lies before the individuals, organisations and government agencies for nations and the world as a whole to protect and enhance biological diversity, while continuing to meet people's need for natural resources. If these goals are not met, the future generation will live in a biologically impoverished world.

The conservation can only be achieved through the concerted efforts of conservation both in nature (in situ) and off site (ex situ). "The conservation of species/populations within their natural habitat is termed in situ, while ex situ efers to conserve species/populations outside their natural habitats".

### *In situ conservation*

In situ conservation, using approaches such as legal protection and habitat management is the most satisfactory, cheapest, and often the only means for maintaining biological diversity. Destruction of natural habitats is the main threat to biodiversity, therefore, it is the necessity to conserve natural habitats. In situ conservation can fulfill this requirement. For many species, in situ conservation is the most appropriate method of conserving gene pool; also maintains evolutionary processes.

The conservation of natural populations includes habitat management, creation of protected areas and by limiting encroachment, overgrazing or drainage. Before appropriate conservation techniques, appropriate data about the vegetation are required. The number of species, populations and individuals, the habitat type, threat to the area, the changing status of each species and soil type must be considered. For the individual species inventory, survey, and monitoring are most important information to be gathered and used. An inventory leads to a field survey to determine the habitat requirement and biological relationships of the species, population sizes and reproductive success. An inventory and survey are usually species-based and can be followed by demographic or ecophysiological studies concentrating on particular populations and individuals. Monitoring over several years reveals whether a population is declining or increasing, and in the form of situation, its cause can be further investigated, and only then appropriate action can be taken.

### *(a) Protected areas*

Protected areas are regarded as ultimate repositories of biological diversity. Under these sites, natural floras and faunas are protected. These sites are established on the principles of in situ conservation. Various rare and endangered plants are conserved in these sites. Currently there are 578 PAs in India covering 1,54,572.80 sq. km area. Under the Wildlife Protection and Indian Forest Act, National Parks (NPs) and Wild life Sanctuaries (WLSs) are recognized in India (Table 6).

**Table 6** Protected areas in Uttarakhand

| Protected Areas | In India | | In Himalaya | | In Uttarakhand | |
|---|---|---|---|---|---|---|
| | Total No. | Area (sq.km) | Total No. | Area (sq.km) | Total No. | Area (sq.km) |
| NPs | 89 | 37,530.76 | 12 | 6498.09 | 06 | 4077.00 |
| WLSs | 489 | 117,042.04 | 56 | 13978.65 | 06 | 2395.52 |

Source: WII, 2002

### (i) National Parks

National parks are given a higher level of protection, considering no grazing is permitted within them and it is specified that no person shall destroy, exploit or remove any wildlife from national parks or destroy or damage the habitat or any wild animal or deprive any wild animal of its habitat within such national park, except under and in accordance with a permit granted by the Chief Wildlife Warden and no such permit shall be granted unless the state government, being satisfied that such destruction, exploitation or removal of wildlife from national park is necessary for the improvement and better management of wildlife therein, authorizes the issue of such permit.

In these sites, certain species are focused for conservation. Here all the human activities are prohibited except tourism. The boundary of a national park is recognized by the state legislation; it is not based on scientific principles. There are 89 national parks in India, out of which 6 are in Uttarakhand (Table 7).

**Table 7** National parks in Uttarakhand

| S. No. | National Parks | Location (District) | Area covered (sq. km) | Importance |
|---|---|---|---|---|
| 1. | Corbett | Nainital and Pauri | 520.80 | Tiger, 50 sp. Mammals, 580 kinds of birds, 25 reptile spp., and crocodile. |
| 2. | Rajaji | Dehradun and Haridwar | 820.00 | Wild elephants, 812 sp. of Angiosperms. |
| 3. | Govind | Uttarkashi | 472.08 | Musk spotted deer, snow leopard, snow cock. |
| 4. | Gangotri | Uttarkashi | 1552.00 | Snow leopard, black bear, brown bear, musk deer, blue sheep, Himalayan tahr, Himalayan monal. |
| 5. | Valley of Flowers | Chamoli | 87.50 | Medicinal and flowering plants. |
| 6. | Nanda Devi | Chamoli | 624.62 | Bharal, musk deer, tatir, serow, snow leopard. |

### (ii) Wildlife Sanctuaries

Sanctuaries are accorded a lesser level of protection; for them certain types of activities might be permitted. In these sites, wildlife conservation is emphasized. People get some condition-based rights for doing some activities. In India there are 489 wildlife sanctuaries, out of which 6 are in Uttarakhand (Table 8). Most of the sanctuaries in India, like other parts of the world, were set up for protecting rare mammals, birds, etc., but rarely if ever to conserve a plant species.

**Table 8** Wildlife sanctuaries in Uttarakhand

| *S. No.* | *Wildlife sanctuaries* | *Location (District)* | *Area covered (sq. km)* | *Importance* |
|---|---|---|---|---|
| 1. | Govind Pashu Vihar | Uttarkashi | 481.00 | Musk spotted deer, snow leopard. |
| 2. | Kedarnath | Chamoli | 957.00 | Snow leopard, snow cock, musk spotted deer. |
| 3. | Mussoorie | Dehradun | 10.82 | Himalayan goat, panther Himalayan bear, mountain quail. |
| 4. | Askot | Pithoragarh | 599.93 | Musk deer, tahr, serow, snow leopard, bharal. |
| 5. | Sonanadi | Pauri | 301.18 | Tiger, cheetal. |
| 6. | Binsar | Almora | 45.59 | Leopard, civet cat, musk deer. |

**Askot Wildlife Sanctuary**

Askot Wildlife Sanctuary (20°46′45″ – 30°27′45″ N and 81°01′53″ – 80°16′25″), located in the north of the Pithoragarh district of state Uttarakhand, covers an area of 599.93 km$^2$ and has a wide altitudinal range (600 – 6905 m). Surrounded by the mountains and forest, this picturesque green Askot Sanctuary is a best dwelling place of wild animals. This sanctuary has been set up primarily with the objective of conserving the musk deer and its habitat.

**Species Found:** Leopard, Black Bear, Spotted Deer, Snow Cock, Kakar, Tahr, Musk Spotted Deer, Chukoras, Pheasant, Serow and a number of species of birds.

The vegetation of the area mainly comprises subtropical, temperate, subalpine and alpine types.

The main forest types found in this sanctuary are listed below:

1. Ban and Moru oak forests
2. Moist deodar forest
3. Temperate mixed coniferous forest
4. Temperate mixed deciduous forest
5. Kharsu oak forest
6. Fir and spruce forest
7. Blue pine forest
8. Dry temperate mixed forest
9. Dry deobar forest
10. Sub-alpine birch fir forest
11. Birch-rhododendron scrub forest
12. Moist alpine scrub forest
13. Dry alpine scrub forest

### *(b) Biosphere reserves*

The word Biosphere means "the sphere or envelope surrounding the earth in which life exists naturally". The concept of biosphere reserves is an approach for maintaining the integrity and biological support system for man and nature throughout the whole biosphere, which represents a natural undisturbed or least disturbed area of representative ecosystems, and may also include centres of endemism and genetic richness. Biosphere reserves are wide ranging ecosystems protected for conserving the balance of nature. As such it involves conservation, restoration and acquisition of knowledge, which can play the role of meeting scientific, economic, educational, cultural and recreational needs. They also represent means for maintaining the gene pools of species of plants, animals and microorganisms in totality by setting aside representative area of wide ranging natural ecosystem throughout the world for conservation and research.

The main objective in the establishment of biosphere reserves is to provide in situ conservation of plants, animals and microorganisms in their totality as part of the wider ecosystem. In India there are twelve biosphere reserves covering 50,225.94 sq. km area (Table 9); out of which one BR is in Uttarakhand, known as Nanda Devi Biosphere Reserve (NDBR).

**Table 9** Comparison of distribution of BRs

| *In India* | | *In Himalaya* | | *In Uttarakhand* | |
|---|---|---|---|---|---|
| *Total No.* | *Area (sq.km)* | *Total No.* | *Area (sq.km)* | *Total No.* | *Area (sq.km)* |
| 13 | 50,225.95 | 05 | 13,570.16 | 01 | 2236.74 |

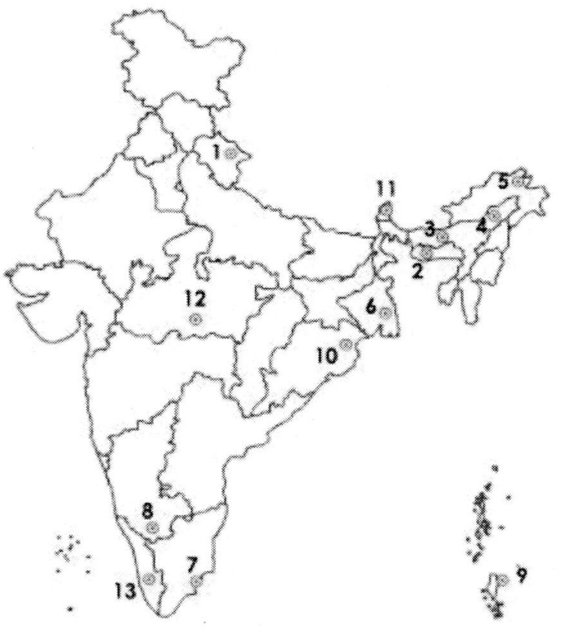


**Fig. 1.** Indian biosphere reserves.

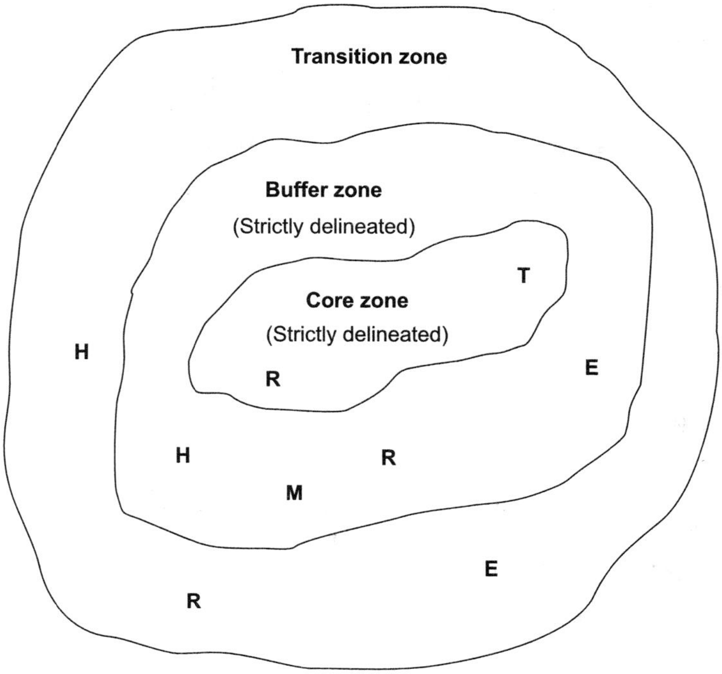


**Fig. 2.** Zones of biosphere reserves.

A biosphere reserve consists of three zone: (1) core zone, (2) buffer zone, and (3) transition zone (Fig. 1.2).

**Core Zone:** Core zone is surrounded by the buffer zone. It contains viable populations of wide ranging species. There is a strict ban on any human activity in this zone. This zone is meant for research, education, demonstration, development of proper techniques for eco-restoration and other manipulation land use.

**Buffer Zone:** Buffer zone is a multiple use area, which acts as a protective barrier for the core zone and may contain a variety of agriculture, horticulture, forestry and other activities to be developed in cooperation with the conservation and research function of the biosphere resources.

**Transition Zone:** Outside the buffer zone, there may be flexible transition zone or area of cooperation, which is considered a zone of cooperation that ties the resources to the surrounding region and thus fulfils the motto 'conservation is development'. People living in this zone pursue forestry, farming, fishing, animal rearing, etc., that is compatible with the conservation of core and buffer zone.

**Nanda Devi Biosphere Reserve (NDBR), Uttarakhand**

It is located in the state of Uttarakhand and is spread over the high altitudes of three districts of Chamoli, Bageshwar and Pithoragarh in Garhwal and Kumaon region. It was established in 1988. It occupies an area of 2236.47 sq. km. It encircles two national parks, namely, Nanda Devi National Park and the world famous Valley of Flowers National Park. Some of the groups of fauna and flora conserved in the NDBR are as follows:

| *Tree species* | 341 |
|---|---|
| Herbs and shrubs species | 552 |
| Grasses | 80 |
| Mammals species | 29 |
| Birds species | 228 |
| Reptiles and amphibian species | 03 |
| Mollusk | 14 |
| Insects | 229 |

Source: HBR, Vol 3, 2001

Various species of animals like snow leopard, musk deer, Himalayan tahr, bharal, black bear, brown bear and different species of birds like monal pheasants, snow cock, golden eagle, black eagle, etc. are also preserved in this biosphere reserve. There are 47 villages under the buffer zone of NDBR. The occupations of residents of these villages are agriculture, medicinal plant cultivation, sheep rearing and tourism.

### (c) Sacred grooves (SG)

Apart from the PA, there is another means for conservation of biological diversity which is completely in the control of local communities, and these communities protect patches of forests by dedicating them to deities or ancestral spirits. Those sites are also conserved due to some religious cause by the local people in a particular area. This mechanism of conservation of biodiversity is known as sacred grooves. It is not a physical unit, but is said to be a multidimensional social unit. Due to religious touch, the local people maintain these sites. Trees like Peepal, Kalpvriksha, Banyan, etc. are considered of great mythological value in our society since ancient times. At present, such places are depleting because of lack of cultural and religious beliefs among the people. This is a challenge for the young generation to maintain their cultural identity in the modern world, so that our traditions remain for long. It is estimated that about 13,720 SGs have been reported from different states of India, which cover an area of about 33,000 ha (0.01%) of the total geographical area of the country.

### *Ex situ conservation*

Ex situ conservation is fundamental for maintaining the wild relatives of crop plants and traditional crop varieties. Whenever species or population is highly threatened, ex situ conservation has a role to play. The rapid loss of natural habitat and the acceleration of climate change means that to protect species for the future, ex situ conservation will become more important. The conservation outside the natural habitat will be necessary to compliment species conservation in the wild. Ex situ activities should be undertaken as far as possible and as appropriate and predominantly for the purpose of complimenting in situ measures. This complimentary rule is becoming highly important because of the present environmental crisis.

### *Conservation in botanical gardens*

Botanic gardens play a pivotal role in conservation, propagation, public education and sustainable utilization of genetic resources. Conservation of many economically important plants in their natural habitat is not always possible or feasible. Genetic erosion has resulted in the decline of population size of many economically important species. This step needs to be taken to collect and maintain the germplasm of these plants in botanical gardens for effective conservation.

Some botanical gardens are fortunate enough to have large natural areas within their properties. In India, about 150 botanical gardens are mentioned under various sectors. Botanical gardens serve as excellent ex situ conservation centres of threatened, rare endemic, ornamental and medicinal plants.

### Zoos

Zoos contribute to the conservation of animal diversity by propagating and reintroducing endangered species. They also act as centres for research to improve the management of captive and wild population. There are more than 800 zoos around the world with about 3000 species of mammals, birds, reptiles, amphibians, etc. Many of them have well developed captive breeding programmes for restoring those species of animals whose number of surviving species is so small that there is no realistic chance of in situ survival. As the number of individuals increases in the breeding, they are selectively released in the wild. Zoos also function for offsite collection to be used to restore depleted populations, and reintroduction of species. Till March, 2006, there were 159 recognized zoos in India.

### Aquaria

Aquaria like botanical gardens and zoos also play a prominent role in restoring degraded ecosystems, reintroducing species in the wild, and restocking of depleted populations of fishes and other aquatic animals. Role of aquaria in captive propagation of threatened freshwater species is significant. The world conservation union (IUCN) is mounting a major effort to develop captive breeding programs for endangered fish species, starting from Lake Victoria, the desert fishes of North America and Appalachian fishes.

### Seed Banks

The most important method of ex situ conservation is through seed banks because they enable the conservation of adequate population of the species banked. For the majority of seed plant species, seed banking works well (70%). The drying and freezing of orthodox seeds enables them to be stored for centuries and remain viable. Seeds of the majority of these species are small and compact and so they occupy little space and can be stored efficiently. Thirty per cent of the species with recalcitrant seeds require other methods of storage or in situ conservation. The future of recalcitrant species is much less secure than those, which have orthodox seeds. Some success is already being achieved through the storage of embryos and tissue through cryogenic storage. The seeds of many rare and endangered plants can be collected and stored in seed banks under low temperature so that the seed will remain viable for a very long period. The seeds thus stored can be utilized when a crisis situation arises and the species can be served.

#### *In vitro conservation*

The storage of plant tissue as callus or in liquid culture can often be used for recalcitrant species. But it is generally of much shorter duration than for seeds and requires much more attention and maintenance. It is also the only way to store species that have lost their means of sexual reproduction and are only propagated vegetatively (e.g., banana). Since in vitro cultures are often stored in condition of slow growth, some somaclonal variation occurs with the disadvantage that these variations are not necessarily appropriate for environmental change. The best way to store tissue, which suspends growth, is through cryogenic storage.

#### *DNA storage*

A more recent method of storing genetic material is to store DNA extract from plants. While this may preserve individual genes, it is presently not possible to regenerate new plant from DNA. However, useful gene can be extracted and used through the techniques of genetic engineering.

#### *Pollen storage*

Another storable plant product is pollen, the male gamete of plants. Since it is a haploid material, it can be used only from crossing the living plant material, so it is not useful as the only way to store germplasm. It could be possible in future to generate haploid plant for pollen. Pollen culture has the disadvantage of storing only paternal material, so the maternal site is not conserved by this method. An advantage is that pollen is relatively easy to store at low cost.

### *Gene bank*

Certain institutions are also actively involved in the conservation of germplasms through gene banks of crops, economically important plants, and domesticated animals. Some such important gene banks are maintained by NBPGR (New Delhi), NBAGR (Karnal), NBFGR (Lucknow), which are set up for ex situ conservation of plants and animals under ICAR, an autonomous body established under Government of India. In addition, the Indian National Gene Bank (INGB) was established for conserving the national heritage of germplasm collection in the form of seeds, vegetative propagation, tissue cell culture, embryo and gametes.

## NEW CONSERVATION STRATEGIES

Sustainable utilization and community involvement requires creating new types of legally recognized reserves.

### Community Reserves

Community reserves will protect landscape, ecosystems and species traditionally or culturally conserved by local communities. These reserves will be limited to private and community land, and not government owned area.

### Community-Oriented Approaches

An important factor contributing to the endangered status of many species is the anthropogenic activity. Populations dependent on local forests also contribute to species endangerment by overexploiting forest produce in the absence of alternative livelihood. There is a need for involving people and their knowledge based protection and management of biodiversity. Local communities who influence conservation, management and utilization of biodiversity are the major stakeholders in its conservation.

### Drawing from Local Values, Knowledge and Experience

There is an abundant evidence that ecosystem people exhibit a number of cultural practices that promote sustainable use of biological resources and conservation of biodiversity. Local communities could help in managing conservation of biodiversity in the areas to meet their bonafied livelihood needs.

### Rendering Civil Society more Responsive

The motivation of communities is possible to protect and conserve biodiversity through control over their resource base, adequate management, incentives to conserve biodiversity and equity in bearing the cost of conservation.

### Harnessing Voluntary Action

Some voluntary field based organizations and NGOs have the flexibility to develop, test and carry out innovative approaches to work with the communities, especially women and the poor. They have a role in policy advocacy information gathering and applied research. They have the ability to raise public awareness and provide technical assistance and other activities. NGOs could complement or supplement government efforts and also plan the very important role of influencing public policies and programmes.

## SCIENCE OF BIODIVERSITY

Biodiversity science is an applied science. It is directed mostly towards the problems associated with the loss of biodiversity. It does not mean that the field does not offer the excitement and joy of doing pure science. In the modern discipline of conservation biology, after more than a decade and a half since the emergence of conservation biology as a distinct area of study, very little work was being done in the field that actually helps in conservation. Much of the research in the field has been basic rather than applied (Ehrenfield, 2000). The difference between basic—pursuing knowledge for knowledge sake, and applied—pursuing knowledge to solve societal problems is not unique to biodiversity science. It is common to many other fields and will remain as long as we face pressing problems that we believe that our science can address.

Work in the applied area of biodiversity, such as assessment and monitoring of biodiversity in the face of global change, restoration, sustainable use of biodiversity, and development of effective conservation approaches cannot be addressed without basic advances in biodiversity science and allied disciplines in social sciences. A theory focuses that if a certain amount of natural habitat is lost, a certain proportion of species will become extinct. The theory does not tell us over what period these species will become extinct; the only certainty is that they will. Not much data is available on the rate of extinction or loss.

Furthermore, some countries have yet to comprehend the vast social, economic, scientific, technological, ecological, and political potential of the resources. If the current situation continues, these countries are likely to be left out in the race to conserve and sustainably utilize their natural resources for the well-being of their people and world at large. The policy makers have to realize that conservation and sustainable utilization of natural resources need to be central to all development planning for the developing countries, because these rely predominantly on natural resources for their well-being. A national commitment is needed, and therefore there is an urgent need for coordination between the various ministries, organizations and institutes.

## REFERENCES

Ambio, (1995). Vol. **24(4):** 251-254.

Ganeshaiah, K.N., Uma Shankar, R. and Bawa, Kamaljit S. (2010). Science and Non Science of Biodiversity and Conservation. *Biotech News* Vol. **5(6):** 216-219.

Khanina, L. (1998). Determining Keystone Species. Conservation Ecology (online) **2(2):** R2. Available from the Internet URL: http://www.consecol.org/journal/vol2/iss2/resp2.

Prance, G.T. (1995). Biodiversity. Encyclopedia of Environmental Biology Vol. 1.

Rosenzweig, M.L. (1995). *Plant Sciences Bulletin* **42:** 48-49.

Stevens, M.H.H., Raikow, D.F., Servedio, M.R., Collins, R., Schumann, T.L., Tipper, A.N., Long, Z.T. and Carson, W.P. (1996). Hutchinson's Chariot: A review of species diversity in space and time.

# 2

# Biodiversity of Phytoplankton, Zooplankton and Benthos in the Bay of Bengal at the East Coast of India, Andhra Pradesh

**Kankal, N.C. and Dhadse, Sharda**

*National Environmental Engineering Research Institute (NEERI), Nehru Marg, Nagpur-440020 (India)*
*Phone: (0712)-2249881; Fax: (0712) 2249878; (Mob.): 09423685941*
Email: nc_kankal@neeri.res.in

## ABSTRACT

The study of phytoplankton, zooplankton and benthos at 8 sampling locations was done around the coastal area of the Bay of Bengal during summer season (April, 2010) to assess the water quality and bio-ecoaesthetic value of the area. Among the major taxonomic groups, bacillariophyceae was the most dominant phytoplankton species. Based on the study of zooplankton, it appeared that the area is mostly preferred by Cladocera (*Daphnias* sp., *Allonella* sp. and *Moina* sp.) followed by Rotifera (*Brachionus* sp. and *Keratella* sp.), Copepoda (*Cyclops* sp.; *Nauplius* larva)) and Protozoa (*Acanthocystis* sp.). Benthos was dominated by Foraminifera (11 sp.) followed by Ostracoda (3 sp.), Polychaeta and Ophiuroidea (1 sp. each). An average count for phytoplankton, zooplanktons, benthos was computed as 165/ml, 1825/m$^3$ and 17750/m$^2$ respectively. Higher diversity indices for these three groups ranged from 1.51 to 2.92 indicating medium impact of pollution on the study area of the Bay of Bengal.

**Keywords:** Water Quality, Pollution, Diversity Index, Bay of Bengal, Sea Coast.

## INTRODUCTION

It is established that the total biological productivity of more fertile parts of the sea is greater than that of agricultural land, and in the sea itself the shorelines form the more fertile and productive parts despite highly complex and variable environmental conditions. Compared to the deep seas, coastal waters are rich in energy and nutrients mostly derived from land drainage and benthic decay of maritime detritus. The photosynthetic productivity of the seas around the coastal water is usually 35 to 75 times that of the open ocean, and coastal benthic production per unit area exceeds that in the water column by a factor of ten. The high energy and nutrients of the coastal waters, specially in the semi-enclosed landlocked bays, imply that they also have (through inputs from land) a great variety of offending substances or pollutants compared to the deep seas which are still, as a whole, relatively free from pollution. Biotic environment comprises microorganisms, phytoplankton, zooplankton, benthos and nekton (NEERI Report, 2010).

Biological productivity of the ocean is influenced by light, nutrients, primary, secondary and tertiary production. Light penetration in the ocean determines the depth of the photosynthetic zone, whereas the nutrients, particularly nitrogen and phosphorus, influence the promotion of productivity. Phytoplankton or planktonic algae give the production at the primary stage of the food chain, while zooplankton and fish are at the secondary and tertiary levels of the food chain. Phytoplankton provides food for herbivores and hence forms a major link in the food chain. Primary productivity in terms of phytoplankton population in the sea is governed by the rate of photosynthesis and availability of nutrients. The energy necessary for the metabolic process is derived from light and is stored as chemical energy in organic compounds. Overall, phytoplankton population density is higher during winter and summer-monsoon and low during the winter-monsoon seasons.

The phytoplankton, zooplankton and benthos constitute important links in aquatic food chain. They are good indicators of water pollution. They play a major role in pollution monitoring studies. These are the basic aspects for environmental impact assessment studies. The phytoplankton and zooplankton are practically suitable choice of indicators of water quality due to ease of sampling, their cosmopolitan distribution and lot of available information on these groups. The monitoring for biological parameters is rapid, inexpensive and reliable, requiring only biological examination of the water samples. The impact of pollution is directly reflected by the survival status of flora and fauna. The biological data gives an overall picture of the subsequent effects of pollution. The study on biological parameters will lead to detecting various factors contaminating or polluting the aquatic environment. With a view to conserve environmental quality and safety of natural flora and fauna, studies of biological aspects of the ecosystem is of priority in environmental impact assessment studies. These are the best indicators of environmental quality. These indicators serve as inexpensive and efficient early warning and control system and are hence included in the study.

Due to Krishnapatanam sea port and fishing activities around the study area, there may be tremendous pollution pressure on the water in the area. The present investigation was carried out to focus attention on biotic faunal features around the study area. The objective of the study is to estimate diversity index of phytoplankton, zooplankton and benthos to identify the level of pollution impacts and bio-ecoaesthetic value in the study area.

## MATERIALS AND METHODS

This study area is the part of the Palar and Pennar basins located in the Bay of Bengal in the area, immediately offshore of the coast located between 12°12′ 37″ to 14°49′39″ N latitudes and 79°58′44″ to 80°05′35″E longitudes to the south-east of Nellore district in Andhra Pradesh. The area of the study is about 9400 $km^2$. Studies were carried out on phytoplankton, zooplankton and benthos to examine the water quality of the coastal area. The water and sediment sampling was carried out during the summer season (April, 2010) in the study area. The sampling programme covered a distance around coastal area in the Bay of Bengal near Nellore, in Andhra Pradesh and Tamil Nadu. The sampling locations were selected for water and sediment quality assessment covering pollution impacted region. Sampling programme was conducted for the period of 15 days.

### Sampling and Analysis

#### *Water*

Physicochemical and biological parameters have been determined to establish the water quality status of the existing offshore, estuary, river and lake water sources. Sampling locations monitoring is shown in Fig. 1 and enlisted in Table 1. Total 8 water samples were collected for which 5 samples were collected from offshore water and 1 each from estuary, Penner river and Pullicat Lake around the study area. All the analysis and estimations were carried out as per the procedure described in the $20^{th}$ edition of Standard Methods, 1998 (APHA, AWWA and WEF). Heavy metals were determined by ICP using JY-24 Model of Jobin Yvon, France. Sodium and potassium were also estimated using Systronic 128 Model Flame Photometer.

#### *Phytoplankton*

The plankton refers to the microscopic organisms floating free or suspended in natural water bodies. The plankton consisting of animal species are called zooplankton, and those consisting of plant species are called phytoplankton. The density of phytoplankton is much higher than that of zooplankton. Therefore, representative water sample of about 100 ml was collected from the water

* Sampling Locations

**Fig. 1.** Study area showing sampling locations.

**Table 1** Sampling locations in the study area

| *S. No.* | *Type* | *Location* | *Depth (m)* | *Latitude* | *Longitude* |
|---|---|---|---|---|---|
| **Offshore Water** | | | | | |
| 1 | Offshore | Surface | 0.20 | 14° 11′ 59.88″ N | 80° 12′ 34.26″ E |
| | | Middle | 9 | | |
| | | Bottom | 20 | | |
| 2 | Offshore | Surface | 0.20 | 14° 15′ 02.94″ N | 80° 30′ 07.2″ E |
| | | Middle | 9 | | |
| | | Bottom | 18 | | |
| 3 | Offshore | Surface | 0.20 | 14° 21′ 21.36″ N | 80° 15′ 40.92″ E |
| | | Middle | 10 | | |
| | | Bottom | 19 | | |
| 4 | Offshore | Surface | 0.20 | 14° 21′ 21.36″ N | 80° 16′ 02.52″ E |
| | | Middle | 13 | | |
| | | Bottom | 26 | | |

**Table 1** Contd.

| *Sr. No.* | *Type* | *Location* | *Depth (m)* | *Latitude* | *Longitude* |
|---|---|---|---|---|---|
| 5 | Offshore | Surface<br>Middle<br>Bottom | 0.20<br>10<br>21 | 14° 37′ 57.78″ N | 80°13′ 29.54″ E |
| **Estuary Water** | | | | | |
| 6 | Estuary (Muttu Kuru) | Surface<br>Middle<br>Bottom | 0.20<br>10<br>19 | 14° 14′ 52.56″ N | 80° 07′ 53.76″ E |
| **River Water** | | | | | |
| 7 | Pennar River | Surface | 0.20 | 14° 28′ 15.95″ N | 79° 59.25′ 25.92″ E |
| **Lake Water** | | | | | |
| 8 | Pullicat Lake | Surface | 0.20 | 13° 25′ 16.02″ N | 80° 19′ 14.35″ E |

body in clean, good quality plastic bottles. This sample was preserved by adding Lugol's Iodine solution @ 1 ml per 100 ml of water sample. Lugol's solution was prepared by dissolving 20 gm potassium iodide (KI) in 50 ml distilled water. This solution was diluted by adding 150 ml distilled water. Then 20 ml glacial acetic acid was added to the solution and stored in a coloured bottle and kept in a cool place. Some phytoplankton organisms are unicellular, whereas others are multicellular or colonial, therefore, the following procedure was adopted to enumerate these algae.

The Lackey Drop (Microtransect) method (Lackey 1938; Edmonds (1969)) is a simple method of obtaining counts of considerable accuracy with samples containing a dense plankton population. The phytoplankton density in clean water is very less, while it is dense in polluted water. Therefore, the phytoplankton in clean water needs to be concentrated before counting for accurate estimation. The water sample was concentrated 15 times by centrifugation at 250 rpm for 15 min. The supernatant water was decanted and the pellet of algal cell suspended in 1 ml drilled water. 0.04 ml of concentrated water sample was pipetted out onto a glass slide and covered with an 18 mm cover slip. Microorganisms were counted in strips on slide from a drop of centrifuged, decanted and concentrated water sample, and number of individuals of each organism was counted in 45 × magnification. Phytoplankton was identified as per the guidelines given in Standard Methods (1998).

The number is expressed per ml of the sample as per the details given below:

- The pipette should be selected such that it gives 25-30 drops/ml.
- If it gives 25 drops/ml, then the volume of one drop is 0.04 ml
- Area of cover slip (18 mm × 18 mm) = 324 $mm^2$
- Area of strip observed (18 × 0.51) = 9.18 $mm^2$
- Number of microscopic fields in 18 mm long cover slip = 35

- Diameter of microscopic field = 18/35 = 0.51 mm
- Area of microscopic field = $3.14 \times (0.51/2)^2 = 0.22\ mm^2$
- Note down an average count of organism per strip
- Calculate the count of algae per microscopic field

$$\text{No./ml} = \frac{Y \times 324 \times 25 \times X}{0.22 \times 15} = Y \times 1964$$

OR

$$\text{No./ ml} = \frac{Y \times \text{Area of cover slip} \times \text{No. of drops/ml} \times \text{Vol. of Concentrated(ml)}}{\text{Area of microscopic field} \times 15}$$

where,

Y- average number of each organism per strip; it is the number of organisms divided by the total number of strips observed for a sample

35 - number of microscopic strips in 18 mm length of cover slip

20 - number of drops making up for 1 ml in a pipette (number can vary according to the type of pipette used)

X - volume of concentrated sample after centrifuging and decanting supernatant

15 - the volume in ml of water used for centrifugation

The diversity is calculated for each community by Shannon Wiener Diversity Index (SWDI). This index was used for the estimation of community diversity (Azeiteiro et al., 1999, 2000; Morgado et al., 2003). It is where proportion is obtained by dividing the number of individuals of species by the total number of individual in a sample for which $\log_2$ proportions were obtained from index table (Shannon, 1948).

$$d = \sum \frac{n}{i = 1n} (\alpha / n).\log_2 (\alpha / n)$$

The method was found to be more reliable for the assessment of phytoplankton species in offshore area. The Palmer's Pollution Index (PPI) was computed to know the level of organic pollution in the water of study area. The ranges of PPI < 15, from 15 – 20, and > 20 indicate absence of organic pollution, presence of moderate and high organic pollutions, respectively.

### *Zooplankton*

Zooplankton density is always less than the phytoplankton density. Therefore around 20-50 L of water is passed through plankton net (mesh size 64 micron) to concentrate zooplankton. The zooplankton thus collected can be preserved immediately with an appropriate quantity of ethanol (70%) or buffered formalin (5%) to make a final concentration. The entire water is centrifuged, decanted and concentrated to make 1 ml volume for observation under S-R (Sedgwick-Rafter) counting cell. The S-R Cell is a rectangular chamber (50 mm long, 20 mm wide, 1 mm deep) having a total area of the bottom as 1000 $mm^2$ and total volume 1000 $mm^3$ or 1 ml. The cell can be easily observed under I0 × magnification. The cover slip is placed diagonally across the cell which helps to prevent the formation of air bubbles in the cell. Do not permit overfilling because this would yield a depth greater than 1 mm and produce an invalid count. Do not permit large air spaces caused by evaporation to develop in the chamber during a lengthy examination. To prevent the formation of air spaces, occasionally place a small drop of distilled water on the edge of cover glass. Before counting, let the S-R cell stand for at least 15 minutes to settle the plankton and count the organisms settled on the bottom. Observe the cell under the microscope and count the organisms present in it.

The zooplankton is counted in 10 × magnification. Its number per $m^3$ is expressed by the following formula:

$$\text{No.}/m^3 = \frac{C \times V_2}{V_1 \times V_3}$$

where,

C - total number of counted individuals of species in a sample
$V_1$ - volume of concentrated sample through plankton net
$V_2$ - centrifuged, decanted and concentrated volume of sample in ml
$V_3$ - volume of grab sample in $m^3$ i.e. 20 L/1000

$$\text{No./m}^3 = \frac{\text{No. of plankton in 1 ml of S} - \text{R cell} \times \text{volume of net concentrated (ml)}}{\text{Volume of water passed through plankton net (litres)}}$$

The organisms were identified with the help of available reference (Kameswara et al., 1987; Krishna, 1986; Kasturirangan, 1963 and Wickstead, 1965).

### *Sediment*

Different types of offshore areas play an important role in the formation of sediment. The study has large deposits of sediments. These sediments account for variations in particle size, sand, soils, clay, sand particles, and silt are important ingredients of soil as they determine the characteristics of sediments. The sediments also contain hydrocarbons, heavy metals and benthos. Representative samples for sediment from depth (5 – 20 m) were collected from offshore, estuary and river surrounding the areas of various distances from the study area. Standard methods have been followed for the analysis of sediment samples. The international pipette method (Black, 1964) was adopted for the determination of particle size distribution. The chemical characteristics of the sediment were determined by preparing sediment extract in distilled water in the ratio 1:1(Jackson, 1967). Organic carbon was determined by Walkley and Black method (1972). Potassium was determined by Flame photometer Method (Jackson ML, 1967). Heavy metals in soil samples were determined by acid digestion and analysis on atomic absorption spectrophotometer (AAS).

### *Benthos*

Benthos is an organism that inhabits the bottom of an aquatic body. Many of them are sessile, while some creep over a burrow in mud. The quality and quantity of the animals found at the bottom are related to the nature of substrate and the depth. Their number and distribution depends upon the physicochemical and biological characteristics of water. For benthos collection, bottom sediment sample was collected by Van Veen grab sampler having an area of 0.03 $m^2$ with the depth of about 15 cm. The sediment samples collected from different locations in the area were passed through 500 micron mesh sieve and again through 45 micron sieve for segregation. The microorganisms retained on both the sieves were preserved immediately with 4% buffered formalin. The diversity is calculated for each community of zooplankton and benthos by using Shannon Wiener Diversity Index (SWDI) (Shannon 1948).

## RESULTS AND DISCUSSION

The physicochemical characteristics (temperature, pH, salinity and DO) of water in the study area are presented in Figure 2. The source of the data is taken from NIOT, Chennai.

The variations in temperature of water body have great bearing upon the biological productivity. The lowest yearly seawater temperature in this region is 20.5°C and maximum is 29.9°C. The mean value of water sample at 3 locations of Nellore coast from surface (1 m depth) and bottom (up to 80 m depth) was recorded as 27.2°C and 23.24°C respectively. Seasonal fluctuations in water temperature distribution play an important role in influencing biological processes. Temperature affects the organisms through direct physiological mechanisms. Temperature related responses vary according to different species, and the repercussions for ecosystem dynamics depend upon food web interactions (Kinne, 1963). Variation in salinity ranges from 31.61–36.04% (parts per thousand) with an average of 33.90%. pH ranges between 7.60 and 8.23, while DO ranges between 2.1–6.2 mg/l at 80 m and 1 m depths respectively. Vertical distribution of DO shows sharp decrease from 10 m down to 50 m depth, and thereafter the values remain almost constant. The water sample of offshore locations was collected from the surface, middle and bottom around the study area during the study period, and the results are presented in Table 2.

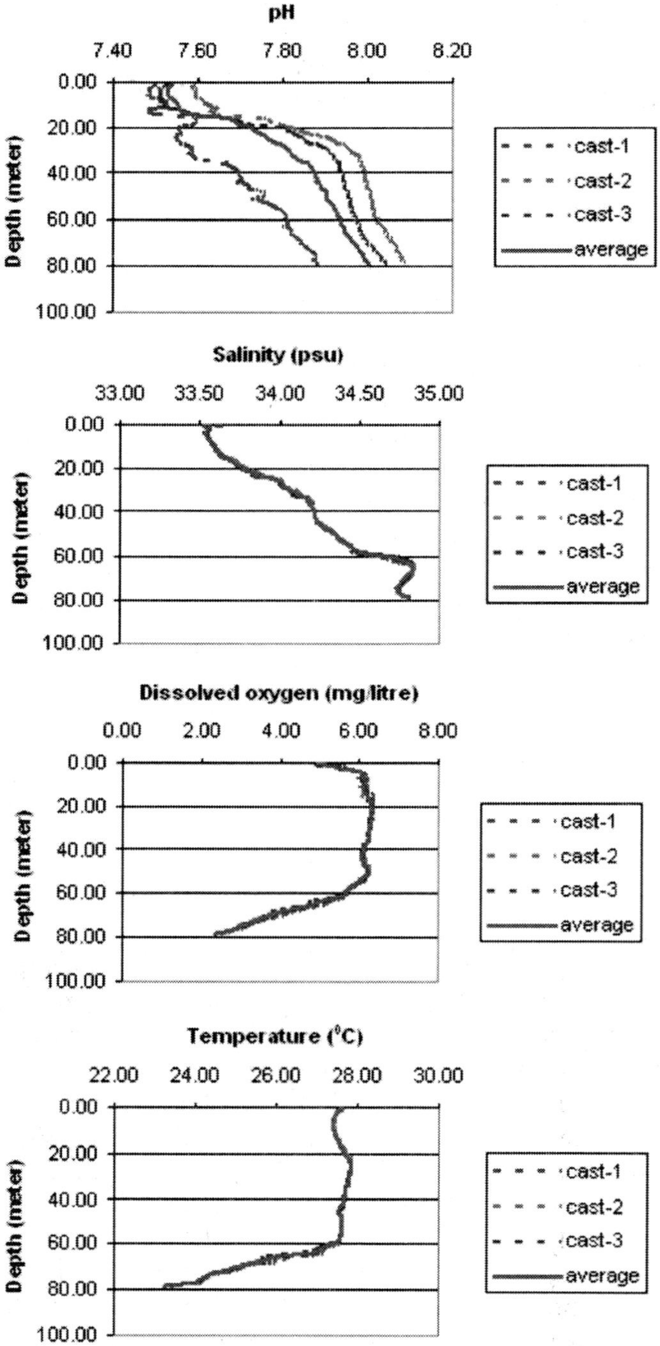


**Fig. 2.** Water quality at various depths (1-80 m).

**Table 2** Coastal water quality, Bay of Bengal

| *Sr. No.* | *Parameter* | *Offshore* | *Estuary* | *River* | *Lake* | *IS:7967/10500 (1976/1991)* |
|---|---|---|---|---|---|---|
| 1. | Temperature °C | 29.0 | 30.0 | 31.5 | 33.5 | 40 |
| 2. | pH | 7.5 | 7.4 | 7.2 | 7.1 | 6.5-9.5 |
| 3. | D.O. mg/l | 2.1-5.8 | 4.2-5.8 | 3.4 | 4.2 | 3.0 |
| 4. | Alkalinity (mg/l) | 134 | 132 | 140 | 125 | – |
| 5. | Chloride (mg/l) | 187318 | 18521 | 23427 | 83 | – |
| 6. | Salinity ($\%_0$) | 33.90 | 33.45 | – | – | – |
| 7. | Nitrate as N (mg/l) | 0.030 | 0.043 | 0.071 | 0.019 | 45 |
| 8. | Total phosphate (mg/l) | 1.62 | 1.23 | 0.87 | 1.38 | 5 |
| 9. | COD (mg/l) | < 5 | < 5 | 19 | 10 | 250 |
| 10. | BOD (mg/l) | < 1 | < 1 | 5 | 3 | 30 |
| 11. | Oil & grease (mg/l) | 0.075 | 0.018 | 0.003 | 0.007 | 10 |
| 12. | Hydrocarbon (mg/l) | 1.078 | 1.558 | 1.808 | 1.687 | – |
| 13. | Ni (mg/l) | ND | 0.001 | 0.03 | 0.04 | – |
| 14. | Cd (mg/l) | 0.0124 | 0.043 | 0.065 | 0.065 | 0.01 |
| 15. | Cr (mg/l) | 0.013 | 0.01 | 0.022 | 0.026 | 0.01 |
| 16. | Cu (mg/l) | 0.005 | 0.011 | 0.014 | 0.015 | 0.05 |
| 17. | Fe (mg/l) | 0.144 | 0.099 | 0.658 | 0.170 | 0.3 |
| 18. | Mn (mg/l) | 0.025 | 0.042 | 0.056 | 0.058 | 0.1 |
| 19. | Zn (mg/l) | 0.916 | 1.222 | 1.888 | 1.890 | 5.0 |
| 20. | Co (mg/l) | 0.013 | 0.025 | 0.065 | 0.098 | – |

The results of physical parameters, viz., pH, temperature, turbidity and TSS during the summer season in the water are found in the range of 7.1–7.5, 29°–33.5°C., <1–2.4NTU and 12–108 mg/l. DO in seawater is an essential requirement for the survival of aquatic organisms. DO values range between 2.1–5.8 mg/l (Avg. 4.1); it is higher than the value (3.1 mg/l) reported by Lokhande et al., 2008 in the water of Thane creek. The average value of DO was recorded to be (4.1 mg/l) in a sample at Bay of Bengal whereas at Ulhas river estuary the value of DO (3.9 mg/l) was reported by Mishra (2002) indicating slight change in water quality due to discharge from small-scale industrial effluents. This is an indication of organic pollution (Goldin et al., 2002). The offshore water quality sources in the study area showed low concentrations of heavy metals except iron, which was recorded higher than the standards at some places. All the values are well below the tolerance limits prescribed under Bureau of Indian Standards for controlling the pollution of marine coastal areas IS:7967 (1976).

## Phytoplankton

During summer season, in offshore water, the Shannon Wiener Diversity Index (SWDI) varies between NIL-2.93, indicating no impact of pollution. The population dynamics of offshore water were estimated by phytoplankton count in number/ml, which was observed in the range of Nil–224 No/ml. Table 3 shows the percent concentration of organisms in this group which reveals highest count for Bacillariophyceae (53%) followed by Chlorophyceae (42%) and Cyanophyceae (9%).

Average count of phytoplankton was computed as 87, 68, 168 and 120 No/ml in the water of offshore, estuary, river and lake was respectively. Phytoplankton was represented by 6 Bacillariophyceae, 5 Chlorpphyceae and 1 Cyanophyceae. No toxic algae were found to be present in the study region. The dominance of Bacillariophyceae and Chlorophyceae as well as low algal count and medium SWDI values (Nil–2.93) indicate medium impact of pollution with medium impact factor. The Palmer Pollution Index (PPI) values were found less than 15 in offshore water, indicating oligotropic in nature or very low pollution, more than 15 in lake water the presence of organic pollution, and more than 21 in river indicated high pollution water bodies. A higher density of phytoplankton was observed at surface water, which gradually decreased while going down below at depth in offshore water due to least photosynthetic activities in deeper layer of water. There were no counts observed at the depth below 20 m.

**Table 3** Biological parameter—phytoplankton

| *Sr. No.* | *Bottle No.* | *Sampling Location* | *Depth (m)* | *Total Phyto-plankton Count (No./ml)* | *% Composition* | | | *SWDI* | *PPI* |
|---|---|---|---|---|---|---|---|---|---|
| | | | | | *Chloro-phyceae* | *Bacillario-phyceae* | *Cyano-phyceae* | | |
| **Offshore Water** | | | | | | | | | |
| 1. | Seawater | Surface | 0.5 | 112 | 50 | 50 | – | 1.97 | 6 |
| | | Middle | 9 | 22 | – | 100 | – | 1 | 6 |
| | | Bottom | 20 | Nil | Nil | Nil | Nil | Nil | Nil |
| 2. | Seawater | Surface | 0.5 | 134 | 33 | 67 | – | 2.58 | 10 |
| | | Middle | 9 | 45 | 25 | 75 | – | 1.50 | 3 |
| | | Bottom | 18 | Nil | Nil | Nil | Nil | Nil | Nil |
| 3. | Seawater | Surface | 0.5 | 112 | 60 | 40 | – | 2.245 | 8 |
| | | Middle | 10 | 45 | 50 | 50 | – | 1 | 3 |
| | | Bottom | 19 | Nil | Nil | Nil | Nil | Nil | Nil |
| 4. | Seawater | Surface | 0.5 | 224 | 50 | 50 | – | 2.245 | 10 |
| | | Middle | 13 | 123 | 36 | 64 | – | 1.97 | 3 |
| | | Bottom | 26 | Nil | Nil | Nil | Nil | Nil | Nil |
| 5. | Seawater | Surface | 0.5 | 190 | 59 | 41 | – | 2.742 | 15 |
| | | Middle | 10 | 45 | 25 | 75 | – | 1.5 | 3 |
| | | Bottom | 21 | Nil | Nil | Nil | Nil | Nil | Nil |
| **Estuary Water** | | | | | | | | | |
| 6. | | Surface | 0.5 | 190 | 35 | 47 | 18 | 2.697 | 12 |
| | Muttu Kuru | Middle | 10 | 45 | 50 | 50 | – | 2 | 4 |
| | | Bottom | 21 | Nil | Nil | Nil | Nil | Nil | Nil |
| **River Water** | | | | | | | | | |
| 7. | Pennar | Surface | 0.5 | 168 | 40 | 53 | 7 | 2.883 | 21 |
| **Lake Water** | | | | | | | | | |
| 8. | Pulicate Lake | Surface | 0.5 | 190 | 41 | 47 | 12 | 2.93 | 18 |

The species of phytoplankton observed during the study period are given below:

| *Chlorophyceae* | *Bacillariophyceae* | *Cyanophyceae* |
|---|---|---|
| *Ankistrodesmus* sp. | *Navicula* sp. | *Osillatoria* sp. |
| *Cosmarium* sp. | *Nitzschia* sp. | – |
| *Scenedesmus* sp. | *Cymbella* sp. | – |
| *Chlorococcum* sp. | *Cyclotella* sp. | – |
| *Chlamydomonas* sp. | *Synedra* sp. | – |
| – | *Gyrosigma* sp. | – |

## Zooplankton

The data on zooplankton community for offshore water quality depth up to 20 m during the summer season are shown in Table 4.

**Table 4:** Biological parameters—zooplankton

| *Sr. No.* | *Bottle No.* | *Sampling Location* | *Depth (m)* | *Total Phyto-Zooplankton/m³* | *% Composition* | | | | *SWDI* |
|---|---|---|---|---|---|---|---|---|---|
| | | | | | *Rotifera* | *Cladocera* | *Copepoda* | *Protozoa* | |
| **Offshore Water** | | | | | | | | | |
| 1. | Seawater | Surface | 0.5 | 1200 | 50 | – | 33 | 17 | 1.916 |
| | | Middle | 9 | 600 | 67 | – | 33 | – | 1.584 |
| | | Bottom | 20 | Nil | – | – | Nil | – | Nil |
| 2. | Seawater | Surface | 0.5 | 1600 | 38 | – | 50 | 12 | 2.250 |
| | | Middle | 9 | 800 | 25 | – | 75 | – | 1.500 |
| | | Bottom | 18 | Nil | Nil | Nil | Nil | Nil | Nil |
| 3. | Seawater | Surface | 0.5 | 1600 | 50 | – | 50 | – | 2.0 |
| | | Middle | 10 | 400 | 50 | – | 50 | – | 1.0 |
| | | Bottom | 19 | Nil | Nil | Nil | Nil | – | Nil |
| 4. | Seawater | Surface | 0.5 | 1200 | 33 | – | 67 | – | 1.916 |
| | | Middle | 13 | 400 | – | – | 100 | – | 1.000 |
| | | Bottom | 26 | Nil | Nil | Nil | Nil | Nil | Nil |
| 5. | Seawater | Surface | 0.5 | 2000 | 40 | 20 | 40 | – | 2.245 |
| | | Middle | 10 | 1000 | 20 | – | 80 | – | 1.522 |
| | | Bottom | 21 | Nil | Nil | Nil | Nil | Nil | Nil |
| **Estuary Water** | | | | | | | | | |
| 6. | Muttu Kuru | Surface | 0.5 | 1600 | 50 | – | 50 | – | 2 |
| | | Middle | 10 | 600 | 67 | – | 33 | – | 1.584 |
| | | Bottom | 21 | Nil | Nil | Nil | Nil | Nil | Nil |
| **River Water** | | | | | | | | | |
| 7. | Pennar | Surface | 0.5 | 2800 | 27 | - | 43 | - | 1.84 |
| **Lake Water** | | | | | | | | | |
| 8. | Pulicate Lake | Surface | 0.5 | 2600 | 46 | 16 | 38 | - | 2.405 |

In offshore water, the Shannon Weiner Diversity Index (SWDI) and zooplankton count varies between Nil–2000 No/m$^3$ and Nil–2.584, respectively. Average zooplankton counts were computed as 900, 1000, 2600 and 2800 No/m$^3$ at offshore, estuary, lake and river water, respectively. Low zooplankton count was also observed at deeper level of sea water due to the absence of phytoplankton at deeper levels and disturbance in food chain. The percent composition reveals highest count for Copepoda (46%), Rotifera (43%), Cladocera (18%) and Protozoa (15%). The zooplankton was represented by 3 Cladocera, 2 each of Rotifera and Copepoda and 1 Protozoa species. Generally, the low tide samples exhibited higher density than high tide samples (Jiyalal et al., 1990). Copepoda was the dominant group in the water of western coast near Thane creek, and it was replaced by Rotifera (Kotangale, 1994; Patil, 2002). In our study it was observed that Cladocera was the dominant group followed by Rotifera, Copepoda and Protozoa in the Bay of Bengal. Tiwari and Nair, 2002, reported the highest numerical abundance of Copepods contributing to the bulk of biomass in Dharamtar creek water samples, which is somewhat different from the present study for the Bay of Bengal water. Usually monsoon and post-monsoon months are the most productive periods for secondary production of zooplankton (Menon et al., 1977) and exhibit higher diversity index value (Srinivasan and Santhanam, 1991).

However, higher species diversity values were recorded in summer as compared to premonsoon and monsoon seasons at Coleroon estuary (Jagadeesan and Avyakannu, 1992). The present study also shows higher diversity index for zooplankton in the water of the Bay of Bengal in summer season. Tolerance of Rotifers to higher salinity was observed by Somani, 2002, and this is also an indication for the Bay of Bengal water quality in respect to zooplankton. The values of diversity index that ranged from nil–2.584 show moderately polluted nature of water in the study area. Above discussions indicate medium water quality with moderate productivity. The list of zooplankton species recorded during the summer season is presented as below:

| *Rotifera* | *Protozoa* | *Cladocera* | *Copepoda* |
|---|---|---|---|
| *Brachionus* sp. | *Acanthocystis* sp. | *Daphnia* sp. | *Cyclops* sp. |
| *Keratella* sp. | – | *Allonella* sp. | *Nauplius larva* |
| – | – | *Moina* sp. | – |

## Sediment Quality

Different types of offshore areas play an important role in the formation of sediment. The study area has large deposits of sediment. The sediments also contain hydrocarbons, heavy metals and benthos, etc. Generally, low organic carbon recorded in the area compared to world average for the shelf sediments can be attributed to the oxidation because of high dissolved oxygen content of bottom water or grazing by zooplankton and other higher organisms. Sediment samples collected from offshore, estuary and river were analyzed for various parameters, viz., pH, electrical conductivity (EC), organic carbon, total nitrogen, phosphorus, potassium and heavy metals (Table 5).

The medium organic and nutrient content of the sediment shows that it is of poor quality. Heavy metal pollution is serious because it can persist for many decades. It creates problems in nutrient utilization for benthic organisms. The benthos count varies between 14400–26400 No/m$^2$ and the Shannon Weiner Diversity Index (SWDI) varies from 1.57–2.75. Foraminifera were the dominant group in the study area. The percent composition reveals highest count for Foraminifera (68%) followed by Ostracoda (27%), Polychaeta (16%) and Ophiuroidea (10%). The count of benthos was observed to be the highest at surface sediment than deeper layers. Benthos were concentrated more in 4–6 mm and decreased progressively in the deeper layers of sediments.

**Table 5** Sediment quality

| *Sr. No.* | *Parameter* | *Offshore* | *Estuary* | *River* |
|---|---|---|---|---|
| 1. | pH | 7.3-8.0 | 7.0-7.8 | 7.6 |
| 2. | EC (mS/m) | 9.7 | 9.2 | 1.41 |
| 3. | Hydrocarbon (gm/g) | 1.038 | 0.155 | 0.333 |
| 4. | Oil & grease (gm/kg) | 1.425 | 1.5 | 1.0 |
| 5. | Organic carbon (%) | 1.626 | 1.193 | 1.56 |
| 6. | Total nitrogen (%) | 0.115 | 0.124 | 0.15 |
| 7. | Total phosphorus (%) | 0.147 | 0.075 | 0.11 |
| 8. | Total potassium (%) | 0.090 | 0.019 | 0.013 |
| 9. | Ni (mg/kg) | 0.098 | 0.0111 | 0.073 |
| 10. | Cd (mg/kg) | 0.018 | 00.017 | 0.020 |
| 11. | Cr (mg/kg) | – | – | – |
| 12. | Cu (mg/kg) | 0.15 | 0.021 | 0.162 |
| 13. | Pb (mg/kg) | 0.050 | – | 0.267 |
| 14. | Fe (mg/kg) | 4.064 | 12.998 | 3.507 |
| 15. | Mn (mg/kg) | 2.873 | 4.002 | 1.773 |
| 16. | Zn (mg/kg) | 0.653 | 0.123 | 0.356 |
| 17. | Co (mg/kg) | 0.046 | 0.052 | 0.020 |

The data on Benthos community collected from offshore, estuary and river sediment samples is presented in Table 6.

**Table 6** Biological parameters—Benthos

| *Sr. No.* | *Sampling Locations* | *Table Benthos/m$^2$* | *% Composition* | | | | *SWDI* |
|---|---|---|---|---|---|---|---|
| | | | *Foraminifera* | *Ostracoda* | *Polychaeta* | *Ophiuroidea* | |
| | | | **Sea Bottom** | | | | |
| 1. | Sea | 14,400 | 50 | 40 | - | 10 | 1.91 |
| 2. | Sea | 16,800 | 57 | 43 | - | - | 2.24 |
| 3. | Sea | 24,400 | 73 | 27 | - | - | 2.23 |
| 4. | Sea | 19,200 | 63 | 25 | 13 | - | 2.75 |
| 5. | Sea | 21,600 | 56 | 11 | 33 | - | 2.50 |
| 6. | Sea | 26,400 | 82 | 9 | 9 | - | 2.36 |
| 7. | Estuary | 16,800 | 71 | 29 | - | - | 1.94 |
| 8. | River | 24,00 | 90 | - | 10 | - | 1.57 |

List of Benthos in the study area is presented below:

| *Sr. No.* | *Family* | *Benthos Species* |
|---|---|---|
| 1. | Foraminifera | Bolivina<br>Globigerina<br>Elphiduim<br>Calcarina<br>Amphistegina<br>Triloculina<br>Rosalina<br>Ammonia<br>Globigerinita<br>Rotalia<br>Triloculina |
| 2. | Ophiuroidea | Brittle star |
| 3. | Ostracoda | Macrocyprina<br>Cytherelloidea<br>Paranesidea |
| 4. | Polychaeta | Polychaete larva |
| 5. | Ophiuroidea | Ophiothrix fragilis |

In the shallow water of river, the benthos count was observed very low due to organic pollution or nutrient enrichment. Benthos showed patchy dispersion, related to the availability of food and predation (Ansari, 1983). Setty (1976) observed abundance of Foraminifera population in the polluted water of Bay of Bengal which ranged from 1.57–2.75 (Avg. 2.187), regarded to be good quality at all the sampling locations including the sediment sample of river. Higher value of benthos count may be due to nutrient enrichment of organic contents (Mishra, 2002). Bivalves was dominant group at Dahanu, Danda and Savta Creek in west coast of India (Kotangale, 1994); it was replaced by Gastropoda. Enrichment of coastal waters due to river flow and land runoff also seems to be one of the factors contributing to the richness of fauna near shore regions in the Bay of Bengal. According to Patil et al. (1975 and 2002), the availability of food and associated chemical changes influence the population dynamics. Thus, the diversity index value for phytoplankton, zooplankton and benthos shows moderately polluted nature of water in the study area.

## ACKNOWLEDGEMENTS

The authors express their sincere gratitude to the Director, NEERI, for his constant encouragement and permission to publish this paper.

## REFERENCES

Ansari, Z.A. and Ingole, B.S. (1983). Meiofauna of some sandy beaches of Andaman Islands. *Indian J. Marine Sciences*. **12:** 245-246.

APHA, (1992). Standard Methods for examination of water and wastewater, APHA, AWWA, WPCE, Washington, DC.

Azeiteiro, U.M., Marques, J.C. and Re, P. (1999). Zooplankton annual cycle in the Mondego river estuary (Portugal). Arquivos do Meseu Bocage, Nova Series. **3:** 239-263.

Azeiteiro, U.M., Marques, J.C. and Re, P. (2000). Zooplankton assemblage in a Portugal: Mondego estuary. Arquivos do Muses Bocago. Nova Series, **3:** 357-376.

Goldin, Q., Athalye, R.P., Somani, V.U., Borkar, M.U. and Patil, N.N. (2002). Present status of water quality of Thane Creek and Ulhas River Estuary. *Proc. The National Seminar on Creeks, Estuaries, and Mangroves: Pollution and Conservation*. 58-62.

Jagadeesan, P. and Ayakkannu, K. (1992). Seasonal variation of benthic fauna in marine zone of Coleroon estuary and inshore waters, south-east coast of India. *Indian J. Marine Sciences.* **21:** 67-69.

Jiyalal, Ram M., Vijayalakshmi, R. Nair and Desai, B.N. (1990). Distribution of phytoplankton of Mithapur (Gujarat). *J. Indian Fisheries Association.* **19:** 49-57.

Kasturirangan, L.R. (1963). A key for the identification of more common planktonic copepoda of Indian coastal water: CSIR New Delhi, India.

Kinne, O. (1963). The effects of temperature and salinity on marine and brackish water animals. *Oceanogr. and Mar. Biol., Ann. Rev.* **1:** 301-340.

Kotangale, J.P., Ghosh, T.K. and Gadkari, A. (1994). Diversity of Zooplankton and Benthos at Dahanu, Danda and Savta Creek in West Coast of India. *Proc. Acad. Environ. Biol.* **3(1):** 1-8.

Krishna, N. (1986). Introduction to planktonology. Himalaya Publishing House, Bombay.

Lackey, J.B. (1938). The manipulation and counting of river plankton and changes in some organisms due to formalin preservation. *Pub. Health Rep.* **53:** 2080.

Lokhande, K.S., Kankal, N.C., Indurkar, M.M. and Wate, S.R. (2008). Zooplankton and Zoobenthos study of Thane Creek, West Cost of India. In: 2nd National Conference of Zoology (NCZ), Ranchi, Jharkhand, 30.

Menon, N.R., Gupta, T.R.C., Hariharan, V., Kotti, R.J. and Shetty, H.P.C. (1977). Marine plankton of Mangalore waters—a pre-pollution assessment. *In: Proc. Symp. warm water Zooplankton*, NIO, Goa. 274-283.

Mishra, V. (2002). Study of macrofauna and mangrove mudflats of Ulhas river estuary, Ph.D. Thesis, University of Mumbai, Mumbai.

Morgado, F.M., Antunes, C. and Pastorinho, M.R. (2003). Distribution and patterns of emergence of superbenthic and pelagic crustaceans in a shallow temperate estuary (Ria de Aveiro, Portugal). *Acta Oecologica,* **24:** 205-217.

National Environmental Engineering Research Institute (NEERI) (2010). Environmental Impact Assessment studies at Bay of Bengal (AP).

Patil H.S., Dodakundi, G.B. and Rodgi, S.S. (1975). Succession in zooplankton and phytoplankton in a sewage stabilization pond. *Hydrobiologia.* **47:** 253-265.

Patil, N.N., Goldin, Q., Somani, V.U., Kurve, P.N. and Athalye, R.P. (2002). Study of zooplankton distribution from Ulhas River Estuary and Thane Creek. In: Goldin Q. (Ed). *Proc. The National Seminar on Creeks, Estuaries, and Mangroves: Pollution and Conservation* 109-115.

Rao, K., Sivadas, P., Narayanan, B., Jayalakshmy, K.V. and Krishna Kutty M. (1987). Distribution of foraminifera in the lagoon of certain islands of the Lakshadweep Archipelago, Arabian Sea India. *Indian J. Marine Sciences.* **16:** 161-178.

Setty, M.G.A.P. (1976). The relative sensitivity of benthonic foraminifera in the polluted marine environment of Cola Bay, Goa. *In:* Proc. VI Ind. Coll. Micropal. Strat. 225-234.

Shannon, C.E. (1948). A mathematical theory of communication. *Bell System Technical Journal*. **27:** 379-423.

Somani, V.U., Patil, N.N., Goldin, Q. and Athalye, R.P. (2002). Study of rotifers in Thane Creek and Ulhas River Estuary. *In:* Goldin Q. (Ed). *Proc. The National Seminar on Creeks, Estuaries, and Mangroves: Pollution and Conservation*. 103-105.

Srinivasan, A. and Santhanam, R. (1991). Tidal and Seasonal variations in zooplankton of pullavazhi brackish water, south-east coast of India. *Indian J. Marine Sciences* **20:** 182-186.

Tiwari, L.R. and Nair, V.R. (2002). Plankton biodiversity of Dharamtar Creek adjoining Mumbai Harbor. *Proc. The National Seminar on Creeks, Estuaries, and Mangroves: Pollution and Conservation*. 96-102.

Weaver, W. and Shannon, C.E. (1948). The mathematical theory of communication. Urbana, Illinois: University of Illinois.

Wickstead, J.H. (1965). An introduction to the study of tropical plankton. Hutchinson Tropical Monograph.

# 3

# Sustainable Development

R.K. KALE AND M.H. FULEKAR

*Central University of Gujarat, Sector-30, Gandhinagar, Gujarat*

**ABSTRACT**

Sustainable development is the development that meets the needs of the present without compromising the ability of future generations to meet their own needs. The goal of sustainable development to enable the people throughout the world to satisfy their basic needs and enjoy a better quality of life without compromising the quality of future generation in the context of environment, society and economics have been discussed. Environmental sustainability, social sustainability and economic sustainability are main themes highlighted in the chapter.

**Keywords:** Sustainable Development, Environment, Economics, Social Sustainability.

## INTRODUCTION

Sustainable development is the development that meets the needs of the present without compromising the ability of future generations to meet their own needs. It contains within it two key concepts:

(i) the concept of needs, in particular the essential needs of the world's poor to which overriding priority should be given; and

(ii) the idea of limitations imposed by the state of technology and social organization on the environment's ability to meet present and future needs".

Sustainable development is often an overused word, but goes to the heart of tackling a number of interrelated global issues such as poverty, inequality, hunger and environmental degradation. The idea of sustainable development grew from numerous environmental movements in earlier decades. Summits such as the Earth summit held in Rio Brazil, 1992, were major international meetings to bring sustainable development to the main stream. The 1992 Rio Earth Summit was attended by 152 world leaders and sustainability was enshrined in agenda 21, a plan of action and a recommendation that all countries should produce national sustainable development strategies. Despite binding conventions and numerous detailed reports, there seems to have been little known about the details to citizens around the world.

*"In many countries – rich and poor – this is often because of a perception that sustainability is expensive to implement and ultimately a break on development occurs. Poor countries for their part usually lack the physical infrastructure, ideas and human capacity to integrate sustainability into their development planning. Besides, they are often quite skeptical about rich countries' real commitment to sustainable development and demand a more equitable sharing of environmental costs and responsibilities. Many people also believe that environmental problems cannot wait still developing countries become richer. There is still no widely shared vision on what sustainable development might mean in practice."*

The millennium ecosystem assessment (MA) was released in March 2005. This 2500 page report was four years in making, drawn upto 1300 researchers from 95 nations over four years and funded by the Global Environment Facility, the United Nations Foundation, the World Bank and various others. The key message from the report included the following points:

- Everyone in the world depends on nature and ecosystem services to provide the conditions for a decent, healthy and secured life.
- Humans have made unprecedented changes to ecosystems in recent decades to meet growing demands for food, freshwater, fibre and energy, which has helped to improve the lives of billions, but at the same time they have weakened nature's ability to deliver other key services such as purification of air and water, protection from disasters, and the provisions of medicines.
- Human abilities have taken the planet to the edge of massive wave of species extinctions, further threatening our own well-being.
- The loss of services defined from ecosystems is a significant barrier to the achievement of millennium development goals to reduce, poverty, hunger and disease.
- The pressure on ecosystems will increase globally in the coming decades due to change in human attitude and action.
- Measures to conserve natural resources are more likely to succeed if local communities are given the ownership of them, share the benefits, and are involved in decisions.
- Even today's technology and knowledge can reduce considerably the human impact on ecosystems. They are unlikely to be deployed fully, however, until

ecosystem services cease to be perceived as free and limitless, and their full value is taken into account.

- Better protection of natural assets will require coordinated efforts across all sections of governments, businesses and international institutions. The productivity of ecosystems depends on policy choices in investment, trader subsidy, taxation and regulation, among others.

(Source: Board of Disasters governing the MA, process, March 2005)

The goal of sustainable development is to enable all people throughout the world to satisfy their basic needs and enjoy a better quality of life without compromising the quality of life of future generations. UNESCO identifies several key themes as critical priorities for planning programmes and activities in support of the decade.

Among the more important are:

- Biodiversity;
- Freshwater Management;
- Environmental Conservation and Protection;
- Rural Transformation;
- Health Promotion;
- Sustainable Production and Consumption;
- Human rights, peace and international understanding; and
- Poverty alleviation and gender equality.

Sustainable development has three components:

### Environment, Society and Economy

If you consider the three to be overlapping circles of the same size, the area of overlap in the centre is human well-being. As the environment, society and economy become more aligned, the area of overlap increases, and so does human well-being. Sustainable development-education has come to be seen as a process of learning how to make decisions that consider the long-term future of the economy, ecology and equity of all communities.

There is a need to develop an appreciative awareness of the interdependence of all elements of the enhancement, including society and economy and encourage a respect and concern for use of natural ecosystems and resources that are regenerative and sustainable.

## ENVIRONMENTAL SUSTAINABILITY

*"Environmental Sustainability" is the development that meets the needs of the present generation without compromising the ability of future generations to meet their own needs".*

Environmental sustainability is partly global economic and social well-being, which envisages the use of natural resources such as forests, land, water and

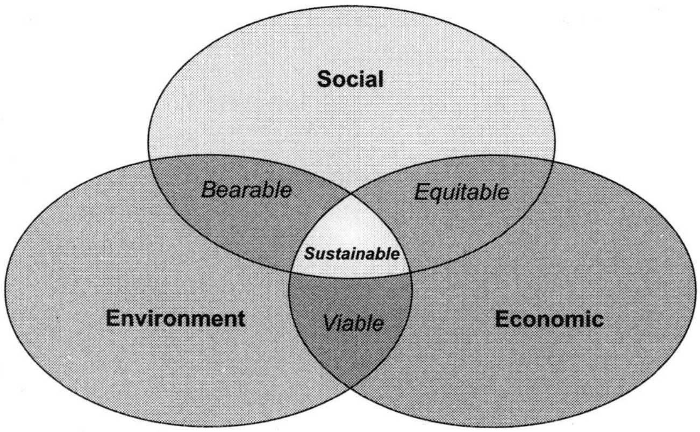


fisheries in a sustainable manner, without causing changes in our natural world.

Environmental problems are commonly viewed as technical problems. Environmental sustainability provides tools for analyzing the human conflict, as well as strategies for social change.

Environmental sustainability has five key areas:

1. **Environmental Protection**

   Environmental protection is responsible for the strategies, processes, systems and guidelines that focus on the protection, environmental stewardship, climate actions, monitoring and assessment, and environmental reporting.

2. **Water Quality**

   Water quality is responsible for sampling, testing, and data management of services that may impact the regional environment. Water quality assurance is also required in quality central services in areas such as portable water, reclaimed water, leachate, storm water, and source water. Water quality is also responsible for developing comprehensive regional standards for environmental testing, data management services and aquatic ecology.

3. **Environmental Partnerships**

   Environmental partnership is responsible for developing and delivering services that transforms behaviour towards the environment. Environmental partnerships uses a comprehensive and consistent social marketing approach to put in place the tools, resources and aids that assist citizens, business es and local government partners in finding solutions to better our environment.

4. **Environmental Resource Management**

   Environmental resource management is responsible for the planning, developing and application of systems that will best ensure an integrated and beneficial use and reuse of our resources, some of which may now be seen as waste. For example, rather than looking at waste to be disposed, there is a need to focus on looking at waste as a community for beneficial reuse. This includes a focus towards zero waste in our landfills, biosolids recycling, methane gas capture and other energy from waste initiative compost and organics processing initiative and other emerging opportunities.

5. **Environmental Engineering**
   Environmental engineering is responsible for all the professional engineering services to the department as well as ensuring that engineering within the entire organization meets engineering standards. Environmental engineering provides leadership to corporate projects, waste management plans, environmental considerations for the emerging opportunities, and landfill projects.

The basic concept for environmental services is to closely align and integrate responsibilities for high performance environmental monitoring, testing, source critical, engineering, social awareness and the 4 R's (reduce, reuse, recycle and recover).

### Environmental Sustainability—Ensure

1. Integrate positive principles of sustainable development into country policies and programming and reverse the loss of environmental resources:
   - The rate of deforestation shows signs of decreasing, but is still alarmingly high.
   - A decisive response to climate change is urgently needed.
   - The unparalleled success of the monitored protocol shows that action on climate change is within our grasp.
2. Reduce biodiversity loss, achieving by 2010, a significant reduction in the rate of loss:
   - The world has missed the 2010 target for biodiversity conservation, with potentially grave consequences.
   - Key habitats for threatened species are not being protected adequately.
   - The number of species extinction is growing by the day, especially in the developing countries.
   - Overexploitation of global fisheries has stabilized, but steep challenges remain to ensure their sustainability.
3. Halve, by 2025, the proportion of the population without sustainable access to safe drinking water and basic sanitation:
   - The world is on track to meet the drinking water target, though much remains to be done in some regions.
   - Accelerated and targeted efforts are needed to bring drinking water to all rural households
   - Safe water supply remains a challenge in many parts of the world
   - With half the population of developing regions without sanitation, the 2015 target appears to be out of reach

- Disparities in urban and rural sanitation coverage remain daunting
- Improvements in sanitation are bypassing the poor

4. By 2020, to have achieved a significant improvement in the lives of at least 100 million slum dwellers
   - Slum improvements, though considerable, are failing to keep pace with the growing ranks of the urban poor
   - Slum prevalence remains high in sub-urban Africa and increases in countries affected by conflict.

### Renewable Energy for Sustainable Future

The largest concern that environmentalists face today is renewable energy generation from natural resources. Renewable energy refers to energy resources that occur naturally and repeatedly in the environment and can be harnessed for human benefit. Renewable energy systems use resources that are constantly replaced and are usually less polluting. The best and immediate sustainable resources for ecofriendly power generation include solar energy, wind energy, hydro-energy, nuclear energy, ocean energy, geothermal energy and biogas. Renewable energy can also be obtained from natural resources such as trees and plants, rivers and even garbage/wastes.

The adverse effects of burning fossil fuels have led to the increasing demand of "Green Energy". There is a need to enhance awareness about green energy to meet the ever increasing demand of energy for sustainable development.

#### *Social sustainability*

*"Social sustainability means responding better to local communities ensuring responses are tailored to local community contexts, and promoting social industries, cohesion and accountability."*

Social sustainability is a concept in community planning which places equal emphasis on creating a community with access to social resources (i.e., human rights, employment, culture, ecology, information and knowledge, etc.) as is now placed on creating a community with a strong infrastructure (roads, water, emergency resources, etc.). It is built on the premise that those social concepts play a strong part in the success and growth of a community. A business built on this theory would encourage community volunteerism in its employees, promote cultural and educational events and use their resources for the growth of the community as a whole.

Social sustainability is one aspect of sustainable development. Social sustainability encompasses human rights, labour rights and corporate government. In common with environmental sustainability, social sustainability is the idea that future generations should have the same or greater access to social resources as the current generation. Social resources include ideas as broad as other cultures and basic human rights. The different aspects of social sustainability are often considered in socially responsible investing (SRI) as social sustainability criteria that are commonly used by SRI. It involves community, diversity, employee relations, human rights, product safety, reporting and governance structure. Social sustainability also involves participation of everyone socially.

Social sustainability is the core element of sustainability. Sustainability essentially means creating and maintaining the quality of life for people. Financial and environmental factors are important; by working towards financial and environmental sustainability we are already working towards social sustainability. The social element of sustainability does have a number of its own distinct criteria. A social sustainable system must:

1. protect mental well-being of all stakeholders;
2. protect physical health of all stakeholders;
3. encourage community;
4. treat all stakeholders fairly; and
5. provide all stakeholders with essential services.

These elements are essential to develop and maintain society.

Social sustainability assessment often includes:

- Social Impact Assessment (SSA)
- Environmental Impact Assessment (EIA) and Strategic Environmental Assessment (SEA)
- Theme Specific Assessment (TSA) with Theoretic Coverage

Social sustainability takes the larger world view into consideration in relation to communities, culture and globalization. Safeguards policy plays an important role in enhancing possible outcome of development prospects—the "benefit side" of the equation. Several policies require meaningful constitution and participation by affected people during the design. Preparation and implementation of bank-finances has often been an issue in projects reviewed by the inspection panel. The appropriate implementation of the banks safeguard policies plays a critical role in ensuring the effectiveness, sustainability and positive development impacts of its projects and programmes.

Sustainability, both social and environmental, is important to understand a system required for economic activity that is comparable with and not destructive of either the ecological web of life or the social web of life of which we are a part, and upon which we depend for our health, well-being and quality of life.

"Human development and the achievement of human potential requires a form of economic activity that is environmentally and socially sustainable in this and future generations".

## Economic Sustainability

*A widely accepted definition of Economic Sustainability is: "Maintenance of Capital, Or Keeping Capital Intact." Sustainable development is a pattern of resource use that aims to meet human needs while preserving the environment so that these needs can be met not only in the present, but also for future generations.*

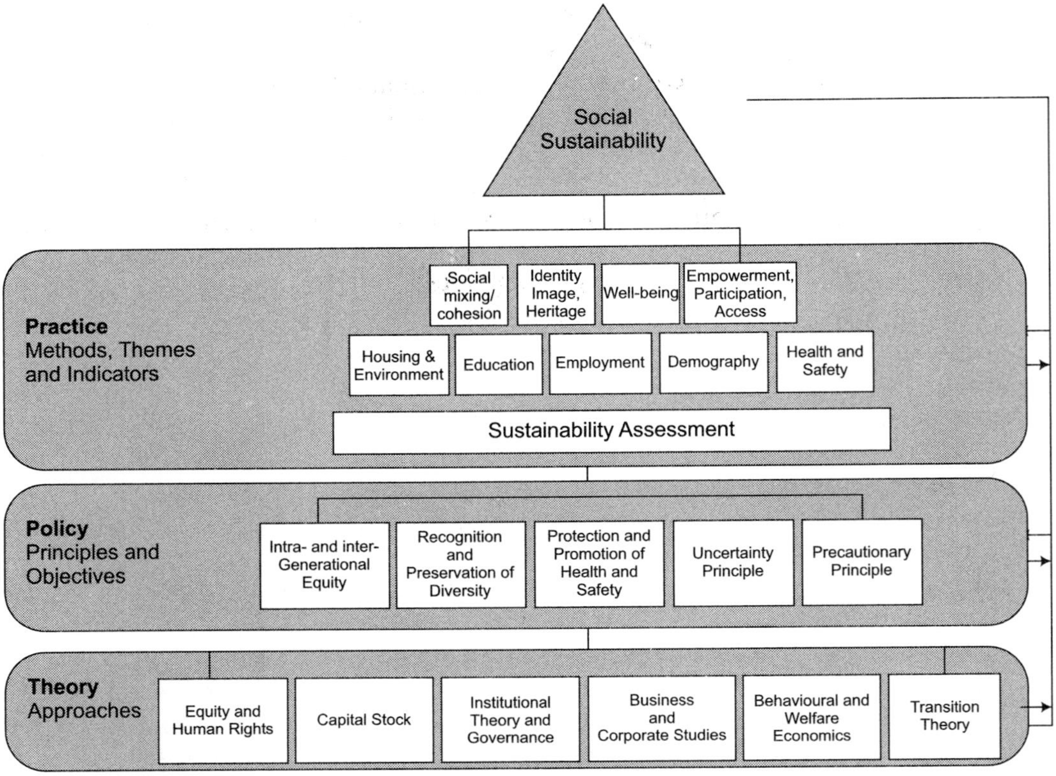


Source: Oxford Brookes University

**Fig. 1.** Social sustainability assessment framework (SSAF).

Economic sustainability is a term used to identify various strategies that make it possible to utilize available resources to best advantage. The idea is to promote the usage of resources that are both efficient and responsible and likely to provide long-term benefits. In the case of a business operation, economic sustainability calls for using resources so that the business continues to function over a number of years and consistently returning profit.

Economic sustainability involves assets and resources so as to identify the amount of return generated by the efficient use of these resources. The idea is to aid in identifying the areas of operation in which resources are utilized in the most efficient manner which might require corrective steps in production and business. The proposed changes, to be operational are considered in terms of their overall effect on the quality production, without any adverse impact on the environment. Economic sustainability encourages responsible use of resources. This involves not only making sure that the business is making a profit, but also that the operation is not creating environmental concerns that could cause harm to the balance of local ecology.

Economic sustainability requires to recognize:

- Strengths
- Opportunities
- Weakness
- Threats

The efficient and responsible use of resources usually discussed in monetary terms is what economic sustainability refers to. Economic sustainability can be applied to individuals, families, small and large business communities, governments or worldwide organizations.

**Ecological economics:** "To create an enduring society, we must devise a system of production and commerce where it is sustainable and restorable. It is believed that all business has a social contact with society and the environment- in effect a fiduciary responsibility and that the concepts of socially responsible business and shareholder democracy can be models for prospering successful business.

1. Economic system is based on a combination of private business, decentralized, democratic cooperatives, publicly owned enterprises, and alternative economic structures. This system puts human and ecological needs alongside profits to measure success, and maintains accommodability to communities.
2. Community-based economy constitutes an alternative to both corporate capitalism and state socialism. It values diversity and decentralization. To accumulate power and wealth is a pernicious characteristic of a civilization. Green revolution advocates that economic relation becomes more direct, more cooperative and more egalitarian. In order to shift to a world with a different way of life characterized by sustainability regionalization, more harmonious balance between the natural ecosphere and the human made technosphere and revival of community life is essential.
3. Greens support a major redesign of eliminate waste. In natural systems somebody's waste is a meal for someone else. Everything recycles; there is no waste. Natural system needs to be assessed for manufacture and production of goods. Durable goods would be designed in such a way that ultimately they can be disassembled and reassembled. Toxics would be safeguarded.
4. Sustaining quality of life, property, environmental health and long term survival demand are new ways of doing business. There is a need to remake commerce to encourage diversity and variety, responding to the local and global conditions. Green support, small business, responsible stakeholder capitalization are broad and diverse forms of economic cooperation. Economic diversity is more responsive than big business to the needs of diverse human populations.
5. There should not be any conflict between the economic growth and environmental protection. Green views economy as a part of the ecosystem not as an isolated subset in which nothing but resources come in and produces products and waste go out. There is an absolute limit to economic growth based on the laws of thermodynamics and principles of ecology.

The goal of economic sustainability is to establish profitability over a long term. A profitable business is much more likely to remain stable and continue to operate from one year to the next. From this perspective, economic sustainability can be seen as a tool to make sure the business does have a future and continues to contribute to the financial welfare of the society.

## REFERENCES

Krishna A. (2007) Understanding Sustainable Development, Social Science Research Network.

Michael C. (2008) Achieving Sustainable Development and Promoting Development Co-operation – Dialogue at ECOSOC, United Nation Publications.

Norton B.G. (2003) Searching for Sustainability, Interdisciplinary Essay in the Philosophy of Conservation Biology, Cambridge University Press.

Polese M. and Stern R.E. (2000) The social Sustainability of Cities: Diversity and the Management of Change. University of Toronto Press.

Taylor T.H. (2000), Alternate Energy Resources, Adam Hilger Ltd. Bristol.

# 4

# Bioprospecting of *Euphorbia nivulia* Buch.-Ham.

**Raghunath T. Mahajan and Shamkant B. Badgujar***

*Department of Biotechnology, KCE Society, Moolji Jaitha College, NH-06, Jalgaon-425001 Maharashtra, India*

**Corresponding author E-mail: sham83badgujar@gmail.com*

**ABSTRACT**

Nature always stands as a golden mark to exemplify the outstanding phenomenon of symbiosis. The history of herbal medicines is as old as human civilization. Many herbal remedies have been employed in various medicinal systems for the treatment and management of different ailments. *Euphorbia nivulia* Buch.-Ham., a member of *Euphorbiaceae* family, is a wild, thorny, xerophytic, succulent plant, commonly used in fencing of the agricultural field and also in dry barren areas. It has different biological activities for the treatment of several ailments of human beings. It possesses antimicrobial, wound healing, haemostatic, larvicidal, insecticidal, nematicidal and cytotoxic activities. Chemically, it contains terpenes, glycoproteins, phytoelements and phytochemicals. This article provides informative data on ethnobotanical importance and bioprospecting of *Euphorbia nivulia* Buch.-Ham.

**Keywords:** *Euphorbia Nivulia,* Biological Activities, Traditional Uses.

## INTRODUCTION

Vegetation is one of the precious gifts of nature where plants are intimately related to human beings. People have been utilizing plants as medicine since the earliest period of civilization. The use of plants for curing human diseases is an ancient practice, in which interest has been revived in modern age (The Wealth of India). People of remote villages and tribal areas are dependent upon the practice of folk medicines (Nadkarni, 1982). In recent times, focus on plant research has increased all over the world and a large body of evidence has been collected to show immense potential of medicinal plants used in various traditional systems (Kirtikar and Basu, 1995). Ethnobotanically, plant latex has a great potential with respect to its medicinal value. Latex has been reported to occur in 12000 plant species belonging to 900 genera. A common feature that can be found in the latex of *Euphorbiaceae* is the presence of noticeable digestive enzyme activity. *Euphorbia* is a large genus consisting of about over 2000 species in the world (http://www.desert-tropicals.com/Plants/Euphorbiaceae/ Euphorbia.html). Approximately 195 species of *Euphorbia* have been recorded from India (Aditya, 2010). The genus includes herbs, shrubs and trees in widely diverse habitats. (Basak et al., 2009). One such plant, *Euphorbia nivulia* Buch.-Ham. invites attention of the researchers worldwide for its biological activities. There is not much literature available on the biological activities of *Euphorbia nivulia* Buch.-Ham. Also, the earlier reviews on euphorbian plants lack satisfactory information regarding its biological activities. The aim of the present review is to provide the updated information on biological uses of *Euphorbia nivulia*. Emphasis is being laid on the areas of most recent interest and those which have not been presented in earlier reports.

## BOTANICAL DESCRIPTION

Kingdom: Plantae, Division: Magnoliophyta, Class: Magnoliopsida, Order: Malpighiales, Family: Euphorbiaceae, Genus: *Euphorbia*, Species: *Euphorbia nivulia* Buch.-Ham.

### Synonyms

Sanskrit: Patrasnuhi; Patta-karie. Hindi: Sehund. Bengali: Sij. Marathi: Sabar; Sabarkanda; Nivdung. Telgu: Akujimudu. Tamil: Akujemudu.

### Habitat

Northern and central India; planted as hedge plant, also wild in arid soils and often planted in dry areas.

### Part Used

Juice of leaves, root bark, stem and latex.

### Species of Euphorbia

Its species are widely distributed in tropical Asia, Africa, Europe and Australia. Other reported species of genus Euphorbia are *Euphorbia acaulis* Rox., *Euphorbia antiquorum* Linn., *Euphorbia clarkeana* Hook f., *Euphorbia cristata* Heyne ex Roth., *Euphorbia cyathophora* Murr., *Euphorbia dracunculoides* Lamk., *Euphorbia fusiformis* Buch.-Ham., *Euphorbia hirta* Linn., *Euphorbia hypericifolia* auct. non Linn., *Euphorbia indica* Lamk., *Euphorbia milii* Desmoul, *Euphorbia neriifolia* auct. non Linn., *Euphorbia nivulia* Buch.-Ham., *Euphorbia notoptera* Boiss., *Euphorbia panchaganiensis* Blatt and McCann., *Euphorbia pilosa* Linn., *Euphorbia prostrata* Ait., *Euphorbia prunifolia* Jacq., *Euphorbia pulcherrima* Willd. ex Klotz., *Euphorbia pycnostegia* Boiss., *Euphorbia resinifera* Berg., *Euphorbia royleana* Boiss., *Euphorbia thomsoniana* Boiss., *Euphorbia thymifolia* Linn. and *Euphorbia tirucalli* Linn (Patil, 2003; Khare, 2007).

## MORPHOLOGICAL CHARACTERS

Tall shrubs or small trees with cylindrical stem and branches. Stipular spines glabrous, straight, paired, often blackish. Leaves appear only during the rainy season, crowded at the end of branches, obovate-oblong or spathulate, glabrous. Cymes - 3- flowered, born from above the leaf scars on the tubercles. Capsules glabrous, trigonous, seeds globose, dorsally lined, smooth. Flowering and fruiting period is March to July (Patil, 2003; Khare, 2007).

## USES OF PLANT

All parts of the plant possess medicinal properties. The juice of the leaf is used as a purgative, diuretic, etc. The paste of the leaf, made with neem oil, is applied externally in rheumatism. Plant latex is used for treating jaundice, dropsy, enlargement of liver and spleen, and applied to hemorrhoids. Coagulated latex is used for bronchitis (Khare, 2004). It is warmed in mustard oil; the resulting mixture is applied in cold and headache. Root bark is boiled in rice water and arrack is given in dropsy. In north-eastern region of India, the leaf juice is utilized in pains and boils by Boro community of Assam (Basumatary et al., 2004). The latex of this plant possesses vesicant, wormicidal and purgative properties (Pullaiah, 2006). Milky juice, i.e., latex of this plant is reported for its bronchodilating activity (Savithramma et al., 2007). Fleshy part of stem is roasted in extremely hot ash for 20-30 minutes, juice is extracted, 1-2 tablespoonful juice (for children 3-7 year aged) per day is a recommended dose for curing cough within 7-10 days (Mahajan and Badgujar, 2008). Leafy latex and root is applied in skin and ear disorders, retention of urine, swelling and worm infection (Britto et al., 2010). Kumar and Chaturvedi (2010) report the application of ethnomedicine derived from the stem of this plant in curing the bone fractures, and antiseptic utility of latex. Although there is a wide range of potential useful medicinal phytoconstituents of the plant, the research in this area is infantile. Some of the experimentally proved biological activities of *Euphorbia nivulia* are given below.

### Antimicrobial

Aqueous alcohol extract of the leaves of *Euphorbia nivulia* is reported to have antimicrobial activity. Nineteen different bacterial strains and two fungal cultures were used for antimicrobial activity. The results indicate that the aqueous alcohol extract is active at concentrations ranging from 3–12.5 μg/ml for Gram-negative and 12.5–200 μg/ml for Gram-positive bacteria, in comparison with known antibiotics such as streptomycin and nystatin (Annapurna et al., 2004). The latex has several triterpenes, which exhibit significant antimicrobial activity against *Staphylococcus aureus* and *Escherichia coli*. This information is given by Khare (2007) in his book, "Indian Medicinal Plants". The remarkable effect on bacterial growth inhibition with increasing concentration of Nivulian, a cysteine protease of *Euphorbia nivulia latex* is reported by us. It shows comparable inhibitory activity against both Gram-positive and Gram-negative bacteria tested. Nivulian had a strong antibacterial activity against *Escherichia coli* and *Staphylococcus aureus,* moderate against *Pseudomonas aeruginosa* and lesser activity against *Klebsiella pneumoniae, Proteus vulgaris* and *Bacillus subtilis* (Badgujar, 2011). The observations made by Nagarathnam et al., 2010, regarding the antimicrobial activity of cysteine protease of plant origin present in *Curcuma longa* are also similar to Nivulian. Another member of the Euphorbiaceae i.e., *Euphorbia prostrata* is also reported for antibacterial activity (Schmelzer and Gurim-Fakim, 2008). Therefore we believe that plant latex may find a position in preparing antimicrobial compounds with a great potential in abatement of infectious diseases in the near future.

### Wound Healing

Recently, Badgujar et al., 2009, reported that the centrifugal fraction of latex of *Euphorbia nivulia* enhanced the rate of wound contraction and period of epithelization in mice. The surface area of latex treated wound was reduced by 93 to 98 per cent on the $18^{th}$ day as compared to control (90 per cent). It is interesting to note that similar type of wound healing activity is present in other members of Euphorbiaceae family, viz., *Euphorbia hirta* and *Euphorbia nerifolia*. Ethanolic extract of the whole plant of *Euphorbia hirta* is reported for burn wound healing purpose (Jaiprakash et al., 2006). Plant latex of *Jatropha curcas* is ethnobotanically reported for curing the wound. Active ingredient i.e., curcain, isolated from the latex of *Jatropha curcas,* has a profound effect on wound healing phenomenon, as is evident from the report of Nath and Dutta (1992). Aqueous extract of latex of *Euphorbia nerifolia* is evaluated for wound healing effect in guinea pig. This extract facilitated the healing process as evidenced by an increase in tensile strength, DNA content, epithelization and angiogenesis (Rasik et al., 1996). Thus, the plant latex contains not only cysteine protease but also other active ingredient responsible for wound healing activity.

### Haemostatic

A significant reduction in bleeding/clotting time in mice is reported by the treatment of Nivulian (Badgujar and Mahajan, 2009). Also, it significantly decreased the coagulation time of whole blood in mice. The Nivulian arrested bleeding from fresh wounds by reducing bleeding/clotting and whole blood coagulation time, which are important indices of haemostatic activity. The reduction in coagulation time of whole blood by the Nivulian indicates that it may also enhance the blood coagulation pathways. Thus, this shrub could be a promising haemostatic agent (Badgujar, 2011). Haemostatic function of plant product is also reported from the latex of Euphorbiaceae member i.e., *Jatropha gossypifolia* (Oduola et al., 2007). Traditional report of hemostatic activity of *Jatropha curcas* is experimentally proved by Osoniyi and Onajobi (2003). Plant latex remarkably reduces the whole blood clotting and bleeding time. However, active ingredient of plant latex of *Euphorbia nivulia* is yet unknown and has to be worked out.

### Larvicidal

The larvicidal activity of centrifugal fraction of latex against fourth instar stage of *Anopheles* sp. is very recently reported by us (Badgujar, 2011). Mortality rate of larvae was high at higher concentration of proteinous fraction of latex. But lower concentration caused several morphological changes in larvae. The Nivulian (centrifugal fraction of latex) had a moderate larvicidal activity as it was 2.5 times less potent than pyrethrin, when fourth instar larvae of *Anopheles* sp. are exposed to both for a period of 24 hrs. The larvicidal activity of ethyl acetate, butanol, and petroleum ether extracts of five species of Euphorbiaceae plants, viz., *Jatropha curcas, Pedilanthus tithymaloides, Phyllanthus amarus, Euphorbia hirta,* and *Euphorbia tirucalli,* is evaluated against the early fourth instar larvae of *Aedes aegypti* L. and *Culex quinquefasciatus* (Say). The larval mortality is observed after 24 hrs. of exposure. The $LC_{50}$ values of petroleum ether extracts of *Jatropha curcas, Pedilanthus tithymaloides, Phyllanthus amarus, Euphorbia hirta,* and *Euphorbia tirucalli* are 8.79, 55.26, 90.92, 272.36, and 4.25 ppm, respectively, against *Aedes aegypti,* and 11.34, 76.61, 113.40, 424.94, and 5.52 ppm, respectively, against *Culex quinquefasciatus.* Of the various ratios tested, the petroleum ether extracts of *Jatropha curcas* and *Euphorbia tirucalli* are more efficient than the other plant extracts (Rahuman et al., 2008). It is, therefore, suggested that plant latex may have an ideal potential larvicidal property against *Aedes aegypti* and *Culex quinquefasciatus.* This may be an ideal ecofriendly approach for the control of dengue vector, *Aedes aegypti,* and the lymphatic filariasis vector, *Culex quinquefasciatus.*

### Insecticidal

The *Euphorbia nivulia* aqueous leaf extract exhibited toxic and insect growth regulatory (IGR) effects on cabbage diamondback moth (DBM), *Plutella xylostella* L. At 96 hrs. after treatment, mortality is observed only in 10 and 20 per cent concentration with a maximum of 10 per cent mortality. At adult emergence, the highest mortality was observed at 2.5 per cent concentration (57.7 % mortality) followed by 10, 20 and 5 per cent concentrations of extract. The mortality is observed to increase from 96 hrs. after treatment till adult emergence in all the concentrations. A change in mortality pattern with variable time is observed such that the initial mortality is largely due to toxic effect, while at the time of adult emergence, the IGR effect was more compared to toxic effects (Uma et al., 2009).

### Nematicidal

The nematicidal effect of Nivulian showed that the reproduction of root knot nematode, *Meloidogyne incognita* in plants grown in soil (treated with Nivulian) is significantly suppressed. However, nematode population declined only in pots treated with Nivulian and standard nematicide, whereas, in untreated-inoculated pots it had almost doubled. It is, therefore, reasonable to believe that plants grown with Nivulian develop certain degree of resistance against nematode attack. It may be due to the latex containing phenolic substances or the absorption of substances liberated during the decomposition of Nivulian by soil microbial flora (Badgujar and Mahajan, 2009). This type of nematicidal property is already reported in some latex bearing plants namely, *Carica papaya, Artocarpus heteropyllus, Ficus carica, Ficus elastica, Ficus glomerata, Ipomoea fistulosa, Nerium odorum* and *Tabernaemontana coronaria* by Siddiqui et al., 1987. Incorporation of chopped shoots of these laticiferous plants significantly suppressed the population of root knot nematode i.e., *Meloidogyne incognita.* Very interestingly, *Euphorbia prostrata* possesses antiparasitic activity against *Meloidogyne incognita.* It is mentioned for its nematicidal potency (Schmelzer and Gurib-Fakim, 2008). Therefore, we believe that the plant latex may occupy the place of biopesticide for controlling the population of root knot nematodes.

### Cytotoxic Activity

The latex of *Euphorbia nivulia* afford ingol diterpenes, 3-acetyl-8-methoxyl-7-angolyl-12-hydroxylingol (6), 3, 12-diacetyl-7-hydroxy-8-methoxylingol (7), and 3,12-diacetyl-7-angolyl-8-hydroxylingol (8) as shown in Figure 1 (Ravikanth et al., 2003) along with three macrocyclic ingol diterpenes derivatives: 3, 7, 12-triacetyl-8-benzoylingol (1), 3, 12-diacetyl-7-angeloyl-8-methoxyingol (2), and 7-angeloyl-12-acetyl-8-methoxyingol (3) and two ingol diterpenes derivatives 3, 12-diacetyl-8-benzoylingol (4) and 3, 12-diacetyl-7-benzoyl-8-nicotinylingol (5) (Ravikanth et al., 2002). The structures of the above natural products are deduced on the basis of spectroscopic analysis (Figure 1). Mostly all diterpenes exhibit cytotoxic activity.

These authors also reported that the ingol derivatives (**1–5**) of *Euphorbia nivulia* latex are reported for prostaglandin E2 (PGE2) inhibitory activity; only isolate **3** ingol derivative showed significant inhibition. These diterpenes are known for their diverse biological activities (Sahai et al., 1981; Khan and Malik, 1990; Fatope et al., 1996). Ingol derivatives of compounds **1–8** are notably reported for their potential anticancer activity (Ravikanth et al., 2003). Of these, compounds **2**, **3**, and **7** showed significant cytotoxic activity against Colo 205, MT2, and CEM cell lines and others showed moderate or no activity. Toxicity potential of compounds **2**, **3**, and **7** is almost same in the three cell lines (Ravikanth et al., 2003).

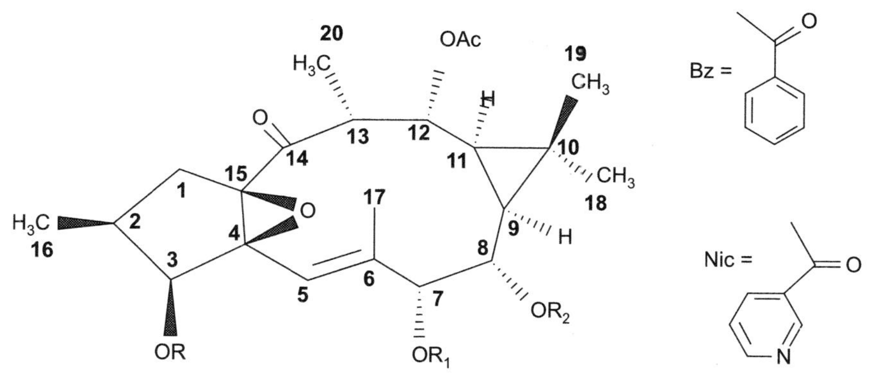


**1** R = $R_1$ = AC; $R_2$ = Bz
**2** R = Ac; $R_1$ = Ang; $R_2$ = Me
**3** R = H; $R_1$ = Ang; $R_2$ = Me
**4** R = Ac; $R_1$ = H; $R_2$ = Bz
**5** R = Ac; = $R_1$ = Bz; $R_2$ = Nic
**6** R = Ac; $R_1$ = Ang; $R_2$ = Me; Ac = H
**7** R = Ac; $R_1$ = H; $R_2$ = Me
**8** R = Ac; $R_1$ = Ang; $R_2$ = H

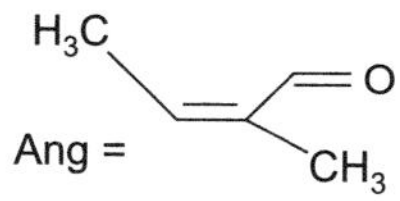


**Fig. 1.** Ingol derivatives of *Euphorbia nivulia* latex.

## CHEMICAL CONSTITUENTS

### Terpenes

The new tetracyclic triterpenes, Cycloart-25-en-2β-ol and cyclonivulinol (Rao et al., 1985) and three ingol diterpenes 3,12-*O*-diacetyl-7-*O*-angeloyl-8-methoxyingol; 3,7,12-*O*-triacetyl-8-*O*-benzoylingol and 7-*O*-angeloyl-8-methoxy-12-*O*-acetylingol (Ravikanth et al., 2003) are identified from the latex of *Euphorbia nivulia.*

### Glycoproteins

High molecular weight lectin (44 kDa) was purified by affinity chromatography from the latex of *Euphorbia nivulia*. It is a type of glycoprotein containing 9% carbohydrate and appears to be a dimer with subunits of $M_r$ of 22 000 Da. Its sedimentation coefficient is 4.8S (Inamdar et al., 1988). Recently, we isolated two glycoproteins from the latex of this plant, viz., Nivulia-II and Nivulian-III (43.42 and 52.96 kDa). Nivulian-II has notable proteolytic activity and it belongs to cysteine protease category (Badgujar and Mahajan, 2010).

### Phytoelements

Phytoelements like Fe (1.484), Cu (0.072), Zn (0.384), Mn (0.173), Mg (0.204), Na (2.085) and Ca (1.031) are detected in ppm quantities from the ash sample of plant latex by using atomic absorption spectroscopy (Badgujar, 2011). All these minerals are also found in latex bearing unripe fruit pulp of *Carica papaya* (Oloyede, 2005) and *Ficus carica* (Badgujar and Mahajan, 2010). Only difference is the per cent occurrence of these minerals with great variation in its content.

## Phytochemicals

The latex of *Euphorbia nivulia* contains phenolic compounds, alkaloids, cynogenic glycosides, terpenes and tannins (Mahajan and Badgujar, 2008). Miscellaneously, it also contains caoutchouc-like substance, fat, albuminoids, hydrolytic enzymes, sugar, pectin, citric, tartaric and mallic acid, dextrin, euphol, nerifloiol etc (Badgujar, 2011). Similar phytochemicals are reported in the plant latex samples of *Carica papaya* and *Hevea brasiliensis* (Ravikanth et al., 2003).

## SUMMARY

In the present global scenario, natural medicines are gaining prominence, because they are economical, easily available and relatively free from side effects. The increased global demand for polyherbal formulations is reflective of positive impact of consolidated efforts aimed at reviving science of phytopharmacy. *Euphorbia nivulia* possesses a wide spectrum of biological activities applicable in both medicine as well as agricultural practices. This plant has a potential for various health related activities such as antimicrobial, antimalarial, cytotoxic, wound healing, haemostatic and many more. As this plant exhibits soil-pest controlling property, it is also used in post-harvest management process. The benefits of the vigorous nature of *Euphorbia nivulia* and its wide geographical distribution offer an opportunity to agro and pharma industries.

## ACKNOWLEDGEMENTS

It is my pleasure to express my thanks to Hon'ble Principal, A.G. Rao, Moolji Jaitha College, Jalgaon, for his moral support and facilitating the use of library during the course of this investigation.

## REFERENCES

Aditya, S. (2010). A revision of genophytic *Euphorbia* species from India. *Euphorbia World*, **6(1):** 18-21.

Badgujar, S.B., Mahajan, R.T. and Chopda, M.Z. (2009). Wound healing activity of *Euphorbia nivulia* Buch.-Ham. latex in mice, *Research Journal of Pharmacology and Pharmacodynamic*, **1(2):** 90-92.

Badgujar, S.B. and Mahajan, R.T. (2010). Characterization of milk clotting cysteine protease of *Euphorbia nivulia* Buch.-Ham. latex. *Green Farming (New Series)*, **1(6):** 645-648.

Badgujar, S.B. and Mahajan, R.T. (2010). Chemical profile of fruit latex of *Ficus carica* L. *Journal of Chemo and Biosphere*, **1(1):** 33-37.

Badgujar, S.B. and Mahajan, R.T. (2009). Evaluation of nematicidal properties of some laticiferous plants. *Green Farming*, **2(10):** 680-684.

Badgujar, S.B. (2011). Proteolytic enzymes of some latex bearing plants belonging to Khandesh region of Maharashtra, Ph.D. Thesis North Maharashtra University. Jalgaon (Maharashtra State), India.

Basak, S.K., Bakshi, P.K., Basu, S. and Basak, S. (2009). Keratouveitis caused by *Euphorbia* plant sap. *Indian journal of Ophthalmology*, **57(4):** 311-313.

Basumatary, S.K., Ahmed, M. and Deka, S.P. (2004). Some medicinal plant leaves used by Boro (tribal) people of Golapara district, Assam. *Natural product Radiance*, **3(2):** 88-90.

Britto, A.J., Sujin, R.M., Mahesh, R. and Dharmar, K. (2010). Ethnomedicinal wisdom of the Manavalakuruchi people in Kanyakumar district, Tamil Nadu. *International Journal of Biological Technology*, **1(2):** 25-30.

Fatope, O.M., Zeng, L., Ohayagha, E.J. and Mclaughlin, L.J. (1996). New 19-acetoxyingol diterpenes from the latex of *Euphorbia poisonii*. *Bioorg. Med. Chem.* **4(10):** 1679-1683.

Inamdar, S.R., Murugiswamy, B. and Madiah, M. (1988). Purification and characterization of a lectin from *Euphorbia nivulia* Buch.- Ham. latex. *International Journal of Protein and Peptide Research*, **35(1):** 35-46.

Jaiprakash, B. and Chandramohan, Reddy, D.N. (2006). Burn wound healing activity of *Euphorbia hirta*. *Ancient Science of Life*, **XXV(3 and 4):** 1-3.

Khan, Q.A. and Malik, A. (1990). A new macrocyclic diterpene ester from the latex of *Euphorbia tirucalli*. *Journal of Natural Product*, **53(3):** 728-731.

Khare, C.P. (2004). Indian herbal remedies: rational western therapy. Ayurvedic and other traditional usage botany. Springer Science, New York, USA, 218-220.

Khare, C.P. (2007). Indian medicinal plants (An illustrated dictionary). Springer Science, New York, USA.

Kirtikar, K.R. and Basu, B.D. (1995). Indian Medicinal Plants, Vol. I-IV, International Book Distributors, Dehradun.

Kumar, G.P. and Chaturvedi, A. (2010). Ethnobotanical observations of Euphorbiaceae species from Vidarbha region, Maharashtra, India, *Ethnobotanical Leaflets*, **14:** 674-680.

Mahajan, R.T. and Badgujar, S.B. (2008). Ethnomedicinal values of laticiferous plants used by the tribal people of North Maharashtra, India. *Research Link.*, **55VII(8):** 20-23.

Mahajan, R.T. and Badgujar, S.B. (2008). Phytochemical investigations of some laticiferous plants belonging to Khandesh region of Maharashtra. *Ethnobotanical Leaflets*, **12:** 1145-1152.

Nadkarni, A.K. (1982). Dr. K.M. Nadkarni's Indian Materia Medica, Bombay Popular Prakashan Pvt. Ltd, Bombay, India.

Nagarathnam, R., Rengasamy, A. and Balasubramanian, R. (2010). Purification and properties of cysteine protease from rhizomes of *Curcuma longa* (Linn.). *Journal Science of Food and Agriculture*, **90:** 97-105.

Nath, L.K. and Dutta, S.K. (1992). Wound healing response of the proteolytic enzyme curcain. *Indian Journal of Pharmacology*, **24:** 114-115.

Oduela, T., Popoola, G.B., Avwioro, O.G. and Oduola, T.A. (2007). Ademosun AA, Lawal MO, use of *Jatropha gossypifolia* stem latex as a haemostatic agent: how safe is it? *Journal of Medicinal Plants Research*, **1(1):** 14-17.

Oloyede, O.L. (2005). Chemical profile of unripe pulp of *Carica papaya*, Pakistan. *Journal of Nutrition*, **4(6):** 379-381.

Osoniyi, O. and Onajobi, F. (2003). Coagulant and anticoagulant activities in *Jatropha curcas* latex. *Journal of Ethnopharmacaology*, **89(1):** 101-105.

Patil, D.A. (2003). Flora of Dhule and Nandurbar Districts. Bishen Singh Mahendra Pal Singh, Dehradun.

Pullaiah, T. (2006). Encyclopedia of world medicinal plants, Volume 2, Regency Publication, New Delhi. 917.

Rahuman, A.A., Gopalakrishnan, G., Venkatesan, P. and Geetha, K. (2008). Larvicidal activity of some Euphorbiaceae plant extracts against *Aedes aegypti* and *Culex quinque fasciatus* (Diptera: Culicidae). *Parasitology Research*, **102(5):** 867-873.

Rasik, A.M., Shukla, A., Patnaik, G.K., Dhawan, B.N., Kulshrestha, D.K. and Srivastava, S. (1996). Wound healing activity of latex of *Euphorbla neriifolia* Llnn. *Indian Journal of Pharmacology,* **28:** 107-109.

Ravikanth, V., Reddy, V.L.N., Rao, T.P., Diwan, P.V., Ramkrishna, S. and Venkateswarlu, Y. (2002). Macrocyclic diterpenes from *Euphorbia nivulia. Phytochemistry,* **59:** 331-335.

Ravikanth, V., Reddy, V.L.N., Reddy, V.A., Ravinder, K., Rao, T.P., Siva Ram, T., Anand Kumar, K., Prakesh, V.D. and Venkateswarlu, Y. (2003). Three new ingol diterpenes from *Euphorbia nivulia*: evaluation of cytotoxic activity. *Chem. Pharm. Bull. (Tokyo),* **51:** 431-434.

Sahai, R., Rastogi, R.P., Jakupovic, J. and Bohlmann, F. (1981). Analgesic and anticancer activities of diterpenes from *Euphorbia* Sp. *Phytochemistry,* **20:** 1665-1667.

Savithramma, N., Sulochana, C. and Rao, K.N. (2007). Ethnobotanical survey of plants used to treat asthma in Andhra Pradesh, India. *J. Ethnopharmacology,* **113(1):** 54-61.

Schmelzer, G.H. and Gurib-Fakim, A. (2008). Plant Resources of Tropical Africa 11 (1): Medicinal Plants 1, Backhuys Publishers, Netherlands, 287.

Siddiqui, M.A., Haseeb, A. and Alam, M.M. (1987). Evaluation of nematicidal properties in some latex bearing plants. *Indian J. Nematology,* **17(1):** 99-102.

The Wealth of India: A Dictionary of Raw Material and Industrial Products–Raw Material Series, Vol. I-XI (Publication and Information Directorate, CSIR, New Delhi, India), 1948-1976.

Uma, M.S., Prasanna, P.M., Manjunatha, Reddy, G.V. and Kumar, A.R.V. (2009). Efficacy of some Euphorbiaceae plant extracts against cabbage diamondback moth *Plutella xylostella* L. *Karnataka. J. Agric. Sci.,* **22(3):** 688-689.

# 5

# Arbuscular Mycorrhizal Fungi (AM Fungi): A Potential Tool for Augmenting Carbon Dioxide Sequestration

**Sibi Balachandran and Seema Mishra***

*SIES Indian Institute of Environment Management, Sri Chandrasekarendra Saraswati Vidyapuram, Nerul, Navi Mumbai-400706*
*E-mail: seemamishra03@gmail.com*

**ABSTRACT**

AM fungi establish mutualistic symbiosis with majority of plant species, acting as a living bridge between soil and plants, where they help in soil fertility improvement, plant nutrition and aboveground plant community composition. Distribution of AM fungi in diverse ecosystem including polluted and degraded sites depicts their survival and adaptation strategy. In view of global environment change, arbuscular mycorrhizal fungi (AM fungi) play a central role in the terrestrial C cycle. The direct and rapid acquisition of recent photosynthate and rapid turnover of hyphal network in the soil by AM fungi indicate their potential in carbon sequestration in soil. Due to the enormous potential of AM fungi in $CO_2$ sequestration and survival under adverse environmental conditions, it could be postulated as a potential tool for the enhanced $CO_2$ sequestration in majority of ecosystems.

**Keywords:** AM Fungi, $CO_2$ Sequestration, Hyphal Turnover, Glomalin.

## INTRODUCTION

Carbon dioxide is essential for photosynthesis in plants and other photoautotrophs, and is also a prominent greenhouse gas. Carbon dioxide efflux from soil or soil respiration is an important component of the global carbon budget that is getting disturbed due to anthropogenic and natural causes. The global $CO_2$ concentrations increased progressively from about 316 ppm in 1960 to 389.78 ppm in 2010 (NOAA, 2011; Figure 1), and it is predicted that there will be a doubling of present concentration of $CO_2$ by the middle of the next century (Rouhier and Read, 1998). This increase in $CO_2$ rate in combination with other greenhouse gases led to an increase in average global temperature, posing a threat to the plant biodiversity and has already affected hydrological, terrestrial and marine ecosystems.

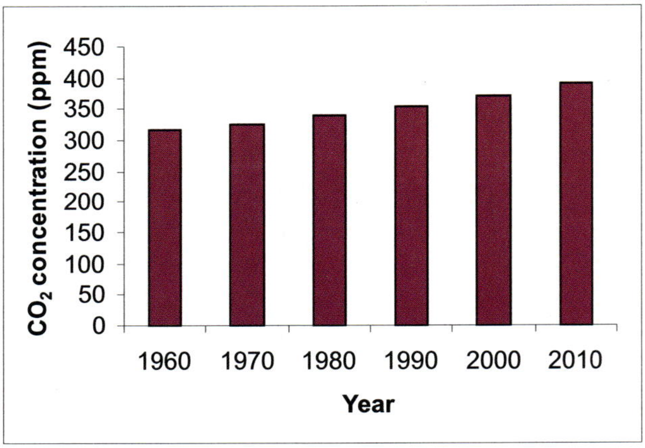


**Fig. 1.** Mean $CO_2$ concentration (ppm) in atmosphere in the last fifty years (Source: NOAA, 2011).

Increase in atmospheric concentration of $CO_2$ and other greenhouse gases has necessitated identification of technological options to sequester $CO_2$ into long-lived pools like terrestrial sequestration (Lal, 2009). Terrestrial sequestration of $CO_2$ involves the transfer of $CO_2$ from the atmosphere into soils and vegetation. Plants fix $CO_2$ and allocate a part of the photosynthate to roots (Rogers et al., 1994) and soil (Jones et al., 1998). It includes both removal of $CO_2$ from the atmosphere and limiting $CO_2$ emission from terrestrial ecosystem into the atmosphere. Carbon in the form of organic matter is a key element to healthy soil. It is estimated that each tonne of soil organic matter releases 3.667 tonnes of $CO_2$, which is lost into the atmosphere. Similarly, the build-up of each tonne of soil organic matter removes 3.667 tonnes of $CO_2$ from the atmosphere. The conversion of natural habitats to cropland and pasture, and unsustainable land practices such as excessive tillage frees carbon from organic matter, releasing it to the atmosphere as $CO_2$. Aerial plant tissue materials usually decompose quickly after plant death, releasing $CO_2$. Root material and root exudates have a much greater resistance to decomposition and therefore a greater potential for long term decomposition. Depletion of organic carbon causes deficit of carbon in the soils. Soils can regain lost carbon by reabsorbing it from the atmosphere. This process is called carbon sequestration (Kumar et al., 2006). Microbes and plant interaction in rhizosphere control the release and sequestration of carbon. Metting et al. (2004) advocated the implementation of proper management practices in terrestrial C sequestration by using technologies like increasing native biodiversity, use of AM fungi, *rhizobium* and soil conditioners for maximization of terrestrial $CO_2$ sequestration (Table 1). AM fungi are primarily responsible for nutrient transfer from soil to plant. This is achieved by the growth of their fungal mycelium within a host root and out into the soil (Dodd et al., 2000). AM fungi form specialized structures inside the root cells called arbuscules, which are believed to be the main site for nutrient transfer between the plant and fungus. AM fungi gain all their carbon for growth from the plant photosynthate in exchange for a supply of mineral elements from the soil. They have significant morphological and biochemical characteristics that induce their potential for $CO_2$ sequestration.

**Table 1** Proposed global carbon sequestration potential from implementing technologies (Source: Metting et al., 2004)

| *Technical Approach* | *Potential Sequestration Pg C/yr* | *Potential success High/ Med/Low (Inverse of Risk)* | *Time to Implement (yr)* |
|---|---|---|---|
| *Crop & forest management* [a] | | | |
| Cropland management | 0.5 - 2 | H | Now |
| Forest management | 1 – 3 | M/H | Now |
| *Underutilized resources* | | | |
| Deserts & saline/ alkaline aquifers [b] | 0 – 1 | L | 25 |
| Native plant biodiversity [c] | N/A | H | Now |
| *Plant biotechnology* [d] | | | |
| Innate productivity | 1 – 6 | L/M | 25 |
| Photoassimilate partitioning | 0 – 2 | L/M | 25 |
| Stress tolerance (salt, drought) | 1 – 3 | M | 10 |
| Nitrogen fixation | 1 – 2 | L/M | 25 |
| Lignin (fiber) biosynthesis | 1 – 2 | L/M | 5 |
| Bio products [e] | N/A | H | 5 |
| *Microbial biotechnology* | | | |
| Soil conditioners | 0 - 0.5 | M | Now |
| Mycorrhizal fungi | 1 – 2 | H | 10 |
| Engineering rhizobial communities | 1 – 2 | H | 25 |
| *Chemical technology* | | | |
| Smart fertilizers | 0 – 1 | M | 10 |
| Soil additives/conditioners | 0 - 0.5 | H | Now |
| Calcium for arid regions | 0 – 2 | L | Now |
| Plant growth regulators | 0 – 1 | H | Now |

[a]Does not include impacts of biotechnology or chemical technology, which are probably additive.
[b]Baseline is unmanaged arid/hyperarid/desert without plant cover. Lower value is periodic desert flooding. Larger value is for highly engineered microalgal mass culture.
[c]Because many scenarios are possible, there is no single comparison base.
[d]All of the plant biotechnology options are intimately linked, so values are probably not additive.
[e]Fossil fuel offsets

## GENERAL CHARACTERISTICS OF AM FUNGI

### Diversity and Taxonomical Features

AM fungi are endophytic and commonly occur in about 80% of plant species. The phylogenetic relationship between origin and diversification of AM fungi and coincidence with vascular land plants was investigated by Simon et al. (1993) by sequencing rDNA genes as a molecular clock to infer dates. The authors estimated the origin of AM-like fungi 353-462 million years ago, which is consistent with the hypothesis that AM were instrumental in the colonization of land by ancient plants. This hypothesis is also supported by the observation that AM fungi can now be found worldwide in the angiosperms, gymnosperms as well as ferns, suggesting that its nature is ancestral (Brundrett, 2002). AM fungi have been placed in glomales order of class Zygomycetes due to the presence of chitin in their cell wall (Bonfante-Fasolo et al., 1990; Weijman and Meuzelaar 1979), presence of nonseptate and coenocytic mycelium and formation of non-motile spores, the chlamydospores (Orlando et al., 2003). Morton and Benny (1990) classified AM fungi into six major genera (Figure 2). The characterization of AM fungi species has been done on the basis of macro and micro-characteristics (Khade, 2008). The macro-characteristics can be used to characterize the visual difference in the gross descriptive morphology of a species (Table 2). The micro-characteristics demonstrate a degree of intra-specific difference in developmental morphology (Mehrotra, 1997).

The distribution of AM fungi is reported from diverse ecosystem, such as grassland (Su et al., 2007), farmland (Li et al., 2007) and forestland (Violi et al., 2008), as well as many stressful environments, viz., mangrove forests (Sengupta and Chaudhury, 2002; Kumar, 2009); polluted mangrove forests (Mishra and Balachandran, 2010); heavy metal contaminated ash ponds (Bedini, 2010 ); over-grazed grasslands (Su et al., 2007), hot and arid forests (Li et al., 2007); and fire affected forests (Knorr et al., 2003).

**Table 2** Macro-and micro-taxonomic characteristics used for the identification of AM fungi

| **Macro-characteristics** | |
|---|---|
| Sporocarp morphology | Size, Shape, Peridium |
| Spore | Colour, Shape, Size |
| Subtending hypha | Shape, Width, Pore occulsion |
| Mycorrhizal anatomy | Hyphal characters<br>Intraradical spores |
| **Micro-characteristics** | |
| Spore wall structure | Colour, Dimension, Number, Type |
| Spore germination | Ornamentation, Direct, Indirect |

### Process and Mechanism of Symbiotic Association of AM Fungi

It is recognized that mycorrhizal association is a natural strategy that most plants have developed in their developmental process (Paul and Clark, 1996). AM fungi are obligate biotrophs; their spores are the only plant-independent phase, and they have an ability to germinate and arrest growth many times if plant derived signals are missing (Requena et al., 2007; Plate 1). The process of the development of AM symbiosis varies with the plant and fungal species involved (Smith and Smith, 1990). AM spores germinate under appropriate water and temperature conditions, hyphae grow for 2-3 weeks and they respond to root present in their vicinity (Requena et al., 2007; Plate 2). Also, root exudates from host root elicit hyphal growth due to the presence of plant phenolic substances, in particular flavonoids (Nair et al., 1991), some lipophylic molecules (Buee et al., 2000) or strigolactones (Akiyama et al., 2005). On contact with host root cell, the fungal hyphal cells form appressoria (Paszkowski, 2006) that develop into haustoria-like structures (arbuscles) to penetrate the root cortical cell walls and form interface with the host cytoplasm (Smith and Read, 1997). Here it undergoes extensive dichotomous branching into a tree-like fungal structure that may entirely fill the cortical region and provide an increased surface area for metabolic exchanges between the plant and the fungus (Plate 3). These arbuscules collapse after some days, leaving an intact cortical cell, which is then able to host another arbuscule. The signals that induce intracellular fungal branching or arbuscule decay have not been identified (Paszkowaski, 2006).

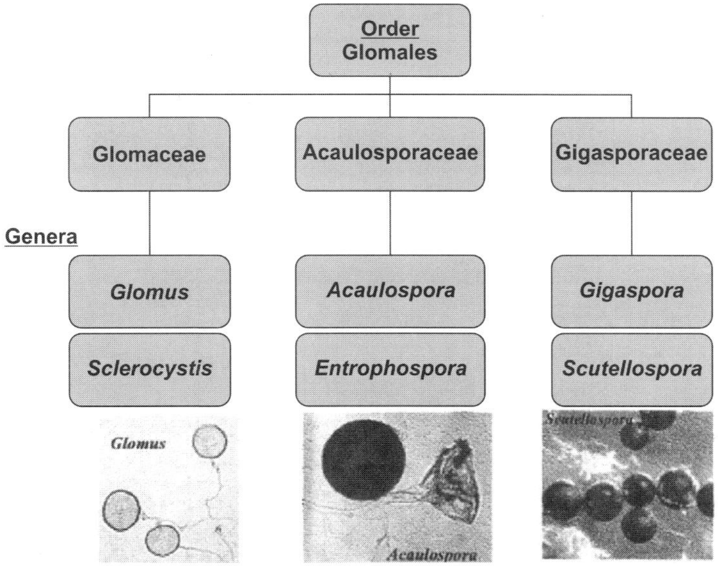


**Fig. 2.** Taxonomy of AM fungi showing spores produced by different genera in soil (Source: Dodd, 2000).

AM fungi produce vesicles which are structures that function as storage (lipids) and reproductive organs (Plate 4). They also interface directly with the soil by producing extraradical hyphae that may extend several centimeters out into the soil, which increases the potential of nutrient uptake, and possibly also water uptake

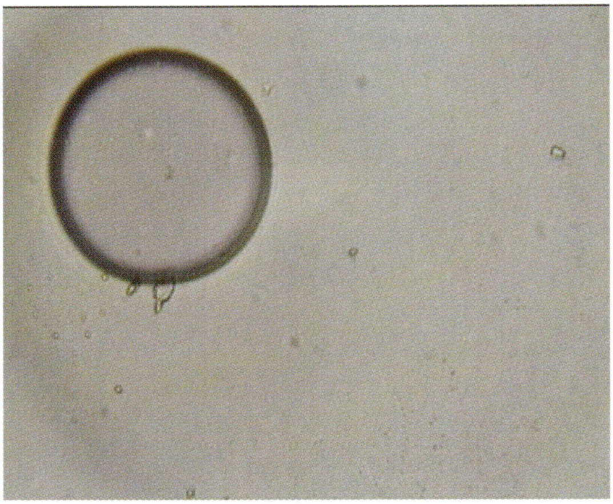

**Plate 1** AM fungal spore of *Gigaspora rosea.*

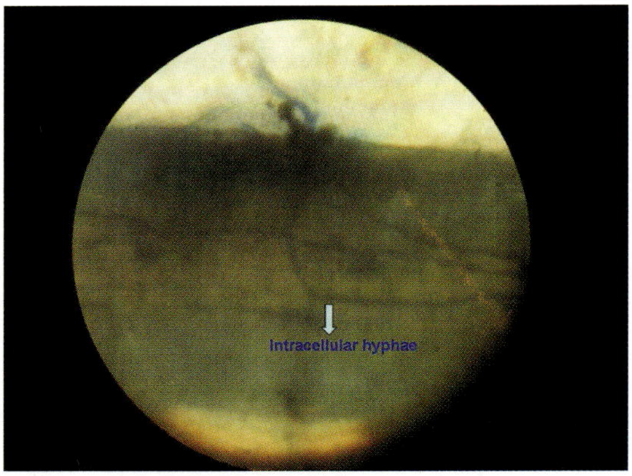


**Plate 2** AM infection in host root showing intracellular hyphae.

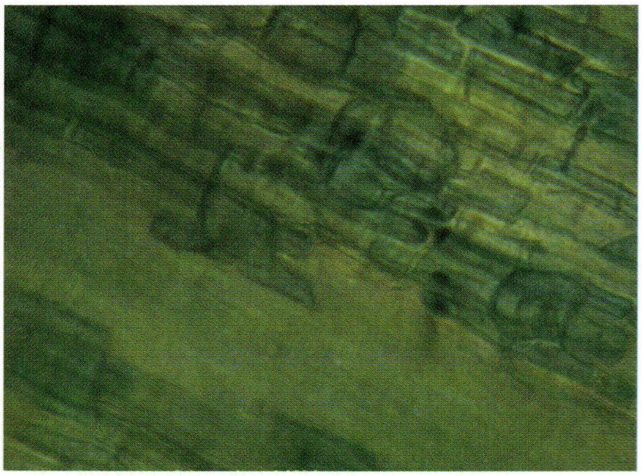

**Plate 3** AM infection showing arbusculae in host root.

**Plate 4** AM infection showing vesicles with intracellular hyphae in host root.

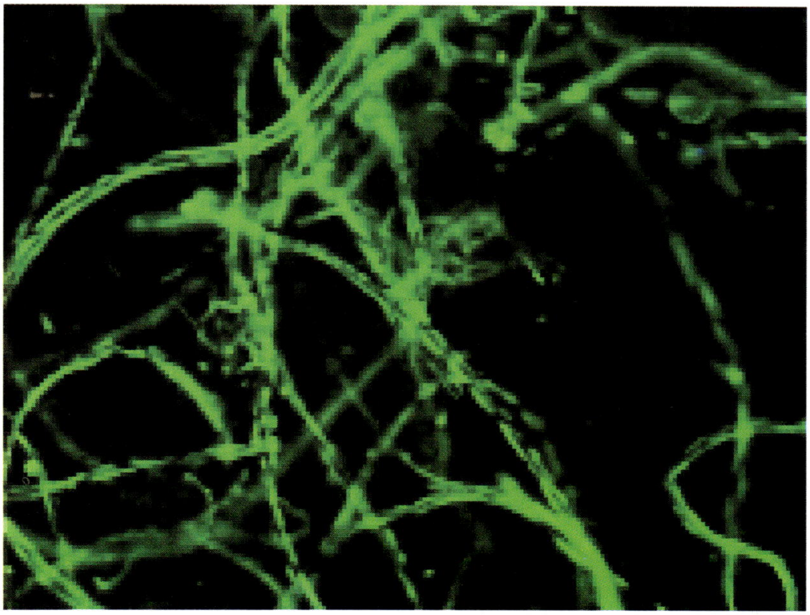

**Plate 5** Showing thin, threadlike strands of mycorrhizal hyphae from pot cultures have an abundant amount of glomalin—seen on their surface as bright green spots here after a laboratory procedure (Source: Wright & Upadhyaya, 1999).

(Auge, 2001). The arbuscules are the key features of AM fungi and are responsible for nutrient exchange (Figure 3). They represent a dead end in the growth of AM fungi (Bonfante and Perotto, 1995) because they finally senescence and collapse after 4–10 days of symbiosis (Sanders et al., 1977). The fungal structures are then degraded completely by the plant cell and the plant cell recovers its original morphology (Jacquelinet-Jeanmougin et al., 1987). This way, cortical cells are able to allow a second fungal penetration and arbuscule formation (Figure 3). During colonization, the fungal arbuscule occupies a major portion of the host cortex cell, but is separated from the cell protoplast by a part of the host plasma membrane, the periarbuscular membrane. This membrane completely surrounds the arbuscule, leading to a fourfold increase of the surface of the plasma membrane. Although it originates from the plant plasma membrane, the periarbuscular membrane exhibits different properties. In particular, phosphate transporters were shown to be located specifically in the periarbuscular membrane (Harrison et al., 2002).

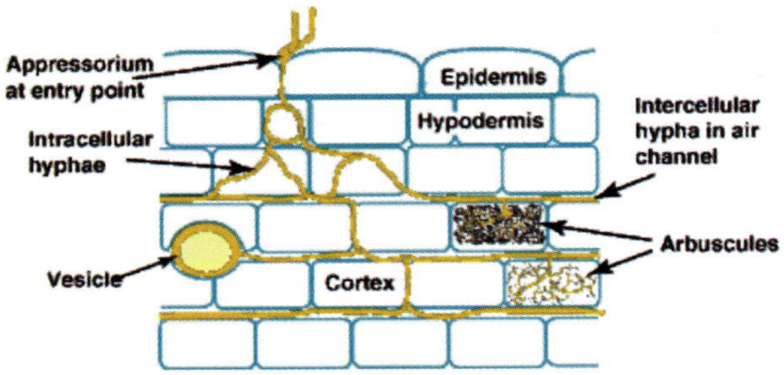


**Fig. 3.** Process of symbiotic association between the plant root and AM fungi (Brundrett et al., 1996).

Moreover, a high amount of H+—ATPase activity (Gianinazzi et al., 1991) accompanied by the highly acidic nature of the space separating plant and fungal plasma membranes has been found (Guttenberg, 2000). The space separating plant and fungal plasma membrane correspond to the apoplastic compartment and represents the symbiotic interface. It has been accepted that the establishment of the symbiosis must be the result of a continuous molecular 'dialogue' between the plant and the fungus, as exerted through the exchange of both recognition and acceptance signals (Azcon-Aguliar and Barea, 1997). In exchange of the nutrients, the fungus is supplied with photosynthates (about 20% of the plant assimilation capacity) and these assimilates are either respired or released into the soil via transfer of sugar from the root to their endosymbiotic fungi, which later translocate this carbon in the form of lipids and sugar into the external mycelium spreading in the soil (Bago, 2003). As a result of higher carbon drain, microbial rhizosphere composition is altered and therefore the term 'mycorrhizosphere' was coined (Rambelli, 1973; Lidermann, 1988). The external mycelium produces chlamydospores and completes its life cycle.

AM fungi are disseminated both actively and passively. Active dissemination is from one root to another by mycelial growth through soil, and passive dissemination is through biotic agents like rodents, worms, insects and birds, or abiotic agents like wind and water.

## SPECIFIC FEATURES OF CARBON SEQUESTRATION IN AM FUNGI

AMF are primarily responsible for nutrient transfer from soil to plant but have other roles such as soil aggregation, protection of plant against drought stress and soil pathogens and increasing plant diversity. This is achieved by the growth of their fungal mycelium within a host root and out into the soil beyond (Dodd et al., 2000). AM fungi have significant morphological and biochemical characteristics that induce their potential for $CO_2$ sequestration (Figure 4).

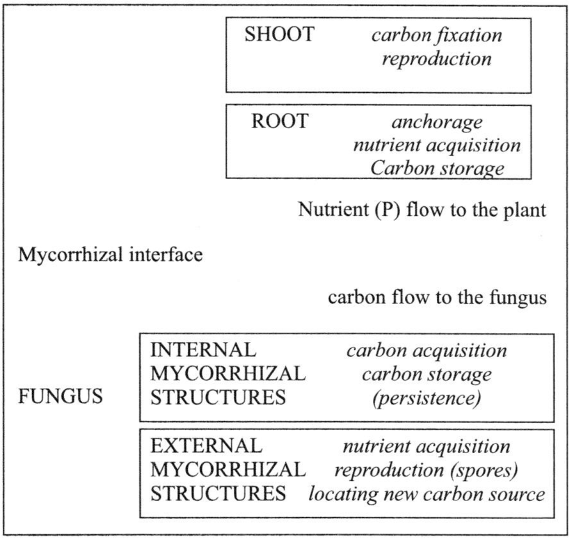


**Fig. 4.** Conceptual representation of an (arbuscular) mycorrhizal association between a plant and a fungus. The lists of attributes given in italics represent key functions for the organism in question (i.e. plant or fungus). In this simplified case of an arbuscular mycorrhiza, the fungus receives all of its carbon from the plant. However, in nature, any given fungus is likely to be receiving carbon from several different plants, which in turn may be in mycorrhizal association with several different fungi. (Source: Staddon P. L. 2005)

### Acquisition of Photosynthates and Rapid Turnover of Hyphae

AM fungi are capable of direct and rapid acquisition of photosynthates from plant cells, which they allocate to intra (arbuscules, vesicles and hyphae) or extra radical structures (mycelium, hyphae etc.) depending upon the environmental conditions. Many of the hyphae network in the soil are short-lived, less than a week (Staddon, 2003). AM fungi biomass, dead or alive in the soil may account for a very large proportion of the soil microbial biomass, and the mycelial network also allows the movement of carbon into the soil bulk. Decomposition of the mycorrhizal hyphae induces increased carbon in the soil. The very short life-span of the hyphal network is rather a beneficial process for the plants as there are no such soil microbes present which can contribute to such a significant amount of soil carbon sequestration (Staddon, 2003).

### Role of Chitin

AM fungi are generally placed in the family of Zygomycetes as this family is characterized by the presence of chitin in their cell walls. The presence of chitin can be used as an indicator of infection of mycorrhizal fungi in plant roots. Chitin estimation indicates both living and dead fungal biomass, and it can thus be used to estimate the total biomass of the mycorrhizal fungi in the roots (Singh, 2005). Studies indicated that chitin added to sand-soil stimulated the root colonization, growth of extraradical mycelium and production of spores of AM fungi. It is a recalcitrant in nature with low decomposition rate that could help in the enhanced sequestration of $CO_2$ in soil.

### Production of Glycoprotein-Glomalin

Glomalin is a glycoprotein produced by hyphae of AM fungi in soil and roots that stores carbon in both its protein and carbohydrate (glucose or sugar) subunits. Glomalin is named after Glomales, the taxonomic order that AM fungi belong to (Wright, 2004). Glomalin is left in the soil as a residue after the AM hyphae die and decompose which constitutes a modest proportion, 0.4–6% of the biomass (Treseder and Mack, 2007). Glomalin contains 30 to 40 percent carbon and also forms clumps of soil granules called aggregates (Plate 5). The fungi use carbon from the plant to grow and make glomalin. In return, the fungal hyphae extend to the reach of plant roots. Hyphae help the plants to take up more water and nutrient. As a plant grows, the fungi move down the root and form new hyphae to colonize the growing roots. When hyphae attached on the roots limit their transportation of nutrients, their protective glomalin moves off into the soil. There it attaches to the particles of minerals (sand, silt, and clay) and organic matter, forming clumps. Higher $CO_2$ levels in the atmosphere stimulate the fungi to produce more glomalin. Studies suggest that glomalin contributes much more nitrogen and carbon to the soil than the hyphae or other soil microbes (Wright and Nichols, 2004).

It has been reported that glomalin content changes with the change in land use, viz., conversion of forest to grassland (Rillig et al., 2002), tillage practices (Wright et al., 1999), and crop rotation practices (Wright and Anderson, 2000). In disturbed soil conditions such as fire affected forest (Knorr et al., 2003), heavy metal polluted ash disposal site (Bedini et al., 2010), and polluted mangrove forests (Balachandran and Mishra, 2011), the presence of glomalin indicates that it is recalcitrant in nature (Table 3). Also, it contributes to the sequestration of heavy metals (Cornenjo et al., 2008) that helps in avoiding direct exposure of plants to toxic compounds and maintaining their growth under stress conditions. In fire affected areas, the concentration of glomalin did not change significantly (Knorr et al., 2003); it may be due to low intensity fire. Also, glomalin is extremely heat stable in nature (Boerner, 2000).

**Table 3** Range of glomalin (mg/g) in soils in different types of ecosystem

| *Type of Ecosystems* | *Range of Glomalin (mg/g)* | *Reference* |
|---|---|---|
| Hawaiian agricultural soils | 60 | Rillig et al., 2001 |
| Fire affected deciduous forest | 2.8-3.5 | Knorr et al., 2003 |
| Tropical rainforest | 2.85 – 6.45 | Violi, 2008 |
| Temperate rainforest | 2 - 15 | Violi, 2008 |
| Polluted ash disposal site | 1.59- 2.27 | Bedini et al., 2010 |
| Polluted mangrove forest of Mumbai | 4.6 – 11.8 | Balachandran and Mishra, 2011 |

### Tripartite Relationship with other Beneficial Microbes

AM fungi interact with other soil microbes like the free-living and symbiotic nitrogen fixers and phosphate solubilisers to improve their efficiency for the biochemical cycling of elements to the host plants. Beneficial effect of microbial colonies may include mobilization of nutrients, phytohormone production and vesicular arbuscular mycorrhizal infection. This relationship is known as tripartite relationship. Intense colonization of microbes in rhizosphere can itself lead to stiff competition between the microbes and the plants for nutrient resources. Chemical molecules of root exudates govern the development of plant-fungal symbiosis and provide powerful signals that alert the mycorrhizal fungi of the presence of a host plant. They are a major localized source of nutritional organic carbon and nitrogen for rhizosphere microorganisms and plant growth (Kumar et al., 2006). Several studies have indicated that dual inoculation of AM fungi and *rhizobium* sequester 10–25% more organic carbon in the soil than control (Mishra et al, 2011; 2009; 2008). Hence, symbiotic association in rhizosphere stimulates the C sequestration, and AM fungi have a more specific role.

## SIGNIFICANCE OF AM FUNGI SYMBIOSIS AND CARBON SEQUESTRATION IN HIGHER PLANT COMMUNITIES

AM fungi diversity is the major factor in the maintenance of plant biodiversity and ecosystem stability and function. The species composition and diversity of AMF communities has the potential to determine plant population and plant community structure. The fact that plant species vary in the degree of response to AMF species has important implications for growth of individual plant species. It plays a vital role in the C sequestration in the plants by helping them to store more amount of carbon into them as AM fungi takes up some of it for their survival. AM fungi also increase carbon allocation in the soil by the production of glomalin. The prospects of AM fungi symbiosis in higher plants could have several direct and indirect benefits (Flow chart 1).

### Enhanced Sequestration of Atmospheric $CO_2$

AM fungi in association with terrestrial forests enhance their capacity for $CO_2$ sequestration as the fungal hyphae can store a large amount of carbon derived from the plants. In other way, it increases the ability of the plant to capture more

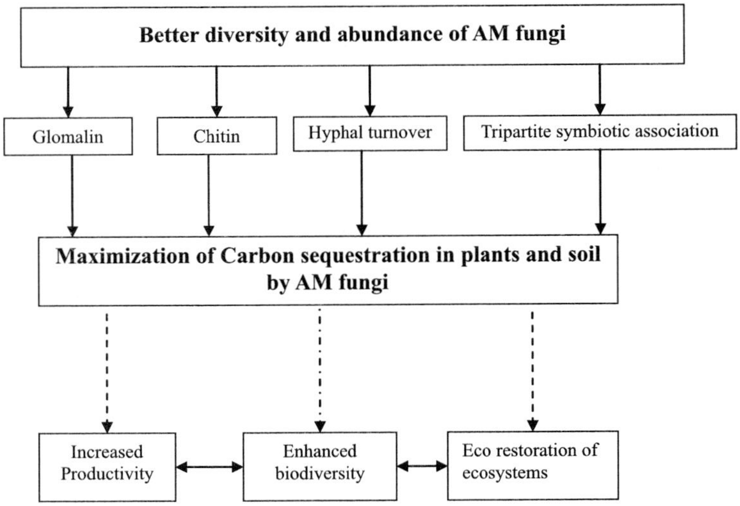


**Flow chart 1: Prospects of AM fungi symbiosis in different ecosystems**



$CO_2$ into its biomass which is ultimately taken up by AM fungi or fixed subsequently in the soil. Tropical forests are more efficient in storing carbon in the living plants and litter instead of transferring carbon to the soil below (Pidgeon, 2009). Coastal ecosystems keep sequestering large amounts of carbon throughout their life cycle. Majority of carbon stays fixed in the soil rather than the plant, and so a relatively small amount is released into the atmosphere when the plant dies. Forests like mangrove ecosystem contribute about half of the total carbon sequestration in ocean sediments, even though they account for less than 2 percent of the ocean surface as coastal vegetation spreads deeper below the ground than it grows above. The diversity of AM fungi is well reported from mangrove forest, and presence of glomalin is also proved (Balachandran and Mishra, 2011).

## Enhanced Turnover of Roots and Nutrients

Fungal hyphae help in the fast turnover of roots leading to sequestration of carbon into the soil. AM fungi play an important role in maintaining carbon exchange between the biosphere and the atmosphere. Live hyphae can turn over rapidly on an average of five to six days (Staddon et al., 2003). This rapid turnover of the hyphae leads to an increase in the carbon fixation in the soil. However, the carbon flow to AM fungi hyphae might be respired back to the atmosphere quickly (Zhu, 2003). But this return of carbon to the atmosphere is a small fraction of the amount that is allocated in the soil after the hyphae perishes. Mycorrhizal hyphae can redistribute the recently fixed carbon away from the roots. Hence, in the presence of mycorrhizal fungi, $CO_2$ sequestration is not restricted only to the rhizosphere.

In a high $CO_2$ concentration, AMF could thus enhance soil C sequestration directly via C allocation to deeper soil (Vicca et al., 2009).

### Increased Productivity of Ecosystems

AM fungi provide phosphorus and other micronutrients as well as help in rapid turnover of roots, which increases the carbon content in the soil ultimately leading to good and increased aboveground productivity (Wright and Nichols, 2004). Mycorrhizal fungi often enhance resource complementarity by providing nutrients that are otherwise inaccessible to plant roots. Enhanced P uptake is one of the mechanisms by which AM fungi can enhance plant productivity, and experiments with single plants or experimental plant communities have shown that AM fungi contribute up to 90% of plant P uptake (Jakobsen et al., 1992; Van der Heijden et al., 2006). Enhanced P uptake is especially important for plant species having high P requirement such as legumes, or under the conditions when plant productivity is strongly determined by P availability, such as in the tropics.

### Eco-restoration of Ecosystems

Plants capable of forming association with AM fungi have been shown to accumulate a considerable amount of trace metals. The interaction between AM fungi and minerals other than phosphorus, particularly trace elements, has been the subject of many recent studies because of the possibility of the beneficial effect of mycorrhizae in improving the tolerance of plants against toxicity. The uptake of trace elements by mycorrhizal plants depends on several factors such as physico-chemical properties of the soil, particularly its fertility level, pH, the host plants and the fungi involved, and above all, the concentration. The survival and adaptation of plants on degraded sites could be easier with the help of AM fungi, and ultimately this symbiosis helps in their eco-restoration.

## EPILOGUE

Significance of microbes, especially AM fungi, in C sequestration is tremendous especially with increasing level of $CO_2$ in the atmosphere. The involvement and active role of AM fungi in C cycle is crucial than any other technologies available. AM fungi acquire recent photosynthate directly from the plant and allocate it to intra- or extraradical structures, depending on the environmental conditions and their nutritional status. The hyphae rapidly turnover in the soil and keep on adding C to it mainly due to the presence of chitin and glomalin. Apart from adding C to the soil, enrichment of other nutrients as P, Zn, Mn, etc., helps in improving the productivity of ecosystem and remediation of heavy metals. This association could be applied to restore degraded ecosystems and entitle it as a potential sink for GHC (greenhouse gas) to enter into the market of carbon credit. However, extensive work is needed in this direction, in view of current and global changes in environment that would affect the functioning of AM fungi in the soil ecosystem.

## ACKNOWLEDGEMENTS

The financial support by Ashoka Trust for Research in Ecology and the Environment (ATREE) sponsored by Sir. Jamshed Ji Tata Trust is gratefully acknowledged.

## REFERENCES

Akiyama, K., Matsuzaki, K. and Hayashi, H. (2005). Plant sesquiterpenes induce hyphal branching in arbuscular mycorrhizal fungi. *Nature* **435:** 824-827.

Auge, R.M., Stodola, A.J.W., Brown, M.S., Tims, J.E. and Sexton, A.M. (2001). Moisture retention properties of a mycorrhizal soil. *Plant Soil* **230:** 87-97.

Azcon-Aguilar, C. and Barea, J.M. (1997). Applying mycorrhiza biotechnology to horticulture: significance and potentials. *Scientia Horticulture* **68:** 1-24.

Bago, B., Pfeffer, P.E., Abubaker, J., Jun, J., Allen, J.W., Brouillette, J., Douds, D.D., Ammers, P.J. and Shachar-hill, Y. (2003). Carbon exports from arbuscular mycorrhizal roots involves the translocation of carbohydrate as well as lipid. *Plant Physiology* **131(3):** 1496-1507.

Balachandran, S. and Mishra, S. (2011). Arbuscular Mycorrhizal Fungi (AMF): Bioindicators in degraded mangrove forest. National Conference on 'Wetlands - Key to the Healthy Environment' organized by *Paryavaran Daxata Manch,* Thane and K. B.P. College, 2nd February 2011.

Bedini, S., Turrini, A., Rigo, A., Argese, E. and Giovannetti, M. (2010). Molecular characterization and glomalin production of Arbuscular mycorrhizal fungi colonizing a heavy metal polluted ash disposal island, downtown Venice. *Soil Biology & Biochemistry* **42:** 758-765.

Bonfante, P. and Perotto, S. (1995). Strategies of arbuscular mycorrhizal fungi when infecting host plants. *New Phytologist* **130:** 3-21.

Bonfante, P., Faccio, A., Perotto, S. and Schbert, A. (1990). Correlation between chitin distribution and cell wall morphology in the mycorrhizal fungus *Glomus versiforme. Mycological Research* **94:** 157-165.

Buee, M., Rossingnol, M., Jauneau, A., Ranjeva, R. and Becard, G. (2000). The symbiotic growth of arbuscular mycorrhizal fungi is induced by branching factor partially purified from plant root exudates. *Molecular Plant- Microbe interaction* **13:** 693-698.

Cerón-Bretón, J.G., Cerón-Bretón, R.M., Rangel-Marrón, M., and Estrella-Cahuich, A. (2010). Evaluation of carbon sequestration potential in undisturbed mangrove forest in Términos Lagoon, Campeche. *Development, Energy, Environment, Economics,* 295-300.

Cornejo, P., Meir, S., Borie, G., Rilling, M.C. and Borie, F. (2008). Glomalin-related soil protein in a Mediterranean ecosystem affected by a copper smelter and its contribution to Cu and Zn sequestration. Science of Total Environment 406, 154-160.

Dodd, J.C., Boddington, C.L., Rodriguez, A., Gonzalez-Chavez, C. and Mansur, I. (2000). Mycelium of arbuscular mycorrhizal fungi (AMF) from different genera: form, function, and detection. *Plant Soil* **226:** 131-151.

Gray, E. and Smith, J.D. (1950). A new aquatic nitrogen fixing bacterium from three Cambridgeshire chalk streams. *Microbiology* **4:** 281-285.

Guttenberger, M. (2000). Arbuscules of vesicular-arbuscular mycorrhizal fungi inhabit an acidic compartment within plant roots. *Planta* **211:** 299-304.

Jacquelinet-Jeanmougin, J., Gianinazzi-Pearson, V. and Gianinazzi, S. (1987). "Endomycorrhizas in the Gentianaceae. II. Ultrastructural aspects of symbionts relationships in Gentiana lutea L. *Symbiosis* **3:** 269-286.

Jakobsen, I., Abbott, L.K. and Robson, A.D. (1992). External hyphae of vesicular-arbuscular mycorrhizal fungi associated with *Trifolium subterraneum* L. I: Spread of hyphae and phosphorus inflow into roots. *New Physiology* **120**: 371-380.

Jones, T.H., Thompson, L.J., Lawton, H.H., Bezemer, T.M., Bardgett, R.D., Blackbum, T.M., Bruce, K.D., Cannon, P.F., Hall, G.S., Hartley, S.E., Howson, G., Jones, C.G., Kampichler, C., Kandeler, E. and Ritchie, D.A. (1998). Impacts of rising atmospheric carbon dioxide on model terrestrial ecosystems. *Science* **280:** 441-443.

Khade, S.W. (2008). Taxonomy of arbuscular mycorrhizal fungi. *Mycorrhiza News* **20(1):** 2-7.

Knorr, M.A., Boerner, R.E.J. and Rilling, M.C. (2003). Glomalin content of forest soils in relation to fire frequency and landscape position. *Mycorrhiza* **13:** 205-210.

Kumar, R., Pandey, S. and Pandey, A. (2006). Plant roots and carbon sequestration. *Current Science* **91:** 7-10.

Kumar, T. and Ghose, M. (2008). Status of arbuscular mycorrhizal fungi (AMF) in the Sunderbans of India in relation to tidal inundation and chemical properties in soil. *Wetland Ecology and Management*. 471-483.

Lal, R. (2009). Sequestering atmospheric carbon dioxide. *Critical Reviews in Plant Sciences* **28:** 90-96.

Li, L.F., Li, T. and Zhao, Z.W. (2007). Differences of arbuscular mycorrhizal fungal diversity and community between a cultivated land, an old field, and a never-cultivated field in a hot and arid ecosystem of southwest China. *Mycorrhiza* **17:** 655-665.

Linderman, R.G. (1992). Vesicular-arbuscular mycorrhiza and soil microbial interactions. *In:* Belhlenfalvay, G.J. and Linderman, R.G. (Eds.) Mycorrhiza in sustainable Agriculture. ASA spec. Publ: Madison, WI, 45-70.

Maia, L.C. (1991). Morphological and ultrastructural studies on spores and germ tubes of selected arbuscular mycorrhizal fungi (Glomales). Gainesville: University of Florida [Doctoral thesis].

Mehrotra, V. (1997). Problems associated with morphological taxonomy of AM fungi. *Mycorrhizal News* **9(1):** 1-10.

Metting, F.B., Smith, J.L., Amthor, J.S. and Izaurralde, R.C. (2001). Science needs and new technology for increasing soil carbon sequestration. *Climate Change* **51:** 11-34.

Mishra, S. and Balachandran, S. (2010). Diversity of Arbuscular Mycorrhizal Fungi (AMF) in degraded mangrove forests of western coast of India. Fourth International Conference on Plants and Environmental Pollution organized by International Society of Environmental Botanists and N.B.R.I., Lucknow from 8-11$^{th}$ December 2010.

Mishra, S., Sharma, S. and Vasudevan, P. (2008). Comparative effect of biofertilizers on fodder production and quality in guinea grass (*Panicum maximum* Jacq.). *J. Science of Food and Agriculture* **88(9):** 1667-1673.

Mishra, S., Sharma, S. and Vasudevan, P. (2009). Effect of single and dual inoculation of *Rhizobium* with AM fungi on nodulation, fodder production and quality in two *Stylosanthes* species. *Biological Agriculture and Horticulture* **26:** 411-421.

Mishra, S., Sharma, S. and Vasudevan, P. (2011). Role of bioinoculants and organic fertilizers in fodder production and quality of leguminous tree species. *J. Environmental Biology* **32(1):** 57-64.

Nair, M.G., Safir, G.R. and Siqueira, J.O. (1991). Isolation and identification of vesicular-arbuscular mycorrhiza—stimulatory compounds from clover (*Trifolium repens*) roots. *Applied Environmental Microbiology* **57:** 434-439.

National energy Technology Laboratory (NETL) (2009). Terrestrial sequestration of Carbon dioxide, USA.

National Oceanic and Atmospheric Administration (NOAA). www. $CO_2$Now.org. 2011.

Paszkowski, U. (2006). A journey through signaling in arbuscular mycorrhizal symbiosis. *New Phytologist* **172:** 35-46.

Paul, E.A. and Clark, F.E. (1996). Soil Microbiology and Biochemistry. San Diego, CA: Academic Press. 340.

Pidgeon, E. (2009). Carbon sequestration by coastal marine habitats: important missing sink. The Management of Natural Coastal Carbon Sinks. IUCN.

Quilambo, O.A. (2003). The vesicular arbuscular mycorrhizal symbiosis. *African J. Biotechnology* **2(12):** 539-546.

Rambelli, A. (1973). The rhizosphere of mycorrhiza. In: *Ecto-mycorrhiza* (Marks, G.L. and Koslowski, T.T. (Eds.). Academic Press, New York. 299-343.

Requena, N., Breuninger, M., Franken, P. and Ocon, A. (2003). Symbiotic status, phosphate and sucrose regulate the expression of two plasma membrane H+--ATPase genes from mycorrhizal fungus *Glomus mossae. Plant Physiology* **132:** 1540-1549.

Rilling, M.C., Wright, S.F. and Eviner, V. (2002). The role of arbuscular mycorrhizal fungi and glomalin in soil aggregation: comparing effects of five plant species. *Plant Soil* **238:** 325-333.

Rogers, H.H., Runion, G.B. and Krupa, S.V. (1994). Plant responses to atmospheric $CO_2$ enrichment with emphasis on roots and the rhizosphere. *Environmental Pollution.* **83:** 155-189.

Rouhier, H. and Read, D.J. (2009). The role of mycorrhiza in determining the response of *Plantago lanceolata* to $CO_2$ enrichment. *New Phytology,* 367-373.

Sanders, F.E., Tinker, B.P., Black, R.L.B. and Palmerly, S.M. (1977). The development of endomycorrhizal root system. I. speed of infection and growth promoting effects with four species of vesicular-arbuscular endophyte. *New Phytol.* **78:** 257-268.

Sengupta, A. and Chaudhari, S. (2002). Arbuscular mycorrhizal relations of mangrove plant community at the Ganges river estuary in India. 169-174.

Simmon, L., Bousquet, J. and Levesque, R.C. (1993). Origin and diversification of endomycorrhizal fungi and coincidence with vascular land plants. *Nature* **363:** 67-69.

Singh, S. (2005). Fungal chitin and its use for estimation of mycorrhizal infection. *Mycorrhizal News* **17(3):** 2-8.

Smith, S.E. and Read, D.J. (1997). Mycorrhizal Symbiosis, 2nd Edn. Academic Press, London.

Smith, S.E., Smith, F.A. and Jakobsen, I. (2003). Mycorrhizal fungi can dominate phosphate supply to plants irrespective of growth response. *Plant Physiology* **133:** 16-20.

Staddon, P.L. (2005). Mycorrhizal fungi and environmental change: the need for a mycocentric approach. *New Phytologist* **167:** 635-637.

Staddon, P.L., Ramsey, C.B., Ostle, N., Ineson, P. and Fitter, A.H. (2003). Rapid turnover of hyphae of mycorrhizal fungi determined by AMS microanalysis of 14C. *Science* **300:** 1138-1140.

Su, Y.Y. and Guo, L.D. (2007). Arbuscular mycorrhizal fungi in non-grazed, restored and over-grazed grassland in the Inner Mongolia steppe. *Mycorrhiza* **17:** 689-693.

Treseder, K.K., Turner, K.M. and Mack, M.C. (2007). Mycorrhizal responses to nitrogen fertilization in boreal ecosystems: Potential consequences for soil carbon storage. *Global Change Biology* **13:** 78-88.

Van der Heijden, M.G.A., Streitwolf-Engel, R., Riedl, R., Siegrist, S., Neudecker, A. and Ineichen, K. (2006). The mycorrhizal contribution to plant productivity, plant nutrition and soil structure in experimental grassland. *New Phytologist* **172:** 739-752.

Vicca, S., Zavalloni, C., Fu, Y.S.H., Voets, L., Herv´e Dupr´e de Boulois, Declerck, S., Ceulemans, R., Nijs, I. and Janssens1, I.A. (2009). Arbuscular mycorrhizal fungi may mitigate the influence of a joint rise of temperature and atmospheric $CO_2$ on soil respiration in grasslands. *International J. Ecology,* 1-10.

Violi, H.A., Barrientos-Priego, A.F., Wright, S.F., Escamill-Prado, E., Morton, J.B., Menge, J.A. and Lovatt, C.J. (2008). Disturbances change arbuscular mycorrhizal fungal phenology and soil glomalin concentration but not fungal spore composition in montane rainforests in Veracruz and Chiapas, Mexico. *Forest Ecology and Management* **254:** 276-290.

Weijman, A.C.M. and Meuzelaar, H.C.L. (1979). Biochemical contributions to the taxonomic status of the Endogonaceae. *Canadian J. Botany* **57:** 284-291.

Wright, S.F. and Anderson, R.L. (2000). Aggregate stability and glomalin in alternative crop rotations for central great plains. *Biology and Fertility of Soils* **31:** 249-253.

Wright, S.F. and Nichols, K.A. (2004). Glomalin: hiding place for a third of the world's stored soil carbon. *Agriculture Research Magazine.*

Wright, S.F., Starr, J.L. and Paltineanu, I.C. (1999). Changes in aggregate stability and glomalin during tillage management transition, *Soil Science Society of American J.* **63:** 1825-1829.

# 6

# Interactions between Arbuscular Mycorrhizal Fungi and N-fixing Bacteria, *Rhizobium*

**Mohammad Miransari***

**Professor of Soil Science, Department of Soil Science, College of Agricultural Sciences, Shahed University, Tehran, Iran*
*E-mail: miransari1@gmail.com, miransari@shahed.ac.ir*

**ABSTRACT**

The interaction between arbuscular mycorrhiza (AM) and soil N-fixing bacteria, specifically *rhizobia*, can affect ecosystem productivity. Hence, evaluating such effects can help in the development of appropriate methods, resulting in higher agricultural production. Unlike the association between AM and the host plant, the symbiosis between rhizobia and the host plant is specific because of the biochemicals communicated between the two symbionts at the onset of symbiosis. However, the interesting point is that there are some common genes controlling the symbiosis processes between AM or bacteria and the host plant. There is also a very specific tripartite interaction among the host plant, arbuscular mycorrhiza and soil rhizobia, which can be beneficial to the host plants, in this case the leguminous family, and the other symbionts. There are also some other important and interesting conclusions regarding this review article as indicated in the conclusion part.

**Keywords:** Arbuscular Mycorrhiza, N-fixation, *Rhizobia*, Tripartite Symbiosis.

## INTRODUCTION

Arbuscular mycorrhiza (AM) are soil fungi developing symbiotic associations with most of the terrestrial plants. In such a mutually beneficial symbiosis, the host plants provide the fungi with necessary hydrocarbons for their growth and development, and the fungi transfer water and nutrients to the host through their extensive network of hypha connecting soil and host roots (Smith and Read, 2008). For the onset and continuation of symbiosis, the presence of the host plant is mandatory, although the fungal spores may germinate in the absence of the host plant (Smith and Read, 2008; Smith et al., 2010). Nitrogen (N-) fixing bacteria, specifically *Rhizobium*, are able to fix atmospheric N, when in symbiosis with their specific host plant. For the onset of symbiosis between the fungi or bacteria and the host plant, signal molecules should be exchanged between the two symbionts activating the genes required for the continuation of symbiosis (Gherbi et al., 2008; Marsh et al., 2008; Oldroyd et al., 2009). Such gene activation results in morphological and physiological alterations in the root host plants and eventually produces root nodules and fungal hypha with the related components (Akiyama and Hayashi, 2006; Artursson et al., 2006; Miransari and Smith, 2007; 2008; 2009; Miransari et al., 2007; 2008; 2009a; b). With respect to the great significance of AM fungi and *Rhizobium* in agriculture and ecosystems, it can be interesting to evaluate how such microbes may interact, especially in the soil rhizosphere, so that appropriate methods economical and environment friendly, can be developed for enhanced soil production. Hence, the focus of this review article is on evaluating such interactions, and suggesting methods that may contribute to the enhancement of such effects for enhanced ecosystem productivity.

## ARBUSCULAR MYCORRHIZAL SYMBIOSIS

Arbuscular mycorrhizal symbiosis is the most common symbiosis on the earth. Arbuscular mycorrhizal fungi are soil fungi developing non-specific symbiosis with a wide range of plants. In this association the host plants provide the fungi with carbon in the exchange of water and nutrients. Such kind of association can considerably affect the ecosystem functioning, as the fungi are able to significantly enhance the water and nutrient uptake of their host plant (Smith and Read, 2008). The fungi spore has to germinate first before the onset of symbiosis. Although the fungal spores are able to germinate in the absence of the host plant, the presence of the host plant is necessary for the symbiosis to proceed. The fungi and the host plant must communicate to develop the symbiosis process. Hence, both symbionts produce signal molecules, activating their symbiotic genes resulting in the development of new tissues for the symbiosis (Akiyama et al., 2006; Badri et al., 2009). Spore growth results in the production of extensive hyphal network and some fungal organelles called arbuscules and vesicles (Harrison et al., 1999). The extensive hyphal network can grow into the soil medium, even beyond the root depletion zones absorbing water and nutrients. This can be very beneficial for the host plant especially under stress, because the enhanced water and nutrient uptake can significantly increase plant growth under stress. In addition, AM fungi are also able to enhance the plant growth through the following:

(1) interacting with other soil microbes including N-fixing bacteria,
(2) improving soil structure,
(3) controlling soil pathogens, and
(4) stress alleviation.

Hence, arbuscular mycorrhizal symbiosis can be very beneficial to plant growth as well as the environment (Miransari et al., 2010a, b, c).

## AM Fungi

AM fungi species belong to the phylum Glomeromycota and order Glomerales. They can form endo- and ecto-mycorrhizal symbiosis with their host plant. The endo-mycorrhizal symbiosis is a more common kind of AM association with their host plants in which the developed fungal hypha grow into the apoplastic space of root cortex forming the highly branched arbuscules and the storage tissue vesicles (Miransari, 2010a, b). The arbuscules serve as the interface of nutrients exchange with the host plant roots and with AM hypha can significantly enhance the absorbing capacity of the host plant for water and nutrients. The vesicles also store tissues, with a high number of cellular vacuoles and can help the host plant under different conditions including stress (Harrison et al., 1999). For example, the vacuoles can store different ions including sodium and chloride and heavy metals under salinity and heavy metal stress, respectively, resulting in the alleviation of stress on plant growth (Miransari 2010a, b).

The functional diversity of AM fungi has also important effects on ecosystem functioning. There have been different species of AM recognized so far, and interestingly it has been indicated that their origin usually does not adversely affect their functioning (Miransari et al., 2007; 2008). Importantly, AM species can determine the structural combination of plant species in the ecosystem and hence can have significant effects on the efficiency of plant growth. This indicates that although the AM-host plant symbiosis is not specific, some AM-host plant combination may be more efficient affecting the host plant and hence ecosystem performance (Hijri and Sanders, 2005).

## Host Plants

There is a wide range, up to 80%, of terrestrial plants hosting AM fungi. In their symbiotic association AM fungi act non-specifically and hence each species can colonize different plant species. However, there are plant species, which are not host to AM species including the species from the Brassicaceae and Caryophyllaceae families, although there are species such as *Biscutella laevigata* and *Thlaspi* spp., which can develop symbiosis association with AM fungi. The reason for the lack of symbiosis development with non-host plants is due to the presence of some compounds in the root exudates of such plants preventing the fungi from developing the symbiosis with the plant (Hildebrandt et al., 2007). Such differences are of particular significance affecting the plant performance, because enhanced nutrient uptake including phosphorus is among the most important effects of AM species on plant growth and crop production. For non-host plants, there are some other strategies rather than inoculating them with AM species which must be used to fulfill optimum conditions for growth and production. In other words, for example, in P deficient soils, other methods of fertilization must be applied for non-host plants. The consideration of non-host plants properties can be useful for determination of more details regarding AM symbiosis.

## Establishment of Symbiosis

For the establishment of AM symbiosis, the presence of both AM fungi and host plant is necessary, because the fungi are obligate autotroph, meaning that they can survive and develop symbiosis in the presence of the host plant. For the onset and hence preceding the symbiosis, there must be some signal communications between the two symbionts. Both symbionts produce some biochemical compounds, which are able to trigger activities at molecular levels including the expression of symbiotic genes (Badri et al., 2009). It has been recently indicated that there is a kind of plant hormone called strigolactones produced by the plant roots, which is able to activate the fungal genes resulting in the production of fungal hypha (Akyama et al., 2006). In addition, this class of plant hormones is able to stimulate the growth of parasitic weeds (Lendzemo et al., 2009). AM fungi can also produce the Myc factors, resulting in physiological and morphological alteration in plant roots. It is interesting that unlike the process of symbiosis in *Rhizobium* and their legume host plant, the AM symbiosis is non-specific indicating that the Myc factors produced (Harrison 2005; Badri et al., 2009) by the fungi are able to trigger morphological changes in the roots of a wide range of host plants. In N-fixing symbiosis by bacteria, the process is specific as there are only some specific signal molecules including flavonoids produced by the legume host plant. For example, geneistin is among the signal molecules produced by soybean roots, triggering genetical activation in *Bradyrhizobium japonicum* (Miransari and Smith 2007; 2008; 2009).

## Diversity and Functions

Various species of AM have been recognized so far. They are categorized based on their morphological and physiological characters. For example, morphological properties of the spore are among the specifications used for the recognition of AM species (Harrison et al., 2005). AM species are also different in the level of their functioning indicating that although they perform similar functions when in symbiosis with their host plant, they differ in the intensity of their functioning. Some AM species are more efficient under different conditions including stress. Such differences are of particular significance as the proper combination of AM-host plant may result in higher efficiency and hence higher plant growth and crop production (Daei et al., 2009). In addition, the origin of AM species usually does not affect their functioning as they can act similarly under different geographical locations. Plant structural combination is also determined by AM diversity, as some AM-host plant symbiosis are more efficient resulting in the establishment of specific plant community under certain conditions (Sanders et al., 1996).

## NITROGEN FIXING SYMBIOSIS

Some bacteria called *Rhizobium*, develop symbiosis with their specific host plants from the leguminous family. In this symbiosis, the bacteria are able to fix the atmospheric nitrogen (N) and convert it into available N ($NH_4$+) for their host plant. The bacteria produce the enzyme, nitrogenase, which can reduce atmospheric $N_2$ into ammonia that is eventually incorporated in the structure of amino acids and proteins. The establishment of symbiosis is detailed in the following (Long 2001).

Through some very interesting and in the meanwhile complicated biochemical dialogue, the bacteria realize the presence of specific host plant roots and chemotactically approach them. There are two main types of biochemicals, specifically signal molecules including *nod* gene inducers and *nod* factors produced by the host plant and rhizobia, respectively. The *nod* gene inducers mostly contain flavonoids, which are able to activate the nodulation genes in rhizobium resulting in the production of *nod* factors. The latter are lipochitooligosaccharide molecules, inducing morphological changes in the host plant roots, especially root hairs, including root hair curling and formation of root nodules (Miransari et al., 2006; Cooper, 2007). Flavonoids can also regulate the translocation of auxin, which can influence the formation of nodules during nodule morphogenesis, and can also stimulate the establishment of tripartite symbiosis between the host plant, rhizobium and AM fungi (Subramanian et al., 2006; Wasson et al., 2006; Cooper et al., 2007).

In brief, the stages related to the development of symbiosis between the rhizobium and the host plant include the root exudation of signal molecules, the gene activation of bacteria by the exudates (Miransari and Smith, 2007; 2008), and production of some biochemicals by the bacteria that can induce morphogenesis and physiological changes in the host roots. This eventually results in the formation of nodules, which are the place of bacterial settlement and hence N fixation (Miransari et al., 2006; 2007; 2008).

### N-fixing Bacteria, *Rhizobium*

Unlike AM fungi, which develop symbiosis with most terrestrial plants, N-fixing bacteria are able to establish symbiotic association with their specific host plant and fix atmospheric N (Parniske, 2000; Sprent, 2001; Scheublin et al., 2004; Miransari and Smith, 2007; 2008). The N-fixing bacterial symbionts, including the genera *Azorhizobium, Bradyrhizobium, Mezorhizobium, Rhizobium and Sinorhizobium*, collectively called rhizobia, are settled in the symbiosome, a plant membrane including compartment (Parniske, 2000; Limpens and Bisseling, 2003). The bacteria act specifically because the *nod* gene inducers, flavonoids, exudated by the legume host plant can trigger the nodulation genes only in the specific *Rhizobium* (Table 1). Similar to AM-plant symbiosis, the bacteria-legume symbiosis also has important agricultural and environmental implications, because it can substantially contribute to available N production and utilization by the plant (Vandermeer, 1989; Cleveland et al. 1999), decreasing the amount of N chemical fertilization (Miransari, 2010c).

**Table 1** *Rhizobium* strains and their specific legume host plant (McInnes et al., 2004).

| *Rhizobial Strain* | *Legume Host Plant* |
|---|---|
| *Rhizobium etli* bv. *phaseoli* | *Phaseolus vulgaris* |
| *Rhizobium gallicum* bv. *phaseoli* | |
| *R. leguminosarum* bv. *phaseoli* | *P. vulgaris* |
| *Sinorhizobium fredii* | |
| *R. leguminosarum* bv. *trifolii* | *Trifolium pratense*<br>T. repens<br>T. subterraneum<br>*T.* spp. |
| *R. leguminosarum* bv. *viciae* | *Lens esculenta*<br>*Pisum sativum*<br>*Vicia* spp. |
| *Sinorhizobium meliloti* | *Medicago sativa*<br>M. lupulina<br>*Melilotus alba* |
| *Bradyrhizobium japonicum* | *Glycine max* |
| *B.* sp. (Cajanus), *B. elkanii* and fast growers | *Cajanus cajan* |
| *B.* sp. (*Amphicarpaea*) | *Amphicarpaea bracteata* |
| *B.* sp. (*Desmodium*) and *B.* sp. (*Macroptilium*) | *Desmodium intortum* and *Macroptilium atropurpureum* |
| *B.* sp. (*Lupinus*) and *B.* sp. (*Macroptilium*) | *Lupinus albus*<br>L. angustifolius<br>M. atropurpureum<br>*Ornithopus compressus* |

## Host Plants

Although atmospheric $N_2$ is fixed by both symbiotic and non-symbiotic (e.g., *Azospirillum* spp., Arzanesh et al., 2010) soil microbes, the most significant part of N fixation is related to symbiotic *Rhizobium*. In legume host plant in symbiosis with *Rhizobium* can fix atmospheric N. Table 1 indicates the legume plants, which are able to develop symbiotic association with their specific *Rhizobium*. The morphological and physiological differences among the different legume species make them specific to *Rhizobium*. In other words, the genetical structure of the legume plant, affecting its functioning, determines which *Rhizobium* bacteria can develop symbiotic association with the legume as the host plant. The root exudates of the legume host plant are able to activate the nodulation genes of only the specific *Rhizobium* (Long, 2001; Cooper, 2007; Badri et al., 2009).

The symbiotic association between *Rhizobium* and legumes results in the fixation of up to 400 kg/ha atmospheric $N_2$. Such high amount of N is of both environmental and economical significance as it reduces the amount of chemical N fertilization (Cleveland et al., 1999). This biological fertilization is very important as far as crop production is concerned as it significantly reduces the amount of N fertilization. (Salvagiotti et al., 2008; Miransari, 2010c; Miransari and Macknezie, 2010a, b).

Because of their N fixing capacity, the legume plants can also be used for fertilizing different crop plants in rotation. They are green fertilization, and after growing in a rotation their fresh tissues are returned to the soil contributing to the soil organic matter as well as soil ammonia and nitrate after being mineralized by the soil microbes (Miransari and Mackenzie, 2010a, b). In addition to being a valuable source of nutrients for soil microbes and crop plants, green fertilization can also improve soil structure and soil water retention capacity by increasing the rate of soil organic matter. Hence, as a host to the *Rhizobium*, legume plants can significantly increase soil efficiency for crop production, while being environment-friendly.

## Establishment of Symbiosis

As previously indicated, unlike the symbiosis process between AM fungi and their host plant, N-fixation by rhizobium and their legume host plants is specific, i.e., only the specific rhizobium bacteria can colonize the host plant root, form nodules and eventually fix atmospheric $N_2$. For the onset of symbiosis, the bacteria has to first realize the presence of the host plant and chemotactically approach it through the production of biochemicals called signal molecules by plant roots (Miransari and Smith, 2007; 2008; 2009). As previously mentioned, the production of signal molecules such as flavonoids by the host plant can express the bacterial *nod* genes resulting in the production of *nod* factors by rhizobium. The *nod* factors, which are lipo-chitooligosaccharides, can induce morphological alteration in plant root hairs such as root hair curling. The bacteria can enter the plant roots through the root hairs and by forming the infection thread eventually reside in the cortex of plant cells and form nodules, which is the place of rhizobium settlement in the form of bacteroid and subsequent N fixation (Long, 2001; Masson-Boivin et al., 2009; Glyanko and Vasileva, 2010).

## Diversity and Functions

The nucleotide sequence of rhizobia differentiate them from each other resulting in their diversity and various functioning processes. *Rhizobium* genes have a plasmid origin, located on their chromosome (Freiberg et al., 1997). They can develop symbiosis with most of the 18,000 species of legumes. Although many nodules properties including their morphology are regulated by the host plant, the *Rhizobium* by itself can also influence the symbiotic process through performing different genetical strategies. The phylogenical, metabolical, and genetical differences of *Rhizobium* make it very diverse. Phylogenically, they belong to two subclasses of proteobacteria; metabolically they perform different metabolical processes, the most important of which is N-fixation; and genetically they include a wide range of bacteria colonizing their specific legume plant. Rhizobia presently include 12 genera and more than 70 species (Masson-Boivin et al., 2009).

## INTERACTIONS BETWEEN AM FUNGI AND N-FIXING BACTERIA

Although N fixation by rhizobial symbiosis supplies the host plant with additional N, high amounts of energy and P must be consumed during the symbiosis process (Scheublin et al., 2004). Different species of AM fungi are able to provide the host plant with various amounts of P (Jakobsen et al., 1992; Ravnskov and Jakobsen, 1995; Smith and Read, 2008). AM fungi are also able to mineralize organic N in the mycorrhizosphere and increase N availability to the plant (Artursson et al., 2006). It is believed that the N-fixing capability of *Rhizobium* may be enhanced when in symbiosis with mycorrhizal host. Accordingly, AM and *Rhizobium* are considered to be synergistic resulting in enhanced colonization rate, nutrient uptake, and plant growth (Artursson et al., 2006).

The beneficial effect of AM fungi on *Rhizobium* P uptake has been proved through providing non-mycorrhizal plants with additional P, resulting in enhanced plant growth comparable with the growth of mycorrhizal plants (Artursson et al., 2006; Smith and Read, 2008). The structure, functioning, and nutritional demand of nodules are different from plant roots. Nodules are produced by cortical cell division, in which rhizobia with high energy and P requirements reside and fix N (Scheublin et al., 2004). Nodule formation also alters physiological properties of the plant and induces plant systematic acquired resistance (Lian et al., 2000; Scheublin et al., 2004), suggesting that the combination of AM species may be different in roots and nodules.

AM communities in legume roots are different from those in non-legume roots. Some AM species are able to develop symbiosis with plants containing high amounts of N. This is also in agreement with the findings that N fertilization can alter the combination of AM species, and that the level of *G. intraradices* with the *Glo8* sequence type increased after fertilization with N and P fertilizers (Johnson, 1993). While N-fixing bacteria utilize dicarboxylates as their source of energy (Udvardi and Day, 1997), AM fungi absorb (Pfeffer et al., 1999) glucose and fructose, just when colonizing plant roots, indicating that hexoses are translocated to the AM fungi (Harrison, 1999). This indicates that energy paths, which eventually produce mass of hypha in AM fungi, can be of higher efficiency compared with the energy paths in *rhizobium*, which eventually produce nodules. It is also interesting to mention that some bacteria, specifically *Paenibacillus* sp., are able to promote the mass production of AM hypha under *in vitro* conditions, which has been attributed to the production of trisaccharide rafinose by the bacteria (Hildebrandt et al., 2007).

Endosymbiosis is accompanied with the production of high amounts of symbiosome membrane. For example, the area of plasma membrane for a nodule cell with bacteroids is about 2800 $mM^2$, however the area for the symbiosome membrane is 21500 $mM^2$ (Roth and Stacey, 1989). Additionally, because different nutrients must be translocated into the cells containing the symbionts, different compositions of proteins including different transporters are located within the symbiosome membrane.

The prebacteroid (Fedorova et al., 1999) and the prearbuscular membranes have ATPase activities, which are not available in pathogenic interactions (Whitehead and Day, 1997). This is another reason for the beneficiary effects of symbiotic *Rhizobium*, relative to the pathogenic microorganisms. Thus, the symbiosome membrane forms two different components, one for the inclusion of the symbionts and one for the control of different proteins compositions and their activities.

While the mutants of *Bradyrhizobium japonicum* are not able to get activated under low oxygen conditions, the bacteria is able to develop a mechanism by which they can be active and fix N (Preisig et al., 1996) under such conditions. Different mutants are categorized based on the lack of ability to synthesize cytochromes, (necessary for bacteroids formation), pass through the infection thread, and transcribe different plant genes necessary for nodulins formation (Ramseier et al., 1991).

AM species increase the efficiency of legume plants through enhancing the uptake of nutrients such as P, Cu, and Zn, which are very important for nodulation and N fixation (Smith et al., 2010). Hence, the specificity between AM fungi and its host plant can be very important, as legumes would develop symbiosis with AM species, which are more efficient (Scheublin et al., 2004). This indicates that host specificity for AM can be of greater importance in a tripartite symbiosis. The tripartite symbiosis between AM, bacteria and legumes is of great significance for both agriculture and ecosystem, and researchers have been trying to find the most efficient combination of AM and bacteria (Scheublin et al., 2004).

Although colonization of legume nodules by AM has been proved in the laboratory, there is very little relevant data under field conditions (Hildebrandt et al., 2007). Compared with plant roots, the combination of AM species is unique for legume nodules. Although the sequence type *Glo8* was found more in legume nodules, the sequence type *Acau5* was only found in legume nodules. It is also worth mentioning that legume roots exclusively contained sequence type 50. These all suggest that legume nodules may have enhancing or inhibiting effects on root colonization of some AM species (Scheublin et al., 2004).

The following reasons indicate the specificity of legume nodules in their association with AM species:

(1) The high tendency of nodules for nutrients such as P, Cu or Zn, might result in the association of some AM species with legume nodules.
(2) High nodule N concentration may be preferable to some AM species (for example, AM species with *Glo8* and *Acau5* sequences).
(3) The similarities between the stages of *Rhizobium* and AM symbiosis, as there are some very important common stages (Provorov et al., 2002; Scheublin et al., 2004; Oldroyd et al., 2009) and hence, legume mutants, which are not able to develop symbioses with rhizobia, may not also develop symbiosis with AM species (Duc et al., 1989).

(4) The development of AM symbiosis is influenced by rhizobial signals (Xie et al., 1995).
(5) The different physiology of nodules and roots and also the alteration of root exudates by rhizobial symbiosis and the induction of systematic acquired resistance in plants.

These all are the likely explanations indicating the different tendency of root nodules for different AM species as all affect the ability of AM species, to colonize legume nodules (Scheublin et al., 2004).

It was found that AM species isolated from the field were able to colonize legume nodules by forming hypha around the nodules and producing hypha and spores in the nodules (Scheublin et al., 2004). Plant host species determine the combination of AM species in legume roots indicating the tendency of different host plants for symbiosis development with different AM species. Some AM species (with *Glo8* and *Glo3* sequences) have the ability to colonize a wide range of host legume plants in both roots and nodules (Opik et al., 2003).

Seed inoculation of legumes with N-fixing bacteria including *Rhizobium* and *bradyrhizobium* can be of practical applications as the method has become a very useful applicable technology in both developed and developing countries (Alves et al., 2003; Miransari and Smith, 2007). The environmental and economical aspects of N fixation are also very significant, e.g., by the application of less N fertilizer and the beneficial interacting effects of bacteria with other soil microorganisms including AM fungi (Miransari, 2010a, b, c). However, it is necessary to indicate how such inoculation can be performed, when AM fungi are also present. In other words, for the production of effective biological fertilizers, it is a must to find the most appropriate combinations of soil microbes under field conditions (Miransari, 2010a, b, c). As previously mentioned, usually legume plants are able to make tripartite symbiosis with N-fixing bacteria and AM fungi. This kind of association can be very beneficial both to the plant and the symbiotic microbes. To make the tripartite symbiosis highly efficient, the interactions between the plants, bacteria and the AM must be exactly elucidated. Researchers have indicated some interesting aspects related to such kind of symbiosis (Antunes et al., 2006 a, b).

However, few researchers have indicated the beneficial effects of AM fungi, both in a non-symbiotic situation and when used as inoculant in a bacteria-legume symbiosis. AM fungi are able to transfer the absorbed N from the soil to the plant. It has been stated that AM fungi are able to mineralize organic N present in organic matter and hence make soil N more available to the bacteria interactive with the fungi. However, the organic matter utilization and mineralization by AM fungi and hence plant and AM nutrient uptake, which is related to the stimulating effects of AM on mineralizing bacteria, has yet to be elucidated (Hodge et al., 2001).

It is important to indicate that how much N is supplied by AM fungi in a tripartite symbiosis (Chalk et al., 2006) as N can significantly influence *Rhizobium*-legume symbiosis (Miransari and Smith, 2007). The network of mycorrhizal hypha is able to transfer N between and within the plants. In addition to the important role of AM fungi in cycling the nutrients (Miransari, 2010a, b, c), their contribution to intensive agriculture should also be clearly indicated (Chalk et al., 2006). Under non-sterile conditions, legume roots can develop symbiosis with AM fungi, which can help plants such as forage and crop legumes absorb N at higher rates. AM fungi and *Rhizobium* can synergistically and significantly affect the symbiotic-related parameters (Chalk et al., 2006).

The different responses of legume species to AM inoculation have also been attributed to different root morphology and architecture and also different dependency on AM (Schoeneberger et al., 1989). Because AM fungi can considerably enhance P uptake, among its very significant effects on plant growth, AM effects on root growth may be the most important one, (Miransari et al., 2007; 2008) affecting root nodules formation and performance. Compared with control, AM inoculation, plant dry matter, fixed N, P and K uptake were significantly increased in different varieties of *Phaseolus vulgaris* and *Vicia faba,* inoculated with a mixture of *R. leguminosarum bv. phaseoli and R. tropici* and also mixed species of *Glomus clarum, G. etunicatum, G. manihotis* and *Gigaspora margarita* (Ibijbijen et al., 1996).

There is little data regarding the effect of AM fungi on micronutrient uptake when in a tripartite symbiosis with *Rhizobium* and legumes. However, AM and *Rhizobium* inoculation increased the uptake of Zn and Cu, hyphal colonization, nodule dry weight, N uptake, fixed N, and P uptake (Antunes 2006a,b). It is also interesting to investigate how AM fungi may affect the uptake of micronutrients, influencing the process of *Rhizobium*-legume symbiosis.

Most research regarding the effects of AM symbiosis on plant growth has been conducted under controlled and greenhouse conditions, and there has been little research regarding the effects of AM symbiosis on plant growth under field conditions (Daei et al., 2009). Hence, it is necessary to develop non-mycorrhizal legume mutants to clearly evaluate the effects of inoculated AM in the presence of indigenous AM on legume plants under field conditions. The mutants that have been developed so far have not been very applicable, because the mutated gene in the mutants ($MYC^-$) controls both the AM symbiosis and nodule formation ($nod^-$) (Chalk et al., 2006).

## Establishment of Dual Symbiosis

The symbiosis of AM and N-fixing bacteria with the host plants is the most important mutual symbiosis, agriculturally and ecologically (Limpens and Bisseling, 2003). As previously mentioned, the symbiotic bacteria make a chemotactic move to the roots and the spores begin to germinate. However, the spores may also germinate in the absence of the host plant, but they are not able to proceed with the following stages of symbiosis, although there are bacteria, which may positively affect such stages in the absence of the host plant (Hildebrandt et al., 2007).

Although because of the differences in host specificity between bacteria and AM fungi, the responses of the two symbionts are different, researchers have found that there are many common stages between bacterial and AM symbiosis with the host plant including the exchange of signal molecules (Limpens and Bisseling, 2003; Badri et al., 2009). For example, there are identical genes in both symbiosis that are activated during the symbiosis (Catoira et al., 2000). In addition, some of the signal molecules necessary for the AM-plant symbiosis have been recently identified (Akiyama and Hayashi, 2006). There should be more than a single gene for the perception of Nod (bacteria) and Myc (AM) factors.

There are some common genetical stages for both bacterial N fixation and AM symbiosis. The extent of plant accommodation for the intracellular settlement of endosymbionts, including AM and bacteria, has yet to be recognized (Parniske, 2000; Marsh et al., 2008; Oldroyd et al., 2009). Accordingly, there are some differences between the bacteria and AM fungi for the preparation of conditions before the onset of symbiosis with their partner. For example, weakening the root cell walls for the passage of bacteria into the roots is not common for AM fungi passage into the roots (Parniske, 2000; Oldroyd et al., 2009).

Because *Rhizobium* and AM fungi result in different responses, including morphological alterations and although both microbes induce similar genes, Nod factors require some extra genes for inducing responses in the host plant. Accordingly, Nod and Myc factors induce genes in different parts of the plant root including epidermis and cortex, respectively, indicating that different plant receptors perceive Nod and Myc factors. Thus, the two factors must result in the activation of different pathways in the plant (Limpens and Bisseling, 2003). Since legumes are able to develop symbiotic association with both N-fixing bacteria and AM, they are ideal for the study of the common stages of symbiosis between the two symbionts (Cook, 1999).

Another very important point is the recognizing role of signal molecules such as flavonoid in the tripartite symbiosis between AM fungi, bacteria, and legume plant (Zhang and Smith, 1995; Davies et al., 2005; Antunes et al., 2006a, b; Miransari and Smith, 2007; 2008; 2009). For example, the signal molecule formononetin is commercially available and has been indicated to be very effective on AM symbiosis (Davies et al., 2005; Chalk et al., 2006).

There are common legume plant genes and biochemical molecules affecting the tripartite symbiosis of the legume host plant with AM and N-fixing bacteria. The presence of a tripartite symbiosis in legumes indicates that legumes have genetical controlling processes that make this kind of symbiosis likely (Borisov et al., 2004). There are some pea (*Pisum sativum* L.) mutants that are not able to develop symbiosis with both AM fungi and N fixing bacteria. There are stages in both symbiosis, controlled by the same genes including hyphal passage into the plant roots by appressorium (Myc1), the development of arbuscule, and AM development rate (Jacobi et al., 2003; Borisov et al., 2004).

## Diversity and Functions

There is not much research regarding AM diversity in legumes and their colonization of legume nodules. Researchers have indicated that AM fungi are able to colonize root nodules in the laboratory (Baird and Caruso, 1994). However, nodule colonization of AM under field conditions and the species of AM, which are able to colonize legumes nodules, have yet to be recognized. Legumes may prefer to develop symbiosis with AM species, which can supply P more efficiently. This can be very advantageous under the conditions, where nutrients are deficient (Schulze, 2004). Different AM species are able to increase nodulation and N fixation differently (Ianson and Linderman, 1993; Scheublin et al., 2004).

Using the same PCR methods, the diversity of AM species ranged from 0.4 to 2.3, using the Shannon index under cultivated field and tropical rainforest conditions (Husband et al., 2002). The parameter most influencing the Shannon index is the specific plant species rather than the number of plant species. Hence, parameters such as the nature of plant functionality (legumes or non-legumes), species of plants, and root components, including roots and nodules, can determine AM communities in plants. More research is required to specify the effects of different AM species combinations on plant and nodule performance, and also on the structure and combination of plants (Scheublin et al., 2004; Meghvansia et al., 2008).

## Beneficial Uses of Dual Symbiosis

The amount of plant photosynthesized C supplied to AM fungi and *Rhizobium* ranges from 4 to 16% of the total fixed C. As a C sink for the host plant, AM fungi and *Rhizobium* stimulate the photosynthesis process in their host plant, although the enhanced plant nutrient uptake by AM and rhizobium can also partially account for such increased photosynthesis (Caschuk et al., 2009). Increased C production as a result of photosynthesis stimulation by microsymbionts can compensate for the plant C, utilized by AM fungi and *Rhizobium*. Interestingly, the dual inoculation with AM fungi and *Rhizobium* resulted in significantly higher rates of photosynthesis as well as nutrient uptake by plant, relative to the single inoculation of each symbiont (Caschuk et al., 2009).

In other words, and particularly for legume plants, the combination of AM fungi and *Rhizobium* is a complementary set, significantly increasing plant growth and crop production through utilization of enhanced nutrient uptake and fixed C by the host plant (Meghvansia et al., 2008; Caschuk et al., 2009). Accordingly, it may be mentioned that there are apparently some regulating processes in the legume plants that can control all the processes related to the tripartite symbiosis between AM fungi, *Rhizobium* and legume host.

It is interesting to mention that AM fungi are able to enhance the performance of inoculated legumes plant with *Rhizobium* under different conditions including stress. Numerous research experiments have illustrated the alleviating effects of AM fungi on plant growth and yield production under stress. The combined effects of AM fungi and *Rhizobium* under stress have also been tested in some experiments, and indicated the enhancing effects of AM fungi on the legume plants (Makoi and Ndakidemi, 2009; Wu and Wong, 2009).

## CONCLUSION

There are some interesting conclusions regarding this review article including: 1) the differences between the efficiency of energy paths in both symbionts, 2) the greater importance of host specificity for AM in the tripartite symbiosis between AM, *Rhizobium* and the host plant, 3) the tripartite symbiosis is very much different under laboratory and field conditions, 4) the genetic combination of nodules can influence AM performance, 5) there are still so many unknown facts regarding the interactive effects of AM, *Rhizobium* and the host plant in a tripartite symbiosis, 6) with respect to the AM ability in providing the host plant with N, it is important to elucidate that how much N is supplied by AM in a tripartite symbiosis as it affects the *Rhizobium*-legume symbiosis performance, 7) development of mutant hosts, which are responsive to *Rhizobium* and not to AM symbiosis is necessary for the precise evaluation of AM symbiosis stages in a tripartite symbiosis, and 8) the role of signal molecules in the tripartite symbiosis must be clearly elucidated.

The interactions between AM, *Rhizobia* and the host plant can affect the agriculture and ecosystem productivity including the practical aspects of biological fertilization production under field conditions. Hence, evaluation of such effects can be very beneficial for the suggestion of methods, which may enhance soil production efficiency. The new insights can always be useful for the development of appropriate agricultural and ecological strategies.

## REFERENCES

Akiyama, K. and Hayashi, H. (2006). Strigolactones: chemical signals for fungal symbiosis and parasitic weeds in plant roots. *Ann. Bot.* **97:** 925-931.

Alves, B.J.R., Boddey, R.M. and Urquiaga, S. (2003). The success of BNF in soybean in Brazil. *Plant Soil* **252:** 1-9.

Antunes, P.M., de Varennes, A., Rajcan, I. and Goss, M.J. (2006a). Accumulation of specific flavonoids in soybean (*Glycine max* (L.) Merr.) as a function of the early tripartite symbiosis with arbuscular mycorrhizal fungi and *Bradyrhizobium japonicum* (Kirchner) Jordan. *Soil Biol. Biochem.* **38:** 1234-1242.

Antunes, P.M., Rajcan, I. and Goss, M.J. (2006b). Specific flavonoids as interconnecting signals in the tripartite symbiosis formed by arbuscular mycorrhizal fungi, *Bradyrhizobium japonicum* (Kirchner) Jordan and soybean (*Glycine max* (L) Merr.). *Soil Biol. Biochem.* **38:** 533-543.

Artursson, V., Finlay, R.D. and Jansson, J.K. (2006). Interactions between arbuscular mycorrhizal fungi and bacteria and their potential for stimulating plant growth. *Environ. Microbiol* **8:** 1-10.

Badri, D., Weir, T., van der Lelie, D. and Vivanco, J. (2009). Rhizosphere chemical dialogues: plant-microbe interactions. *Curr. Opin. Biotechnol.* **20:** 642-650.

Baird, L.M. and Caruso, K.J. (1994). Development of root nodules in *Phaseolus vulgaris* inoculated with *Rhizobium* and mycorrhizal fungi. *Internation. J. Plant Sci.* **155:** 633-639.

Borisov, A.Y., Danilova, T.N., Koroleva, T.A., Naumkina, T.S., Pavlova, Z.B., Pinaev, A.G., Shtark, O.Y., Tsyganov, V.E., Voroshilova, V.A., Zhernakov, A.I., Zhukov, V.A. and Tikhonovich, I.A. (2004). Pea (*Pisum sativum* L.) regulatory genes controlling the development of nitrogen-fixing nodule and arbuscular mycorrhiza: a review of basic and applied aspects. *In Russian, Abstract in English Prikl. Biokhim. Mikrobiol.* **13:** 137-144.

Catoira, R., Galera, C., De Billy, F., Penmetsa, R.V., Journet, E.P., Maillet, F., Rosenberg, C., Cook, D., Gough, C. and Denarie, J. (2000). Four genes of *Medicago truncatula* controling components of a Nod factor transduction pathway. *Plant Cell* **12:** 1647-1666.

Chalk, P.M., Souza, R., Urquiaga, S., Alves, B.J.R. and Boddey, R.M. (2006). The role of arbuscular mycorrhiza in legume symbiotic performance. *Soil Biol. Biochem.* **38:** 2944-2951.

Cleveland, C.C., Townsend, A.R., Schime, D.S., Fisher, H., Howarth, R.W., Hedin, L.O., Perakis, S.S., Latty, E.F., Von Fischer, J.C., Elseroad, A. and Wasson, M.F. (1999). Global patterns of terrestrial biological nitrogen ($N_2$) fixation in natural ecosystems. *Global Biogeochemistry Cycles* **13:** 623-645.

Cook, D.R. (1999). *Medicago truncatula*—a model in the making! *Curr. Opin. Plant Biol.* **2:** 301-304.

Cooper, J.E. (2007). Early interactions between legumes and rhizobia: disclosing complexity in a molecular dialogue. *J. Appl. Microbiol.* **103:** 1355-1365.

Daei, G., Ardakani, M., Rejali, F., Teimuri, S. and Miransari, M. (2009). Alleviation of salinity stress on wheat yield, yield components, and nutrient uptake using arbuscular mycorrhizal fungi under field conditions. *J. Plant Physiol.* **166:** 617-625.

Davies, Jr. F.T., Calderon, C.M., Huaman, Z. and Gomes, R. (2005). Influence of a flavonoid (formononetin) on mycorrhizal activity and potato crop productivity in the highlands of Peru. *Sci. Horticul.* **106:** 318-329.

Duc, G., Trouvelot, A., Gianinazzi-Pearson, V. and Gianinazzi, S. (1989). First report of non-mycorrhizal plant mutants (*myc-*) obtained in pea (*Pisum sativum* L.) and faba bean (*Vicia faba* L.). *Plant Sci.* **60:** 215-222.

Fedorova, E., Thomson, R., Whitehead, L.F., Maudoux, O., Udvardi, M.K. and Day, D.A. (1999). Localization of H+-ATPase in soybean root nodules. *Planta* **209:** 25-32.

Freiberg, C., Fellay, R., Bairoch, A., Broughton, W., Rosenthal, A. and Perret, X. (1997). Molecular basis of symbiosis between *Rhizobium* and legumes. *Nature* **387:** 391-401.

Gherbi, H., Markmann, K., Svistoonoff, S., Estevan, J., Autran, D., Giczey, G., Auguy, F., Peret, B., Laplaze, L., Franche, C., Parniske, M. and Bogusz, D. (2008). SymRK defines a common genetic basis for plant root endosymbiosis with arbuscular mycorrhiza fungi, rhizobia, and Frankia bacteria. *Proc. Nat. Acad. Sci. USA* **105:** 4928-4932.

Glyanko, A. and Vasileva, G. (2010). Reactive oxygen and nitrogen species in legume-Rhizobial symbiosis: A Review. *Appl. Biochem. Microbiol.* **46:** 15-22.

Harrison, M.J. (1999). Biotrophic interfaces and nutrient transport in plant fungal symbioses. *J. Exp. Bot.* **50:** 1013-1022.

Harrison, M.J. (2005). Signaling in the arbuscular mycorrhizal symbiosis. *Annu. Rev. Microbiol.* **59:** 19-42.

Hijri, M. and Sanders, I.R. (2005). Low gene copy number shows that arbuscular mycorrhizal fungi inherit genetically different nuclei. *Nature* **433:** 160-163.

Hildebrandt, U., Ouziad, F., Marner, F.J. and Bothe, H. (2007). The bacterium *Paenibacillus validus* stimulates the growth of arbuscular mycorrhizal fungus *Glomus intraradices* up to the formation of fertile spores. *FEMS Microbiol Lett.* **254:** 258-267.

Hodge, A., Campbell, C.D. and Fitter, A.H. (2001). An arbuscular mycorrhizal fungus accelerates decomposition and acquires nitrogen directly from organic material. *Nature* **413:** 297-299.

Husband, R., Herre, E.A., Turner, S.L., Gallery, R. and Young, J.P.W. (2002). Molecular diversity of arbuscular mycorrhizal fungi and patterns of host association over time and space in a tropical forest. *Mol. Ecol.* **11:** 2669-2678.

Ianson, D.C. and Linderman, R.G. (1993). Variation in the response of nodulating pigeon pea (*Cajanus cajan*) to different isolates of mycorrhizal fungi. *Symbiosis* **15:** 105-119.

Ibijbijen, J., Urquiaga, S., Ismaili, M., Alves, B.J.R. and Boddey, R.M. (1996). Effect of arbuscular mycorrhizal fungi on growth, mineral nutrition and nitrogen fixation of three varieties of common beans. *New Phytol.* **134:** 353-360.

Jacobi, L.M., Zubkova, L.A., Barmicheva, E.M., Tsyganov, V.E., Borisov, A.Y. and Tikhonovich, I.A. (2003). *Mycorrhiza* **13:** 9-16.

Jakobsen, I., Abbott, L.K. and Robson, A.D. (1992). External hyphae of vesicular-arbuscular mycorrhizal fungi associated with *Trifolium subterraneum* L. I. Spread of hyphae and phosphorus inflow into roots. *New Phytol.* **120:** 371-380.

Johnson, N.C. (1993). Can fertilization of soil select less mutualistic mycorrhizae? *Ecol. App.* **3:** 749-757.

Kaschuk, G., Kuyper, T., Leffelaar, P., Hungria, M. and Giller, K. (2009). Are the rates of photosynthesis stimulated by the carbon sink strength of rhizobial and arbuscular mycorrhizal symbiosis? *Soil Biol. Biochem.* **41:** 1233-1244.

Lendzemo, V., Kuyper, T. and Vierheilig, H. (2009). Striga seed-germination activity of root exudates and compounds present in the stems of Striga host and nonhost (trap crop) plants is reduced due to root colonization by arbuscular mycorrhizal fungi. *Mycorrhiza* **19:** 287-294.

Long, S. (2001). Genes and signals in the *Rhizobium*-legume symbiosis. *Plant Physiol* **125:** 69-72.

Lian, B., Zhou, X., Miransari, M. and Smith, D.L. (2000). Effects of salicylic acid on the development and root nodulation of soybean seedlings. *J. Agron. Crop Sci.* **185:** 187-192.

Limpens, E. and Bisseling, T. (2003). Signaling in symbiosis. *Curr. Opin. Plant Biol.* **6:** 343-350.

Makoi, J. and Ndakidemi, P. (2009). The agronomic potential of vesicular-arbuscular mycorrhiza (VAM) in cereal-legume mixtures in Africa. *Afric. J. Microbiol. Res.* **3:** 664-675.

Masson-Boivin, C., Giraud, E., Perret, X. and Batut, J. (2009). Establishing nitrogen-fixing symbiosis with legumes: how many rhizobium recipes? *Trend Microbiol.* **17:** 458-466.

Marsh, J.F., Gobbato, E., Schultze, M. and Oldroyd, G.E.D. (2008) Ram1 and Ram2: signaling specificity in mycorrhizal symbiosis. Abstract Book of 8th European Nitrogen Fixation Conference, Gent, Belgium, August 30-September 3, 2008, p. 26.

Meghvansia, M., Prasad, K., Harwani, D. and Mahna, S. (2008). Response of soybean cultivars toward inoculation with three arbuscular mycorrhizal fungi and *Bradyrhizobium japonicum* in the alluvial soil. *Europ. J. Soil Biol.* **44:** 316-323.

McInnes, A., Thies, J.E., Abbotta, L.K. and Howieson, J.G. (2004). Structure and diversity among rhizobial strains, populations and communities—a review. *Soil Biol. Biochem.* **36:** 1295-1308.

Miransari, M., Smith, D.L., Mackenzie, A.F., Bahrami, H.A., Malakouti, M.J. and Rejali, F. (2006). Overcoming the stressful effect of low pH on soybean root hair curling using lipochitooligosaccahrides. *Commun. Soil Sci. Plant Anal.* **37:** 1103-1110.

Miransari, M. and Smith, D.L. (2007). Overcoming the stressful effects of salinity and acidity on soybean [*Glycine max* (L.) Merr.] nodulation and yields using signal molecule genistein under field conditions. *J. Plant Nut.* **30:** 1967-1992.

Miransari, M., Bahrami, H.A., Rejali, F., Malakouti, M.J. and Torabi, H. (2007). Using arbuscular mycorrhiza to reduce the stressful effects of soil compaction on corn (*Zea mays* L.) growth. *Soil Biol. Biochem.* **39:** 2014-2026.

Miransari, M. and Smith, D.L. (2008). Using signal molecule genistein to alleviate the stress of suboptimal root zone temperature on soybean-*Bradyrhizobium* symbiosis under different soil textures. *J. Plant Interact* **4:** 287-295.

Miransari, M., Bahrami, H.A., Rejali, F. and Malakouti, M.J. (2008). Using arbuscular mycorrhiza to reduce the stressful effects of soil compaction on wheat (*Triticum aestivum* L.) growth. *Soil Biol. Biochem.* **40:** 1197-1206.

Miransari, M. and Smith, D. (2009). Alleviating salt stress on soybean (*Glycine max* (L.) Merr.) - *Bradyrhizobium japonicum* symbiosis, using signal molecule genistein. *Europ. J. Soil Biol.* **45:** 146-152.

Miransari, M., Rejali, F., Bahrami, H.A. and Malakouti, M.J. (2009a). Effects of soil compaction and arbuscular mycorrhiza on corn (*Zea mays* L.) nutrient uptake. *Soil Till Res.* **103:** 282-290.

Miransari, M., Rejali, F., Bahrami, H.A. and Malakouti, M.J. (2009b). Effects of arbuscular mycorrhiza, soil sterilization, and soil compaction on wheat (*Triticum aestivum* L.) nutrients uptake. *Soil Till Res.* **104:** 48-55.

Miransari, M. (2010a). Contribution of arbuscular mycorrhizal symbiosis to plant growth under different types of soil stresses. Review article. *Plant Biol.* **12:** 563-569.

Miransari, M. (2010b). Arbuscular mycorrhiza and soil microbes, In Mycorrhizal Biotechnology, editors: Thangadurai, D., Busso, C.A., Hijri, M. CRC Press, 226p, USA.

Miransari, M. (2010c). Biological Fertilization. In: Current Research, Technology and Education Topics in Applied Microbiology and Microbial Biotechnology Editor: Méndez-Vilas A. Formatex Research Centre, Spain.

Oldroyd, G.E.D., Harrison, M.J. and Paszkowski, U. (2009). Reprogramming plant cells for endosymbiosis. *Science* **324:** 753-754.

Opik, M., Moora, M., Liira, J., Koljalg, U., Zobel, M. and Sen, R. (2003). Divergent arbuscular mycorrhizal fungal communities colonize roots of *Pulsatilla* spp. in boreal Scots pine forest and grassland soils. *New Phytol.* **160:** 581-593.

Parniske, M. (2000). Intracellular accommodation of microbes by plants: a common developmental program for symbiosis and disease? *Curr. Opin. Plant Biol.* **3:** 320-328.

Pfeffer, P.E., Douds, D.D., Becard, G. and Shachar-Hill, Y. (1999). Carbon uptake and the metabolism and transport of lipids in an arbuscular mycorrhiza. *Plant Physiol.* **120:** 587-598.

Preisig, O., Zufferey, R., Thöny-Meyer, L., Appleby, C.A. and Hennecke, H. (1996). A high-affinity cbb(3)-type cytochrome oxidase terminates the symbiosis-specific respiratory chain of *Bradyrhizobium japonicum*. *J. Bacteriol.* **178:** 1532-1538.

Provorov, N.A., Borisov, A.Y. and Tikhonovich, I.A. (2002). Developmental genetics and evolution of symbiotic structures in nitrogen-fixing nodules and arbuscular mycorrhiza. *J. Theoretic Biol.* **214:** 215-232.

Ramseier, T.M., Winteler, H.V. and Hennecke, H. (1991). Discovery and sequence analysis of bacterial genes involved in the biogenesis of *c*-type cytochromes. *J. Biol. Chem.* **266:** 7793-7803.

Ravnskov, S. and Jakobsen, I. (1995). Functional compatibility in arbuscular mycorrhizas measured as hyphal P transport to the plant. *New Phytol.* **129:** 611-618.

Roth, L.E. and Stacey, G. (1989). Bacterium release into host-cells of nitrogen fixing soybean nodules, the symbiosome membrane comes from 3 sources. *Europ. J. Cell Biol.* **49:** 13-23.

Salvagiotti, F., Cassman, K., Specht, J., Walters, D., Weiss, A. and Dobermann, A. (2008). Nitrogen uptake, fixation and response to fertilizer N in soybeans: A review. *Field Crop Res.* **108:** 1-13.

Sanders, I.R., Clapp, J.P. and Wiemken, A. (1996). The genetic diversity of arbuscular mycorrhizal fungi in natural ecosystem: A key to understanding the ecology and functioning of the mycorrhizal symbiosis. *New Phytol.* **133:** 123-134.

Scheublin, T.R., Ridgway, K.P., Young, P.W. and van der Heijden, M.G.A. (2004). Nonlegumes, legumes, and root nodules harbor different arbuscular mycorrhizal fungal communities. *Appl. Environ. Microbiol.* **70:** 6240-6246.

Schoeneberger, M.M., Volk, R.J. and Daveym, C.B. (1989). Factors influencing early performance of leguminous plants in forest soils. *Soil Sci. Soc. Am. J.* **53:** 1429-1434.

Schulze, J. (2004). How are nitrogen fixation rates regulated in legumes? *J. Plant Nut. Soil Sci.* **167:** 125-137.

Smith, S.E. and Read, D.J. (2008). Mycorrhizal symbiosis, 2nd ed. Academic Press, London, United Kingdom.

Smith, S.E., Facelli, E., Pope, S. and Smith, A. (2010).Plant performance in stressful environments: interpreting new and established knowledge of the roles of arbuscular mycorrhizas. *Plant Soil* **326:** 3-20.

Sprent, J.I. (2001). Nodulation in legumes. Royal Botanical Gardens, Kew, United Kingdom.

Subramanian, S., Stacey, G. and Yu, O. (2006). Endogenous isoflavones are essential for the establishment of symbiosis between soybean and *Bradyrhizobium japonicum Plant J.* **48:** 261-273.

Udvardi, M.K. and Day, D.A. (1997). Metabolite transport across symbiotic membranes of legume nodules. *Ann. Rev. Plant Physiol. Plant Mol. Biol.* **48:** 493-523.

Vandermeer, J.H. (1989). The ecology of intercropping. Cambridge University Press, New York, N.Y.

Whitehead, L.F., Day, D.A. (1997). The peribacteroid membrane. *Physiol Plant* **100:** 30-44.

Wu, F. and Wong, M. (2009). Dual inoculation with an arbuscular mycorrhizal fungus and *rhizobium* to facilitate the growth of alfalfa on coal mine substrates. *J. Plant Nut.* **32:** 755-771.

Xie, Z., Staehelin, C., Vierheilig, H., Wiemken, A., Jabbouri, S., Broughton, W.J., Vogeli-Lange, R. and Boller, T. (1995). Rhizobial nodulation factors stimulate mycorrhizal colonization of nodulating and non-nodulating soybeans. *Plant Physiol.* **108:** 1519-1525.

Zhang, F. and Smith, D.L. (1995). Preincubation of *Bradyrhizobium japonicum* with genistein accelerates nodule development of soybean (*Glycine max.* (L.) Merr.) at suboptimal root zone temperatures. *Plant Physiol.* **108:** 961-968.

# 7

# Biotechnology: Applications for Conservation and Sustainable Use of Plants of Fragile Arid Ecosystems

**Narpat Singh Shekhawat[1], Mahendra Phulwaria[1], Harish[2], Amit Kumar Gupta[3], Kheta Ram[1], Smita Shekhawat[1], Vibha JB[1], Manoj Kumar Rai[1], Vinod Kataria[1], Gurdeep Kaur[1], Ashok Kumar Patel[1], Jitendra Singh Rathore[3] and R. P. Singh[4]**

[1]*Biotechnology Unit, Department of Botany, Jai Narain Vyas University, Jodhpur-342001, India*

[2]*Department of Botany, Mohanlal Sukhadia University, Udaipur-313001, India*

[3]*Department of Botany and Biotechnology, L.M. College of Science and Technology, Jodhpur-342001, India*

[4]*Shri B.R.M. Govt. College, Nagaur-341001, India*

**ABSTRACT**

Arid and dry lands are significant and important components of earth's ecosystems as these are inhabited by sizeable human population and harbour considerable biodiversity providing ecosystem services. In the era of climate change and global warming, there is greater need for understanding dryland ecosystems. The Indian Thar Desert and Aravalli ecosystems harbour unique life systems that experience types of recurring/non-recurring and biotic/abiotic stresses and these have acquired mechanisms of survival under the fragile climatic conditions. Plants and microbes of these regions produce biomass that support other life forms and bear traits/characters of biological importance and curiosity. Multidisciplinary research and prioritization of research areas are need of the time for sustainability of these arid ecosystems. Development and application of modern methods of biology/biotechnology and information technology may help in amelioration and sustainable utilization of natural resources. We have used tools of biotechnology for characterization, conservation and propagation of a number of plant species that contribute in biomass production in the arid regions. Tissue culture/micropropagation processes have been defined for trees and shrubs that are ecologically and economically important and also for rare/endemic and threatened taxa. This has been possible with the support from various government agencies that provided funds under R and D projects. The methods developed for in vitro culture establishment, and multiplication, in vitro and ex vitro rooting of in vitro produced shoots have enabled field transfer of cloned plants. Thus, tissue culture and molecular biological methods for characterization are valuable as non-GM biotechnology for sustainable utilization of plant resources of arid regions. There is further scope for use of instruments of genomics, proteomics, epigenetics and other emerging sciences. This review summarizes the biotechnological research done for the plants of arid regions of India.

## INTRODUCTION

In July, 2008, the National Institutes of Health (NIH), National Science Foundation (NSF), and Department of Energy (DOE) of the USA asked the National Research Council's Board on Life Sciences, to convene a committee to examine the current status of biological research and recommend how best to capitalize on recent technological and scientific advances that have allowed biologists to integrate biological research findings, collect and interpret vastly increased amounts of data, and predict the behavior of complex biological systems. A committee of 16 experts from the fields of biology, engineering and computational science undertook to delineate those scientific and technological advances and come to a consensus on how the best capitalize on them. The report authored by the Committee on a New Biology for the 21st Century concluded that biological research has indeed experienced extraordinary scientific and technological advances in recent years. The essence of the New Biology, as defined by the committee, is integration - re-integration of the many sub-disciplines of biology, and the integration into biology of physicists, chemists, computer scientists, engineers, and mathematicians to create a research community with the capacity to tackle a broad range of scientific and societal problems. Integrating knowledge from many disciplines will permit deeper understanding of biological systems, which will both lead to biology-based solutions to societal problems and also feed back to enrich the individual scientific disciplines that contribute new insights. The New Biology is not intended to replace the research that is going on now; that research, much of it fundamental and curiosity-driven by individual scientists, is the foundation on which the New Biology rests and on which it will continue to rely (http://www.nap.edu/catalog/12764.html).

Three major challenges for humankind in the $21^{st}$ century are food, energy, and environment, particularly climate change and environmental degradation due to pollution and habitat loss. In a word: sustainability (http://www.aspb.org/

plantsummit) food and energy because essentially all of our food and the great majority of our energy are produced by photosynthetic plants. Climate, pollution, and habitat are of major concern as plants are major players in determining our climate. Agricultural expansion is a major factor in habitat encroachment and pollution of waterways by fertilizer application and runoff. Besides their unique relevance to our lives, plants themselves are unique in their biology. Fundamental biology of plants is similar to human. McGary (2010) suggested that plants and people share at least 48 functional modules; sets of genes that act in common to produce a phenotype. They use the same genetic code, share many homologous genes and even regulatory mechanisms, and they share the same basic biochemical pathways and even fundamental processes in cell biology - their form and life style are fundamentally different. The issues of climate change and pollution are not independent as these affect plant performance and their habitats.

Plant scientists have great challenges of providing solutions for these major challenges. However, beyond these applications of science, we also believe in the intrinsic value of exploring how organisms work as a valuable human quest and a worthy goal. Plants are one of the fascinating and important organisms on earth.

As early as 100-300 AD Hindu philosophers first pondered the nature of reproduction and inheritance and later on during 1000 AD they observed that certain diseases may "run in the family." Moreover, they came to believe that children inherit all their parents' characteristics. "A man of base descents can never escape his origins," went the laws of Manu. Studies of inheritance and genetics have been subjects of curiosity and part of human education, development of society and evolution of cultures. During history humans learned to select and grow the plants with bigger and sweeter fruits and tried to breed only healthy animals. The earliest farmers developed improved species of plants and animals by cross pollination or cross breeding. Use of tools and techniques for utilization of biological systems has been practiced for thousands of years. It dates back to when man first learned how to plant crops or breed animals. The discovery of fermentation started new beginning. The natural processes involving microorganisms were understood and utilized for the production of beverages, food and medicine. The method included making cheese and yogurt from milk; converting fruit juices into wine; leavening bread; and making beer, which the Egyptians recorded as a prescription medicine in 1600 B.C. Archeologists reveal that these early practices has were around between 5,000 to 10,000 BC.

While some of the richest ideas in the life sciences have been developed without application of specialized technology - the theory of natural selection and evolution stands out as a prominent example - technology is more often a primary driver of new understanding. The concept of modern cell theory and technological advances/ experimental research has enabled the scientists to reveal secrets of cell/life and utilize cell and tissues/organism for getting services and products for welfare of society.

## CELL THEORY AND BIOTECHNOLOGY

The rejection of philosophy of spontaneous generation of life forms by Luis Pasteur during 19$^{th}$ century has been key factor in our understanding of life. Pasteur's concepts established foundation of modern biology. The concept of origin of life and evolution of species has revolutionized our thinking about living organisms. The cell as an integral component of living organisms has been based on metaphor since its inception, and the history of cell theory has continued to rely on metaphor and analogy. In the nineteenth century, cells were most popularly conceived either as building stones or elementary autonomous organisms from which larger organisms are composed. With advances in physiology and the rise of modern biochemistry in the early twentieth century, the chemical factory or laboratory became the dominant metaphor for this biological unit. Today in the twenty-first century, the metaphorical imagery has become a reality, with cells acting as chemical factories for the synthesis of commercially valuable bio-products. A dominant metaphor in the second half of the nineteenth century described the body as a 'society' or 'state' of cells (Zellenstaat). Cells were 'citizens' arranged into separate classes or professions according to their functions, together making up the 'economy of the organism'. There are hints of this metaphor in the writings of Matthias Schleiden (1804–1881) and Theodor Schwann (1810– 1882), who are credited with founding the cell theory in the 1830s. In biology the term is "cell" entered the scientific lexicon in the 17$^{th}$ century and attained prominence (notoriety as described by some critics of cell theory) in the early 19th century under the aegis of the "Cell Theory", which defined cell as the fundamental unit of all life forms (Welch and Clegg 2010). The proposition that all living things are composed of small fundamental vital units, otherwise known as cells, ranks next to Darwin's theory of evolution as one of the great unifying theses of biological science. Today, it is not at all unusual to run across some version of the following description of what scientists understand a cell to be:

'All living things are made from cells, the chemical factories of life. Cells act as chemical factories, taking in materials from the environment, processing them, and producing "finished goods" to be used for the cell's own maintenance and for that of the larger organism of which they may be part. In a complex cell, materials are taken in through specialized receptors ("loading docks"), processed by chemical reactions governed by a central information system ("the front office"), carried around to various locations ("assembly lines") as the work progresses, and finally sent back via those same receptors into the larger organism. Far from being a shapeless blob of protoplasm, the cell is a highly organized, busy place, whose many different parts must work together to keep the whole functioning' (cited by Reynolds 2007).

The modern tenets of the Cell Theory include; (i) all living organisms are composed of cell or cell products and the cell is structural and functional unit of living things, (ii) all energy flow (metabolisms and biochemistry) of life occurs within cells, (iii) all cells come from pre-existing cells by division, and (iv) hereditary of an organism of which the cell is a component, is present in a cell which is passed from cell to cell during division.

The cell theory, when it was proposed envisioned the theoretical and experimental foundation of modern biology and biotechnology. Technological advancements in any biological provide the means to test hypotheses formulated from less complete or less precise information and technology also provides new information to generate fresh hypotheses. Certainly, the most influential advance in modern biology has been uncovering the molecular basis of inheritance and biological information transfer, namely the discovery and structure of DNA and its relation to RNA and proteins. This advance depended critically on advanced technologies such as x-ray diffraction to test hypotheses generated from chemical analysis, and in turn uncovered vast new fields of information that continue to revolutionize the course of biological research. The central dogma is also a prime example of how discoveries made possible by technology themselves engender powerful new technologies, perhaps the most prominent being the possibility to "design" genes by fusing sequences from different parts of a genome or different organisms and to introduce these designer genes into living organisms from bacteria to humans.

## BIOTECHNOLOGY; GMO/NON-GMO AND USES

By the end of nineteenth century, the work of Gregor Mendel on traits and inheritance promoted the concept of the science through selective breeding. In 1919, a Hungarian engineer named Karl Ereky coined the word biotechnology. During that time, all techniques were referred to as production with the help of living organisms. UN Convention on Biological Diversity has come up with one of many definitions of biotechnology: "Biotechnology means any technological application that uses biological systems, living organisms, or derivatives thereof, to make or modify products or processes for specific use." Biotechnology in simple terms is defined as the "utilization of biological processes, be they microbial, plant, animal cells or their constituents, for the provision of goods and services" for society. This in fact is convergence of tools and techniques evolved and applied during several decades of research and investigations in efforts to understand biology of living organisms. Emergence and development of molecular biology and genetic engineering provided new paradigms to life sciences and new principles of engineering and technology for applications. The newly emerged fields of genome-sequencing, -editing and -imprinting, epigenetics, reverse genetics, genomics, proteomics, bioinformatics and other omics have changed pitch of Biotechnology.

There are many major areas of biotechnology and its applications. Some of these are bioprocessing technology, monoclonal antibodies, cell culture, recombinant DNA technology, cloning, protein engineering, biosensors, nanobiotechnology and microarrays. Cell culture is one of the areas meaning growing cells outside of a living organism. These include plant cell culture, insect cell culture and animal/ mammalian cell culture. Cloning became well known after the cloning of Dolly the sheep years back. Identification, characterization and culture of stem cell and cloning technology actually allows for the generation of genetically identical molecules, plants, cells or animals.

Biotechnology is also categorized like red, white, green, and blue and multicolor biotechnology. Red biotechnology involves medical processes such as getting organisms to produce new drugs, or using stem cells to regenerate damaged human tissues and perhaps re-grow entire organs. White (also called gray) biotechnology involves industrial processes such as the production of new chemicals or the development of new fuels for vehicles. Green biotechnology applies to agriculture and involves such processes as the development of pest-resistant plants or the accelerated evolution of disease-resistant animals. Blue biotechnology, rarely mentioned, encompasses processes in marine and aquatic environments, such as controlling the proliferation of noxious water-borne organisms.

Modern biotechnology has expanded in sophistication, scope, and applicability. The 1970s pioneered modern biotechnology. Herb Boyer and Stanley Cohen demonstrated the generation of human genes with the aid of bacteria in 1973. There is intense debate around the applications of modern biological research/ biotechnology to agriculture, and food production, genetic modification (GM) and the novel crops those results from their applications.

In the recent past due to or/and in order to meet regulatory/legal and ethical requirements, biotechnologies are being classified as involving Non-GMO/Cisgenic or Targeted Genetic modifications (TagMO)/GMO. A term genetic modification is frequently used in association with biotechnology. This sometimes is quite misleading terminology as "modification" means change and conventional breeding is constantly changing to produce new varieties. "Genetic modification" only refers to the recombinant DNA plant breeding technique and the GM problems occur because consumers reject this type of plant breeding technique. "GM" or "GE" or "transgenics"/"transplastomic' are where individual gene constructs are transferred from one organism to another and the most common application uses cross kingdom breeding. For example, bacterial and virus genes are used in GM canola to produce chemical resistance or insect resistance Bt Cotton. Alternatively, other varieties (non-GM) types of plant breeding are accepted by consumers, have no labeling legislation, no contamination issues, no liability issues, etc.

Consumers are accepting thoroughout the world the non-GM plant breeding including (i) Cloning of cell, tissue and organs culture and plant generation for micropropagation and mass multiplication and for production of secondary metabolites (ii) anther and pollen culture for haploid production, and their large scale utilization in crop breeding and improvement (iii) Embryo rescue and its culture for hybrid production using classical breeding approaches (iv) Somaclonal, protoclonal and gametoclonal variations as sources of new sources of desired variants (v) Random mutagenesis and Cell Tissue Selection for useful mutants production and Colchicine treatment for production and utilization of polyploidy crop plants (vi) Protoplast culture and fusion for production of somatic hybrids/ partial hybrids and cybrids and of course (vii) Grafting.

Plant tissue and cell culture technology is used worldwide and it is a non-GMO biotechnology. Also there are other alternative approaches offered for plant breeding and mechanical improvements that are achieving similar aims to those promised by recombinant DNA. Molecular markers and DNA bar-coding are becoming fast track research and being utilized in applied and basic sciences for as part of new biology. These are actually non-GM and genomic applications of biotechnology. Due to the advances in this non-GM technique, plant breeders have far more accuracy and speed in selecting for desired traits. Biotechnology also includes non-GM/O approaches yet it is very common to have promotion of GM under the guise of biotechnology.

## TRANSGENESIS VERSUS CISGENESIS

In the case of transgenesis, the transferred gene usually derives from an alien species that is neither the recipient species nor a close, sexually compatible relative. In other words, transgenesis can extend the gene pool of the recipient species. Such a novel gene might provide the target plant with a new trait that neither occurs in the recipient species in nature nor can be introduced through traditional breeding. This novel trait might affect the fitness of the recipient species in various ways; a change in fitness can then spread through gene flow between a GM crop and its wild relatives (den Nijs et al. 2004), potentially creating shifts in natural vegetation. Consequently, lawmakers and regulatory authorities have paid much attention to the safety of deliberate releases of transgenic crops into the environment and have put in place biosafety frameworks to control this risk.

In a cisgenic plant, the gene of interest, together with its promoter, has been present in the species or in a sexually compatible relative for centuries. Therefore cisgenesis does not alter the gene pool of the recipient species and provides no additional traits. No changes in fitness occur that would not happen through either traditional breeding or natural gene flow. Similarly, cisgenesis carries no risks—such as effects on non-target organisms or soil ecosystems, toxicity or a possible allergy risk for GM food or feed—other than those that are also incurred by traditional breeding. This is the fundamental difference between cisgenesis and transgenesis. Consequently, the deliberate release and market introduction of cisgenic plants is as safe as the release and market introduction of traditionally bred plants. On the issue of safety, regulators could treat cisgenic plants the same as conventionally bred plants (Schouten et al. 2006; Jacobsen and Schouten 2008, 2009). Production of cisgenic barley with improved phytase activity (Holmes et al. 2011) and development of cisgenic apple (Vanblaere et al. 2011) are reported recently. Of the new wave of technologies, targeted genetic modifications (TagMo), are being developed as these have great potential to modify native plant genes in directed and targeted ways. TagMo enables more sophisticated modifications of plant genomes without using recombinant DNA technology, but similar to classical breeding or mutagenesis, which is not formally regulated. It has been argued that TagMo might fall outside of existing regulatory definition.

## PLANT BIOTECHNOLOGY: THE BASICS DEVELOPMENTS, HISTORY AND THE PRESENT

One of the stellar achievements of twentieth century plant biology was the genetic transformation of somatic cells enabling the regeneration of whole plants that were stably transformed and capable of transmitting the inserted genetic material to subsequent generations. This achievement grew out of three independent lines of research initiated early in the twentieth century: plant tissue culture, regeneration of plants from single somatic cells, and the study of crown gall disease. The early discoveries made in these areas represent a combination of basic scientific research and technological innovations and led to the development of genetically transformed crop species expressing traits unobtainable by conventional breeding. Each of these fields can be traced back to a single research publication (Haberlandt 1902; Smith and Townsend 1907; White 1939) that later came to be considered as the foundation of the field. It is instructive to follow how these three fields were established, progressed, converged, and finally coalesced. Early workers, of course, could not know the ultimate way in which their discoveries would be applied to modern plant biotechnology. Instead, they posed questions that were of interest to them and to the scientific communities of which they were a part, and some of their discoveries led in directions that would later prove to be productive in the progression to biotechnology. This essay is intended to provide a review of the crucial discoveries that ultimately led to modern plant biotechnology and show how they contributed to this progression. Despite continued efforts by Haberlandt's collaborators and others, no significant progress on cultures derived from single cells was made for 56 years when W.H. Muir, Albert Hildebrandt, and Albert Riker at the University of Wisconsin investigated this question (Muir et al. 1958). They used tissue cultures of tobacco and carrot and crown gall cultures of grape, marigold, periwinkle, and sunflower. By testing the capacity for growth in liquid culture media, they identified several that produced large numbers of single cells. Single cells were then transferred by micropipette to filter paper placed on nurse cultures of the same or other species growing on an agar medium. Those of tobacco, marigold, sunflower, and grape divided and produced macroscopic cultures, some of which were transferred through 25 or more agar subcultures without diminution of growth rate. Marigold clones consistently produced roots, and hybrid tobacco clones produced shoots on media containing adenine and kinetin.

Subsequently, Vimla Vasil and Hildebrandt (Vasil and Hildebrandt 1965a, 1965b) transferred single tobacco hybrid cells to a drop of culture medium on a microscope slide that could be observed and photographed repeatedly under phase contrast microscopy. Cells were observed to divide to form a filament and subsequently a microcallus mass that was transferred to an agar medium for further growth, where roots and leafy shoots were differentiated. Rooted shoots were transferred to soil where they produced buds and flowers. Thus, these studies demonstrated that plantlets derived from single cultured cells had the capacity to produce whole plants. However, they did not prove that the whole plants were the direct product of a single cell, rather than the product of a tissue mass within which somaclonal or other genetic changes might have taken place during growth to produce a chimeric tissue mass.

The final step proving Haberlandt's prediction that "one could successfully cultivate artificial embryos from vegetative cells" came from the research of Dietlinde Backs-Hüsemann and Jakob Reinert in Berlin. They cultured single carrot cells from suspension cultures on microscope slides where they could be observed and photographed repeatedly. Isolated cells divided to form a mass of embryogenic and parenchyma cells, and the embryogenic cells developed into heart-shaped and torpedo-shaped embryos with recognizable cotyledons, hypocotyls, and radicles (Backs-Hüsemann and Reinert 1970)

The in vitro culture of plant tissue and cells has a long history and has contributed much to the contemporary biotechnology (Sussex 2008; Vasil 2008).While an understanding of somatic embryo induction is starting to emerge, details of the signaling mechanisms involved remain poorly understood ( Braybrook and Harada 2008; Wang et al. 2011).

Grant opportunities in plant sciences to address the grand challenges facing the planet are reviewed by Frommer (2010). Deal (2011) and Sinha (2011) have described accelerating discovery through technology development and plant developmental biology in the post-genomic era respectively. Jenkins et al. (2011) have suggested/proposed the use of synthetic biology approaches to "re-construct" plant feedstock for optimal processing of biomass for non-food applications and considered plants as biofactories for the future.

## ARID REGIONS AND THEIR PLANT RESOURCES

Considerable and significant portions of the earth are arid and drylands (Reynolds et al. 2007; Schimel 2010). These make a key terrestrial biome and are socio-economically important. The dry lands are characterized by low, unpredictable and highly variable precipitation (rainfall). These experience intense sun (solar radiations), extreme temperatures and high evaporations (Maestre et al. 2012). Fernando T. Maestre et al. (2012) have reviewed probable/expected impacts of global environmental change on drylands and identified key under-studied areas that need more attention and research. They have suggested that (i) our ability to predict and understand global climate change impact dry and arid lands can be achieved by explicitly considering how the responses of individuals, populations and communities in turn affect ecosystem services and (ii) future research should explore linkages between these responses and their effect on key components like water and climate while taking care of human development and well-being. However, situation become highly complicated in the arid regions Indian Thar Desert and neighboring Aravallies that are densely populated. The state of Rajasthan occupies the major aridlands of India and as on date, this is the largest state of the Union of India, covering about 11 percent (3, 42,274 sq km) of total are of India. The western and northwestern Rajasthan has international boundary with Pakistan. The north-east and eastern parts are covered by the states of Punjab, Haryana and Uttar Pradesh (UP). The state of Madhya Pradesh (MP) and Gujarat are the South-Eastern and South-Western neighbors respectively. It is located between 23°3′ and 30°12′ N Latitude and 69°30′ and 78°17′ E longitude. The elevation of the land surface varies from 214 to 1375 m.

The Aravalli mountain range – the oldest folded mountain in the world, intersect the state end to end north-east to south-west diagonally. The western three-fifth part is sandy (The Thar) arid plain called Marusthalli (Arid land), has shifting and stabilized sand-dunes. This desertic part has climate characterized by extremes of temperature, low rainfall (342 mm mean annual), high evapo-transpiration, low relative humidity and high wind velocity (10.7 km/hr average annual). A number of salt lakes and a water bodies exist in this region. This region attracts a large number of tourists and is home for Wildlife.

A considerable part of Rajasthan is in the east of Aravalli, occupied by isolated chain(s) of rocks and mountains with undulating tectonic plains, alluvial plains or agricultural lands. Vindhyan hills also enter in the south-east and spread out west-ward. The climate in the east of Aravalli is semi-arid with mean annual rainfall of 850 mm. This region has several rivers and man-made lakes. The Aravalli range itself is third important physiographic region of Rajasthan that runs 550 km from Khetri (Jhunjhunu) in North east to Khed Braham in South-west in the State. The climate of Aravalli range is semi-arid except Mt. Abu where annual rainfall varies from 700-1250 mm per year.

The variability in the climatic, edaphic and topographic conditions offers natural conditions for growth and evolution of diversity of microbes and plants/vegetation. A variety of tree, shrubs, under-shrubs, perennial woody herbs, and grasses/sedges exist in the region that produce biomass and support life-system. *Ephedra foliata* (Oont Phog) is, the only native Gymnosperm. The vegetation of state of Rajasthan is mixture of Indian, Perso-Arabian, Indo-Malayan and general. Shetty and Singh (1987-1993) enlisted plant species that are endemic to the desert zones and the Mt. Abu. About 65 species are the probable wild relatives of crop plants/fruit crops. About 275 species are cultivated for grains, fruits, vegetables, fodder, avenue and ornamental plants. Rajasthan is major producer of seed spices, Bajra, moth and Guar and provide habitat for numerous medicinal and herbal plants.

Plants microbes and animals experience a variety of abiotic stresses during their life cycles. These have however survived and evolved while interacting with harsh environmental conditions. Plants and associated microbes bear unique abilities to survive the hostilities. The plants of these regions are sources of biomass and energy to support life. These provide food, fodder, fuel, timber and number of products useful for human health and economy and industry. The biological and ecological importance of plants of fragile ecosystems are being realized with greater extent in era of global warming/climate change, as the study of biology of these plants may provide answers to some of the challenges being faced. Many of these plants species yield secondary metabolites, gums and bioactive products under useful for human well being.

Recurring and non-recurring stresses and rising biotic and abiotic pressures have caused habitat destruction/disturbance in the aridlands and are adversely affecting diversity and productivity of the plants and microbes of these systems. Fast growing and superior looking/high yielding plants are cut and used in preference ruthlessly; thus, threatening/eliminating germplasms of elite /superior plants. Many endemic and rare plants species are under threat and some are endangered. A few plant species are at verge of extinction. These plants have constrained reproduction and have slow growth and poor propagation. *Calligonum polygonoides,* an ecologically important woody shrub, a source of fodder (even food/source of neuraceuticals) fuel and timber has been eliminated over larger areas. This plant species covered in the recent past the sand dunes in the desert. *Ceropegia species, Glossonema varians, Caralluma edulis* and many others at the verge of disappearances. Similar is situation with many woody plants and trees. There is also need for generation of awareness among people and politicians/administration. Conservation and propagation of the Aravallis and the Thar Desert need technological inputs of conventional non-conventions methods of plant propagation of improvements. Development of tools of modern biology/biotechnology and bioinformatics and their applications are needed for understand biology of plant resources and their sustainable utilization.

## BIOTECHNOLOGICAL TOOLS FOR CONSERVATION, PROPAGATION AND IMPROVEMENT OF PLANTS OF RAJASTHAN

At the Department of Botany, Jai Narain Vyas University, Jodhpur for last 40 years individuals and laboratories have been working on development of modern biology/biotechnology for applications in characterization, conservation, propagation and sustainable utilization. We worked on tree species, medicinal and herbal plants, rare and threatened taxa and crop plants that are adapted to the harsh environmental conditions of Rajasthan. We, in our endeavors got support in form projects/funds by the University Grants Commission (UGC) of India, Department of Biotechnology (DBT), Department of Science and Technology (DST) Department of Environment and Forests & Wildlife (DOEN) and other agencies like Council of Scientific and Industrial Research (CSIR), the State Government of Rajasthan and some international agencies. We developed and defined tissue culture/micropropgation protocols for a number of plants species. We initiated research work on molecular characterization and DNA bar-coding desert of desert plants and allied systems. We have contributed a few reviews/technical papers in the field of plant biotechnology (Arya and Shekhawat 1986; Shekhawat et al. 1995; 1998; Kackar et al. 2009; Rai et al. 2011; Gupta et al. 2011; Gehlot, et al. 2012; Shekhawat et al. 2012c). The research work done applying biotechnological tools on various plant species of Rajasthan is enlisted in Table 1 and Figures 1-10.

**Table 1** Tissue culture and molecular marker studies in a number of plants: work done during last 30 years in the laboratory

| *Plant Species* | *Achievements* | *References* |
|---|---|---|
| *Acacia senegal* | Somatic embryogenesis from cotyledon explants | Rathore et al. (2012a) |
| *Aloe vera* | Plantlet regeneration from callus cultures of selected genotype<br>Genetic fidelity of tissue culture raised plants by ISSR and RAPD | Singh et al. (2009); Rathore et al. (2011a)<br>Rathore et al. (2011b) |
| *Anogeissus acuminata* | Multiple shoots and *in vitro* rooting* | Rathore et al. (1993) |
| *Anogeissus latifolia* | Multiple shoots and *in vitro* and *ex vitro* rooting*<br>Isolation of DNA and sequencing of barcoding gene (*rbcl*) | Shekhawat et al. (2000)<br>Gupta et al. (2011) |
| *Anogeissus pendula* | Multiple shoots and *in vitro* rooting*<br>Isolation of DNA and sequencing of barcoding gene (*rbcl*) | Joshi et al. (1991)<br>Gupta et al. (2011) |
| *Anogeissus rotundifolia* | Multiple shoots and *in vitro* rooting*<br>Isolation of DNA and sequencing of bar coding gene (*rbcl*) | Singh et al. (1997)<br>Gupta et al. (2011) |
| *Anogeissus sericea* | *In vitro* propagation* | Kaur et al. (1992a) |
| *Arnebia hispidissima* | Micropropagation and production of alkannin from callus and cell suspension culture | Shekhawat and Shekhawat (2011) |
| *Calotropis procera* | Development of laticifer cells in callus cultures | Dhir et al. (1984) |
| *Caralluma edulis* | *In vitro* micropropagation | Kaur et al. (1992b) |
| *Capparis decidua* | *In vitro* clonal propagation** | Deora et al. (1995) |
| *Cenchrus setigerus and C. ciliaris* | Plant regeneration from cultured immature inflorescences | Kackar and Shekhawat (1991a) |
| *Celastrus paniculatus* | In vitro propagation and rooting of microshoots<br>Analysis of genetic stability in tissue culture raised plants by RAPD | Arya et al. (2003)<br>Phulwaria et al. (2013a) |
| *Ceropegia bulbosa* | Plant regeneration from callus culture | Phulwaria et al. (2013b) |
| *Chloris virgata* | *In vitro* clonal propagation | Dagla et al. (2005a) |
| *Citrus limon* | Micropropagation of mature tree** | Rathore et al. (2007) |

**Table 1** Contd.

| *Plant Species* | *Achievements* | *References* |
|---|---|---|
| *Commiphora wightii* | Micropropagation** | Yusuf et al. (1999) |
| *Eulophia nuda* | *In vitro* tuberization and *ex vitro* rooting | Panwar et al. (2012) |
| *Haloxylon recurvum* | *In vitro* shoot multiplication | Dagla et al. (2005b) |
| *Lasiurus scindicus* | Plant regeneration | Kackar and Shekhawat (1989) |
| *Lawsonia inermis* | Multiple shoots and *in vitro* and *ex vitro* rooting** | Ram et al. (2011) |
| *Leptadenia reticulata* | Multiple shoots and *in vitro* and *ex vitro* rooting** | Arya et al. (2003); Rathore et al. (2013) |
| *Maerua oblongifolia* | Shoot multiplication and rooting** | Rathore et al. (2005) |
| *Macuna pruriens* | Somatic embryogenesis and plant regeneration | Vibha et al. (2009) |
| *Maytenus emarginata* | *In vitro* clonal propagation** | Rathore et al. (1992a) |
| *Momordica dioica* | Plantlet regeneration of female plant from nodal shoot segment | Shekhawat et al. (2011) |
| *Pennisetum glaucum* | Plant regeneration from peduncle explant | Shekhawat et al. (2012a |
| *Prosopis cineraria* | *In vitro* clonal propagation** | Shekhawat et al. (1993) |
| *Prosopis juliflora* | Bioinformatics analysis of distribution of microsatellite markers (SSRs)/Single Nucleotide Polymorphism (SNPs) in expressed transcripts | Sablok and Shekhawat (2008) |
| *Psidium guajava* | Genetic fidelity of somatic embryo derived plants by ISSR and SSR markers | Rai et al. (2012) |
| *Rauvolfia serpentina* | Multiple shoots and *in vitro* and *ex vitro* rooting** | Shekhawat et al. (2012b) |
| *Salvadora persica* | Multiple shoots and *ex vitro* rooting** | Phulwaria et al. (2011a) |
| *Syzygium cuminii* | *In vitro* clonal propagation** | Rathore et al. (2004) |
| *Tecomella undulata* | *In vitro* clonal propagation** | Rathore et al. (1991) |
| *Terminalia arjuna* | Isolation of DNA and sequencing of barcoding gene (*rbcl*) | Gupta et al. (2011) |
| *Terminalia bellirica* | Multiple shoots and *in vitro* and *ex vitro* rooting**<br>Isolation of DNA and sequencing of barcoding gene (*rbcl*) | Phulwaria et al. (2011b)<br>Gupta et al. (2011) |
| *Terminalia cattapa* | Multiple shoots and *ex vitro* rooting**<br>Isolation of DNA and sequencing of barcoding gene (*rbcl*) | Phulwaria et al. (2011c)<br>Gupta et al. (2011) |
| *Terminalia chebula* | Isolation of DNA and sequencing of barcoding gene (*rbcl*) | Gupta et al. (2011) |
| *Trigonella* | Isolation, culture and shoot regeneration | Shekhawat and |
| *foenumgraecum* | From mesophyll protoplast | Galston (1983a) |

**Table 1** Contd.

| *Plant Species* | *Achievements* | *References* |
|---|---|---|
| *Urochloa panicoides* | Plant regeneration by cultured immature inflorescence | Kackar and Shekhawat (1991b) |
| *Withania coagulans* | *In vitro* micropropagation | Rathore et al. (2012b) |
| *Vigna aconitifolia* | Variation amongst protoplast-derived plants<br>Isolation, culture, and regeneration from leaf protoplasts | Arya et al. (1990)<br>Shekhawat and Galston (1983b) |
| *Zizyphus* sp. | Clonal propagation** | Rathore et al. (1992b) |

* Juvenile explants
** Mature explants

## MOLECULAR CHARACTERIZATION, TISSUE CULTURE AND MICROPROPAGATION OF DESERT PLANTS AND ALLIED SPECIES

We worked on extraction and analyses of genomic DNA of plants and microbes (Sablok and Shekhawat 2008; Sablok et al. 2009; Gupta et al. 2011; Gehlot et al. 2012) of the region Micropropagation as non-GM biotechnology has been developed for application in large-scale plant production and conservation of germplasms of threatened and rare taxa.

We have worked on cloning of mature trees of *Capparis decidua, Anogeissus sericea* subspecies *nummularia, Calligonum polygonoides, Maytenus emarginata, Salvadora oleoides, Salvadora persica, Prosopis cineraria, Tecomella undulata* and a number of woody shrubs and herbs. It has been possible to define highly efficient and reproducible protocols for cloning of mature tees of *Capparis decidua, Citrus limon, Maytenus emarginata, Maerua oblongifolia Salvadora oleoides, S. persica, Terminalia bellirica, Terminalia catappa* and number of medicinal, herbal plants and millet/grasses. In case of *Anogeissus* species and *Capparis spinosa,* establishment of cultures from adult/mature trees has been difficult; therefore seedling derived juvenile explants were used for culture initiation. *Anogeissus* species have poor seed viability and propagation in nature is extremely slow and difficult. All the species of *Anogeissus* are ecologically and economically important and there is need for development of technology for their propagation and conservation. Hence, the development of in vitro technology for large-scale production even if derived from seedling (juvenile) explants is significantly important contribution in tree biotechnology (Shekhawat et al. 2012c).

Our major contributions have been in defining conditions for *in vitro* and *ex-vitro* rooting of in vitro generated and amplified shoots of a number of woody plants and shrubs. Ex-vitro rooting of micropropagated shoots is significant as this save time and resource and it use easy to acclimatize the plants for field transfer.

**Fig. 1.** a. A tree of *Anogeissus acuminata*. b. *A. latifolia*. c. *A. pendula*. d. *A. sericea*

**Fig. 2.** a. Fruits of *Anogeissus sericea*. b. Viable seeds of *A. sericea*. c. Cotyledonary node-derived multiple shoots of *A. sericea* on nutrient medium. d. Hardening and acclimatization of micropropagated plantlets *Anogeissus* species. e. Tissue culture raised plants of *Anogeissus* species being supplied to farmers. f. Tissue culture raised plant of *A. sericea* (subsp. *nummularia*).

**Fig. 3.** A lopped tree with juvenile shoots of *A. sericea* subsp. *nummularia*. b. nodal explants from mature tree of *A. sericea* subsp. *nummularia*, exhibiting bud break in culture. c. Multiplication of shoots of *A. sericea* subsp. *nummularia* by repeated transfer of mother explants. d. *ex vitro* rooting micropropagated shoots of *A. sericea* subsp. *nummularia*.

**Fig. 4.** a. Nodal shoot explants of mature tree of *Capparis decidua*. b. Amplification of shoots derived from mature explants. c. cloned trees of *Capparis decidua* growing in the field.

**Fig. 5.** Nodal explants of mature tree of *Maytenus emarginata*. b. *In vitro* multiplication of shoots of *M. emarginata* initiated from mature explants. c. micropropagated plants of *M. emarginata* being acclimatized in nursery. d. cloned trees of *M. emarginata* in field.

**Fig. 6.** a. explants harvested from mature tree of *Prosopis cineraria*. b and c. multiplication of shoots in culture from nodal shoot explants of mature tree of *Prosopis cineraria* (Repeated transfer of mother explants). d. ISSR profile pattern of different collections of *Prosopis cineraria*.

**Fig. 7.** a. Tree of *Tecomella undulata* (Teak of Marwar). b. *In vitro* shoot multiplication from nodal explants of mature tree of *T. undulata*. c. *ex vitro* rooting of *in vitro* raised shoots of *T. undulata* d. Hardening of cloned plant of *T. undulata*.

**Fig. 8.** a. Micropropagation of shoots of *Chlorophytum* species (Safed Musli), a medicinal plant in liquid culture. b. cloned plants of safed musli in nursery.

**Fig. 9.** a. Amplification of shoots of *Celastrus peniculatus* derived from nodal explants of mature plants. b. cloned plants of *C. peniculatus* in nursery.

**Fig. 10.** a. *In vitro* multiplication of shoots of *Caralluma edulis* (a rare and endemic species of Rajasthan). b. tissue culture raised plants of *C. edulis.*

## ACKNOWLEDGEMENTS

We are grateful to Department of Biotechnology (DBT) and Department of Science and Technology (DST), Government of India and University Grants Commission (UGC), New Delhi for providing funds for establishment of laboratory and greenhouse infrastructure under different programs of support used for present research. We also thank to the Head, Department of Botany, J.N.V. University for providing facilities. Authors (M. Phulwaria, Harish, M. K. Rai) wish to acknowledge the support of the University Grants Commission (UGC), New Delhi for the award of Post Doctoral Fellowship and Dr. DS Kothari Post Doctoral Fellowship. Financial assistance provided by Council of Scientific & Industrial Research (CSIR), New Delhi, to Kheta Ram and Ashok Kumar Patel are gratefully acknowledged.

## REFERENCES

Arya, H.C. and Shekhawat, N.S. (1986). Clonal multiplication tree species in the Thar Desert through tissue culture. *Forest Ecology and Management* **16:** 201-208.

Arya, I.D., Arya, S., Rao, D.V. and Shekhawat, N.S. (1990). Variation amongst protoplast-derived moth bean, *Vigna aconitifolia* plants. *Euphytica* **47:** 33-38.

Arya, V. and Shekhawat, N.S. (2004). Cloning of *Rauvolfia serpentina* - an endangered Medicinal Plant. *Journal of Sustainable Forestry* **20(1):** 53-65.

Arya, V., Shekhawat, N.S. and Singh, R.P. (2003). Micropropagation of *Leptadenia reticulata* - an endangered multipurpose medicinal plant. *In Vitro Cellular Developmental Biology-Plant* **39:** 180-185.

Arya, V., Singh, R.P. and Shekhawat, N.S. (2002). A Micropropagation protocol for mass multiplication and off-site conservation of *Celastrus paniculatus* - a vulnerable medicinal plant of India. *Journal of Sustainable Forestry* **14:** 107-120.

Backs-Hüsemann, D. and Reinert, J. (1970). Embryobildung durch isolierte Einzelzellen aus Gewebeculturen von Daucus carota. *Protoplasma* **70:** 49–60.

Bates, G.W., Gaynor, J.J. and Shekhawat, N.S. (1983). Fusion of plant protoplasts by electric fields. *Plant Physiology* **72:** 1110-1113.

Bhati, R., Shekhawat, N.S. and Arya H.C. (1992). *In vitro* regeneration of plantlets from root segments of *Aegle marmelos. Indian Journal of Experimental Biology* **30:** 844-845.

Braybrook, S.A. and Harada, J.J. (2008). LECs go crazy in embryo development. *Trends in Plant Science* **13(12):** 624-630.

Dagla, H.R. and Shekhawat, N.S. (2005b). *In vitro* multiplication of *Haloxylon Recurvum* (Moq.) - a plant for saline soil reclamation. *Journal of Plant Biotechnology* **7(3):** 1-5.

Dagla, H.R. and Shekhawat, N.S. (2005a). Plant regeneration from tissue culture of *Chloris virgata*: A salt-tolerant desert grass. *Indian Journal of Biotechnology* **4:** 400-403.

Deal, R.B. (2011). Grand challenges: accelerating discovery through technology development. *Frontiers in Plant Science* **2:** 1-3.

denNijs, H.C.M., Bartsch, D. and Sweet, J. (2004). Introgression from genetically modified plants into wild relatives. Wallingford, UK: CABI

Deora, N.S. and Shekhawat, N.S. (1995). Micropropagation of *Capparis decidua* (Forsk.) Edgew. - A tree of arid horticulture. *Plant Cell Reports* **15:** 278-281.

Dhir, S.K., Shekhawat, N.S., Purohit, S.D. and Arya, H.C. (1984). Development of laticifer cells in callus cultures of *Calotropis procera* (Ait.) *R.Br. Plant Cell Reports* **3:** 206-209.

Frommer, WB. (2010). Grand opportunities in plant science to address the grand challenges facing the planet. *Frontiers in Plant Science* 1, 1-3.

Gehlot, H.S., Panwar, D., Tak, N., Tak, A., Sankhla, I.S., Poonar, N., Parihar, R., Shekhawat, N.S., Kumar, M., Tiwari, R., Ardley, J., James, E.K. and Sprent J. I. ( 2012). Nodulation of legumes from the Thar desert of India and molecular characterization of their rhizobia. *Plant and Soil* **357**: 227–243.

Gupta, A.K., Harish, Rai, M.K., Phulwaria, M. and Shekhawat, N.S. (2011). Isolation of genomic DNA suitable for community analysis from mature trees adapted to arid environment. *Gene* **487**:156-159.

Haberlandt, G. (1902). Kulturversuche mit isolierten Pflanzenzellen. Sitzungsber. Akad. Wiss. Wien. Math.-Naturwiss. *Kl., Abt. J.* **111:** 69-92.

Holme, I.B., Dionisio, G., Brinch-Pedersen, H., Wendt, T., Madsen, C.K., Vincze, E. and Holm, P.B. (2011). Cisgenic barley with improved phytase activity. *Plant Biotechnology Journal* **10(2):** 237-47.

http://www.nap.edu/catalog/12764.html

http://www.aspb.org/plantsummit

Jacobsen, E. and Schouten, H.J. (2008). Cisgenesis, a new tool for traditional plant breeding, should be exempted from the regulation on genetically modified organisms in a step by step approach. *Potato Research* **51**:75-88.

Jenkins, T., Bovi, A. and Edwards, R. (2011). Plants: biofactories for a sustainable future? *Philosophical Transactions of The Royal Society B Biological Sciences* **369:** 1826-1839.

Joshi, R., Shekhawat, N.S. and Rathore T.S. (1991). Micropropagation of *Anogeissus pendula* Edgew.- An arid forest tree. *Indian Journal of Experimental Biology* **29:** 615-618.

Kackar, A. and Shekhawat, N.S. (1989). Regeneration of *Lasiurus scindicus* from tissue culture. *Annals of Botany* **64:** 455-458.

Kackar, A. and Shekhawat, N.S. (1991a). Plant regeneration from cultured immature inflorescences of *Cenchrus setigerus* and *C. ciliaris. Indian Journal of Experimental Biology* **29:** 62-64.

Kackar, A. and Shekhawat, N.S. (1991b). Plant regeneration by cultured immature inflorescence of *Urochloa panicoides* (Beauv.). *Indian Journal of Experimental Biology* **29:** 331-333.

Kackar, A., Rathore, M.S. and Shekhawat, N.S. (2009). Spotlight on botanics in India: characterization of Thar Desert plants. *Biotechnology Journal* **4:** 310.

Kaur, G., Rathore, T.S., Rama Rao, S. and Shekhawat, N.S. (1992b). *In vitro* micropropagation of *Caralluma edulis* (Edgew.) Benth. & Hook. f. - A rare edible plant species of Indian Desert. *Indian Journal of Plant Genetic Resources* **5:** 51-56.

Kaur, G., Singh, R.P., Rathore, T.S. and Shekhawat, N.S. (1992a). *In vitro* propagation of *Anogeissus sericea. Indian Journal of Experimental Biology* **30:** 788-791.

Maestre, F.T., Salguero-Gómez, R. and Quero J.L. (2012). It is getting hotter in here: determining and projecting the impacts of global environmental change on drylands. *Philosophical Transactions of The Royal Society B Biological Sciences* **367:** 3062-3075.

*McGary,* K.L. et al. *(2010).* Systematic discovery of nonobvious human disease models through orthologous phenotypes. *PNAS* **107:** 6544-6549.

Muir, W.H., Hildebrandt, A.C. and Riker, A.J. (1958). The preparation, isolation, and growth in culture of single cells from higher plants. *American Journal of Botany* **45:** 589-597.

Panwar, D., Ram, K., Harish, and Shekhawat, N.S. (2012). In vitro propagation of *Eulophia nuda* Lindl., an endangered orchid. *Scientia Horticulturae* **139:** 46-52.

Phulwaria, M., Rai, M.K., Harish, Gupta, A.K., Ram, K. and Shekhawat, N.S. (2011b). An improved micropropagation of *Terminalia bellirica* from nodal explants of mature tree. *Acta Physiologiae Plantarum* **34:** 299-305.

Phulwaria, M., Rai, M.K., Patel, A.K., Kataria, V. and Shekhawat, N.S. (2013a). A genetically stable rooting protocol for propagating a threatened medicinal plant - *Celastrus paniculatus. AoB Plants* (in press)

Phulwaria, M., Shekhawat, N.S., Rathore, J.S. and Singh, R.P. (2013b). An efficient in vitro regeneration and ex vitro rooting of *Ceropegia bulbosa* Roxb. – a threatened and pharmaceutical important plant of Indian Thar Desert. *Industrial Crops and Products* **42:** 25-29.

Phulwaria, M., Kheta, R., Gahlot, P. and Shekhawat, N.S. (2011a). Micropropagation of *Salvadora persica*-a tree of arid horticulture and forestry. *New Forests* **42:**317-327.

Phulwaria, M., Ram, K., Harish, Gupta, A.K. and Shekhawat, N.S. (2011c). Micropropagation of mature *Terminalia catappa* (Indian almond) - a medicinally important forest tree. *Journal of Forest Research* **17:** 202-207,

Rai, M.K., Shekhawat, N.S., Harish, Gupta, A.K., Phulwaria, M., Ram, K. and Jaiswal U. (2011). The role of abscisic acid in plant tissue culture – a review on the recent progress. *Plant Cell Tissue and Organ Culture* **106:** 179-190.

Rai, M.K., Phulwaria, M., Harish, Gupta, A.K., Shekhawat, N.S. and Jaiswal, U. (2012). Genetic homogeneity of guava plants derived from somatic embryogenesis using SSR and ISSR markers. *Plant Cell Tissue and Organ Culture* **111:** 259-264.

Ram, K. and Shekhawat, N.S. (2011). Micropropagation of commercially cultivated Henna (*Lawsonia inermis*) using nodal explants. *Physiology and Molecular Biology of Plants* **17(3):** 281-289.

Rathore, J.S., Rathore, M.S., Singh, M., Singh, R.P. and Shekhawat, N.S. (2007). Micropropagation of mature tree of *Citrus limon. Indian Journal of Biotechnology* **6(2):** 239-244.

Rathore, J.S., Rai, M.K. and Shekhawat, N.S. (2012a). Induction of somatic embryogenesis in gum arabic tree [*Acacia senegal* (L.) Willd.]. *Physiology and Molecular Biology of Plants* **18:** 387-392.

Rathore, M.S., Dagla, H.R., Singh, M. and Shekhawat, N.S. (2008). Rational development of *in vitro* methods for conservation, propagation and characterization of *Caralluma edulis. World Journal of Agricultural Science* **4(1):** 121-124.

Rathore, M.S., Chikara, J. and Shekhawat, N.S. (2011a). Plantlet regeneration from callus cultures of selected genotype of *Aloe vera* L.- an ancient plant for modern herbal industries. *Applied Biochemistry and Biotechnology* **163:** 860-868.

Rathore, M.S., Chikara, J., Mastan, S.G., Rahman, H., Anand, K.G.V. and Shekhawat, N.S. (2011b). Assessment of genetic stability and instability of tissue culture-propagated plantlets of *Aloe vera* L. by RAPD and ISSR markers. *Applied Biochemistry and Biotechnology* **165:** 1356-1365.

Rathore, M.S., Shekhawat, S., Kaur, G., Singh, R.P. and Shekhawat, N.S. (2012b). Micropropagation of vegetable rennet (*Withania coagulans* [stocks] dunal) - a critically endangered medicinal plant. *Journal of Sustainable Forestry* **31:** 727-746.

Rathore, J.S., Rathore, M.S. and Shekhawat, N.S. (2005). Micropropagation of *Maerua Oblongifolia*-a liana of arid areas. *Phytomorphology* **55(3& 4):** 241-247.

Rathore, M.S., Rathore, M.S. and Shekhawat N.S. (2013). Ex vivo implications of phytohormones on various in vitro responses in *Leptadenia reticulata* (Retz.) Wight. & Arn.-an endangered plant. *Environmental and Experimental Botany* **86:** 86-93.

Rathore, T.S., Deora, N.S. and Shekhawat, N.S. (1992a). Cloning of *Maytenus emarginata* (Willd.) Ding Hou - a tree of the Indian desert, through tissue culture. *Plant Cell Reports* **11:** 449-451.

Rathore, T.S., Deora, N.S., Shekhawat, N.S. and Singh, R.P. (1993). Rapid micropropagation of *Anogeissus acuminata. Biologia Plantarum* **35:** 381-386.

Rathore, T.S., Singh, R.P. and Shekhawat, N.S. (1991). Clonal propagation of desert teak (*Tecomella undulata*) through tissue culture. *Plant Science* **79:** 217-222.

Rathore, T.S., Singh, R.P., Deora, N.S. and Shekhawat, N.S. (1992b). Clonal propagation of *Zizyphus* species through tissue culture. *Scientia Horticulturae* **51:** 165-168.

Rathore, V., Shekhawat, N.S., Singh, R.P., Rathore, J.S. and Dagla, H.R. (2004). Cloning of adult trees of Jamum (*Syzygium cuminii*). *Indian Journal of Biotechnology* **3:** 241-245.

Rathore, M.S. and Shekhawat, N.S. (2009). Micropropagation of *Pueraria tuberosa* (Roxb.Ex Willd.) and determination of puerarin content in different tissues. *Plant Cell Tissue and Organ culture* **99:** 327-334.

Reynolds, A. (2007). The cell's journey: from metaphorical to literal factory. *Endeavour* **31:** 65-70.

Reynolds, J.F. et al. (2007). Global desertification: building a science for dryland development. *Science* **316:** 847-851.

Rose, R.J. and Nolan, K.E. (2006). Genetic regulation of somatic embryogenesis with particular reference to *Arabidopsis thaliana* and *Medicago truncatula. In Vitro Cellular Developmental Biology - Plant* **42:** 473-481.

Sablok, G., Gahlot, P., Gupta, A.K., Pareek, K. and Shekhawat, N.S. (2009). Extraction of PCR-usable DNA using modified CTAB method from the Resource Trees adapted to arid environment sustainability. *Plant Omics Journal* **2(3):**103-109.

Sablok, G. and Shekhawat, N.S. (2008). Bioinformatics analysis of distribution of microsatellite markers (SSRs) / single nucleotide polymorphism (SNPs) in expressed transcripts of *Prosopis juliflora*: frequency and distribution. *Journal of Computer Science and Systemic Biology* **1:** 087-091.

Schimel, D.S. (2010). Drylands in the earth system. *Science* **327(5964):** 418-419.

Schouten, H.J., Krens, F.A. and Jacobsen, E. (2006). Do cisgenic plants warrant less stringent oversight? *Nature Biotechnology* **24(7):**753

Shekhawat, M.S., and Shekhawat, N.S. (2011). Micropropagation of *Arnebia hispidissima* (Lehm).DC. and production of alkannin from callus and cell suspension culture. *Acta Physiologiae Plantarum* **33:** 1445-1450.

Shekhawat, M.S., Shekhawat, N.S., Harish, Ram, K., Phulwaria, M. and Gupta, A.K. (2011). High frequency plantlet regeneration from nodal shoot segment culture of female *Momordica dioica* (Roxb.). *Journal of Crop Science and Biotechnology* **14 (2):** 133-137.

Shekhawat, N.S., Mohnot, S., Phulwaria, M., Harish, and Shekhawat, S. (2012a). Micropropagation of *Salvadora oleoides*-an oil yielding tree of arid forests. *Journal of Sustainable Forestry* **31:** 620-632.

Shekhawat, N.S., Phulwaria, M., Harish, Rai, M.K., Kataria, V., Shekhawat, S., Gupta, A.K., Rathore, N.S., Vyas, M., P.athore, N., JB, V., Choudhary, S.K., Patel, A.K., Lodha, D. and Modhi, R. (2012). Bioresearches of fragile ecosystem/ Desert. *Proceeding of the National Academy of Sciences, India Section B: Biological Sciences* **82:** 319-334.

Shekhawat, S., Shekhawat, N.S., Choudhary, S.K., Harish, and Kataria, V. (2012b). Peduncle, a potential source of competent cells for plant regeneration in Pearl Millet (*Pennisetum glaucum*). *Phytomorphology* **62:** 1-11.

Shekhawat, N.S. and Kackar, A. (1987). Protoplasts of *Prosopis cineraria* (L.) Duce: Isolation and culture. *Nitrogen Fixing Tree Research Reports* **5:** 51-53.

Shekhawat, N.S. and Galston, A.W. (1983a). Mesophyll protoplasts of Fenugreek (*Trigonella foenumgraecum*). Isolation, culture and shoot regeneration. *Plant Cell Reports* **2:** 119-121.

Shekhawat, N.S. and Galston, A.W. (1983b). Isolation, culture, and regeneration of moth bean *Vigna aconitifolia* leaf protoplasts. *Plant Science Letters* **32:** 43-51.

Shekhawat, N.S., Johri, B.M. and Srivastava, P.S. (1995). Morphogenesis and plant tissue culture. In: Johri, B.M. (ed.) Botany in India- History and Progress. Oxford & IBH Publishing Co. Pvt. Ltd., New Delhi, pp. 307-356.

Shekhawat, N.S., Mughal, M.H., Johri, B.M. and Srivastava, P.S. (1998). Indian contribution to plant tissue and organ culture. In: Srivastava P.S. (ed.). Plant Tissue Culture and Molecular Biology: Application and Prospects, Narosa Publishing House, New Delhi, pp. 751-811.

Shekhawat, N.S., Rathore, T.S., Singh, R.P., Deora, N.S. and Rao, S.R. (1993). Factors affecting *in vitro* clonal propagation of *Prosopis cineraria*. *Plant Growth Regulation* **12:** 273-280.

Shekhawat, N.S., Yadav, J., Arya, V. and Singh, R.P. (2000). Micropropagation of *Anogeissus latifolia* (Roxb. Ex DC) Wall. Ex Guill. & Perr.-a tree of fragile ecosystems. *Journal of Sustainable Forestry* **11:** 83-96.

Shetty, B.V. and Singh, V. Flora of Rajasthan, 3 Vols. Botanical Survey of India, Calcutta. 1987-1993.

Singh, M., Rathore, M.S., Panwar, D., Rathore, J.S., Dagla, H.R. and Shekhawat, N.S. (2009). Micropropagation of selected genotype of *Aloe vera* L. - An ancient plant for modern industry. *Journal of Sustainable Forestry* **28:** 935 -950.

Singh, R.P. and Shekhawat, N.S. (1997). Micropropagation of *Anogeissus rotundifolia* Blatt. and Hallb. - An endemic and rare tree of the Thar Desert. *Journal of Sustainable Forestry* **4:** 159-170.

Sinha, N.R. (2011). Plant development biology in the post-genomic era. *Frontiers in Plant Science* **2:** 1-2.

Smith, E.F. and Townsend, C.O. (1907). A plant-tumor of bacterial origin. *Science* **25:** 671-673.

Sussex, I.M. (2008). The scientific roots of modern plant biotechnology. *The Plant Cell* **20:** 1189-1198.

Vanblaere, T., Szankowski, I., Schaart, J., Schouten, H., Flachowsky, H., Broggini, G.A. and Gessler C. (2011). The development of a cisgenic apple plant. *Journal of Biotechnology* **154(4):** 304-11.

Vasil, I.K., (2008). A history of plant biotechnology: from the Cell Theory of Schleiden and Schwann to biotech crops. *Plant Cell Reports* **27:** 1423-1440.

Vasil, V., and Hildebrandt, A.C. (1965a). Growth and tissue formation from single, isolated tobacco cells in microculture. *Science* **147:** 1454–1455.

Vasil, V., and Hildebrandt, A.C. (1965b). Differentiation of tobacco plants from single, isolated cells in microcultures. *Science* **150:** 889–892.

Vibha, J.B., Choudhary, K., Singh, M., Rathore, M.S. and Shekhawat, N.S. (2009). An efficient somatic embryogeneis system for velvet bean [*Macuna pruriens*(L.)]: a source of anti Parkinson's drug. *Plant Cell Tissue and Organ Culture* **99:** 319-325.

Wang, X.D., Nolan, K.E., Irwanto, R.R., Sheahan, M.B. and Rose, R.J. (2011). Ontogeny of embryogenic callus in *Medicago truncatula*: the fate of the pluripotent and totipotent stem cells. *Annals of Botany* **107:** 599-609.

Welch, G.R. and Clegg, J.S. (2010). From protoplasmic theory to cellular systems biology: A 150-year reflection. *American Journal of Physiology – Cell Physiology* **298:** C1280–C1290.

White, P.R. (1939). Potentially unlimited growth of excised plant callus in an artificial nutrient. *American Journal of Botany* **26:** 59-64.

Yusuf, A., Rathore, T.S. and Shekhawat, N.S. (1999). Micropropagation of *Commiphora wightii* (Arnott) Bhandari- A threatened medicinal plant of semi-arid region. *Indian Journal Plant Genetic Resources* **12:** 371-375.

# 8

# Advancement in Environment Biotechnology: Applications of Bioinformatics in Bioremediation

**JAYA SHARMA* AND M.H. FULEKAR****

*Department of Life Sciences, University of Mumbai, Santacruz (East). Mumbai-400098*
*School of Environment and Sustainable Development, Central University of Gujarat Gandhinagar-382030, Gujarat*

**ABSTRACT**

Bioinformatics incorporates the development to store and search data and statistical tools and algorithms to analyze and determine relationships between biological data sets, such as macromolecular sequences, structures, expression profiles and biochemical pathways. Bioinformatics is the focus on cellular and molecular levels of biology. Biology and computers are becoming close cousins who are mutually respecting, helping and influencing each other and synergistically merging more than ever (Fulekar, 2008). The huge data from biology mainly in the form of DNA, RNA and protein sequences is putting heavy demand on computers and computational scientists. Bioinformatics has taken on a new glittering by entering the field of bioremediation. Bioinformatics is the application of computer sciences and related technology to the industries for using the huge available database for computational biology. Computational biologists are those who are specialized in using computational tools and computer systems to solve the problems of biology in the area of bioinformatics (Westhead, 2003). The major branches of bioinformatics are genomics, proteomics, biological databases, data mining, molecular phylogenetics, microarray informatics and systems biology, which play a vital role in understanding bioinformatics and its applications. The roles of bioinformatics related tools are described for bioremediation of hazardous wastes to develop environmental clean-up technology.

**Keywords:** Environment, Bioinformatics, Genomics, Proteomics, Bioremediation.

## INTRODUCTION

Bioremediation offers many possibilities from bioinformatics point of view, still slightly explored. This discipline requires the integration of huge amount of data from various sources: chemical structure and reactivity of organic compounds sequence, structure, function of proteins, comparative genomics, environmental microbiology and so on. An accumulation of huge amount of data on individual genes and proteins allowed the first studies of biology from a 'systems' perspective (Alves et al., 2002; fraser et al., 2002). The bioinformatics resources devoted to bioremediation is still scarce. Some interesting projects are being carried out to organize and store this huge amount of information relating to this subject. The microorganisms display a remarkable range of contaminant degradable ability that can efficiently restore natural environmental conditions (Ideker et al., 2001; Jeong et al., 2001).

The availability of genomic and proteomic data and improved bioinformatics and biochemical tools have raised the expectation of humanity to be able to control the genetics by manipulating the existing microbes. Bioinformatics analysis will facilitate and quicken the analysis of systematic level of behavior of cellular processes and understanding the cellular processes in order to treat and control microbial cells as factories. For the last decades, bioinformatics technology has been developed to identify and analyze various components of cells such as genes and protein function, interactions and metabolic and regulatory pathways. The next decade will belong to understanding cellular mechanism and cellular manipulation using the integration of bioinformatics. The major impact of bioinformatics has been in automating the microbial genome sequencing, the development of integrated databases over the Internet, and analysis of genomes to understand gene and genome functions.

Advances in the molecular biology technologies are making a global gene expression profile possible; genome-wide analysis of DNA (genomics), RNA expression (transcriptomics) and protein expression (proteomics) as well as protein-protein interaction (interactomics) create the opportunity to systematically study the physiological expressions of such organisms. Attempts are made to interpret some of the areas of genomics and proteomics, which have been employed for bioremediation studies. The studies on bioinformatics data obtained, viz., various sources are interpreted by combining these techniques and applying them towards future studies of active bioremediation to develop environmental clean-up technology.

## GENOMICS AND PROTEOMICS

The biology of the past century was, undoubtedly, reductionist. Its general trend was expressed in the appearance and burst of scientific disciplines, such as biochemistry, biophysics, molecular biology, molecular genetics, virology, cell biology, and bioorganic chemistry. All these disciplines are united by an aspiration to explain the phenomena of life through the description of the properties of molecules, first of all, biopolymers, which constitute the living organisms. In current science, all these directions are often united under the general heading "physicochemical biology", while classical biological disciplines, such as zoology, botany, embryology, etc., are referred to as "general biology." Certain events have recently taken place in biology that show the conditionality of this division and suggest a new general approach called integratism. Unlike reductionism, integratism tends not to divide the biological phenomenon into components but to provide an integral picture of the entire phenomenon on the basis of the properties of its components. Naturally, the volume of knowledge required for the realization of this integral approach is much larger than for the reductionist approach. Where can this huge volume of additional information required for the new integrative strategy of life cognition be obtained from? The answer is genomics, proteomics, and bioinformatics. The deciphering of genomes gave rise to a new biological discipline, genomics, which was divided, in turn, into comparative and functional branches.

There is a significant difference between genomics and proteomics: the genome is stable for a given species, since mutations occur very rarely, while the proteome is different not only for different cells of the same organism but also for the same cells with respect to its state (division, dormancy, differentiation, etc.). Therefore, thousands upon thousands of proteomes in even the simplest organism, not to mention humans, are very difficult to study. However, new strategies, such as biological microchips, phage display, mass-spectrometry, and some others, have recently produced a new methodical foundation of genomics and proteomics. Therefore, the coming decade will be a time of new methodical thinking, which will lead to a substantial revision of many general biological concepts.

## GENOMICS

Genomics is the study of complete set of genetic information—all the DNA in the organism. Combinations of genes, often interacting with environmental factors, ultimately determine the physical characteristics of an organism. Once the genome has been sequenced, positions that define the features of biological importance must be identified and annotated. When the newly identified gene has a close relative already in a DNA database, gene finding is straightforward. The genes tend to be simple, uninterrupted open reading frames that can be translated and compared with the database. However, the discovery of new genes without close relatives is more problematic.

### Genomics Applied in Bioremediation

The new scientific discipline of genomics is providing insights into some key microbiology involved in metal bioremediation. Microbial genome sequencing has, for the first time, made accessible all the components needed for both the

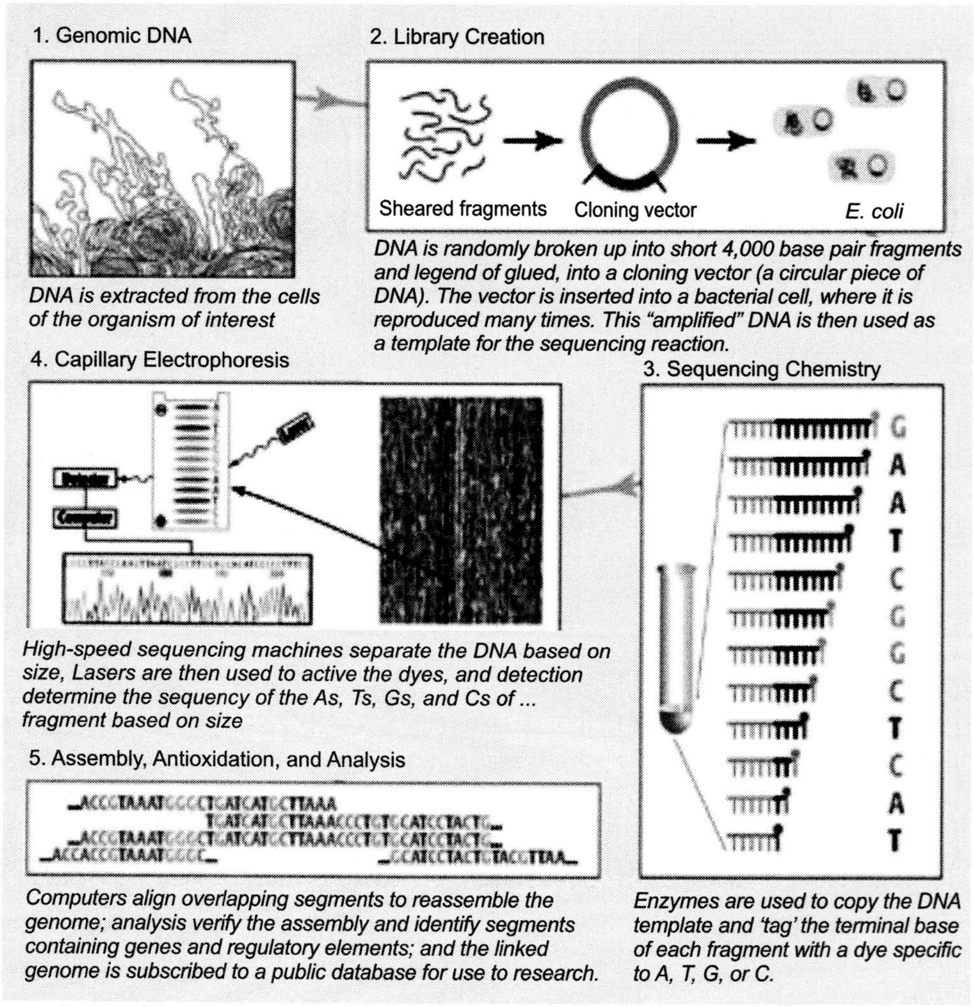


**Fig. 1.** Sequencing genomes. While the number of chromosomes, genes, and base pairs in the genomes of different organisms vary, their fundamental structures are very similar, and the techniques for sequencing and studying them are the same. Whole genome shotgun sequencing is used to determine the order of the bases of an entire genome. (Graphic courtesy of DOE Joint Genome Institute, Walnut Creek, CA.)

elaboration and the functioning of a cell. Associated with other global methods such as protein and mRNA profiling, genomics has considerably extended our knowledge of physiological processes and their diversity not only in human, animal and plant pathogens but also in environmental isolates. At a higher level of complexity, the so-called meta-approaches have recently shown great promise in investigating microbial communities. Combined with classical methods of physico-chemistry and microbiology, these endeavors should provide us with an integrated view of how microorganisms adapt to particular ecological niches and participate in the dynamics of ecosystems (Bertin, 2008).

Genomics is a recent conceptual approach for the biological study of microorganisms, which relies on the analysis of the complete genetic information they contain. In the microbial field, most of the studied genomes belong to the Proteobacteria, which represent the majority of known bacteria in taxonomy, pathology or biotechnology. The knowledge of the complete genome sequence of numerous organisms has led to profound changes in investigation methods, aiming to characterize genes whose function was not identified by conventional strategies or to study stress responses because of their multifactorial character. The use of global analysis methods, usually called large-scale or high-throughput technologies, have spread to many laboratories. Instead of studying individual genes, proteins or metabolic products, the integral profile of organisms can now be established. In particular, the differential analysis in various growth conditions of the whole protein content ('proteome'), transcripts ('transcriptome') or metabolites ('metabolome') allows the simultaneous quantification of gene expression or the accumulation of the corresponding products in an organism (Delneri et al., 2001; Kahn, 1995; Velculescu et al., 1997). Integrative biology, which aims at including all physiological processes in a bioinformatics predictive approach, will also benefit from these expanded genomic possibilities (Bonneau et al., 2006). Most of the sequencing efforts were initially focused on human pathogens, such as bacteria and parasites, and later on higher eukaryotes, to establish a basis for a better understanding of physiological processes in human beings. More recently, there has been a growing interest in microorganisms isolated from various environments. The objective is to exploit available methodologies to identify specific properties that allow these organisms to grow in diverse environments, including extreme ecological niches. These studies should also lead to a better understanding of ecosystems themselves. Moreover, they should lead to the identification of novel functions that may be exploitable for biotechnological applications, in particular the bioremediation of contaminated environments. The wide availability of these data to the scientific community and their comparison offer unprecedented opportunities for studying how the components of a living organism, also the organisms themselves, function together and respond to environmental challenges. A deep knowledge of the elements involved, their spatial and temporal distribution, and the metabolic pathways they belong to, would allow an integrated picture to be drawn of the biological processes under study and could lead to an optimal use of microorganisms' properties, favoring the desired effects.

Whatever the organisms under study, i.e., pathogenic microbes and environmental isolates explored in the first genomic programmes or more recently, the results of genome sequencing are always of importance, as they allow organisms to be examined in a global way. In many cases, these data have greatly increased the previously accumulated knowledge of the organisms, e.g., secondary metabolite biosynthesis in *Pseudomonas fluorescens* (Paulsen et al., 2005) or the acquisition of virulence factors in *Yersinia enterocolitica* (Thomson et al., 2006). When genomic approaches are applied to organisms that have been the subject of a limited number of studies, the result obtained is often remarkable.

The ease and availability of genomic data has created a new level of understanding this system. Bioremediation capabilities of the microbial population can be analyzed not only by physiological parameters, but also by the use of genomics tools, and efficient remediation strategies can be planned. PCR and DNA- or oligonucleotide-based microarray technology is a powerful functional genomics tool that allows researchers to view the physiology of a living cell from a comprehensive and dynamic molecular perspective (Kapley and Purohit, 2003).

If environmental genomic approaches are to have a real impact on optimizing bioremediation strategies, studies will need to advance beyond the sequencing of a few isolated segments of genomic DNA. Such studies can provide relatively limited information on the physiology of the organisms being investigated.

The steadily decreasing cost and increasing speed of DNA sequencing has now made it feasible to consider sequencing billions of bases of environmental genomic DNA. This offers the possibilities of assembling environmental sequences into complete or nearly complete, genomes of the individual microorganisms that predominate in the environment of interest, especially in some instances of

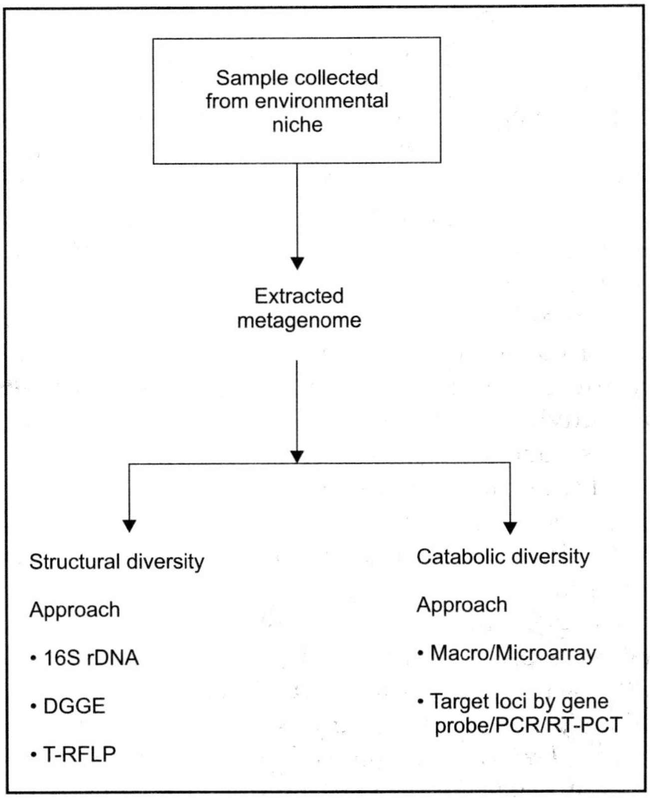


**Fig. 2.** General approach for analysis of metagenome in bioremediation.

bioremediation in which a particular genus of microorganism predominates. Parallel environmental proteomics studies are also imaginable. A global analysis of which genes are being expressed under various environmental conditions in contaminated environment will reveal metabolic status of microorganisms and indicate environmental modifications possibly as simple as addition of trace nutrient, which can accelerate bioremediation (Lovely, 2003).

## PROTEOMICS

It is a new branch of science dedicated to all the proteins expressed by a cell and how these proteins change under different growth conditions. The terms 'proteomics' and 'proteome' were introduced in 1995 , which is a key post genomic feature that emerged from the growth of large and complex genome sequencing datasets. Proteomic analysis is particularly vital because the observed phenotype is a direct result of the action of the proteins rather than the genome sequence. Traditionally, this technology is based on highly efficient methods of separation using two-dimensional polyacrylamide gel electrophoresis (2-DE) and modern tools of bioinformatics in conjunction with mass spectrometry (MS). However, 2-DE has been considered to be a limited approach for very basic and hydrophobic membrane proteins in compartmental proteomics. In bioremediation, the proteome of the membrane proteins is of high interest, specifically in polycyclic aromatic hydrocarbon biodegradation. The improvements in 2-DE for use in compartmental proteomics have been made by introducing an alternative approach for multidimensional protein identification technology (MudPIT) (Santos, 2004).

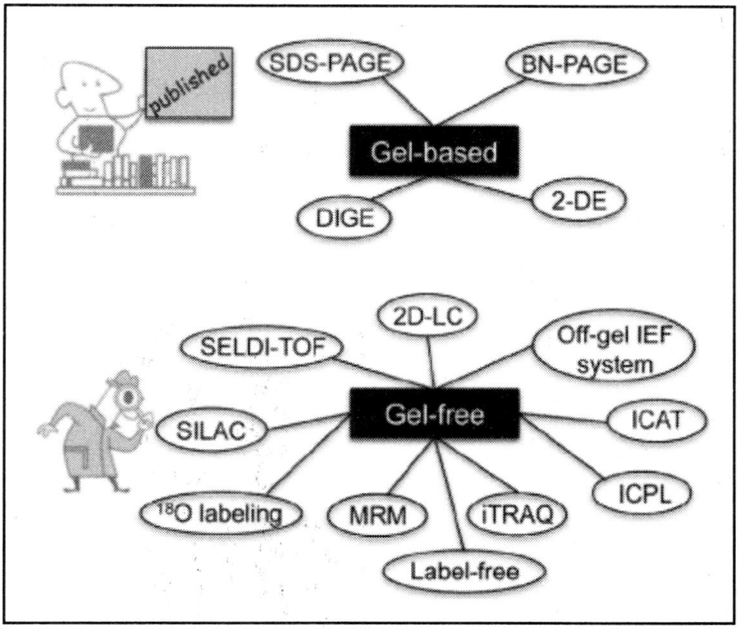


**Fig. 3.** Proteomics approaches for bioremediation.

## Proteomics Applied in Bioremediation

Elevated concentration of metals is known to trigger adverse redox-reactions in the cell such as generation of toxic hydroxide radicals in the reaction (Brown et al., 1992). The hydroxide radicals thus formed are highly reactive molecules and play a role in a number of deleterious reactions such as peroxidation of lipids, which cause membrane disruption and the oxidation of proteins, causing their inactivation (Elisa et al., 2000). This can lead to alteration in structure and/or function of these molecules. Bacteria and higher organisms have developed resistance mechanisms to toxic metals to make them innocuous. Organisms respond to heavy metal stress using different defense systems, such as exclusion, compartmentalization, making complexes and the synthesis of binding proteins such as metallothioneins (MTs) or phytochelatins (PCs). Organisms with unique abilities of metal absorption, accumulation or resistance can be searched for among naturally occurring organisms. Alternatively, these systems can be utilized in engineering bacteria for the remediation of polluted waters and soils. The introduction and/or over expression of metal binding proteins have been widely exploited to increase the metal binding capacity, tolerance or accumulation of bacteria. In bacteria, various peptides consisting of metal-binding amino acids (mainly histidine and cysteine residues) have been studied for enhanced heavy metal accumulation by bacteria.

Efforts have been made to understand the mechanisms by which cells react when exposed to exogenous metals. One of the approaches is the use of recently developed technology of DNA micro-array as a high throughput method for global analysis of gene expression to explore the cellular response to metal toxicity. However, it does not correlate well with the relative protein abundance in the cell, and also the post-translational modifications of proteins are not taken into consideration (Krizek et al., 2003). The other approach is proteomic (Nouwens et al., 2000; Aebersold and Matthias, 2003), which involves the global analysis of the differentially expressed proteins along with their post-translational modifications.

## Proteomics using Mass Spectrometry

Initial proteomics efforts relied on protein seperation using 2D gel electrophoresis, with subsequent mass spectrometric identification of protein spots. This technology can easily yield information about how the concentration of proteins varies over a course of a cell's lifetime. The rapid developments in mass spectrometry have shifted the balance to direct mass spectrometric analysis, and further developments will increase sensitivity, robustness and data handling. Proteomic analysis coupled with mass spectrometry, however, have several limitations. An inherent limitation of this approach is the depth of coverage, which is necessarily constrained to the most abundant proteins in the sample. Indeed, all proteins of an organism cannot be visualized whatever the detection method used because some proteins are present and effective at very low levels. Abundance levels of proteins in a yeast cell for instance can vary from one to a million per cell (Ghaemmaghami et al.,

2003), and it is of course difficult to monitor over such a large dynamic range. Furthermore, some proteins are quite labile while others are not. Moreover, membrane proteins are usually difficult to detect on 2D gels because of a low solubility level. Nevertheless, a global or partial proteomic map (cytoplasmic, membrane or extracellular fraction) of several organisms has been drawn. These bacteria are often model organisms, e.g., *B. subtilis* (Hecker & Volker, 2004) or human pathogens, e.g., *Mycobacterium tuberculosis*, the aetiological agent of tuberculosis, or *Helicobacter pylori*, responsible for stomach ulcer (Jungblut, 2001). Recently, the physiology of an increasing number of environmental isolates has been studied by a proteomic approach. For example, arsenic metabolism and the adaptation to cold have been investigated in *Herminiimonas arsenicoxydans* (Muller et al., 2007) and the archaeon *Methanococcoides burtonii* (Saunders et al., 2005), respectively. Similarly, several proteomic approaches addressed the adaptive response of plants interacting with mycorrhizae or rhizobacteria (Kim et al., 2004). The development of interactive databases, such as SWISS-2DPAGE (http://expasy.org/ch2d/) or InPact (http://inpact.u-strasbg.fr/~db/index.php), should greatly facilitate the exploitation of these data. Finally, using specific staining procedures, a 2D map of iron-metalloproteins has been recently drawn for the acidophilic archaeon *Ferroplasma acidiphilum* (Ferrer et al., 2007). The results suggest that the high content of metalloproteins in this organism may not only reflect its current iron-rich habitat, but may also indicate that it represents a relic of early life on Earth, where metals were more prevalent than today due to widespread volcanic and hydrothermal activities.

### Structural Proteomics with Special Reference to NMR Spectroscopy and X-ray Crystallography

A full description of cell behaviour necessitates structural information at the level not only of all single proteins but also of all salient protein complexes and the organisation of such complexes at a cellular scale. This all-encompassing structural endeavour spans several orders of magnitude in measurement scale and requires a battery of structural techniques, from x-ray crystallography and nuclear magnetic resonance at the protein level, to electron microscopy of mega-complexes and electron tomography for high–resolution visualization of the entire cellular milieu. Microbial cells usually contain several thousand genes. Many of them code for proteins with an enzymatic activity, which therefore are suspected to generate hundreds of metabolic reactions. The constant progress in techniques of analytical chemistry, such as NMR, Fourier transform mass spectrometry (FT-MS) or high-resolution liquid chromatography (HPLC), will lead to the global identification of various categories of metabolites present in a cell, such as saccharides (glycome), lipids (lipidome) and amino acids (Hollywood et al., 2006). These techniques will be necessary to build in dynamics of protein interactions.

Protein structure can be determined using some of the techniques of structural proteomics, X-ray crystallography and NMR spectroscopy. Rapid advancement in cryo-electron microscopy is also enabling the determination of gross features of protein structure at a low resolution. NMR can deliver strong results in several areas of structural biochemistry. A wide range of experiments can be performed to determine structure-function relationships, to find binding partners with their specific binding sites, to investigate dynamics of proteins (Renner et al., 2001), or to distinguish multiple conformations (Muhlhahn et al., 1998), to compare apo and holo forms of proteins and map the binding sites of their cofactors (Wijesinha-Bettoni et al., 2001), or to determine pKa values of ionizable groups (Fielding, 2000). A series of spectra taken under different conditions may be used to monitor aggregation and formation of amyloid fibrils (Zurdo et al., 2001), to determine KD values of binding partners, or to track hydrogen exchange with real-time NMR in proteins dissolved in D2O (Canet et al., 2002). The ability to detect ligands binding only very weakly to target molecules has made NMR also increasingly important in drug discovery (Diercks et al., 2001; Pellecchia et al., 2002).

## One-Dimensional NMR

It is a simple proton experiment, the most basic spectrum in NMR spectroscopy that can be acquired in a short time (usually not longer than a few minutes) for samples as dilute as 0.01 mM, contains a great amount of information.

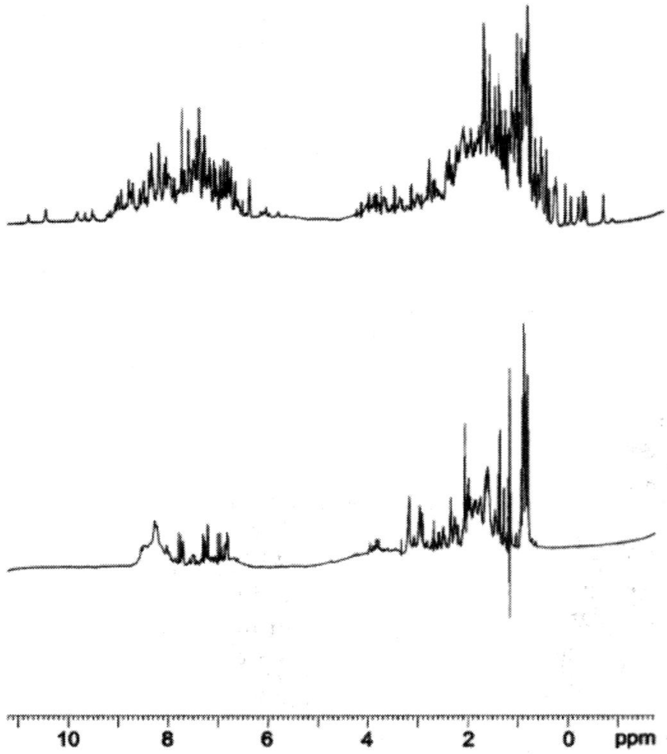


**Fig. 4.** Characterization of protein structures by one-dimensional NMR spectroscopy.

## Two-Dimensional NMR

Because of the greatly improved resolution of two-dimensional experiments, they are frequently used for screening and binding studies. The simplest and most powerful among them is the heteronuclear single-quantum coherence (HSQC) experiment. In an impressive large-scale approach, Yee et al. (Yee et al., 2002) have recently investigated more than 500 proteins from five different organisms, using 15N-HSQC experiments to screen for those proteins amenable to NMR structure analysis. This spectrum is the first step in any structure elucidation as it maps the backbone amide groups of a protein to their proton and nitrogen frequencies. A whole set of three-dimensional spectra later used to assign the NMR signals to their respective amino acid residues is based on the HSQC experiment.

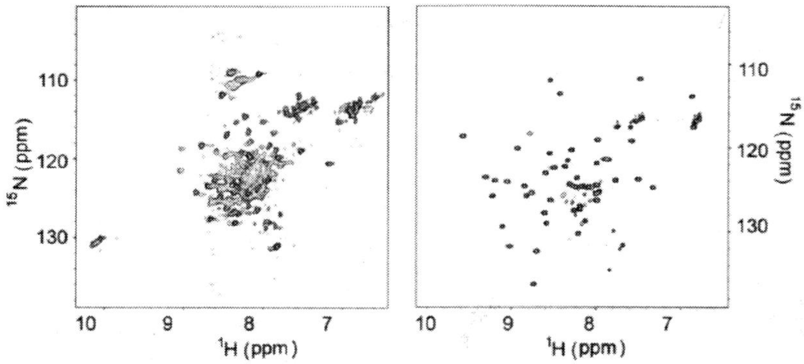


**Fig. 5.** $_{15}$N-HSQC spectra of unfolded and folded proteins.

The left panel shows a 15N-HSQC spectrum of a partially unstructured protein fragment of 80 amino acid residues. All signals cluster around a 1H frequency of 8.3 ppm. Also, the signal dispersion in the 15N dimension is limited. The broad unresolved signals in the middle of the spectrum indicate either aggregation in the sample or conformational heterogeneity on a ms-_s time scale (both cases are unfavorable for NMR studies). The signal at 10 ppm is not diagnostic for a folded protein but stems from the side chain amide group of a tryptophan residue. The right panel shows the spectrum of a folded, 55-residue-long construct of the IGFBP-5 protein. The peaks show a large signal dispersion in both dimensions.

NMR can serve as a powerful tool in structural proteomics, as a method for screening proteins for sample conditions that favor crystallization or that are candidates for the NMR solution structure determination (Rehm et al., 2002).

## APPLICATIONS OF GENOMICS AND PROTEOMICS

Bioinformatics, despite a young field, has helped both fundamental microbiology and biotechnology through the development of algorithms, tools, and discoveries

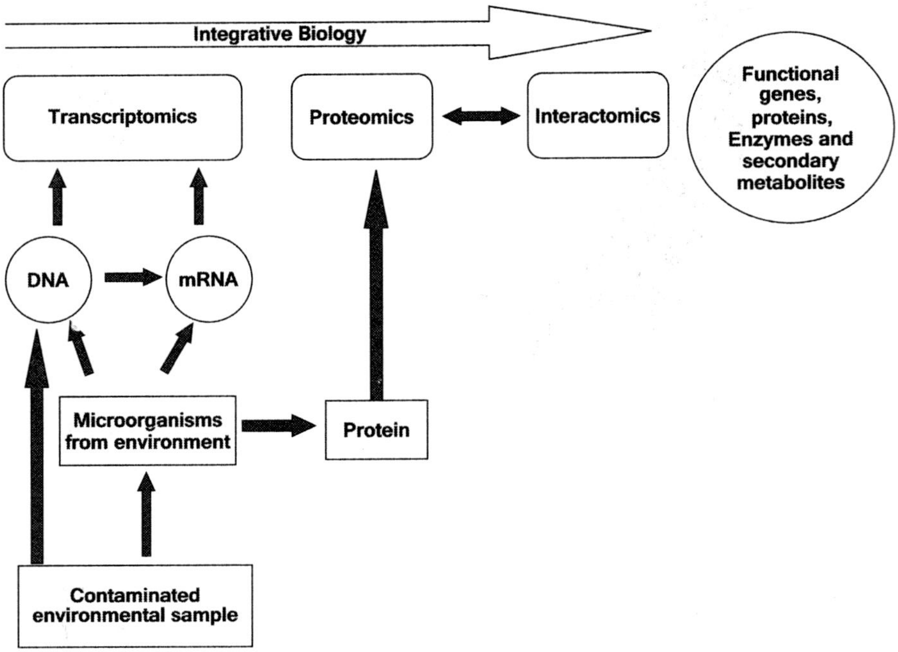


**Fig. 6.** Approaches of bioinformatics for environmental clean-up.

refining the abstract model of microbial cell functioning. The major impact of bioinformatics has been in automating the microbial genome sequencing, the development of integrated databases over the Internet, and analysis of genomes to understand gene and genome functioning. Bioinformatics techniques are critically dependent on the knowledge derived from wet laboratories and the available algorithms and tools. Both the resources have limited capability of handling a vast amount of data to interpret genomics and proteomics with so many unknowns. The progress in bioinformatics and wet laboratory techniques has to remain interdependent and focused, complementing each other for their own progress of biotechnology. The application of genomic tools in the identification of microbial community has led to the discovery of unique bacteria that were not accessible by culture-based techniques. Nucleic acid extraction from target niches and amplification of the 16S rRNA gene by polymerase chain reaction (PCR) has proved extremely useful in assessing the microbial community (Malik et al., 2008; Rani et al., 2008). Changes in the population dynamics can also be assessed by analyzing the microbial diversity at different time points or after treatment/change in carbon source of the niche (Kapley et al., 2007).

In addition to the purely practical significance for health sciences (prognostication, diagnosis, genotherapy, and pharmacology), the new knowledge will especially affect developmental biology. It will be possible to obtain integral and, at the same time, sufficiently detailed notions of individual cells, from the egg and spermatozoon of the differentiated cells. This will allow us for the first time to follow the interaction of individual cells at different developmental stages, which was always a dream for developmental biologists. The triad genomics-proteomics-bioinformatics will also markedly affect the classical biological disciplines, such as the theory of evolution and taxonomy. It is evident that biologists in the new century will be able to approach so far unsolved problems, such as carcinogenesis and aging, with levels of knowledge and information that exceed by several orders of magnitude, those that biologists in the last century had at their disposal. The new century will bring a new integral biology, whose contours have been outlined but whose effects on natural science and the fate of humankind are still difficult to fully comprehend. Genomics and proteomics operate with such huge arrays of structural information that an acute requirement in bioinformatics has arisen where new information is collected, sorted, described, analyzed, and processed. In model organisms, such procedures might, for example, enable the relationship between a mutation and a specific biosynthetic pathway to be established. In this respect, they would play an important role in functional genomics, which aims to attribute functions to the numerous hypothetical proteins identified in sequenced organisms (Kell et al., 2005). What is applicable to model organisms such as *E. coli* and *S. cerevisiae* will hopefully be used in the future to characterize environmental isolates or microbial communities in their environment. Moreover, the relationship between metabolites and proteins, between proteins and nucleic acids, and between proteins and proteins would constitute another development in the understanding of molecular interactions in a cell (Pu et al., 2007).

## 16SRDNA TECHNIQUE AND PHYLOGENETIC ANALYSIS FOR IDENTIFICATION OF POTENTIAL STRAINS

The rRNA-based analysis is a central method in microbiology used not only to explore microbial diversity but also to identify new strains. Bacterial species have at least one copy of the 16S rRNA gene containing highly conserved regions together with hypervariable regions. The use of 16S rRNA gene sequences to identify new strains bacteria is gaining momentum in recent years.

Kebria et al., isolated and characterized a novel *Bacillus* strain which is capable of degrading diesel fuel. Phenotype and phylogeny analysis of this strain was characterized, and it was identified as diesel degrading bacteria based on gram staining, biochemical tests, and 16S rRNA gene sequence analysis. These results indicate that this new strain was *Bacillus* sp. and could be considered *Bacillus cereus* with 98% 16 S rRNA gene sequence similarity. The results indicate that native strains have great potential for in situ remediation of diesel-contaminated soils in oil refinery sites (Kebria et al. 2009).

## METAGENOMICS

Metagenomics is the study of metagenomes, genetic material recovered directly from environmental samples. The broad field may also be referred to as environmental genomics, ecogenomics or community genomics. This relatively new field of genetic research enables the studies of organisms that are not easily cultured in a laboratory as well as studies of organisms in their natural environment.The extensive suite of molecular-based approaches developed over the past decade has enabled the field of metagenomics, the study of uncultured microorganisms. Paramount to metagenomic analysis is the use of high-throughput DNA sequencing technologies, which with the advent of low cost next-generation methods is transforming metagenomics. The application of metagenomics, to both global environments and microbes associated with a living host, has facilitated the study of the functional ecology of environmental microorganisms. Novel functional genes and environmental functional signatures can be retrieved using metagenomics, and these can form the basis of hypothesis driven analyses of uncultured microorganisms. As with any technology, the daunting task is to understand and apply the growing number of metagenomics sequences in the context of microbial ecology and evolution.

## IDENTIFICATION OF GENES AND PROTEINS INVOLVED IN BIODEGRADATION

One major application of functional genomics and proteomics is to discover novel genes/proteins involved in the degradation of xenobiotics such as aromatic hydrocarbons and transformations of heavy metals. The fundamental strategy in a functional genomics approach is to expand the scope of biological investigations from studying a single gene or protein to studying all the genes or proteins simultaneously in a systematic fashion. The classic approach to assess gene function is to identify which gene is required for a certain biological function at a given condition through gene disruption or complementation. With the combination of technologies, such as transcriptomics and proteomics complementing traditional genetic approaches, the detailed understanding of gene functions becomes feasible. The biochemical potential of natural microflora can be assessed by monitoring the available catabolic gene pool using target catabolic genes, or a target genotype can be tracked in environmental niches using PCR. For example, phenol degrading genotype was monitored in environmental samples by designing primers from the dmpN and pheA operons (Kapley et al., 2000). Moharikar et al. (2003) have also demonstrated the tracking of catabolic gene loci *phe*B, *xyl*E, tod-isp, *bed*A and *nah*G responsible for the enzymes catechol 1,2-dioxygenase, catechol 2,3-dioxygenase, toluene dioxygenase iron-sulphur protein component, benzene dioxygenase and naphthalene dioxygenase in activated biomass of effluent treatment plant treating wastewater generated at refineries. The natural degradative potential of a niche can be determined using tools like PCR or DNA hybridization, and bioremediation strategies can be planned for more effective degradation. If the target gene is present in low numbers, enrichment for degradative genotype can be an option (Sutherland, 2000), or in the absence of target gene, bioaugmentation can be an alternative, as demonstrated in the treatment of wastewater containing hydrocarbons, where the bioaugmentation of the *alk*B genotype was shown to improve the efficiency of treatment.

Several recent functional genomics studies using proteomics approaches provide a framework and starting point to define the function of a particular protein. Furthermore, partial protein sequences from high-resolution, 2-D gels and electrospray MS of protein complexes can be used to unambiguously assign a protein to be encoded by a specific gene in microorganisms with complete genome sequences.

Ying-Tsong Chen et al., successfully did the sequencing and analysis of the large virulence plasmid pLVPK of *Klebsiella pneumoniae* CG43, which shows the presence of genes of heavy metals resistance. Many genes for the maintenance of heavy-metal ion homeostasis have been identified in bacteria. Three physically linked gene clusters were identified in the pLVPK that are related to metal resistance phenotype in *K. pneumoniae*. These gene clusters include homologs of the lead resistance genes pbrRSABC of *Ralstonia metallidurans* CH34 (Borremans et al., 2001), the copper-resistance genes pcoEABCDRS of *E. coli* plasmid pRJ1004 (Brown et al., 1995), and the silver-resistance gene cluster silCBAPsilRSE of *S. enterica* serovar Typhimurium (Gupta et al., 1999). Copper resistant gene pcoA was isolated and identified by Piater et al., 2008. This was isolated first time from *Proteus mirabilis*, a Gram-negative soil bacterium. First, South African mine isolates were screened for resistance to copper, and an isolate showing the highest resistance was identified as *Proteus mirabilis*. 16SrDNA sequencing and PCR techniques were used to isolate the potential microorganism and pcoA gene responsible for bioremediation. Similarly, LIU Xue-duan 2010, reported isolation, characterization and extraction of *mer* gene of $Hg^{2+}$ resisting strain D2 using 16S rDNA sequence technique and PCR amplification technique. Trajanovska et al., 1997, detected heavy metal ion resistance genes in Gram-positive and Gram-negative bacteria isolated from a lead-contaminated site.

Polymerase chain reaction, DNA-DNA hybridization and DNA sequencing were used to explore the nature of genetic systems responsible for metal resistance in eight isolates. Specific DNA sequences could be amplified from the genomic DNA of all the isolates using primers for sections of the *mer* (mercury resistance determinant on the transposon Tn*501*) and *pco* (copper resistance determinant on the plasmid pRJ1004) genetic systems.

Gade et al., (2007) have done a characterization of the proteins of bacterial strain isolated from contaminated site involved in heavy metal resistance. Proteomic approach involving two-dimensional polyacrylamide gel electrophoresis (2D PAGE) and mass spectrometry (MS) was used to identify the differentially expressed proteins under heavy metal stress. Two of the differentially expressed proteins were identified to be l-isoaspartate protein carboxymethyltransferase type II and DNA gyrase A. To our knowledge, this is for the first time that *K. pneumoniae* has been reported to be present in a metal contaminated site and l-isoaspartate protein carboxymethyltransferase type II protein to be over expressed under heavy metal stress. Further, a research done by Surbhi Sharma et al. (2006) shows the role of proteins in resistance mechanism of *Pseudomonas fluorescens* against heavy metal induced stress with proteomics approach. The aim of the study was to elucidate the differential protein expression of this bacterium when exposed to heavy metal stress, using two-dimensional electrophoresis. The proteins spo VG and enolase showed upregulation during the bacterial exposure to lead and copper. Some proteins like xylosyltransferase, ORF 18 phage phi KZ, OMP H1 and translational elongation factor EF-Tu appeared only during their exposure to cobalt.

Sharma and Fulekar have isolated copper resistant gene pcoA from the multiple metal resistant potential strain *Klebsiella pneumoniae* MHF ENV III. The strain was isolated using 16SrDNA technique from small scale metal industries and showed bioremediation of heavy metals (Cu, Cd, Fe) upto a concentration of 100 ppm.

**Table 1** Isolated and identified genes using functional genomics tools

| *Metal* | *Gene* | *Resistant Mechanism* |
|---|---|---|
| **Hg2+** | *mer* | $Hg^{2+}$ and organomaterials are enzymatically detoxified |
| **$AsO_4^{3-}$, $AsO^{2-}$,$SbO^+$** | *ars* | Arsenate is enzymatically reduced to arsenite by ArsC, arsenite and antimonite are pumped out by the membrane protein ArsB. |
| **$Cd^{2+}$** | *cada* | Pumped out from Gram-positive bacteria by a P-type ATPase |
| **$Cd^{2+}$,$Zn^{2+}$,$Co^{2+}$ and $Ni^{2+}$** | *czc* | Pumped out from Gram-negative bacteria by a three-polypeptide membrane complex that functions as a divalent cation |
| **$Cu^{2+}$** | *cop* | Plasmid $Cu^{2+}$ resistance results from four polypeptide complex consisting of an inner membrane protein, an outer membrane protein and two periplasmic copper proteins |
| **$CrO_4^{2-}$** | *chr* | Results from a single membrane polypeptide that causes reduced net cellular uptake |
| **$Pb^{2+}$** | *pum* | Lead resistance appears to be due to an efflux ATPase in Gram-negative and accumulation of intracellular Pb3(PO4)2 in Gram-positive bacteria |

## BIOINFORMATICS TOOLS: IDENTIFYING GENE FUNCTION (SEARCHING AND ALIGNMENT)

After identifying the ORFs (open reading frames), the next step is to annotate the genes with proper structure and function. The function of the gene has been identified using popular sequence search and pair-wise gene alignment techniques. The four most popular algorithms used for functional annotation of the genes are BLAST (Altschul et al., 1990) and its variations (Altschul et al., 1997), dynamic programming technique, Smith-Waterman alignment (Waterman, 1995) and its variations, indexing based scheme, FASTA and its variations, and BLOCKS that uses multiple sequence alignment of conserved domains to identify motifs—characterizing patterns of proteins.

**BLAST (Basic Local Alignment Search Tool)**

Search is based upon expanding multiple probable seed points (longer than four nucleotides) that match (with the help of scoring matrices such as BLOSUM or PAM (Mwangi and Siggia, 2003) to identify the largest matching nonrandom segment. Scoring matrices have positive match-value for the amino acids that have common biochemical or biophysical properties and negative match-values if the amino acids do not share biophysical or biochemical properties. Substitution matrices such as BLOSUM (BLOcks SUbstitution Matrix) have been derived by statistically comparing the frequency patterns of the amino acids occurring in conserved domains of protein families. Nucleotide sequences use a nucleotide matrix for scoring that penalizes non-matching positions. BLAST algorithm has near linear time complexity, and the current implementations are fast. However, in order to enhance computational efficiency, BLAST algorithm uses most probable combinations of nucleotide seeds to index the sequences in the database sacrificing some accuracy.

BLAST algorithm has gone through many improvements in heuristics to improve the execution speed, accuracy, and the dependence on predefined scoring matrices. Two major improvements are the use of two or more hits within a matching region before extending the high scoring segment, and the use of multiple iteration of matching to derive a position specific scoring matrix to be used in place of predefined biochemical matrix. PSI-BLAST (Position Specific Iterative BLAST) is a popular implementation of BLAST that uses both these improvements. The use of two hits improves the execution efficiency in the segment extension, and the use of position specific matrix improves the search for weakly homologous sequences in evolutionary distant species. Position specific matrix is built by deriving multiple sequence alignment of the best matching segments and analyzing the frequency of the amino acid substitutions in the matching segments.

Dynamic algorithms such as Smith-Waterman and other indexing schemes are more accurate for pair-wise gene alignment. The alignment of gene-pairs using dynamic algorithms is based upon incremental matching by maximizing the sum of the score of the best alignment of the preceding subsequences and the score of matching the current amino-acid characters (or nucleotide characters). The mismatches in amino acid sequences are penalized using scoring matrices such as BLOSUM or PAM; the nucleotide sequences use a nucleotide matrix for scoring that penalizes non-matching positions. A gap is inserted to show the insertion and deletion of nucleotides (or amino acids). Gaps are not part of a substitution matrix, and are provided as parameters by users. The presence of a gap also results into score penalty. There are two major types of protein (or gene) alignments using dynamic programming: global and local. In global alignment, the amino acid (or nucleotide) characters are placed to maximize the overall score. In contrast, local alignment finds the segment with the maximum score, and the segments with negative scores are ignored. For comparing amino acid sequences from evolutionary distant organisms, local alignment is preferred to take care of large-scale amino acid variations. Global alignment fares well when small amount of random mutation is involved. Due to the pair-wise comparison of all characters in an amino acid sequence to identify best matching subsequence, all dynamic programming techniques have quadratic time complexity making them less suitable for large-scale pair-wise genome comparisons unless preprocessed by BLAST to remove dissimilar genes.

### Multiple Sequence Alignment

Techniques (Chhena et al., 2003) compare multiple homologous genes (genes that have similar sequences) to derive conserved segments and evolutionary tree. The technique uses the integration of pair-wise alignment between two homologs and the notion of distance between two nucleotide sequences or between two amino acid sequences. The notion of distance can be derived either as an edit distance—number of mismatches derived after pair-wise alignment of two sequences—or as the evolutionary distance between two microorganisms given by an evolutionary tree. The technique is based upon progressive pair-wise comparison to make intermediate alignments between the nearest neighbors—homologs having shortest distance—and has been implemented as a greedy algorithm. ClustalX is a popular multiple sequence alignment technique that has been used to identify conserved portions in a gene, and to develop new evolutionary trees.

A major source of problem in the above sequence comparison techniques is the assignment of user defined equal weight to indels (gaps) that undermines the importance of a specific amino acid(s) or a group of amino acid. Another minor problem is the presence of repeat characters in the sequences as the repeat characters only show the functional or structural separation of the component units within a gene, and cannot be mixed with other amino acid characters.

Multiple sequence comparison techniques such as BLOCK (Henicoff et al., 1995) have been used to identify the conserved subsequences in very similar gene sequences, and are good to derive motifs. Motifs—a set of unique subsequences characterizing a protein—have been found very useful in identifying genes with the same functionality. Motifs are derived by identifying the conserved subsequences of the functionally equivalent genes from multiple organisms after aligning the sequences.

Protein domain is the basic unit of protein function and is associated with a unique pattern (possibly one) of folding (alpha helix, beta sheet or their variations) at the structure level. The researchers have used multiple sequence alignment and HMM to identify the regions that are individually homologous to each other in multiple homologous genes. These regions are probable domains. Currently, there are many domain-related databases such as PRODOM, Pfam and SMART. Pfam is a database of multiple alignments of protein domains or conserved protein regions. The alignments represent some evolutionary conserved structure that has implications for the protein's function. Profile hidden Markov models (profile HMMs) built from the Pfam alignments are useful for automatically recognizing that a new protein belongs to an existing protein family, even under weak homology. Currently Pfam is derived automatically by cluster analysis of PRODOM database.

### Bioinformatics Tools for Structure Prediction—3D Structure Modeling and Docking

A protein may live under one or more low free-energy conformational states depending upon its interaction with other proteins. Under a stable conformational state, certain regions of the protein are exposed for protein-protein or protein-DNA interactions. Since function is also dependent upon exposed active sites, protein function can be predicted by matching the 3D structure of an unknown protein with the 3D structure of a known protein [Baker and Sali, 2000; Whisstock and Lesk, 2003). However, 3D structures from X-ray crystallography and NMR spectroscopy are limited. Thus, there is a need for alternate mechanism to match genes. Generally, there is a close correspondence between the gene sequence and 3D structure. In such cases, sequence matching is sufficient for function annotation. However, many times multiple sequences map to the same 3D structure; the lack of matching of amino acid sequences does not exclude same 3D structure. In such cases, matching 2D structure (Thommas et al., 1996)—patterns of alpha helix and beta sheets—and 3D structures is needed to verify the function of the newly sequenced protein.

There are two major approaches to model 3D structure of a protein: (i) sequence homology-based prediction and (ii) *ab initio* (or *de novo*) method. The sequence homology approach uses sequence alignment to identify the best matching 3D structure for different components: conserved portion, loop portion and side chains from the database, and threads them to predict the overall 3D structure. The *ab initio* method is based upon the energy minimization principle, and predicts the structure from the sequence alone. Recent advances in *ab initio* methods integrate the biochemical and biophysical properties such as folding of beta sheets and the information of hydrophobic regions to achieve better accuracy.

Docking is a term used to identify best matches between 3D structures of two molecules (receptor and ligand) that bind to each other by simulating interacting surfaces and free energy minimization at the domain level. Docking problem requires modeling of surfaces using spheres (or grids) and identifying the best match that will fit two surfaces without excessive intersection. Many times biochemical information such as binding sites is provided. There are three major problems in docking: (i) for multi-domain proteins conformation may change during docking, (ii) docking algorithms have high computational overhead that makes large-scale modeling quite slow, and (iii) docking algorithms suffer from over-prediction that results in a high number of false positives.

## FUTURE NEEDS

Towards a deep understanding of bioremediation, new techniques in molecular biology, particularly genetic engineering, transcriptomics, proteomics and interactomics offer remarkable advances-promise as tools to study the mechanisms involved in regulation of mineralization pathways. The application of these techniques is still in its infancy, but the amount of data that is continuously being generated, today's genomics and proteomics technocrats need to be organized in a stepwise manner within informative databases. The strategies need to be refined in which transcriptomics, proteomics data are combining together in order to understand the mineralization process in a meaningful way. These techniques show great promise in their ability to predict organisms' metabolism in contaminated environments and to predict the microbial-assisted attenuation of contaminants to accelerate bioremediation.

## REFERENCES

Aebersold, R. and Matthias, M. (2003). Mass spectrometry-based proteomics. *Nature* **422:**198-207.

Altschul, S.F., Gish, W., Miller, W., Myers, E.W. and Lipman, D.J. (1990). Basic alignment search tools. *J. Molecular Biology*, **215:** 403-410.

Altschul, S.F., Madden, T.L., Schaffer, A.A., Zhang, J., Zhang, Z., Miller, W. and Lipman, D.J. (1997). Gapped BLAST and PSI-BLAST: A new generation of protein database search programs. *Nucleic Acids Research*, **15:** 3389-3402.

Alves, R., Chaleil, R.A.G. and Sternberg, M.J.E. (2002). Evolution of enzymes in metabolism: a network perspective. *J. Mol. Biol.* **320:** 751-770.

Baker, D. and Sali, A. (2001). Protein structure prediction and structural genomics. *Science.* **294:** 93-96.

Bar, C., Patil, R., Doshi, J., Kulkarni, M.J. and Gade, W.N. (2007). Characterization of the proteins of bacterial strain isolated from contaminated site involved in heavy metal resistance—A proteomic approach. *J. biotechnology*, **128:** 444-451.

Bertin, P.N., Médigue, C. and Normand, P. (2008). Advances in environmental genomics: towards an integrated view of microorganisms and ecosystems. *Microbiology*, **154:** 347-359.

Bonneau, R., Reiss, D.J., Shannon, P., Facciotti, M., Hood, L., Baliga, N.S. and Thorsson, V. (2006). The Inferelator: an algorithm for learning parsimonious regulatory networks from systems-biology data sets de novo. *Genome Biol.* 7.R36.

Borremans, B., Hobman, J.L., Provoost, A., Brown, N.L. and van der Lelie, (2001). Cloning and functional analysis of the pbr lead resistance determinant of *Ralstonia metallidurans* CH34. *J. Bacteriol*, **183:** 5651-5658.

Brown, N.L., Barrett, S.R., Camakaris, J., Lee, B.T. and Rouch, D.A. (1995). Molecular genetics and transport analysis of the copper-resistance determinant (pco) from *Escherichia coli* plasmid pRJ1004. *Mol. Microbiol*, **17:** 1153-1166.

Brown, N.L., Rough, D.A. and Lee, B.T.O. (1992). Copper resistance determinants in bacteria. *Plasmid*, **27:** 41-51.

Canet, D., Last, A.M., Tito, P., Sunde, M., Spencer, A., Archer, D.B., Redfield, C., Robinson, C.V. and Dobson, C.M. (2002). Local cooperativity in the unfolding of an amyloidogenic variant of human lysozyme. *Nat. Struct. Biol.* **9:** 308-315.

Chenna, R., Sugawara, H., Koike, Lopez, T.R., Gibson, T.J., Higgins, D.J. and Thompson, J.D. (2003). Multiple sequence alignment with the clustal series of programs. *Nucleic Acids Research*. **31:** 3497-3500.

Delneri, D., Brancia, F.L. and Oliver, S.G. (2001). Towards a truly integrative biology through the functional genomics of yeast. *Curr. Opin. Biotechnol*, **12:** 87-91.

Diercks, T., Coles, M. and Kessler, H. (2001). Applications of NMR in drug discovery. *Curr. Opin. Chem. Biol.* **5:** 285-291.

Elisa, C., Jordi, T. and Joaquim, R. (2000). Oxidative stress in bacteria and protein damage by reactive oxygen species. *Int. Microbiol*, **3:** 3-8.

Ferrer, M., Golyshina, O.V., Beloqui, A., Golyshin, P.N. and Timmis, K.N. (2007). The cellular machinery of *Ferroplasma acidiphilum* is iron-protein-dominated. *Nature* **445:** 91-94.

Fielding, L. (2000). Determination of the association constants (Ka) from solution NMR data. *Tetrahedron* **56:** 6151-6170.

Fraser, H.B., Hirsh, A.E., Steinmetz, L.M., Scharfe, C. and Feldman, M.W. (2002). Evolutionary rate in the protein interaction network. *Science*, **296:** 750-752.

Fulekar, M.H. (2008). Bioinformatics-Application in Life and Environmental Science.Springer Publication.

Gupta, A., Matsui, K., Lo, J.F. and Silver, S. (1999). Molecular basis for resistance to silver cations in *Salmonella*. *Nat. Med.* **5:** 183-188.

Hecker, M. and Volker, U. (2004). Towards a comprehensive understanding of *Bacillus subtilis* cell physiology by physiological proteomics. *Proteomics*, **4:** 3727-3750.

Henikoff, S., Henikoff, J.G., Alford, W.J. and Pietrokovski, S. (1995). Automated construction and graphical presentation of protein blocks from unaligned sequences. *Gene*. **163:** GC17-26.

Hollywood, K., Brison, D.R. and Goodacre, R. (2006). Metabolomics: current technologies and future trends. *Proteomics* **6:** 4716-4723.

http://elm.eu.org *webcite*). Pfam [16,63] (and http://pfam.cgb.ki.se*webcite*

Ideker, T., thorsson, V., Ranish, J.A., Charistmas, R., Buhler, J., Eng, J.K., Bumgarner, R., Goodlett, D.R., Aebersold, R. and Hood, L. (2001). Integrated genomic and proteomic analysis of a systematically perturbed metabolic network. *Science*, **292:** 929-934.

Jeong, H., Mason, S.P., Barabasi, A.L. and Oltvai, Z.N. (2001). Lethality and centrality in protein networks. *Nature*, **411:** 41-42.

Jungblut, P.R. (2001). Proteome analysis of bacterial pathogens. *Microbes Infect.* **3:** 831-840.

Kahn, P. (1995). From genome to proteome: looking at a cell's proteins. *Science* **270:** 369-370.

Kapley, A., Prasad, S. and Purohit, H.J. (2007a). Changes in microbial diversity in fed-batch reactor operation with wastewater containing nitroaromatic residues. *Bioresour. Technol.* **98:** 2479-2484.

Kapley, A. and Purohit, H.J. (2000). Tracking of phenol degrading genotype. *Env. Sci. Pollut. Res.* **16:** 89-90.

Kell, D.B., Brown, M., Davey, H.M., Dunn, W.B., Spasic, I. and Oliver, S.G. (2005). Metabolic footprinting and systems biology: the medium is the message. *Nat. Rev. Microbiol.* **3:** 557-565.

Kim, S.T., Kim, S.G., Hwang, D.H., Kang, S.Y., Kim, H.J., Lee, B.H., Lee, J.J. and Kang, K.Y. (2004). Proteomic analysis of pathogen responsive proteins from rice leaves induced by rice blast fungus, *Magnaporthe grisea*. *Proteomics* **4:** 3569-3578.

Krizek, B.A., Prost, V., Joshi, R.M., Stoming, T. and Glenn, T.C. (2003). Developing transgenic *Arabidopsis* plants to be metal-specific indicators. *Environ. Toxicol. Chem.* **22:**175-181.

Lovely, D. (2003). Cleaning Up With Genomics: Applying Molecular Biology To Bioremediation. *Nature reviews,* **1:** 35-44.

Malik, S., Beer, M., Megharaj, M. and Naidu, R. (2008). The use of molecular techniques to characterize the microbial communities in contaminated soil and water. *Environ. Int.* **34:** 265-276.

Moharikar, A., Kapley, A. and Purohit, H.J. (2003). Detection of dioxygenase genes present in various activated sludge. *Env. Sci. Pollut. Res.* **10:** 373-378.

Muhlhahn, P., Zweckstetter, M., Georgescu, J., Ciosto, C., Renner, C., Lanzendorfer, M., Lang, K., Ambrosius, D., Baier, M. and Kurth, R., et al. (1998). Structure of interleukin 16 resembles a PDZ domain with an occluded peptide binding site. *Nat. Struct. Biol.* **5:** 682-686.

Muller, D., Médigue, C., Koechler, S., Barbe, V., Barakat, M., Talla, E., Bonnefoy, V., Krin, E., Arsène-Ploetze, F. and other authors. (2007). A tale of two oxidation states: bacterial colonization of arsenic-rich environment. *PloS Genet,* 3: e53.

Mwangi, M.M. and Siggia, E.D. (2003). Genome wide identification of regulatory motifs in *Bacillus subtilis. BMC Bioinformatics,* **4:** 18-20.

Nouwens, A.S., Cordwell, S.J., Larsen, M.R., Molloy, M.P., Gillings, M., Willcox, M.D.P. and Walsh, B.J. (2000). Complementing genomics with proteomics: the membrane subproteome of *Pseudomonas aeruginosa* PAO1. *Electrophoresis* **21:** 3797-3809.

Paulsen, I.T., Press, C.M., Ravel, J., Kobayashi, D.Y., Myers, G.S., Mavrodi, D.V., DeBoy, R. T., Seshadri, R., Ren, Q. and other authors. (2005). Complete genome sequence of the plant commensal *Pseudomonas fluorescens* Pf-5. *Nat. Biotechnol,* **23:** 873-878.

Pellecchia, M., Sem, D.S. and Wüthrich, K. (2002). NMR in drug discovery. *Nat. Rev. Drug Discov.* **11:** 211-219.

Piater, L.A., Ojo, A.O. and VanHeerden, E. (2008). Identification and initial characterization of a copper resistant South African mine isolate. *African Journal of Microbiology Research,* **2:** 281-287.

Pu, S., Vlasblom, J., Emili, A., Greenblatt, J. and Wodak, S.J. (2007). Identifying functional modules in the physical interactome of *Saccharomyces cerevisiae. Proteomics,* **7:** 944-960.

Rani, A., Porwal, S., Sharma, R., Kapley, A., Purohit, H.J. and Kalia, V.C. (2008). Assessment of microbial diversity in ef?uent treatment plants by culture dependent and culture independent approaches. *Bioresour Technol* **99:** 7098-7107.

Rehm, T., Huber, R. and Holak, T.A. (2002). Application of NMR in Structural Proteomics: Screening for Proteins Amenable to Structural Analysis. *Structure,* **10:** 1613-1618.

Renner, C. and Holak, T.A. (2001). NMR N-15 relaxation of the insulin-like growth factor (IGF)-binding domain of IGF binding protein-5 (IGFBP-5) determined free in solution and in complex with IGF-II. *Eur. J. Biochem.* **268:** 1058-1065.

Santos P.M., Benndorf D. and Sa-Correia I. (2004). Insights into *Pseudomonas putida* KT2440 response to phenol-induced stress by quantitative proteomics. *Proteomics,* **4:** 2640-2652.

Saunders, N.F.W., Goodchild, A., Raftery, M., Guilhaus, M., Curmi, P.M.G. and Cavicchioli, R. (2005). Predicted roles for hypothetical Bertin, P.N., Médigue, C. and Normand, P. proteins in the low-temperature expressed proteome of the Antarctic archaeon Methanococcoides burtonii. *J Proteome Res.* **4:** 464-472.

Shuker, S.B., Hajduk, P.J., Meadows, R.P. and Fesik, S.W. (1996). Discovering high-affinity ligands for proteins: SAR by NMR. *Science* **274:** 1531-1534.

Stoll, R., Renner, C., Zweckstetter, M., Brüggert, M., Ambrosius, D., Palme, S., Engh, R.A., Golob, M., Breibach, I. and Buettner, R., et al. (2001). The extracellular human melanoma inhibitory activity (MIA) protein adopts an SH3 domain-like fold. *EMBO J.* **20:** 340-349.

Sutherland, T.D., Horne, I., Lacey, M.J., Harcourt, R.L., Russell, R.J. and Oakeshott, J.G. (2000). Enrichment of an endosulfan-degrading mixed bacterial culture. *Appl. Environ. Microbiol.,* **66:** 2822-2828.

Thomas, D.J., Casari, G. and Sander, C. (1996). The prediction of protein contacts from multiple sequence alignments. *Protein Engineering,* **9:** 941-948.

Thomson, N.R., Howard, S., Wren, B.W., Holden, M.T., Crossman, L., Challis, G.L., Churcher, C., Mungall, K., Brooks, K. and other authors. (2006). The complete genome sequence and comparative genome analysis of the high pathogenicity *Yersinia enterocolitica* strain 8081. PLoS Genet 2.e206.

Trajanovska, S., Britz, M.L. and Bhave, M. (1997). Detection of heavy metal ion resistance genes in Gram-positive and Gram-negative bacteria isolated from a lead-contaminated site biodegradation. **8:** 113-124.

Velculescu, V.E., Zhang, L., Zhou, W., Vogelstein, J., Basrai, M.A., Bassett, D.E. Jr., Hieter, P., Vogelstein, B. and Kinzler, K.W. (1997). Characterization of the yeast transcriptome. *Cell,* **88:** 243-251.

Waterman, M.S. (1995). Introduction to Computational Biology: Maps, Sequence, and Genomes.Chapman & Hall, London.

Whisstock, J.C. and Lesk, A.M. (2003). Prediction of protein function from protein sequence and structure. *Q. Rev. Biophysics.* **36:** 307-340.

Wijesinha-Bettoni, R., Dobson, C.M. and Redfield, C. (2001). Comparison of the structural and dynamical properties of holo and apo bovin lactalbumin by NMR spectroscopy. *J. Mol. Biol.* **307:** 885-898.

Yee, A., Chang, X., Pineda-Lucena, A., Wu, B., Semesi, A., Le, B., Ramelot, T., Lee, G.M., Bhattacharyya, S. and Gutierrez, P., et al. (2002). An NMR approach to structural proteomics. *Proc. Natl. Acad. Sci. USA* **99:** 1825-1830.

Zurdo, J., Guijarro, J.I., Jimenez, J.L., Saibil, H.R. and Dobson, C.M. (2001). Dependence on solution conditions of aggregation and amyloid formation by an SH3 domain. *J. Mol. Biol.* **311:** 325-340.

# 9

# Replacement of Pine by Oak in Central Himalaya and its Implications in Forest Management

**S.C. Garkoti**

*School of Environmental Sciences, Jawaharlal Nehru University, New Delhi-110067*
*E-mail: sgarkoti@yahoo.com*

## ABSTRACT

*Quercus leucotrichophora* (Indian white oak) is widely distributed forest forming species between 1200-2200 m asl in central Himalaya. The white oak is one of the most important species that fulfils human needs in the region by providing various ecosystem services such as provisions for fodder, fuel wood, nutrient subsidy to cropping and conservation of soil and water. Chronic disturbance during last few decades due to increased demand to exploit these ecosystem services, white oak forests in the central Himalaya have degraded severely and were invaded by *Pinus roxburghii* (chir pine). Present study describes the re-establishment and regeneration of white oak in two fire disturbed sites previously invaded by chir pine. It was found that fire damages to chir pine in moist valley sites allow re-establishment of white oak. In both the sites of the present study, white oak has expanding population structure. Once the site comes under white oak, the soil nutrient quality improves (as evidenced from the soil N) which in turn facilitates the arrival of its several associates in the community.

**Keywords:** White Oak, Chir Pine, Ecosystem Services, Regeneration, Central Himalaya.

## INTRODUCTION

Disturbance regimes are key factors influencing the dynamics of most terrestrial ecosystems including forest ecosystems (White, 1979; Pickett and White, 1985; Clinton and Boring, 1994). Timings of disturbance in natural ecosystems have significant influence on the establishment of communities in future (Singh and Singh, 1992). Establishment of plant communities in forest gaps created in disturbed sites depends on various factors along the complex path defined by species wise variations in availability of seeds, soil moisture and nutrients, and temperature. These factors may vary seasonally as well as annually, hence favour or disfavour one or the other forest forming species depending on which year or season the disturbance has occurred, frequency, magnitude and extend of disturbance. Although centres of oaks in terms of species richness lie somewhere else in different parts of the world, but a few species found in the Himalayas are one of the largest forests forming species in the world (Singh et al., 2000). Himalayan oaks contribute to a greater proportion of ecosystem services than most of the other species in the region (Singh, 2002).

Ecological literature during the last three decades from the central Himalaya is full of the concern that fire adapted chir pine (*Pinus roxburghii*) is replacing ecologically and socially valuable white oak (*Quercus leucotrichophora*) forests, because the later is subjected to various degrees of anthropogenic disturbances like lopping, browsing, and chronic forest fires (Singh et al., 1984; Singh et al., 1997; Singh 1998; Semwal and Mehta, 1996; Semwal et al., 2003). In central Himalaya, white oak (*Quercus leucotrichophora*), a climax species, is widely distributed between 1200 and 2200 m asl (Troup, 1921). White oak shares its dominance in this part of Himalaya with Chir pine (*Pinus roxburghii*), a successional species (Vishnu Mittre, 1974). Studies so far indicate that due to anthropogenic activities such as forest fires, chronic disturbances in terms of continuous lopping of white oak trees for fodder, chir pine has been spreading greatly at the expanse of white oak in the region (Singh et al., 1984, Upreti et al. 1985, Singh and Singh 1992, Singh 1998). In most of the villages in Indian central Himalaya, despite all technological advancements, people still depend heavily on forest resources for their subsistence living. People collect fuel wood from the forests for cooking, fodder for cattle, forest litter for livestock beddings and various other species as food and medicine. Among all the species, oak is most preferred because it produces high quality fuel wood. Leaves of white oak are perennial source of fodder and also used as cattle bedding, which ultimately after decomposition reach crop fields and replenish soil fertility that has lost after cropping every season. Studies show that the chronic disturbance in white oak forests because of high demand for various purposes resulted in the degradation of these forests and hence made them susceptible to fire damages that spread periodically from adjacent chir pine forests. Being a successional species, pine is not only fire adapted but also a fire promoter species. Fire causes extensive damages to nearby degraded white oak, especially on dry south facing slopes, and has resulted in the encroachment of *Pinus roxburghii* in oak forests (Singh et al. 1984, Singh and Singh 1992). Fire in chir pine forests occur at the intervals of every 3-5 years and eliminates seedlings and saplings of not only chir pine but also other co-occurring species (Tewari 1982, Semwal and Mehta 1996). As chir pine is a successional, fire adapted species, fire favours chir pine to occupy the burnt sites (Singh and Singh, 1992). It has, however, been observed that fire damages during certain particular years result into paradoxically vigorous regeneration of white oak in chir pine occupied sites. Probably combination of habitat characteristics and other environmental factors like soil moisture and temperature and biological factors like mast seeding in white oak have favours the regeneration and establishment of the white oak in some sites after the disturbance. In central Himalaya, in general, people's perception towards white oak has been more favourable compared to chir pine. People protect regenerating oak from fire and animal grazing to ensure sustainable supplies of goods and said services from white oak forests.

Present study was conducted in two sites with the objective to understand regeneration of oak in the sites previously encroached by pine and to suggest management implications of oak re-establishment in Himalaya, which is not only favoured by the people in central Himalaya, but is also known as the species that conserves water and soil (Negi et al. 1996, Negi 2006).

In dry slopes, especially in mountain ridges, once the area comes under pine, it reduces the available N pool in the soil through its immobilization in decomposing pine litter by microbes (Singh et al. 1984). As white oak has comparatively higher demand for N and other nutrients, it is not able to reinvade the areas from where it is displaced by the pine. However, in several sites in mountain valley areas, oak is reinvading pine forests and successfully encroaching in pine regime. Although replacement of white oak by pine, which is common in central Himalaya, has been reported by various ecologists, there is no published record that describes the reverse (Singh and Singh 1992).

## MATERIAL AND METHODS

### Study Area

The forest sites studied were located at 29° 20′11.6° N 80° 01′05.2° E (oak I) and at 29° 19′06.9 ° N 80° 01′26° E (oak II) at 1849 m and 1653 m asl, respectively. In the past both the sites were under pine forest with a few scattered oak trees. The climate is temperate with an annual average rainfall of 1430 mm. In all aspects in ridge sites pine occurs even above the present study site elevations. In valley sites, wherever sites were protected from disturbance for some period, white oak was able to re-establish itself in sites that were encroached by pine and is now apparently a dominant species. Monsoon rains in this part of Himalaya occur from late June/ early July to the middle of September. The year is divisible into three main seasons; cool and usually dry winters (October–March), warm and dry summer (April–June), and warm humid rainy season (July–September).

### Methods

After a preliminary reconnaissance, two representative oak regeneration sites were selected for detailed study. Both the sites were in oak–pine transition zone and were surrounded by old-growth pine forests. Each oak regeneration forest was studied by using ten, 10 m × 10 m randomly placed quadrats. Species richness, number of individuals of different species and circumference at breast height (cbh) of the individuals were recorded. Seedlings number was counted species wise.

## RESULTS AND DISCUSSION

In both the sites, total five different tree species were recorded. In oak I site, pine was represented by only few seedlings, while in oak II site, adult individuals of pine were also present (Figures 1 and 2). In both the forests, tree basal area was higher for *Q. leucotrichophora*; 10.4 $m^2 ha^{-1}$ (86.9 %) and 28.4 $m^2 ha^{-1}$ (85.3 %) in oak I and oak II, respectively (Table 1). *Quercus leucotrichophora* was the dominant species in both the forests followed by *Myrica nagi*.

**Table 1** Vegetation and soil characteristics of two different age regenerating oak (*Quercus leucotrichophora*) forests of central Himalaya

| *Categories* | *oak I (16-years-old)* | *oak II (34-years-old)* |
|---|---|---|
| Elevation (masl) | 1849 | 1653 |
| Aspect | North | South-west |
| Tree density (trees $ha^{-1}$) | 1280(950) | 1100(660) |
| Tree basal area ($m^{-2}$ $ha^{-1}$) | 12.0 (10.4) | 33.3(28.3) |
| Soil N (%) | 0.41 ± 0.03 | 0.48 ± 0.03 |
| Available P (%) | 0.024 ± 0.002 | 0.030 ± 0.002 |
| Soil bulk density (g $cm^{-3}$) | 0.741 ± 0.05 | 0.631 ± 0.12 |
| Soil pH | 5.77 ± 0.11 | 5.36 ± 0.17 |
| Soil moisture (%) | 28.31 ± 2.13 | 31.46 ± 2.69 |
| Soil temperature (°C) | 8.4 | 9.2 |

Values in parentheses are for *Q. leucotrichophora*.

### *Size class distribution in tree species*

In general, two population structure categories have been derived from the size class distribution of tree species (Figures 1 and 2): (a) expending or stable with many seedlings and gradual decline with increase in size class (e.g., *Q. leucotrichophora, R. arboreum*). This structure reflects continuous establishment of population of species. (b) Youthful with seedlings and saplings or young trees (e.g., *Q floribunda, Pyrus pesia, Ficus nimalia, L ovalifolia*). Such species could be recent invaders and may increase their dominance in the community later. Most of these species were absent in adjacent pine forest (Garkoti pers. Communication). It seems, reinvasion of the site by *Q. leucotrichophora* has ameliorated the habitat condition (e.g., N and moisture), which allowed these associates of white oak to occupy the available niches in the community.

In both the forests of present study, *Q. leucotrichophora* has expanding population structure. As the disturbance in the form of forest fire and lopping was excluded for some time in the past, the oak was able to re-establish and expand.

According to Singh et al. (1984), pine has greater nutrient conserving ability than oak and creates N shortage in the soil that makes it difficult for oak to reinvade pine encroached sites. Present study addressed the question: which conditions favour oak to re-establish in these sites? The study observes that after the fire removes the pine, oak is able to reoccupy the sites in only moist valley areas. The mountain ridge areas remain under pine. It seems that nutrient required by the oak would be flowing down to valley sites from the mountain slopes through surface and subsurface runoff during rainy seasons. Once the oak gets established, it further improves the nutrient status of the habitat and makes gradual opening for its associates from the adjacent forests.

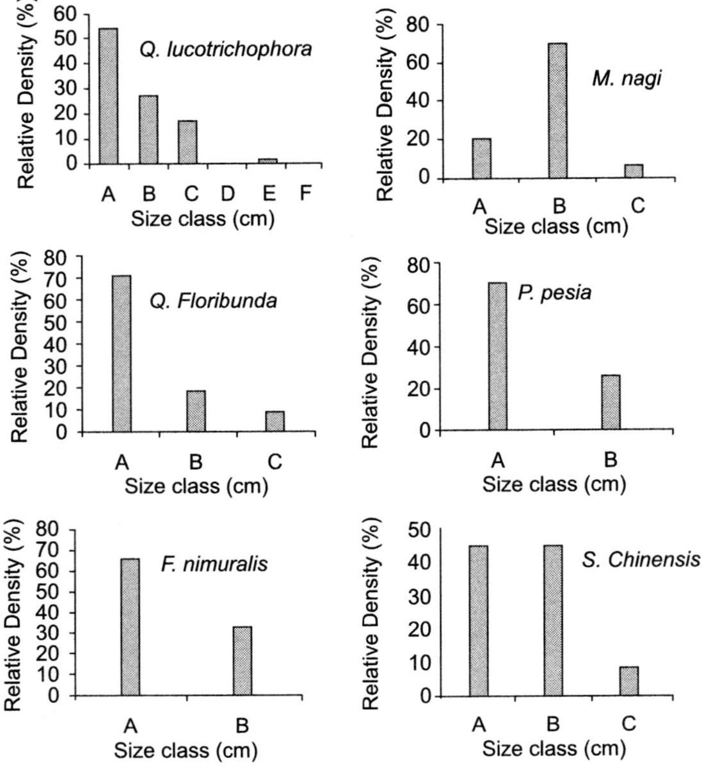


**Fig. 1.** White oak regenerating forest site I. A – H represent girth classes of individuals of the species in cm. (a) seedlings, (b) 10-30, (c) 30.5-60, (d) 60.5-90, (e) 90.5-120, (f) 120-150, (g) 150.5-180 and, H > 180.

Oak has comparatively higher demand for N than the pine (Singh and Singh 1992). In ridge area that was occupied by pine, the dynamic available N pool in the soil at any point of time is very low and N immobilization by microbes in decomposing pine litter further increases the N shortage in the soil (Singh et al. 1984). Because of this reason, oak is in general not able to reoccupy the habitats in ridge areas. However, the wash off of nutrients from the slopes that brings soil and nutrients improves the moisture as well as nutrient status in valley areas. Once established, white oak is able to maintain high soil fertility through return of nutrient rich litter and rapid mineralization of nutrients (Singh and Singh 1992). Total soil N was 0.48 and 0.41 % in valley areas, and 0.38 and 0.39% in ridge areas, (at 0-10 and 10-20 cm soil depths, respectively). If fire occurs at the time when other environmental factors are in favour of oak, it regenerates vigorously (Negi et al. 2002) and paradoxically outcompetes pine from the site. Existing remnant trees facilitate to fill the gap created due to the killing of trees by fire (Teegalpalli et al. 2009).

## MANAGEMENT IMPLICATIONS

Present study indicates that even a short period of protection (e.g., 5–8 years) is sufficient to permit the oak to re-establish and regenerate and to check the invasion of pine in valley areas in central Himalaya. It is important to identify such sites in

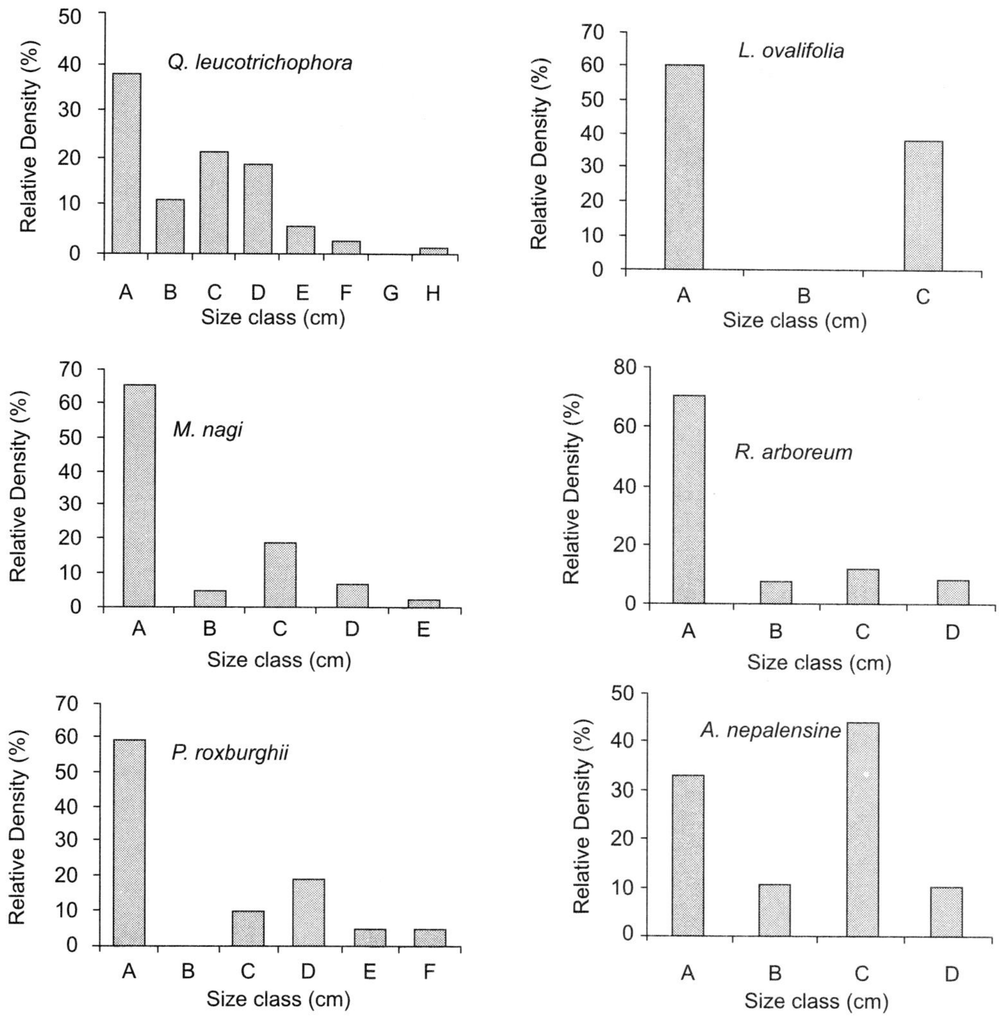


**Fig. 2.** White oak regenerating forest site II in India central Himalaya A – H represent girth classes of individuals of the species in cm. (a) seedlings, (b) 10-30, (c) 30.5-60, (d) 60.5-90, (e) 90.5-120, (f) 120-150, (g) 150.5-180 and, H > 180.

the region. Conservation of such areas not only helps people directly by supplying the goods, but also indirectly through increase in the biodiversity and conservation of soil, nutrients, moisture and other valueable ecosystem services.

## REFERENCES

Clinton, B.D. and Boring, L.R. (1994). Regeneration patterns in canopy gaps of mixed oak forests of the southern Appalachians: Influences of topographic position and evergreen understory. *Am Midl. Nat.* **132:** 308-319.

Garkoti, S.C. (2011). Fine root dynamics in three central Himalayan high elevation forests ranging from closed canopied to open-canopied tree line vegetation. *Journal Forest Research,* **16:** 136-143.

Metj, J.J. (1997). Vegetation dynamics of several little disturbed temperate forests in east central Nepal. *Mountain Research and Development,* **17:** 33-51.

Negi, A.S., Negi, G.C.S. and Singh, S.P. (1996). Establishment and growth of *Quercus floribunda* after a mast year. *J. Vegetation Science* **7:** 550-564.

Negi, G.C.S. (2002). Hydrological research in the Indian Himalayan mountains: Soil and water conservation. *Current Science,* **83:** 974-980.

Negi, G.C.S. (2006). Leaf and bud demography and shoot growth in evergreen and deciduous trees of central Himalaya, India. *Trees – Structure and Functions,* **20:** 416-429.

Pickett, S.T.A. and White, P.S. (1985). The ecology of natural disturbance and patch dynamics. Academic Press, New York, 472p.

Semwal, R.L. and Mehta, J.P. (1996). Ecology of forest fires in chir pine (*Pinus roxburghii* Sarg.) forests of Garhwal Himalaya. *Current Science,* **70:** 426-425.

Semwal, R.L. Chatterjee, S., Punetha, J.C., Pradhan, S., Dutta, Pijush, Soni, S., Sharma, G, Singh, V.P. and Malayia, A. (2003). Forest fires in India. Lessons from case studies. World Wide *Fund for Nature-India,* pp. 52.

Singh, J.S., Rawat Y.S. and Chaturvedi, O.P. (1984). Replacement of oak with pine in the Himalaya affects the nitrogen cycle. *Nature,* **311:** 54-56.

Singh J.S. and Singh S.P. (1992). Forests of Himalaya: structure, functioning and impact of man. Gyanodaya Prakashan, Nainital.

Singh, S.P. (1998). Chronic disturbance, a principal cause of environmental degradation in developing count. *Environmental Conservation* **25:** 1-2.

Singh, S.P., Rawat, Y.S. and Garkoti, S.C. (1997). Failure of brown oak (*Quercus semecarpifolia*) to regenerate in Central Himalaya: a case of environmental semisurprise. *Current Science,* **73:** 371-374.

Singh S.P. (2002). Balancing the approaches of environmental conservation by considering ecosystem services as well as biodiversity. *Current Science,* **82:** 1331-1335.

Teegalapalli, K., Hiremath, A.J. and Jathanna, D. (2010). Patterns of seed rain and seedling regeneration in abandoned agricultural clearings in a seasonally dry tropical forest in India. *J. Tropical Ecology,* **26:** 25-33.

Tewari, J.C. (1982). Vegetation analysis along altitudinal gradients around Nainital. Ph.D. Thesis, Kumaun University Nainital, India. 570 p.

Thadani, R. and Ashton, P.M.S. (1995). Regeneration of banj oak (*Quercus leucotrichophora* A. Camus)in the central Himalaya. *Forest Ecology and Management,* **78:** 217-224.

Troup, R.S. (1921). The silviculture of Indian trees, Vols. I-III. Oxford, Clerandon Press, 1195 pp.

Upreti, N., Tewari, J.C. and Singh, S.P. (1985). The oak forests of Kumaun Himalaya (India): composition, diversity, and regeneration. *Mountain Research and Development,* **5:** 163-174.

Vetaas, O.R. (2000). The effect of environmental factors on the regeneration of *Quercus semecarpifolia* Sm. in Central Himalaya, Nepal. *Plant Ecology,* **146:** 137-144.

Vishnu-Mittre, (1974). Late quaternary paleobotany and Palynology in India: an appraisement. In Vishnu-Mittre (ed.). Late quaternary vegetation development in extra European areas, Birbal Sahani Institute of Paleobotany, Lucknow, pp. 16-51.

White, P.S. (1979). Patterns, process and natural disturbance in vegetation. *Botanical Review,* **450:** 229-299.

# 10

# The Bioelectromagnetics: Present Explorations

N.K. TIWARI[1], R.A. KHAN[2] AND M.Y. KHAN[3]

[1,2]Institute of Engineering and Technology Department of Conputer Science and Engineering, Shri Ramswaroop Memorial University, Barabanki-225003, UP, India

[3]Department of Biotechnology, School of Bioscience and Biotechnology, Babasaheb Bhimrao Ambedkar University, Lucknow (Central University) Raebareli Road, Lucknow-226025, UP, India

[1]corresponding author: Neeraj Kumar, RA-UPCST (Young Scientist Scheme)
Email: neerajmtech@gmail.com

**ABSTRACT**

Bioelectromagnetics is the study of interaction of biological tissues with electromagnetic radiation (EMR). The BioEMR is a recently developed area of research that is being discussed globally by innovators on its health and environment challenges. In the modern age of communication we are living in the sea of electromagnetic radiation due to growing exposure of electromagnetic radiation through the newly developed electronic gadgets and wireless communicating devices. The cellular phones have played a major role in increasing the EMR exposure. Now we are forced to think again of the possibilities of health hazards through the EMR exposure particularly cellular phone EMR. The close proximity of the cellular phone to the users body has raised a serious concern in the society. Although researches in this area of bioelectromagnetic had initiated and placed the safely limits for EMR in 60-70s decay but in present we have to arrange the EMR exposure guidelines and safetly patterns for human health & environment. The bioelectromagnetics is a newly exposed area of exploration by which we are trying to explain the ideas and innovations from basic to advance science of natural and applied electromagnetic fields in biology and medicine. The possibility of adverse effects of exposure to electromagnetic fields has become a major issue of health policies, both at the level of individual countries and

**Fig. 1.** Demonstration of EMR surrounding the human body.

internationally. At present, a number of scientists, are conducting research in the area of bioelectromagnetics. These studies focus on two main aspects of human health: therapeutic applications and potentially harmful effects of exposure to electromagnetic fields. They have began discussion over the latest findings on biological effects of electromagnetic fields including dosimetry, interaction mechanisms, health effects of cellular communication systems and medical applications. Cellular telephones have been used throughout the world within a short period of time. These are widely used by the majority of adults, teenagers and even the children, who are using them out of inquisitiveness and interest. Although there is no clear evidence to show the harmful physiological effects of Global System for Mobile communications (GSM) at the levels used by cellular telephones, there is a widespread public concern that there may be potential for harm or radiation toxicity. Therefore, it is appropriate to conduct sensitive studies to allay any public concern. Such studies should employ methodology that is able to detect very small effects with appropriate statistical power. The World Health Organization (WHO) is the United Nations scientific health body, which has been assessing health risks for over 50 years. With its mandate to provide national authorities with sound advice based on the best available science, and using methodology that has been recognized and accepted worldwide, WHO is the logical agency that conducts health risk assessments of new and emerging technologies, especially those that emit electromagnetic fields (EMF). Biological effects and possible health consequences of exposure to EMF need to be assessed according to an appropriate set of guidelines. In its constitution, WHO defines health as the state of complete physical, mental and social well-being and not merely the absence of disease or infirmity. It is thus important to identify whether EMF-induced biological effects can be considered a hazard to human health. This chapter is an attempt to serve as a vehicle for rapid dissemination of information about basic facts required to understand the bioelectromagnetics, toxic potential of EMR, mechanisms involved in the interaction of human body with EMR, present scenario of research and possibilities of health hazards, and regulatory challenges to minimize the EMR exposure.

## INTRODUCTION

Every second, every minute, every hour and every day, you are being zapped with the dangerous and damaging form of pollution on earth. This modern day threat is something you cannot see, hear, smell, feel, or taste. It is in your home, workplace and school. It is over your head, under your bed and even right next to your ear. We are exposed to electromagnetic field (EMF) in two ways, natural and man-made, but man-made EMF poses greater threat. It is produced mainly from electricity power lines, microwave ovens, TV screens, security devices, mobile phones and home appliances that run on electricity. EMF is everywhere in modern world and lurk unseen around us. Microwave and related radiation is estimated to be up to 1 billion times as great as that which naturally exists in the environment today. Yet most people do not even know what the term EMF means. EMF is the dark side of technology that contributes to increased diseases and deaths everywhere. How much do you know about it? How are you affected? And why electromagnetic pollution becomes a growing health concern?

### Electromagnetic Radiation Exposure

EMF is invisible lines of force that occur when electrical current flows through a cable or an electrical device. It is made up of two fields, electric and magnetic; each having the potential to cause harmful side effects. The combined field emanates outwards and affects everything it touches. All electrical devices create electromagnetic field when in use. The power needed to operate the device (voltage) produces electric fields; while current flow produces the magnetic fields. Intensity of the field varies with the voltage. Higher voltage creates stronger electric fields. Primary transmission towers such as cables on high towers generate more intense electromagnetic field (Augner et al., 2009). Interaction between the magnetic field and the flowing electricity produces electromagnetic radiation, EMR. Trees, buildings and even human skin can shield or weaken the electric field. Its exposure can sometimes be felt because it can make skin or hair tingle. However, we cannot easily block magnetic fields. You cannot feel or avoid magnetic field. It travels through most materials in its path including humans, animals, plants, roofs, bricks, glass and even concrete. That is why most recent studies focused on health effects of magnetic fields. They are also easier to measure than the electric fields. Electromagnetic pollution is caused by electromagnetic field or EMF radiation. EMF is a combination of electric and magnetic fields that occur in our environment, both naturally and artificially. Electromagnetic radiation is an invisible health hazard spreading out of control worldwide. It is an environmental threat as dangerous as tobacco and asbestos, and as serious as toxic wastes and global warming. Artificial EMF poses a greater risk to human health, sources being power lines, mobile phone base stations, numerous modern appliances powered by electricity, batteries and motor. Exposure to high levels of human-made EMF radiation may damage brain cells and DNA, cause cancer, leukemia in children, brain tumor in adults, and adversely affect the central nervous, cardiovascular and immune systems. Some effects are linked to electromagnetic pollution like decreased testosterone levels in men, miscarriages in pregnant women, birth defects in babies, Alzheimer's disease, cataracts, depression and suicides, chronic fatigue and others (Eberhardt et al., 2008). Quite a few studies have suggested that the radiation emitted by a cell phone can interact with brain activity and behavior (Eulitz, et al., 1998; Freude, et al., 1998; Freude et al., 2000), gene expression and DNA (Koivisto et al., 2000a; Koivisto et al., 2000b; Krause et al., 2000a; Krause et al., 2000a), cell growth, proliferation and tumors (Koivisto et al., 2000a; Edelstyn et al., 2001; Lee et al., 2001; Smythe, et al., 2003; Trosic, et al., 2006; Gorlitz et al., 2005), hormones, proteins and enzymes (Trosic et al., 2004; Mashevich et al., 2003; Tice et al., 2002; Tkalec et al., 2005; Lonn et al., 2004 & 2005; Christensen, et al., 2004; Cook et al., 2003; Warren et al., 2003), but some are not support for the same (Kahn et al., 2003; Hardell et al., 2002; Auvinen et al., 2002).

World Health Organization (WHO) states that electromagnetic fields of all frequencies represent one of the most common and fastest growing environmental influences, about which anxiety and speculation are spreading. Before understanding the bio-effects of the EMF radiation or bioelectromagnetics, we have to know the basic facts related to this emerging issue. Technically, electromagnetic radiation or EMR is the energy that radiates. This energy takes the form of waves and spreads outwards from its source in concentric circles. It is very much like the pattern of ripples when you drop a pebble into calm water. EMR passes through space at the speed of light, about $3 ´ 10^8$ meters per second. It is the interaction of electric and magnetic fields in it with matter that determines EMR effects. There are many forms of electromagnetic radiation, each having different wavelengths and frequency. Light, for example, is a very high frequency EMR, while radio is lower frequency EMR. The electromagnetic spectrum includes several different classes of radiation: low frequency, radio waves, microwaves, infrared, visible light, ultraviolet light, X-rays and gamma rays. Wave frequency is what differentiates one class of radiation from another. In modern homes and offices, anything electrical like food processors, refrigerators, microwave ovens, vacuum cleaners, etc. generate electromagnetic radiation. We are all exposed to brief doses of EMR when we use a hair dryer or make a call with mobile phone. Power lines and mobile phone masts, however, are constantly operational sources of EMR. They emit EMR ceaselessly and steadily, creating a field of low-frequency radiation. Electromagnetic pollution is a term given to all the man-made electromagnetic fields (EMFs) of various frequencies, which fill our homes, workplaces and public spaces. When we call something in our environment a pollutant, we are implying that it is somehow harmful to nature and ourselves. There is a plenty of scientific evidence to support the idea that some types of electromagnetic radiation are harmful to us and nature. When we refer to electromagnetic pollution, we are generally speaking of frequencies below visible light waves. Of course, X-rays and gamma rays (which oscillate faster than the visible light) are highly dangerous, but we have not filled our homes and workplaces with these rays. We know they were dangerous, so we have been careful. We did not realize that low frequency electromagnetic radiation (including the 50/60Hz frequency commonly used for power supplies) was dangerous to our health, so we have not been very careful about that. Unfortunately, electromagnetic pollution is all around us. Some main culprits are cell phones or mobile phones, computers, electrical appliances, electronic equipment, radio and TV transmitters, microwave ovens, high and low voltage power lines, information networks, cars, motor cycles, buses, trains, planes, etc.

Almost every new invention adds to the pollution. In fact, collectively, we have all been adding to EMF pollution for over 100 years. The rate of increase is rising exponentially. Electromagnetic pollution has now reached the critical level at which it can seriously damage our health. It is true that fixing the world's EMF pollution is a big challenge. We have to avoid becoming the victims of EMF pollution. We have to find out the solution to reduce electromagnetic exposure for ourselves, our families and others around us. Before beginning the discussion over the health effects of electromagnetic, we should know the EMF factors which may be responsible for belongings. Some effects depend on the following aspects:

### EMF intensity

Exposure to strong EMFs may cause damage, even though the exposure is of short duration. In one study, researchers asked pregnant volunteers to wear a device which measured the highest intensity EMF in a 24 hour period. Results indicated that higher peak EMFs correlated with higher rates of damage.

### Cumulative EMF exposure

During a normal day a person may be exposed to EMFs of various frequencies. For example, these could arise from electric shavers and hair dryers, traveling by car, bus or train, domestic appliances such as heaters, ovens and microwaves, neon lights, household wiring, power lines, cell phone carrying and use, to name just a few of the more common sources. The combination of these exposures may overwhelm the body defenses and repair mechanisms.

### EMF exposure duration

Many studies have indicated that various health effects are only noticed after many years of exposure to electromagnetic pollution, for example, high voltage power lines, or cell phones.

### EMF transience

It could be more biologically stressful to be exposed to the fluctuating EMF which arises from an electrical appliance that cycles on and off (for example, a photocopier or printer) than a similar appliance that remains steady all day.

### *EMF frequency*

It is not yet known with certainty which electromagnetic frequencies cause which health effects, but it does appear that different frequencies cause different effects.

### *Superimposed signals*

An electromagnetic wave can be modulated in various ways to carry a signal, which might be digital or analogue. Where the wave is used for communication (e.g., radio, television, mobile telephony, etc.), it is normal to have a signal imposed on a carrier frequency. There is an evidence that in some cases the signal component can be more harmful than the carrier wave EMF itself.

## RADIATION TOXICITY

Earlier we have discussed that radiation is energy travelling through the space in the form of waves or particles. It occurs naturally and has always been around. We have evolved with it and are bombarded with it in one form or another everyday of our lives—from the earth, space and even within our own bodies. Some experts suggest a little radiation is a good thing and we all know of its uses in medical science to combat and diagnose some illnesses. Radiation is a subject most of us know very little about, especially its side effects.

Radiation toxicology is the study of adverse effects of radiation on living organisms. It is a multidisciplinary science borrowing freely from several of the basic sciences. The cytopathologic consequences of radiation exposure are similar to those induced in other types of cellular injury. Radiation induced cell changes may result in death of the organism, death of the cells, modulation of physiological activity, or cancers that have no features distinguishing them from those induced by other types of cell injury. Electromagnetic radiation is divided into non-ionizing and ionizing radiation according to the energy required to eject electrons from molecules (Sanders et al., 1986).

## IONIZING RADIATION

Ionizing radiation is a radiation that has sufficient energy to dislodge electrons from atoms, which are the basic building blocks of matter. It thus changes the chemical properties of the substance it interacts with i.e., X-rays, gamma rays, alpha particles, beta particles, cosmic rays. The term *radioactivity* generally refers to the release of ionizing radiation. More specifically, it is the property of certain unstable elements to change and emit radiation. Ionizing radiation may cause damage to the living tissues. However, different types of radiation such as alpha, beta, gamma and X-ray differ in their ability to cause tissues damage.

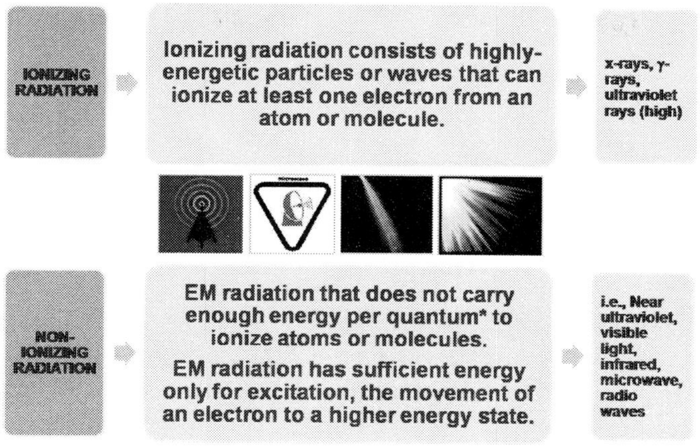


**Fig. 2.** Major classification of radiation.

## NON-IONIZING RADIATION

Non-ionizing radiation is radiation without enough energy to ionize the atoms. Examples of this radiation and devices using it are ultraviolet, visible light, lasers, infrared, microwave, radio frequency, extremely low frequency, acoustic, and ultrasound.

## NATURAL OR BACKGROUND RADIATION

Radioactive material is found throughout the nature. It occurs naturally in soil, water, and vegetation. All living things are constantly bombarded by radiation from space known as cosmic radiation. Everyone has radioactive materials inside their bodies from birth. The common background radiation level in the environment is largely due to natural resources of radiation (Shivaprashad et al., 2008). Background radiation is naturally and inevitably present in our environment. The Earth is naturally radioactive. We can find some radioactive substances like uranium and thorium in rocks such as granite, sandstone and limestone. As the rocks weather, they form soils which also contain traces of radioactive materials. Then, the soils are washed into streams and rivers. So we are exposed to background radiation directly from rocks, soils, streams and trees. Air contains two radioactive gases, radon and thoron. These gases come from uranium and thorium in Earth's crust. As they seep into air, they disperse with low concentration. But when they enter a building through floor and building materials, the concentration increases unless the building is well ventilated. Our exposure to the radiation depends on (i) the altitude above the sea level where we live, (ii) the materials used in the construction of our building, (iii) the way building is constructed and ventilated, and (iv) the composition of underlying rocks. The biggest single source of natural background radiation is cosmic radiation, coming from the sun and outer space. If you live in mountains, you get more cosmic radiation than at sea level because there is less air to act as shield. At the cruising height of a jet, you get over 100 times more radiation than at sea level. A British study (Source: www. coribarena. com) estimates that an average dope of radiation 0.25 mSv is received at sea level. We get an extra 0.01 mSv each year for every 30 meters above the sea level. For every hour spent in flight, we get an extra 0.04 mSv. Building materials made from Earth's natural resources like clay, sand and gravel used for bricks and concrete, and from wood, are also a source of natural background radiation. The radiation dose from building is 0.14 mSv, wooden house gives 0.3 mSv and a house made mainly from concrete, stone or brick gives 0.35 mSv. Food and drink: naturally radioactive materials in the Earth's crust are taken up by plants and animals. They dissolved in water too. So everything we eat and drink is also slightly radioactive. No one can avoid natural background radiation. Humans have lived with it in the atmosphere since beginning of time. Background radiation in our everyday environment does not cause widespread harm.

## ARTIFICIAL OR MAN-MADE RADIATION

Modern Scientific and technological activities also contribute to the radiation in the environment. Radiation can also be produced artificially by humans. For example, X-rays in hospitals are produced electrically without using radioactive substances. The same is true for radio waves that transmit our radio and TV programs through air. It has been established that out of the total radiation dose that the world population receives, about 96.1 per cent is from natural resources and the remaining 3.9 per cent from man-made sources (Chougankar et al., 2003). There is no data available that how much electromagnetic radiation (EMR) level in the environment of the study area has increased in the last decade. Man-made electromagnetic radiation can be harmful. The sources are: (i) underground and overhead electric power cables, (ii) cooking devices, (iii) televisions, (iv) radios, and (v) mobile phones. We see using part of electromagnetic spectrum that our eyes detect as visible light. Radiation is commonly used to diagnose ailments. Some people are treated with artificial radiation to cure disease. For example, medical and dental X-rays check hidden problems, and a multitude of products that make communication convenient such as telephones and computers are made possible with radiation. However, various sicknesses and even death can occur in people exposed to certain levels and types of radiation. Ionizing radiation is known to be carcinogenic, or cancer-causing. Large doses of ionizing radiation exposure, very much larger than the background levels, can also cause a measurable increase in cancers and leukemia in later life. Animal and plant studies indicate that it can also cause genetic mutations that affect future generations. At very high levels, radiation exposure can cause sickness and death within weeks. To measure long term radiation exposure risks, experts use a standard unit of measurement called Sieverts, or more commonly thousandths of Sieverts or millisieverts (mSv). These units measure quantities called equivalent dose to individual tissues or effective dose to the whole body.

## WIRELESS COMMUNICATIONS: NEW THREAT OF EMR

Wireless communication has played a vital role for being a progressive tool of Information Technology. In the race to adopt new technologies, we often ignore their ill effects and do not realize this. Presently people are adopting the wireless communication devices for varied purposes. They feel comfortable when they communicate with others through wireless communication devices in the air instead of wire. Most of these devices work in the radio frequency/microwave range. Radio frequency fields may cause thermal effects at high exposure levels. Mobile phone or cellular phone is a synonym of the wireless communication technology. Why is mobile phone called cellular phone or cell phone? A mobile phone system requires a number of mobile phone towers. Each mobile phone tower sits in the middle of a geographical area known as ***cell***. That is why mobile phones are also known as ***cellular*** phones or cell phones (Tiwari et al., 2007)

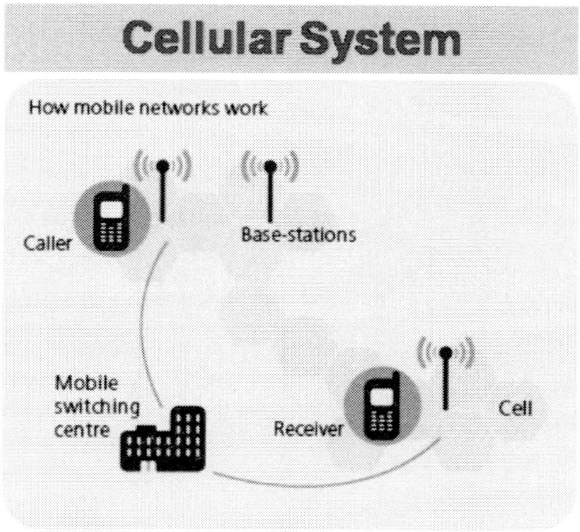


**Fig. 3.** Demonstration of cellular network how it work.

The cordless telephone handset contains a small radio transmitter and receiver that is communicated, via its antenna, with the base unit that is connected to standard telephone line. Speaking into the handset's microphone modulates the internal radio transmitter, which sends the message to the base unit by high frequency radio waves. The base unit's transmitter sends the incoming messages to the handset's receiver and the handset's internal electronics converts this radio signal into audio output in the handset's earphone. Radio frequency waves constitute a range of what is called the electromagnetic (EM) spectrum, which is divided into eight ranges according to the wave's frequency, associated energy and other characteristics. Electromagnetic wave energy is measured in electron volts (eV). The energy of all electromagnetic waves is directly proportional to their frequency.

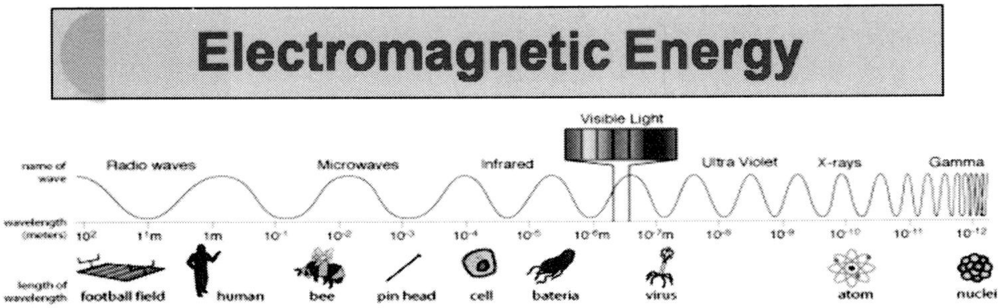


| S. No. | Frequency Spectrum | Frequency range(≈) | Corresponding Energy(≈) |
|---|---|---|---|
| 1 | Power line frequency | Few to about 30 KHz | near zero to $1.2 \times 10^{-10}$ eV |
| 2 | Radio and television broadcasting frequencies | 30 KHz to 300 KHz | $1.2 \times 10^{-10}$ eV to $1.2 \times 10^{-6}$ eV |
| 3 | Cellular telephone systems | 900 MHz | $3.5 \times 10^{-6}$ eV* |
| 4 | Microwave frequencies | 300 MHz to $3 \times 10^{11}$ Hz | $1.2 \times 10^{-6}$ eV to $1.2 \times 10^{-3}$ eV |
| 5 | Infrared frequencies | $3 \times 10^{11}$ to $4.3 \times 10^{14}$ Hz | $1.2 \times 10^{-3}$ eV to 1.8 eV |
| 6 | Visible frequencies | $4.3 \times 10^{14}$ Hz to $7.5 \times 10^{14}$ Hz | 1.8 eV to 3.1 eV |
| 7 | Ultraviolet frequencies range | $7.5 \times 10^{14}$ Hz to $3 \times 10^{17}$ Hz | 3.1 eV to 1.2 keV* |
| 8 | X-ray frequencies | $3 \times 10^{17}$ Hz to $3 \times 10^{19}$ Hz | 1.2 keV to 120 keV |

* Electromagnetic radiation with energies in below of 10 eV is called Non-ionizing radiation.

**Fig. 4.** Chart of frequency spectrum and corresponding energy.

Power line frequencies range from a few to about 30 KHz with corresponding energies from near zero to1.2 × $10^{-10}$ eV. Radio and television broadcasting frequencies range from about 30 KHz to 300 KHz with corresponding energies from 1.2 × $10^{-10}$ eV to 1.2 × $10^{-6}$ eV. Microwave frequencies range from about 300 MHz to 3 × $10^{11}$ Hz with corresponding energies from 1.2 × $10^{-6}$ eV to 1.2 × $10^{-3}$ eV. Infrared frequencies range from about 3 × $10^{11}$ to 4.3 × $10^{14}$ Hz with corresponding energies from 1.2 × $10^{-3}$ eV to 1.8 eV. This frequency is associated with radiant heat. If the exposure is too great, burning may result. Visible frequencies range from about 4.3 × $10^{14}$ Hz to 7.5 × $10^{14}$ Hz with corresponding energies from eV to 3.1 eV. Ultraviolet frequencies range from about 7.5 × $10^{14}$ Hz to 3 × $10^{17}$ Hz with corresponding energies from 3.1 eV to 1.2 keV. There is an additional hazard associated with the ultraviolet radiation above 10eV. Electromagnetic radiation with energies in excess of 10eV is called Ionizing radiation, which means that such radiation has sufficient energy to remove electrons from the outer orbits of atoms and can break the bonds between the atoms in molecules. X-ray frequencies range from about 3 × $10^{17}$ Hz to 3 × $10^{19}$ Hz with corresponding energies from 1.2 keV to 120 keV, well into the ionizing radiation range. These EM waves are produced in medical and industrial X-ray machines and also in the emissions from some radioactive substances. Gamma ray frequencies range from about $3 \times 10^{19}$ Hz on up with corresponding energies from 120 keV on up. These energies are usually measured in millions of electron volts, MeV. These waves are produced from some radioactive materials. Ionizing radiation can be damaging to tissue. For example, it can break the bonds of the DNA molecules in the nucleus of a living cell. This may lead to the cell's death or perhaps, start of a cancer. The focus here is on radio frequency (RF) waves, the energy of which is far below the 10 eV threshold radiation. For example, cellular telephone systems operate in the 800 to 900 MHz range, and these radio waves have energies of about $3.5 \times 10^{-6}$ eV. Radio waves pass through most matter, including living tissue, with very little being absorbed. It may be, therefore, absorbed in very small fraction by the living tissue. The RF energy absorbed in the tissue is converted into heat to raise the temperature of that tissue. The rise in temperature is approximately proportional to the quantity of radiation absorbed, and this, of course, will depend upon the intensity of the incident radiation. The microwave oven is a good example of the use of intense radiofrequency energy to raise the temperature and cook a toast. The intensity of electromagnetic radiation is a measure of power on a surface area and its unit is watts per square meter. Another term that needs explanation is SAR, Specific Absorption Rate, which is a measure of the rate at which the body absorbs radiofrequency energy, is measured as power incident on one unit of mass, W/kg.

## MECHANISM INVOLVED IN HUMAN BODY INTERACTION TO CELLULAR PHONE'S EMR

Electric currents exist naturally in the human body and are an essential part of its physiology. All nerves relay their signals by transmitting electric impulses. The effects of external electromagnetic fields on the human body depend on the

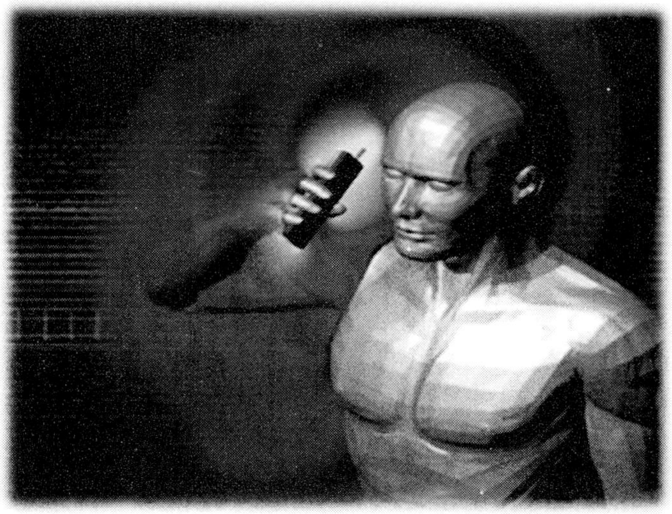

**Fig. 5.** Imagination of Human body interaction to cellular Phone's EMR. Source: http: luismmx.bbgspot.com/2012/04/cellphone-radiation

electromagnetic field frequency and their magnitude or strength. The frequency simply describes the number of oscillations or cycles per second. At low frequencies, electromagnetic field passes through the body, while at radiofrequencies, the fields are partially absorbed and penetrate only a short depth into the tissue. Low frequency electric fields influence the distribution of electric charges at the surface of conducting tissues and cause electric current to flow in the body. Low frequency magnetic field induces circulating currents within the human body. The strength of these induced currents depends on the intensity of the outside magnetic fields and the size of the loop through which the current flows. When sufficiently large, these currents may cause simulation of nerves and muscles. At radiofrequencies, the fields penetrate a short distance into the human body. The energy of these fields is absorbed and transformed into the moving molecules, which result in the rise of temperature. The levels of radiofrequency field to which people are normally exposed in our environment are lower to produce significant heating affect. Some scientists previously believed that the only way that EM radiation could produce damaging effects was if the radiation was sufficiently intense to cause a heating effect on the tissues. This theory has now been roundly discredited by many studies in which biological effects have been observed at intensities far too small to cause any measurable heating effect. The mechanisms by which electromagnetic radiation may affect disease processes are not yet fully understood, but promising candidates include the follows:

### DNA Damage

Our cells have mechanisms to repair DNA damage to a limited extent, but it appears that EMFs can overwhelm these mechanisms. Damaged DNA is implicated in several disease processes, including various types of cancer. German research group Verum studied the effect of radiation on human and animal cells. After being exposed to

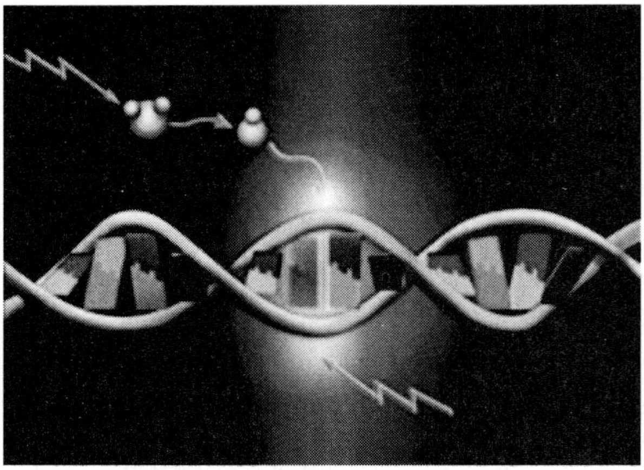

**Fig. 6.** Explicit image of DNA damage.

cell phone frequencies the cells showed increased breaks in their DNA. These DNA breaks could not always be repaired by the cells. The damage would therefore be passed on to future cells which could predispose them to becoming cancerous (USA Today, 2004).

### Interference with Melatonin Production

Electromagnetic radiation appears to interfere with the production of melatonin, a hormone that is produced normally in the body. Low melatonin levels have already been linked to several diseases, including cancers. Recent research indicates that seratonin production can also be affected by EMFs.

### Interference with Cellular Communication

Our body cells communicate internally and externally by means of electrical signals. These signals can be altered by EMF radiation (which generates electrical currents within the body) causing changes in both cellular activity and cellular structures.

## BIOLOGICAL EFFECTS OF EMR

There are many ways to classify health effects of electromagnetic radiation. Here we distinguish between biological changes and pathological effects substantiated by epidemiological studies. Some of the clinicians and scientists are calling for severe restrictions on wireless technology because they are seeing such a dramatic increase in certain diseases and symptoms in their patients and believe that the association of cellular phone EMR is linked directly or indirectly to the diseases including attention deficit disorder, extreme fluctuations in blood pressure, heart arrhythmias, heart attacks and strokes, Alzheimer's Diseases, epilepsy, leukemia, and brain tumors. In addition, this association is also linked to the symptoms such as headaches, migraines, chronic fatigue, agitation, sleep disorders, tinnitus, nervous and connective tissue pains of unexplained origin, and susceptibility to infection. Now, this is a big issue of the possibilities to the cell phone EMR to health hazards mainly through cancer, genetic damage, brain and other neurological problems, birth defect, increased blood pressure, impaired learning and short-term memory, disturbed sleep, headache, stress, etc. But there are controversies about the adverse effects of cellphone radiations on human health in a few research studies (Kundi, 2010). The list of EMF health effects presented here is in fact just a small sample from the mountain of research that has been reported in scientific literature. Here are some of the biological changes caused by electromagnetic radiation, as observed and reported in various studies. In Figure 7, the absorption of radiofrequency radiation is clearly demonstrated. RFR absorption into the brain was greater in adult than the child (Dubey et al., 2010).

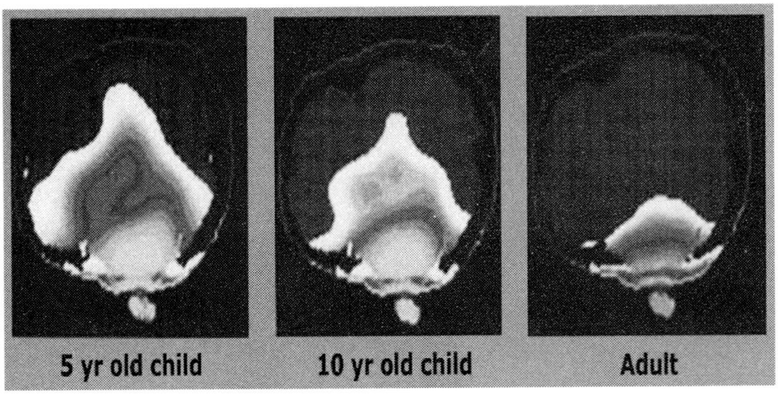


**Fig. 7.** Illustration of absorption of RFR into the brain (source: Dubey et al., 2010).

### Effects on Human Skin Cells

In the most comprehensive analysis of its type published to date, scientists at Pacific Northwest National Laboratory have found that exposing human skin tissue cells, or fibroblasts, to low doses of ionizing radiation regulates phosphorylation of proteins that are involved in a wide range of biological processes. In short, the body can sense low doses of radiation and activate the cell signaling pathways needed to respond to any induced cellular damage (PNNL, 2010).

### Excited Brain Cells

Researchers from Fatebenefratelli Hospital in Isola Tiberina found that the electromagnetic field emitted by cell phones can cause some cells in the brain's cortex (adjacent to the side of phone use) to become excited for about an hour, while others become inhibited (Ferreri et al., 2006).

### Brain Cell Damage

A study of the effects of cell phone frequencies (applied at non-thermal intensity) on rat brains showed damage to the neurons (brain cells) in various brain parts, including the cortex, hippocampus and basal ganglia (Salford et al., 2003).

### Lymphatic Cancer and Bone Marrow Cancer

Researchers from the University of Tasmania and the University of Bristol studied the records of 850 patients who had been diagnosed with lymphatic and bone marrow cancers. They concluded that people living within 300 metres of a high voltage power line for extended periods (particularly in childhood) were up to five times more likely to develop these diseases later in life (Lowenthal et al., 2007).

## MAJOR STUDY AND ANALYSIS RELATED TO CELLULAR PHONE RADIATION

In present scenario, a number of studies show possible health implications through the extensive use of cellular phones. A number of significant studies are described here, which have been conducted on humans and animal models.

### STUDY 1: Occupational Exposure to Radio Frequency/Microwave Radiation and the Risk of Brain Tumors

A population-based, case-control study including 381 meningioma cases, 366 glioma cases, and 1,494 controls aged 30-69 years was performed in three German regions (Dubey et al., 2010). An exposure matrix for occupational activity was constructed by using information on RF/MW-EMF exposure collected in a computer-assisted personal interview. "High" exposure was defined as an occupational exposure that may exceed the RF/MW-EMF exposure limits for the general public recommended by the International Commission on Non-Ionizing Radiation Protection. Multiple conditional logistic regressions were performed separately for glioma and meningioma. No significant association between occupational exposure to RF/MW-EMF and brain tumors was found. For glioma, the adjusted odds ratio for highly exposed persons compared with persons not highly exposed was 1.21 (95% confidence interval: 0.69, 2.13); for meningioma, it was 1.34 (95% confidence interval: 0.64, 2.81).

### STUDY 2: Prenatal and Postnatal Exposure to Cell Phone Use and Behavioral Problems in Children

The association between prenatal and postnatal exposure to cell phones and behavioral problems in young children was examined (Divan et al., 2008). Mothers were recruited to the Danish National Birth Cohort early in pregnancy. When the children of those pregnancies reached 7 years of age in 2005 and 2006, mothers were asked to complete a questionnaire regarding the current health and behavioral status of children, as well as past exposure to cell phone use. Mothers evaluated the child's behavior problems using the Strength and Difficulties Questionnaire. Mothers of 13,159 children completed the follow-up questionnaire reporting their use of cell phones during pregnancy as well as current cell phone use by the child. Greater odds ratios for behavioral problems were observed for children who had possible prenatal or postnatal exposure to cell phone use. After adjustment for potential confounders, the odds ratio for a higher overall behavioral problems score was 1.80 (95% confidence interval = 1.45-2.23) in children with both prenatal and postnatal exposure to cell phones. Exposure to cell phones prenatally-and, to a lesser degree, postnatally-was associated with behavioral difficulties such as emotional and hyperactivity problems around the age of school entry. These associations may be noncausal and may be due to unmeasured confounding.

### STUDY 3: Enhancement of Chemically Induced Reactive Oxygen Species Production and DNA Damage in Human SH-SY5Y Neuroblastoma Cells by 872MHz RFR

The objective of the study was to investigate the effects of 872MHz radiofrequency (RF) radiation on intracellular reactive oxygen species (ROS) production and DNA damage at a relatively high SAR value (5W/kg) (Luukkonen et al., 2009). The experiments also involved combined exposure to RF radiation and menadione, a chemical inducing intracellular ROS production and DNA damage. The production of ROS was measured using the fluorescent probe dichlorofluorescein, and DNA damage was evaluated by the Comet assay. Human SH-SY5Y neuroblastoma cells were exposed to RF radiation for 1h with or without menadione. Control cultures were sham exposed. Both continuous waves (CW) and a pulsed signal similar to that used in global system for mobile communications (GSM) mobile phones were used. Exposure to the CW RF radiation increased DNA breakage ($p < 0.01$) in comparison to the cells exposed only to menadione. Comparison of the same groups also showed that ROS level was higher in cells exposed to CW RF radiation at 30 and 60min after the end of exposure ($p < 0.05$ and $p < 0.01$, respectively). No effects of the GSM signal were seen on either ROS production or DNA damage. The results of the present study suggest that 872MHz CW RF radiation at 5W/kg might enhance chemically induced ROS production and thus cause secondary DNA damage.

### STUDY 4: To Assess the Human Sperm Motility through the EM Radiation Emitted by Cellular Phone (GSM 900) in Normal Environmental Conditions

A total number of 27 males were enrolled in the study (Osman et al., 2006). The semen sample obtained from each participant was divided equally into two parts. One of the specimens was exposed to EMR emitted by an activated 900 MHz cellular phone, whereas the other was not. The concentration and motility of the

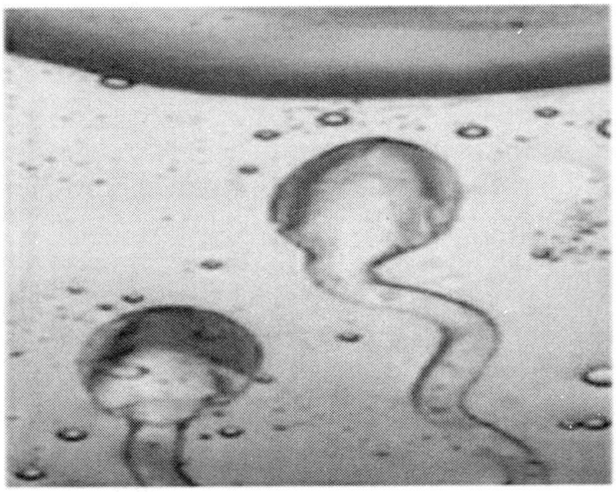

**Fig. 8.** Illustration of human sperm motility.

specimens were compared to analyze the effects of EMR. Assessment of sperm movement in all specimens was performed using four criteria: (A) rapid progressive, (B) slow progressive, (C) nonprogressive, (D) no motility. Statistically significant changes were observed in the rapid progressive, slow progressive and no-motility categories of sperm movement. EMR exposure caused a subtle decrease in the rapid progressive and slow progressive sperm movement. It also caused an increase in the no-motility category of sperm movement. There was no statistically significant difference in the sperm concentration between two groups. These data suggest that EMR emitted by cellular phone influences human sperm motility. In addition to these acute adverse effects of EMR on sperm motility, long-term EMR exposure may lead to behavioral or structural changes of the male germ cell. These effects may be observed later in life, and they are to be investigated more seriously.

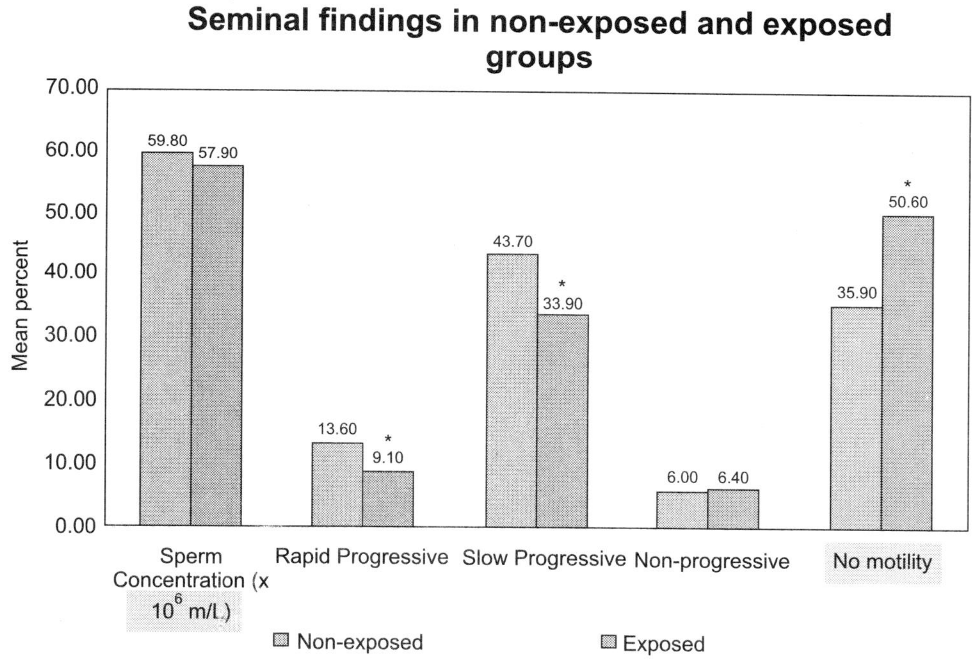


**Fig. 9.** Graphical representation of outputs of study 4.

## STUDY 5: Prevalence of Headache among Extensive and Normal Cellular Phone Users

'Headache' is the most prevalent symptom among cellular phone users as has been found in a survey study in Singapore (Chia et al., 2000). Study shows significant increase in the prevalence of headache with increasing duration of cellular phone usage. Another study in Turkey (Balicki et al., 2005) also supported for the same.

We conducted a survey among the 188 cellular phone users in Lucknow city of India to investigate the possibility of cellular phone usage and its association with headache. All subjects (77.7% male and 22.3% female) were enquired about general profile through a well designed questionnaire (age, sex, income, education etc.), calls detail and general health (disease, never smoker or smoker). The subjects belonged to 14 - 62 years age range (mean age ± SD; 29.1 ± 9.5). Study revealed that 11.2% subjects exhibited 'headache' symptom reporters, which are lower than the studies mentioned in the literature. Further the prevalence of headache was compared in into high user (HU>500 hours use) and low user (LU<500 hours use). Headache was reported by 7.01% subjects in LU, whereas 12.97% subjects reported in HU. Headache was reported by 7.01% subjects in LU, whereas 12.97% subjects reported in HU. Study concluded that the symptoms for headache were found higher among HUs in comparison to LUs (Kumar et al., 2009). It means possibility of headache symptoms may increase in high users. They utilize cellular phone above 30 minutes per day upto three years or more.

### STUDY 6: Neurobehavioral Effects in Mice through the Acute Exposure of Cellular Irradiation

In this study we assessed the spontaneous locomotors activity (SLA) of the mice after the acute exposure of cellular telephone (CDMA) (Kumar et al., 2010). 10 mice 30-days-old (average weight 30gm) were grouped as control and EMR and they were habitualised for one week prior to beginning the experiment in well-designed circular cellphone exposure (CCPE) cage. CCPE cage was specially designed for EMR exposure through the cellular phone over the mice in the natural environment, and it was placed inside the exposure room where other electric appliances did not exist. In the cage, cellular phone was placed in the center and mice were allowed to move around the device within 10 cm. radius. 3 hours/day

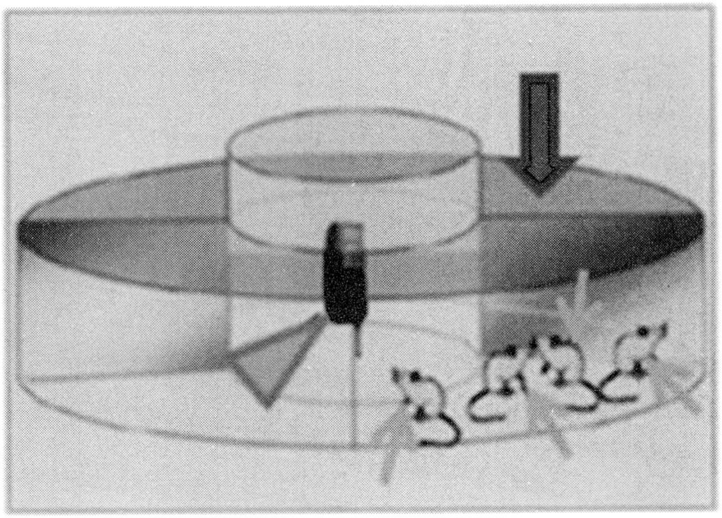

**Fig. 10.** Explicit image of CCPE cage.

exposure was applied to the mice of EMR group by the cellular phone in CCPE cage. Locomotor activity was assessed using computerized Optovarimax animal activity monitor, following the Agrawal et al (Agrawal et al., 2004). In this study, sub parameters such as distance traveled, time resting, time moving and stereotypic time were measured with the help of the activity monitor. No significant changes were found in the EMR exposed mice in comparison to control mice.

## REGULATORY RESPONSIBILITIES TO MINIMIZE THE EMR EXPOSURE

Due to idespread concerns of the public, a number of expert groups and health agencies around the world have reviewed the scientific literature related to possible health effects of radiofrequency energy used by wireless communication devices. Currently many International efforts are underway for final assessment on health effects associated with cell phone radiation. World Health Organization recommended using precautionary principles with the purposes of decreasing the risk. Some countries have developed a precautionary approach regarding cell phone use among children and have recommended that children should use these devices as little as possible (NRPB, 2004). Russian National Committee of Non-Ionizing Radiation Protection (RNCNIRP) recommended to limit the use of Cell phones by children and teenagers under 16 years old (MMF 2003). The Specific Absorption Rate (SAR), which measures the rate at which the human body absorbs radiation, and its maximum levels for modern handsets have been set by governmental regulatory agencies in many countries. The possible hazards of radiofrequency radiation energy have been debated for many years, and exposures limits for radiofrequency energy have been in place in the United States since the 1960s, and most other countries have adopted similar limits.

The International Expert Group on Mobile Phones (IEGMP) was formed in 1999 by the United Kingdom Government to analyze the possible health impacts of mobile phones. The committee's report concluded: there is an evidence that at the frequencies used in the cell phone technology, children will absorb more energy per kilogram of body weight from an external electromagnetic field than adults. A one-year-old could absorb around double, and a five-year-old around 60% more than an adult. Committee also recommends that the mobile phone industry should refrain from promoting the use of mobile phones by children. The National Radiological Protection Board (NRPB), merged with Health Protection Agency (HPA), in April, 2005, concluded that the available evidence does not suggest that there are adverse health effects from exposures to radiofrequency fields below the guideline levels. The Food and Drug Administration (FDA) is the principal agency that follows the issue of possible health effects of radiofrequency energy. According to FDA, there is no hard evidence of adverse effects on the general public. With regards to the safety and use of cell phones by children, the scientific evidence does not show a danger to users of wireless communication devices including children (FDA, 2005). The International Commission on Non-Ionizing Radiation Protection (ICNIRP) establishes exposure guidelines for radiofrequency energy, which have been adopted by many countries throughout the world. Five members of ICNIRP have given epidemiological review. They concluded that the results of epidemiologic studies give no consistent or convincing evidence of a fundamental relation between radiofrequency exposure and any adverse health effects. The Health Council of Netherlands (HCN) published a series of reports on the health effects of electromagnetic fields on human health starting in 2001. According to report of HCN 2002 concludes – It is unlikely from a development point of view that major changes in brain sensitivity to electromagnetic fields still occur after the second year of life. There is no reason to recommend that mobile phone use by children should be limited as far as possible (Gezond-heidsraad et al., 2002). The Australian Radiation Safety and Nuclear safety Agency (ARPANSA) is an advisory body on the topic of possible health and environmental effects of radiation. Agency states-At present, there is insufficient evidence in the science to substantiate the hypothesis that children may be more vulnerable to radiofrequency emissions from mobile phones than adults.

Now at the national level, responsible authorities are also taking this important issue seriously. They are trying to minimize the possible effects from the cellular phone over dose EMR exposure. Union Health Ministry of India & Indian Council of Medical Research (ICMR) are now approaching for quality research to look at the effect of the cellular phone exposure on neurological disorders like cognitive impairment, depression and sleep-related problems (Lantow et al., 2006). The Uttar Pradesh Council of Science and Technology (UPCST) has begun the study at Babasaheb Bhimrao Ambedkar Central University, Lucknow, to envisage the symptoms and sensations due to cellular telephone usage among the urban and rural population of Uttar Pradesh.

## SUGGESTIVE SAFETY MEASURES—HOW TO REDUCE CELL PHONE HEALTH HAZARD

Expert groups have suggested that the radiation from mobile phones greatly exceeds the safe levels even at normal use; even a short call can have harmful effects. Some of the information available support that microwave radiation from mobile and cordless phones causes neurological problems and other health disturbances (Dubey et al., 2010). The following tips may play a role to reduce exposure to electromagnetic energy from cell phone, but these steps are only for suggestions and not recommended by the authorities.

a) Limit the use of cell phones to essential calls and keep calls short.
b) Children should be allowed to use a cell phone only in cases of emergency. Because of their developing skulls, the radiation can penetrate much more deeply.
c) Wear an air tube headset (not regular wired headset). The regular wired headset has been found to intensify radiation into the ear canal. The wire not only transmits the radiation from the cell phone but also serves as an antenna, attracting EMFs from the surroundings.
d) The best choice is using speaker-phone handset.

e) Do not put the cell phone in a pocket or a belt while in use or while it is on. The body tissue in the lower body area has good conductivity and absorbs radiation more quickly than the head. One study shows that men who wear cell phones near their groin could have their sperm count dropped by as much as 30%.
f) Moving your phone 20cm away from your head reduces radiation.
g) If using the phone without a headset, wait for the call to connect before placing the phone next to the ear.
h) Do not use the cell phone in enclosed metal spaces such as vehicles or elevators, where devices may use more power to establish connection.
i) Do not make a call when the signal strength is 1 bar or less, which means the phone must work harder to establish a connection.
j) Purchase a phone with a low SAR. Most phones have a SAR level listed in its instruction manual. The SAR level is a way of measuring the quantity of RF energy that is absorbed by the body.
k) Use a scientifically validated EMF protection device. There are advanced technologies available nowadays that strengthen the bioenergy field and immune system against the effects of EMF.
l) Use text instead of talk.
m) Have a preference to use landlines.
n) Keep cell phone off most of the time. Let people leave messages and then call them back from a landline.
o) Limit the use of cell phones in rural areas.

## REFERENCES

Agrawal, A.K., Chaturvedi, R.K., Shukla, S., Seth, K., Chauhan, S., Ahmad, A. and Seth, P.K. (2004). Restorative potential of dopaminergic grafts in presence of antioxidants in rat model of Parkinson's disease. *J. Chem. Neuroanat.* **28:** 253-264.

Augner, C., Florian, M., Pauser, G., Oberfeld, G. and Hacker, G. W. (2009).GSM base stations: short-term effects on well-being. *Bioelectromagnetics,* **30(1):** 73-80.

Auvinen, M., Hietanen, R., Luukkonen, R. and Koskela, S. (2002). Brain tumors and salivary gland cancers among cellular telephone users. *Epidemiology,* **13:** 356-359.

Balikci, K., Ozcan, I.C., Balik, D.T. and Balik, H.H. (2005). A survey study on some neurological symptoms and sensations experienced by long-term users of mobile phones. *Pathology Biology,* **53(1):** 30-34.

Chia, S.E., Chia, H.P. and Tan, J.S. (2000). Prevalence of headache among handheld cellular telephone users in Singapore: A Community Study. *Environmental Health Perspectives,* **108(11):** 1059-1062.

Chougankar, M.P., Eppen, K.P. and Ramchandran, T.V. (2003). Profiles of doses to population living in the high background radiation areas in Kerala. *J. Environ. Radioact,* **71:** 275-297.

Christensen, H.C., Schüz, J., Kosteljanetz, M., Poulsen, H.S., Thomsen, J. and Johansen, J. (2004). Cellular telephone use and risk of acoustic neuroma. *Am. J. Epidemiol,* **159:** 277-283.

Cook, Woodward, A., Pearce, N. and Marshall, C. (2003). Cellular telephone use and time trends for brain, head and neck tumours. *NZ Med. J.* **116:** 1175, U457.

Divan, H.A., Kheifets, L., Obel, C. and Olsen, J. (2008). Prenatal and postnatal exposure to cell phone use and behavioral problems in children. *Epidemiology.* **19(4):** 523-529.

Dubey, R.B., Hanmandlu, M. and Gupta, S.K. (2010). Risk of brain tumors from wireless phone use. *J. Computer Assisted Tomography,* **34(6):** 799-807.

Eberhardt, J.L., Persson, B.R., Brun, A.E., Salford, L.G. and Malmgren, L.O. (2008). Blood-brain barrier permeability and nerve cell damage in rat brain 14 and 28 days after exposure to microwaves from GSM mobile phones. *Electromagn. Biol. Med.* **27(3):** 215-229.

Edelstyn, N. and Oldershaw, A. (2001). The acute effects of exposure to the electromagnetic field emitted by mobile phones on human attention. *NeuroReport.* **13:** 119-121.

Eulitz, C., Ullsperger, P., Freude, G. and Elbert, T. (1998). Mobile phones modulate response patterns of human brain activity. *Neuro Report,* **9:** 3229-3232.

FDA. (2005). FDA response to NRPB report on mobile phones and health. Cell phone facts-consumer information on wireless phone. *Food and Drug Administration.*

Ferreri, F., Curcio, G., Pasqualetti, P., Gennaro, L.D., Tech, R.F. and Rossini, P.M. (2006). Mobile phone emissions and human brain excitability. *Annals of Neurology,* **60(2):** 188-196.

Freude, G., Ullsperger, P., Eggert, S. and Ruppe, I. (1998). Microwaves emitted by cellular phones effects humans slow brain potentials. *Bioelectromagnetics,* **19:** 384-387.

Freude, G., Ullsperger, P., Eggert, S. and Ruppe, I. (2000). Microwaves emitted by cellular telephones affect human slow brain potentials. *Eur. J. Appl. Physiol.* **81:** 18-27.

Gezond-heidsraad. (2002) Mobile telephones: an evaluation of health effects. The Hague: Health Council of Netherlands, Publication no. 2002/01E

Gorlitz, B.D., Muller, M., Ebert, S., Hecker, H., Kuster, N. and Dasenbrock, C. (2005). Effects of 1-Week and 6-Week Exposure to GSM/DCS Radiofrequency Radiation on Micronucleus Formation in B6C3F1. Mice. *Radiat Res.* **164(4):** 431-439.

Hardell, L., Hallquist, A., Hansson, M.K., Carlberg, M., Pahlson, A. and Lilja, A. (2002). Cellular and cordless telephones and the risk for brain tumours. *Europ. J. Cancer Prevent,* **11:** 377-386.

Kahn, O'Brien, D.F., Kelly, P., Phillips, J.P., Rawluk, D., Bolger, C. and Pidgeon, C.N. (2003). The anatomical distribution of cerebral gliomas in mobile phone users. *Ir. Med. J.* **96(8):** 240-242.

Koivisto, M., Krause, C., Revonsuo, A., Laine, M. and Hamalainen, H. (2000b). The effects of electromagnetic field emitted by GSM phones on working memory. *NeuroReport,* **11:** 1641-1643.

Koivisto, M., Revonsuo, A., Krause, C., Haarala, C., Sillanmaki, L., Laine, M. and Hamalainen, H. (2000a). Effects of 902 MHz electromagnetic field emitted by cellular telephones on response times in humans. *NeuroReport,* **11:** 411-415.

Krause, C., Sillanmaki, L., Koivisto, M., Haggqvist, A., Saarela, C., Revonsuo, A., Laine, M. and Hamalainen, H. (2000a). Effects of electromagnetic field emitted by cellular phones on the EEG during a memory task. *NeuroReport,* **11:** 761-764.

Krause, C., Sillanmaki, L., Koivisto, M., Haggqvist, A., Saarela, C., Revonsuo, A., Laine, M. and Hamalainen, H. (2000b). Effects of electromagnetic fields emitted by cellular phones on the electroencephalogram during a visual working memory task. *Int. J. Radiat. Biol.* **76:** 1659-1667.

Kumar, N., Khan, R.A. and Sharma, V.P. (2010). Neurobehavioral effects in mice through the acute exposure of cellular irradiation. *J. Neurochemistry,* **115(1):** 76.

Kumar, N., Sharma, V.P., Mathur, N., Khan, M.Y. and Khan, R.A. (2009). Prevalence of Headache among extensive and normal cellular phone users. *J. Neurochemistry,* **110(2):** 228.

Kundi, M. (2010). The controversy about a possible relationship between mobile phone use and cancer. *Cien saude Colet.* **15(5):** 2415-2430.

Lantow, M. Schuderer, J., Hartwig, C. and Simko, M. (2006). Free Radical Release and HSP70 Expression in Two Human Immune-Relevant Cell Lines after Exposure to 1800 MHz RFR. *Radiat Res.* **165(1):** 88-94.

Lee, T.M.C., Ho, S.M.Y., Tsang, L.Y.H., Yang, S.Y.C., Li, L.S.W. and Chan, C.C.H. (2001). Effect on human attention of exposure to the electromagnetic field emitted by mobile phones. *NeuroReport.* **12:** 729-731.

Lonn, S., Ahlbom, A., Hall, P. and Feychting, M. (2004). Mobile phone use and the risk of acoustic neuroma. *Epidemiology,* **15(6):** 653-659.

Lonn, S., Ahlbom, A., Hall, P. and Feychting, M. (2005). Long-term mobile phone use and brain tumor risk. *Am. J. Epidemiol.* **161(6):** 526-535.

Lowenthal, R.M., Tuck, D.M. and Bray, I.C. (2007). Residential exposure to electric power transmission lines and risk of lymphoproliferative and myeloproliferative disorders: a case-control study. *Internal Medicine J.,* **37(9):** 614-619.

Luukkonen, J., Hakulinen, P., MŠki-Paakkanen, J., Juutilainen, J. and Naarala, J. (2009). Enhancement of chemically induced reactive oxygen species production and DNA damage in human SH-SY5Y neuroblastoma cells by 872MHz RFR. *Mutat Res.* **662(1-2):** 54-58.

Mashevich, M., Folkman, D., Kesar, A., Barbul, A., Korenstein, R., Jerby, E. and Avivi, L. (2003). Exposure of human peripheral blood lymphocytes to electromagnetic fields associated with cellular phones leads to chromosomal instability. *Bioelectromagnetics,* **24:** 82-90.

MMF. (2003). Unofficial translation by Mobile Manufacturers Forum of the Russian standard, Hygienic requirements for the sitting and operation of land mobile radio communication equipment. SanPiN. *2.1.8/2.24,* 1190-1203.

NRPB. (2004). Mobile phones and Health 2004. National Radiological Protection Board, **15:** 5.

Osman, E., Emin O., Ibrahim, Y., Tayfun, K., Emin, A., Gokhan, K., Hasan, C.I., Mehmet, K.I. and Ahmet, F.P. (2006). Effects of electromagnetic radiation from a cellular phone on human sperm motility: An *In vitro* Study. *Archives of Medical Research,* **37(7):** 840-843.

PNNL. (2010). Pacific Northwest National Laboratory (U.S. Department of Energy's), Richland, Washington. Biological Sciences Division Research Highlights. December 10. (http://www.pnl.gov/science/highlights/highlight.asp?id=875)

Salford, L.G., Brun, A.E., Eberhardt, J.L., Malmgren, L. and Persson, B.R.R. (2003). Nerve cell damage in mammalian brain after exposure to microwaves from GSM Mobile Phones. *Environmental Health Perspectives,* **111(7):** 881-883.

Sanders, L. (1986). Toxicological aspects of energy production Radiological health. Battelle Press. *Columbus Ohio.* 253-284.

Shiva Prasad, N.G., Nagaiah, N.G.V., Ashok, and Karunakara, N. (2008). Concentrations of 226Ra, 232Th and 40K in the Soils of Bangalore Region, India. *Health Physics – A J. Radiation Safety,* **34(3):** 264-271.

Smythe, J.W. and Costall, B. (2003). Mobile phone use facilitates memory in male, but no female subjects. *NeuroReport.* **14:** 243-246.

Tice, R.R., Hook, G.G., Donner, M., McRee, D.I. and Guy, A.W. (2002). Genotoxicity of radiofrequency signals. Investigation of DNA damage and micronuclei induction in cultured human blood cells. *Bioelectromagnetics,* **23:** 113-126.

Tiwari, N., Shukla, K., Ojha, C.S., Pant, A.B. and Sharma, V.P. (2007). Cell Shocked. Science Reporter - publication of National Institute of Science communication And Information Resources (NISCAIR) CSIR. **44(1):** 1-5.

Tkalec, M., Malaric, K. and Pevalek-Kozlina, B. (2005). Influence of 400, 900, and 1900 MHz electromagnetic fields on Lemna minor growth and peroxidase activity. *Bioelectromagnetics,* **26(3):** 185-193.

Trosic, Busljeta, I. and Modlic, B. (2004). Investigation of the genotoxic effect of microwave irradiation in rat bone marrow cells: *in vivo* exposure. *Mutagenesis,* **19(5):** 361-364.

Trosic, I., Busljeta. (2006). Erythropoietic dynamic equilibrium in rats maintained after microwave irradiation. *Exp Toxicol Pathol.* **57(3):** 247-251.

USA Today. (2004). Mobile phone radiation harms DNA, new study finds. December 21. (http://www.usatoday.com/tech/wireless/phones/2004-12-21-mobilerads_x.htm)

Warren, H.G., Prevatt, A.A., Daly, K.A. and Antonelli, P.J. (2003). Cellular telephone use and risk of intratemporal facial nerve tumor. *Laryngoscope,* **113(4):** 663-667.

## SUPPORTIVE WEBSITES

1. www.newscientist.com
2. www.alternative-magnetic-therapy.com
3. www.pnl.gov
4. www.pubmed.com
5. www.google.com
6. www.sciencedirect.com
7. www.emfacts.com
8. www.bioelectromagnetics.com

# 11

# Antioxidant Enzymes: A Mode of Defence Mechanism during the Process of Phytoremediation of Heavy Metals

ANWESHA M. BHADURI AND M.H. FULEKAR

*Environmental Biotechnology Laboratory, Department of Life Sciences, University of Mumbai, Mumbai-400098*

*School of Environment and Sustainable Development, Central University of Gujarat, Gandhinagar, Gujarat, India*

**ABSTRACT**

Phytoremediation is a cost effective, environment-friendly, aesthetically pleasing approach for remediation of heavy metals. In the process of phytoremediation plants experience various stresses resulting due to toxic effects of heavy metals which leads to overproduction of reactive oxygen species (ROS) in plants ultimately resulting in oxidative stress. These, in turn differentially affect other plant processes that lead to loss of cellular homeostasis. Plants possess homeostatic mechanisms that allow them to keep correct concentrations of essential metal ions in cellular compartments and to minimize the damaging effects of an excess of nonessential ones by mechanisms of stress avoidance. A common feature to combat stress factors is synchronized function of efficient enzymatic and non-enzymatic antioxidant defense systems that helps alleviating cellular damage by limiting ROS. Thus, careful investigation of the mechanism of tolerance of heavy metal at physiological and genetic level is essential since the use of naturally occurring metal tolerant plants which will accelerate the process of transmitting this technology from experimental place to field and will result in developing a better phytoremediation technique.

**Keywords :** Heavy Metals, Phytoremediation, Antioxidant Enzymes, Reactive Oxygen Species (ROS).

## INTRODUCTION

The outset of the industrial revolution resulted in drastic environmental changes mainly originating from point sources such as emission, effluents and solid discharge from industries, vehicle exhaust and metals from smelting and mining, and non-point sources such as soluble salts (natural and artificial), use of insecticides/ pesticides, disposal of industrial and municipal wastes in agriculture, and excessive use of fertilizers (McGrath et al., 2001; Schalscha and Ahumada, 1998). At the outset of industrial revolution the amount of wood, coal and other fuels that were being burnt increased massively. The resulting pollutants although natural products were provided in excess to natural rates of formation which in combination with their innate recalcitrance to degradation meant that these compounds accumulated in environnment (Singer et al., 2003). Eventually these pollutants are posing an ever increasing and unremitting stress in all forms of life as they affect the quality of pedosphere, hydrosphere, atmosphere, lithosphere and biosphere. Although each source of contamination has its own damaging effects on plants, animals and ultimately on human health, but those that add heavy metals to soils and water are of serious concern due to their persistence in the environment as unlike many substance they are not biodegradable, but are only transformed from one oxidation state or organic complex to another (Garbisu and Alkorta 2001; Gisbert et al. 2003). Currently attempts are made to remediate the heavy metals with conventional remediation technologies such as solidification and stabilization; soil flushing; electro kinetics; chemical reduction/oxidation; soil washing; low temperature thermal desorption; incineration; vitrification; pneumatic fracturing; excavation/ retrieval; and landfill disposal. However, they experience different efficiencies for different metals and they can be very expensive especially if large volumes, low metal concentration and high standards of cleaning are required and may have side effects which ultimately will fail to uproot the main concern of heavy metal removal. Moreover these technologies do not eradicate the problem; they merely transfers it to the future generations. Evidently, there is an urgent need for alternative, affordable and efficient methods to clean-up heavily contaminated industrial areas. Thus the potential role of bioremediation, particularly by higher plants, has achieved significant interest for detoxifying the pollutants.

Plant-based bioremediation technologies for cleaning polluted soils and waters have been collectively termed as phytoremediation. This refers to the use of green plants and their associated micro biota for in-situ treatment of contaminated soil, sediments and ground water. Biologically-based remediation strategies, including phytoremediation where plants are used to extract, sequester, and/or detoxify pollutant are gaining interest in scientific community as one of the successful, non-intrusive, aesthetically pleasing, socially accepted technology to remediate polluted soils and water bodies (Garbisu et al 2002). Moreover it has been estimated to be four to 1000 times cheaper, on a per volume basis, than current non-biological technologies. (Sadowsky, 1999). It is not only efficient in remediating high concentrations of heavy metal but also a broad range of contaminants including numerous metals and radionuclides, various organic compounds such as chlorinated solvents, BTEX, PCBs, PAHs, pesticides/insecticides, explosives, nutrients, and surfactants. (Miller, 1996).

Numerous plant species have been identified and tested for their traits and mechanisms of metal uptake (Lone et al. 2008). Plants are unique organisms equipped with remarkable metabolic and absorption capabilities, as well as transport systems that can take up nutrients or contaminants selectively from the growth matrix, soil or water, and it has been proved in many studies that different plants have different abilities to uptake and withstand high levels of pollutants. This can be attributed to the fact that higher plants survive in a constantly fluctuating environment, which has driven the evolution of a highly flexible metabolism and growth (basic growth curve of slow-fast-slow) and development necessary for their sessile lifestyle (Pitzschke and Hirt,2006,Liu and Bush 2006, Buchanan and Luan 2005, Kwon et al 2006). Therefore, some of the plant species developed tolerant races that can survive and thrive on metal concentrated environment seemingly by adapting mechanisms that may also be involved in the general homeostasis and constitutive tolerance to essential metal ions found in all plants (Hall, 2002). The stress incurred as a result of excessive amount of any heavy metals in plants tissues shows toxicity symptoms which may be attributed to a range of interactions at the cellular or molecular level. Plants have a range of potential mechanisms at the cellular level that might be involved in detoxification and tolerance to heavy metal stress. The cellular state is defined as the total active level that regulates cells to adapt to their surroundings in a narrow sense (Zeng and Wang, 2001; Halliwell, 2006; Coupe et al. 2006). Redox state is the important part of the above cellular state and one of the most important indicators for evaluating the stress experienced by plants. (Zeng and Wang, 2001; Coupe et al. 2006; Fey et al. 2005). In order to improve a plant's protection during the process of phytoremediation for better results, it is important to understand the mechanisms contributing to stress tolerance. A universal feature of any environmental stress or the defense mechanism is the ability to produce toxic oxygen derivatives. For quite a long period, ROS is considered mainly as dangerous molecule, whose levels need to be maintained as low as possible. Now studies are proving this view to be true to a certain extent only. It has been realized that reactive oxygen species play important roles in the plant's defense system against any biotic and abiotic stress (Bolwell et al., 2002). ROS are yield of normal metabolic processes which are produced only in low levels. But in biological systems, increasing the synthesis of ROS is one of the initial responses to different stress factors (Singh and Sinha, 2005). Plants have complex ROS scavenging mechanisms at the molecular and cellular levels to combat the stress incurred by the system. These mechanisms inhibit or slow the oxidation of biomolecules and oxidative chain reactions (Michalak, 2006), as a result decreasing the cellular oxidative damage and increasing resistance to heavy metals, which is one of the major abiotic stresses. (Rastgoo and Alemzadeh 2001).

ROS are partially reduced forms of atmospheric oxygen, and under normal conditions their production in plant cells is tightly controlled by the scavenging system. ROS can oxidize biomolecules such as DNA, proteins and lipids, creating oxidative injury that results in the reduction of plant growth and development (Hernandez-Jimenez et al. 2002). Since the half-lives of ROS are extremely short, their stable end products of oxidative damage to cellular macromolecules can be used for oxidative stress monitoring (Orhanl et al. 2004). Plants possess a number of antioxidant molecules and enzymes that protect them against the oxidative damage to control the level of oxidative damage resulting due to increase in ROS levels. The ROS species can react with cellular components (lipids, proteins, nucleic acids) and cause lipid peroxidation, membrane damage and inactivation of enzymes, thus affecting many physiological processes as well as cell viability (Ruciñiska-Sobkowiak 2010). Studies in plant tolerance mechanism have established the fact that in any biotic or abiotic stress such as metal toxicity, drought, and temperature fluctuations, higher activities of antioxidant enzymes and higher contents of non-enzymatic constituents are important for plants to endure the stress. These were originally thought to function as osmotic buffers; on the other hand, apart from the osmotic adjustment, they also seem to play a key role in maintaining the natural state of macromolecules, probably by scavenging ROS (Xiong and Zhou 2002). So it can be affirmed that plants have evolved a complex array of mechanisms to maintain low ROS level and avoid the detrimental effects of excessively high ROS concentrations. Tolerance and protective mechanisms have evolved through the antioxidant network to scavenge free radicals and peroxidase generated during various metabolic reactions. Strategies to minimize oxidative damage are a universal feature of plant defense responses against the oxidative damage. There is a good evidence that the alleviation of oxidative damage and increased resistance to environmental stresses is often correlated with an efficient antioxidative system (Cakmak et al 1993).The protection against oxidative stress is achieved by the production of enzymatic antioxidants comprised of superoxide dimutase (SOD), peroxidase (POD) and catalase (CAT), while glutathione, carotenoids and ascorbate represent non enzymatic components (Hall 2002; Caregnato et al. 2008). Above all, interplay between different ROI-producing and ROI scavenging mechanisms can change drastically depending upon the physiological condition of the plant and

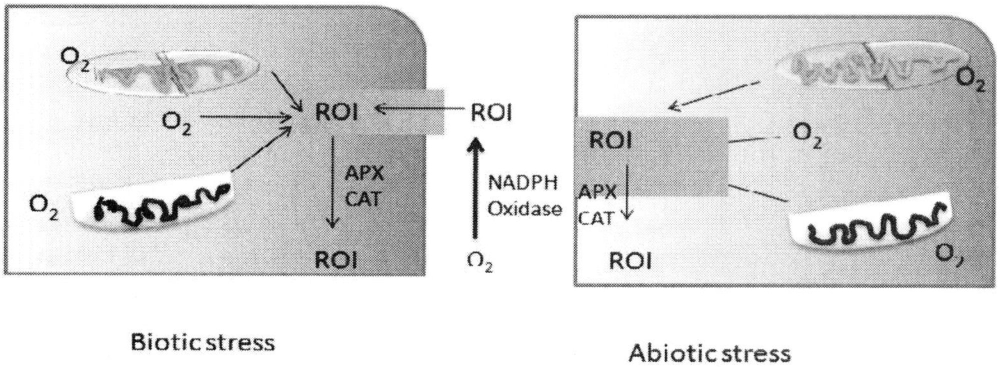


**Fig. 1.** Differences in the steady state level between biotic and abiotic stress. ROI = Steady state level of ROI.

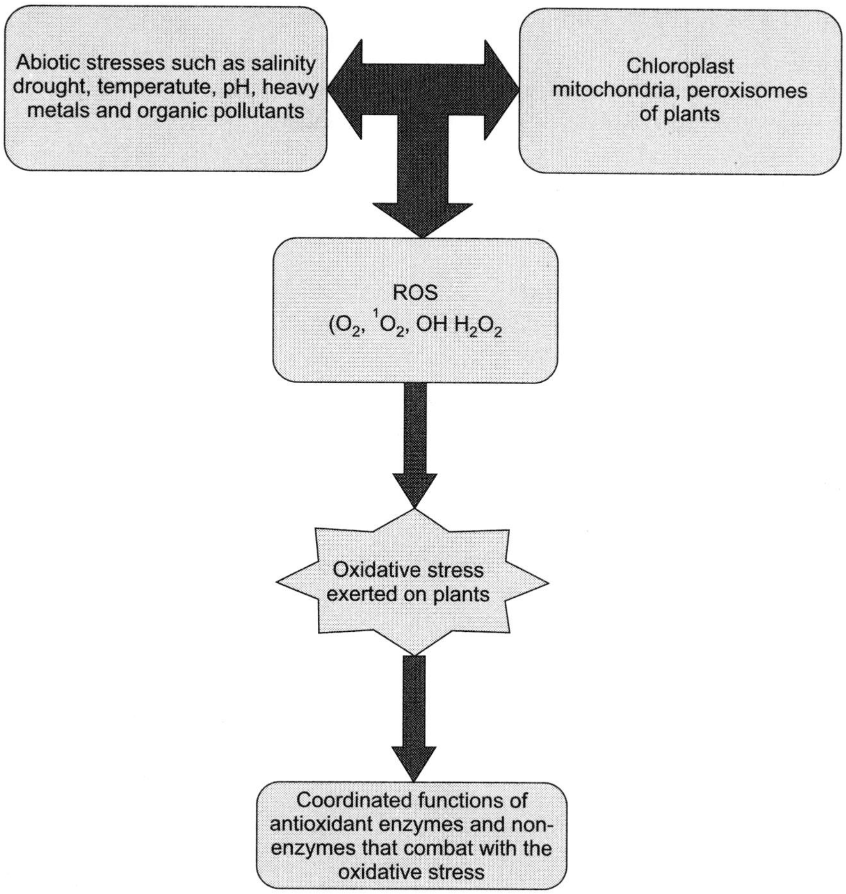


**Fig. 2.** Flow diagram of the interplay of oxidative stress and antioxidant enzymes.

the integration of different environmental, developmental and biochemical stimuli. (Mittler, 2002).

## MODE OF ACTION OF ROS IN PLANTS DURING HEAVY METAL STRESS

Heavy metals are elements having atomic weight between 65.54 and 200.59 and a specific gravity greater than 4 (Kennish, 1992). Among these metals, Fe, Mo and Mn are important as micronutrients. Zn, Ni, Cu, V, Co, W, and Cr are toxic elements with high or low importance as trace elements. As, Hg, Ag, Sb, Cd, Pb, and?U have no known function as nutrients and seem to be more or less toxic to plants and microorganisms (Breckle, 1991; Nies, 1999). Micronutrients are known as essential as they are required not only in small amount for normal plant growth and development, but also constituents of many enzymes and other proteins. However elevated concentrations of both essential and non essential heavy metal in the soil and water can lead to toxicity, thereby damaging the normal metabolic processes of the plants. The reason why metals may be toxic is that they can replace essential metals in pigments or enzymes disrupting their function. Thus, metals render the land unsuitable for plant growth and destroy the biodiversity (Ghosh and Singh, 2005). High concentrations of heavy metals in soil interfere with the metabolic functions in plants, including physiological and biochemical processes, inhibition of photosynthesis, respiration and degeneration of main cell organelles, eventually leading to death of plants (Garbisu and Alkorta 2001; Schmidt 2003). Although heavy metals have proved to be detrimental to plants, phytoremediation demonstrated itself as one of the best alternatives to non-biological degradation of metals. During the period 1997 to 2002, as many as 600 studies reported that the addition of a plant to a contaminated medium (e.g., soil, wetland and hydroponics) enhances the removal and/or transformation of a pollutant. However, the mechanisms to tolerate certain pollutant by certain species of plants are often left unresolved.

Phytoremediation is an effective and affordable technological solution used to extract or remove inactive metals and metal pollutants from contaminated soil and water. This technology is environmental friendly and potentially cost effective (Tangahu et al, 2011). At the end of the 19th century, *Thlaspi caerulescens* and *Viola calaminaria* were the first plant species documented to accumulate high levels of metals in leaves by Hartman in 1975. The idea of using plants to extract metals from contaminated soil was reintroduced and developed by Utsunamyia in the year 1980 and Chaney (1983), and the first field trial on Zn and Cd phytoextraction was conducted by Baker and his team in the year 1991. The plants suitable for metal uptake or phytoextraction should be remarkably resistant plants and organisms that constitute the plant's rhizosphere. Moreover there should be certain addition properties like (ii) accumulating reasonably high levels of the metal, (iii) rapid growth rate (iv) produce reasonably high biomass in the field, and (v) profuse root system (Garbisu et al 2002). Such plants are uncommon, but about 400 hyperaccumulator species (plants able to accumulate huge amounts of heavy metals in their tissues) have been identified.

Even though metal uptake is subjective to the availability of metals, which is in turn determined by both external (soil associated) and internal (plant associated) factors, but the sensitivity of plants to heavy metals and potential of plant for accumulation depends on an interrelated network of physiological and molecular mechanisms such as:

- uptake and accumulation of metals through binding to extracellular exudates and cell wall constituents;
- efflux of heavy metals from cytoplasm to extranuclear compartments including vacuoles;
- complexation of heavy metal ions inside the cell by various substances, for example, organic acids, amino acids, phytochelatins and metallothioneins;

- accumulation of osmolytes and osmoprotectants and induction of antioxidative enzymes; and
- activation or modification of plant metabolism to allow adequate functioning of metabolic pathways and rapid repair of damaged cell structures. (Cho et al. 2003).

Plants possess homeostatic mechanisms that allow them to keep correct concentrations of essential metal ions in cellular compartments and to minimize the damaging effects of an excess of nonessential ones. One of the common consequence of most abiotic and biotic stresses is that they result, at some stage of stress exposure, in an increased production of reactive oxygen species (Polle and Rennenberg, 1993). The successive reduction of molecular oxygen to $H_2O$ yields the intermediates $O_2^{\bullet -}$, $HO^{\bullet}$ and $H_2O_2$, which are potentially toxic, because they are relatively reactive compared with $O_2$. Under physiological conditions $.O_2$ - is not very reactive against the biomolecules of the cells. Likewise $H_2O_2$, which is relatively stable and not very reactive, electrically neutral ROS, but is very dangerous because it can pass through cellular membranes and reaches cell compartments far from the site of its formation (Wojtaszek 1997; Leonardo et al., 2004). As a result a large number of biological molecules such as carbohydrates, unsaturated lipids, proteins and nucleic acids are the target of these reactive species, and the damage may occur in different ways in several sites of the molecule. Among all the macromolecules it is observed that proteins are the most abundant cellular targets of the oxidative species, more than DNA and lipids, making up 68% of the oxidized molecules in the cell (Rinalducci et al., 2008). If uncontrolled reaction proceeds which ultimately are capable of interacting with nuclear proteins and DNA resulting in unspecific oxidation of proteins and membrane lipids causing oxidative degeneration of biological macromolecules To minimize the damaging effects resulting due to $H_2O_2$ thus the scavenging of $H_2O_2$ in cells is significant to avoid oxidative damage.

The aim of this chapter is to understand the mode of action leading to heavy metal toxicity in living cells, therefore we need to understand the chemical properties of these metals and how they change the biochemical process of the plants. The toxic metals or heavy metals that are the core of this chapter are mostly transition metals with an incompletely filled δ-orbital present as cations under physiological conditions. The physiological redox range of aerobic cells stretches from -420mV to +800mV. Therefore, heavy metals of biological significance can be divided into two groups of redox active and inactive metals. Metals with lower redox potentials than those of biological molecules cannot participate in biological redox reactions. Autoxidation of redox metals such as $Fe^{2+}$ or $Cu^{+}$ results in $O_2^{\bullet -}$ formation and subsequently $H_2O_2$ and $OH^{\bullet}$ production via Fenton-type reactions. Cellular injury by this type of mechanism is well-known for iron (Halliwell and Gutteridge, 1986; Imlay et al., 1988), copper (Li and Trush, 1993 *b*) as well as other metals (Lund et al., 1993; Shi and Dalal, 1993). Metal coordination chemistry and their redox properties have provided them with an added advantage that these metals could escape out of the control mechanism such as transport, homeostasis, compartmentalization and binding to designated cell constituents. Although, this process does not occur on a regular basis, but such an action by metals could lead to malfunctioning of cells and eventually toxicity. Heavy metal toxicity comprises inactivation of biomolecules by either blocking essential functional groups or by displacement of essential metal ions (Goyer, 1997). As mentioned above, autooxidation of redox-active heavy metals and generation of reactive radicals resulted in cellular damage like depletion of enzyme activities, damage to lipid bilayer and DNA (Stohs and Bagchi, 1995) by the Fenton reaction causing cellular injury (Stohs and Bagchi, 1995) when metal toxicity stress point is reached at the toxic threshold level of the metal in the tissues of the plants. The metal accumulation and its toxic effects in plants can be evaluated. Streit and Strumn in 1993 classified the exchange of chemicals between soil and plants. They divided the most common methods of assessing metal toxicity to plants from soil into three categories in the conditions of closed system:

(i) monitoring of the presence or absence of specific plant ecotypes and/or plant species (indicator plant),
(ii) measurements of metal concentration in tissues of selected species (accumulative bioindicators),
(iii) recording of physiological and biochemical responses (biomarkers) in sensitive bioindicators.

These reactive radical species include wide variety oxygen-, carbon-, sulfur- and nitrogen- radicals, originating not only from superoxide radical, hydrogen peroxide, and lipid peroxides, but also in chelates of amino acids, peptides, and proteins complexed with the toxic metals. The unifying factor in determining toxicity for all these metals is the generation of reactive oxygen and nitrogen species. The toxic manifestations of these metals are caused primarily due to imbalance between pro-oxidant and antioxidant homeostasis which is termed as oxidative stress. Metals have high affinity for thiol groups containing enzymes and proteins, which are responsible for normal cellular defense mechanism (Flora et al., 2008). Above certain point of tolerance, the physiological state of the cells of plant will be irreversibly changed. This change is reflected by an increase in the activity of certain enzymes defined as enzyme induction or inhibition of enzyme activity in the plants as a result of tolerance or protective mechanism.

Two mechanisms of enzyme inhibition predominate during the process of metal uptake:

I. Binding of the metal to sulphydryl groups, involved in the catalytic action or structural integrity of enzymes.
II. Deficiency of an essential metal in metalloproteins or metalprotein complexes eventually combined with substitution of the toxic metal for the deficient element.

Different plants absorb toxic and non-toxic metals from soil and water to varied extent and accumulate in different body parts (Chamber and Sidle, 1991). Although the morpho-physiological responses of plants may vary according to the nature and the dose of the pollutant and the species, at biochemical and molecular levels, there appears to be a similarity among different pollutants and also among most of the environmental stresses which can be classified as:

1. Increase in antioxidant enzymes and metabolites.
2. Induction of protection-related secondary metabolite genes. If the plants are able to express these responses adequately, pollution-induced "visible" or "hidden" damages do not occur. However, if these protective responses are inadequate and unable to cope with the incidence and the dose of the pollutant, the injury occurs. These responses, therefore, can be compared to immune responses in animals, which of course are evoked in response to pathogens (Srivastava 1999).

It is observed that an active antioxidative metabolism does not represent a metal tolerance mechanism, but many a times it is beneficial for plant performance under heavy metal stress. Eventually scientists reported that the protective mechanisms of plants under heavy metals stress act in cells and tissues in the following ways:

- They transfer electrons directly in single-electron reactions, which generate free radicals. The so-called transition metals (Fe, Cu, Mn, etc.), which have unpaired electrons in their orbitals, accept and donate single electrons, thus promoting monoelectron transfers to $O_2$ and generally ROS interconversion and oxireduction phenomena.
- Metals disturb metabolic pathways, especially in the thylakoid membrane, which also results in increased formation of free radicals and reactive oxygen species.
- In addition, heavy metals mainly inactivate the antioxidant enzymes (peroxidases, catalases, superoxide dismutases) responsible for free radical detoxification, although peroxidases may also be activated due to metal stress.
- Finally, heavy metal accumulation results in the depletion of low molecular weight antioxidants, such as glutathione, which is consumed under phytochelate formation.
- The activation of the cellular antioxidant metabolism belongs to the general stress responses induced by heavy metals (Dietz et al. 1999 and Shaw, 1995).

Irrespective of the production pathway, ROS are highly cytotoxic and their level within the plant cells must be controlled by enzymatic and non-enzymatic antioxidant defense systems. Close contact between the pollutant and the detoxifying enzymes of plants that are localized in the cytosol of living cells are the necessary prerequisite for successful uptake by plants. The presence and activity of these enzymes is crucial for a potential metabolization and further degradation of the chemicals under consideration. Conjugation to biomolecules is regarded as a beneficial detoxification reaction.

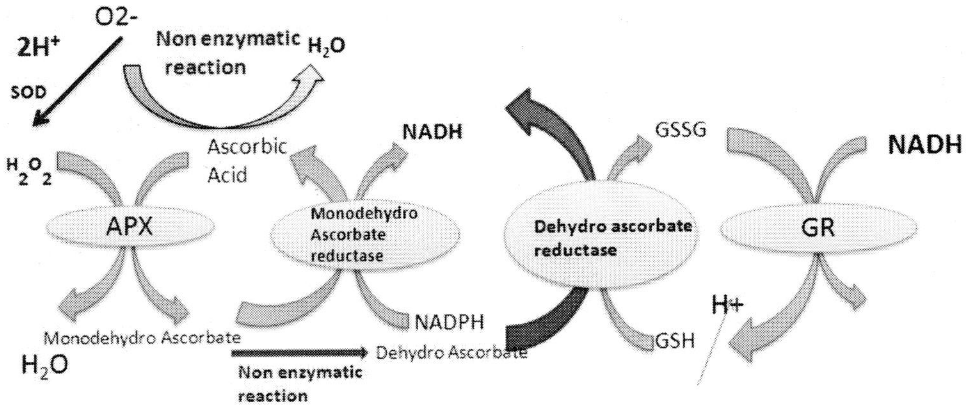


**Fig. 3.** Asada–Halliwell pathways of $H_2O_2$ scavenging and ascorbic regeneration involving various antioxidant enzymes.

## INTERACTION OF ANTIOXIDANT ENZYME AND OXIDATIVE STRESS

A variety of environmental stresses like soil salinity, drought, extremes of temperature and heavy metals are known to cause oxidative damage to plants either directly or indirectly by triggering an increased level of ROS (Malecka 2001; Shah et al 2001). Plants are sessile organisms and have only limited mechanisms for stress avoidance. They need flexible means for adjusting to changing environmental conditions. In order to improve a plant's protection, it is important to understand the mechanisms contributing to stress tolerance (Schutzendubel and Polle, 2002). Tolerance to heavy metals in plants may be defined as the ability to survive in a soil that is toxic to other plants, and is manifested by an interaction between a genotype and its environment (Macnair et al., 2000), although frequently the term is used more widely to include changes that may occur experimentally in the sensitive response to heavy metals. There is an evidence that the tolerance of plants is correlated with increasing amounts of antioxidants and increasing activity of radical scavenging enzymes. Generally a plant's cells try to keep the concentration of ROS at the possible low level because they are more reactive than molecular oxygen ($O_2$) (Wojtaszek 1998) and they react with almost every organic constituent of the living cell. The high reactivity of ROS is based on the specificity of their electronic configuration (Rama devi and Prasad 1998). The most popular ROS are $.O^{2-}$ *-superoxide radical,* $H_2O_2$ *–hydrogen peroxide,* and .OH -hydroxyl radical originating from one, two or three electron transfers to dioxygen ($O_2$). Transition metals cause oxidative injury to plant tissues, but a literature survey did not provide evidence that this stress could be alleviated by increased levels of antioxidative systems. The reason may be that transition metals initiate hydroxyl radical production, which cannot be controlled by antioxidants. Exposure of plants to non-redox reactive metals also resulted in oxidative stress as indicated by lipid peroxidation, $H_2O_2$ accumulation, and an oxidative burst (Schützendübel and Polle 2002). These cytotoxic properties of ROS explain the evolution of complex arrays of nonenzymatic and enzymatic detoxification mechanisms in plants. Oxidative stress is essentially a regulated process; the equilibrium between the oxidative and antioxidative capacities determines the fate of the plant. Under non stressful conditions, the antioxidant defense system provides adequate protection against active oxygen and free radicals. In response, the capacity of the antioxidative defence system is increased (Srivastava, 1999). Plant cells require at least two mechanisms to regulate their intracellular ROI concentrations by scavenging of ROIs: one that is required to enable the fine modulation of low levels of ROIs for signaling purposes, and one that will enable the detoxification of excess ROIs, especially when plants encounter any type of biotic or abiotic stress.

In addition, the types of ROIs produced and the balance between the steady-state levels of different ROIs can also be important. These are determined by the interplay between different ROI-producing and ROI scavenging mechanisms, and can change drastically depending upon the physiological condition of the plant and the integration of different environmental, developmental and biochemical stimuli (Mittler, 2002).

The term antioxidant generally refers to a broad class of compounds that protect cells from damage otherwise caused by exposure to certain highly reactive compounds (Salt 2004). An antioxidant is a molecule capable of inhibiting the oxidation of other molecules. Oxidation is a chemical reaction that transfers electrons from a substance to an oxidizing agent. The free radicals produced during oxidation reaction can then start chain reactions that damage cells. Hence antioxidant enzymes play a crucial role which terminates these chain reactions by removing free radical intermediates, and inhibit other oxidation reactions. They do this by being oxidized themselves, so antioxidants are often reducing agents such as thiols, ascorbic acid or polyphenols (Sies 1997). The antioxidant enzymatic systems that counteract the ROS generated in stress induced plants include a variety of scavengers, such as SOD, APX, GPX, GST, and CAT and non-enzymatic low molecular metabolites, such as ASH, GSH, a-tocopherol, carotenoids and flavonoids (Mittler et al 2004).

In addition, proline can now be added to an elite list of non-enzymatic antioxidants that microbes, animals, and plants need to counteract the inhibitory effects of ROS (Chen and Dickman 2005). Plant stress tolerance may therefore be improved by the enhancement of *in vivo* levels of antioxidant enzymes.

The above said antioxidants found in almost all cellular components demonstrate the importance of ROS detoxification for cellular survival (Gill and Tuteja, 2010)

## Role of Antioxidant Enzymes

The phytoremedial potential of plants can be assessed with tolerance mechanisms for toxic metals that allows plant to combat increased ROS levels during abiotic stress conditions. Moreover, in other circumstances, plants appear to purposefully

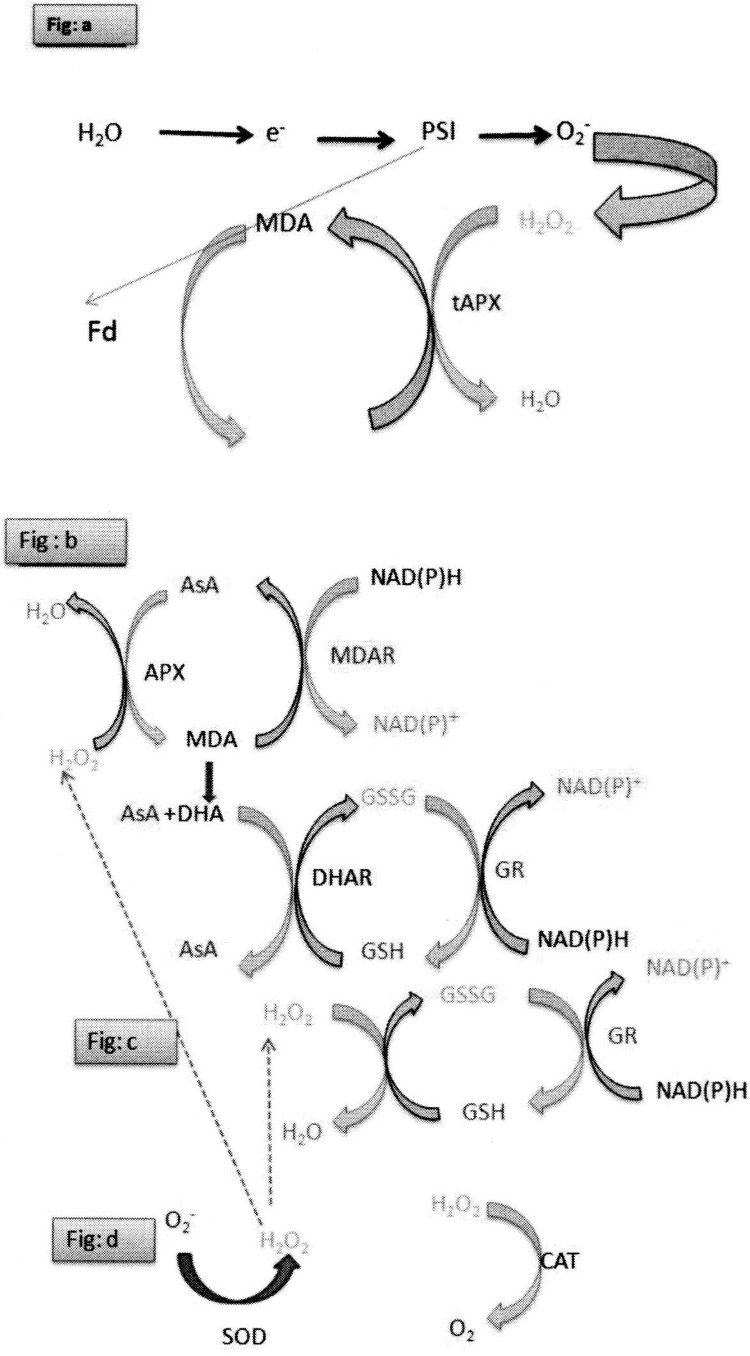


**Fig. 4.** Pathways for reactive oxygen species.

generate ROS as signaling molecules to control various processes including pathogen defense, programmed cell death, and stomatal behavior (Apel and Hirt, 2004).

Plant damage occurs when the capacity of antioxidant processes and detoxification mechanisms is lower than the amount of ROS production. In plants, ROS are continuously produced predominantly in chloroplasts, mitochondria and peroxisomes. Therefore production and removal of ROS must be strictly controlled (Apel and Hirt, 2004).

The oxidative damage to cellular components is limited under normal growing conditions due to efficient processing of ROS through several enzymes and redox metabolites. The major ROI-scavenging pathways of plants include SOD, found in almost all cellular compartments, the water-water cycle in chloroplasts, the ascorbate-glutathione cycle in chloroplasts, cytosol, mitochondria, apoplast and peroxisomes, glutathione peroxidase (GPX; and CAT in peroxisomes. The finding of the ascorbate-glutathione cycle in almost all cellular compartments tested to date, as well as the high affinity of APX for $H_2O_2$, suggests that this cycle plays a crucial role in controlling the level of ROIs in these compartments. By contrast, CAT is only present in peroxisomes, but it is indispensable for ROI detoxification during stress, when high levels of ROIs are produced.In addition, oxidative stress causes the proliferation of peroxisomes. The balance between SOD and APX or CAT activities in cells is crucial for determining the steady-state level of superoxide radicals and hydrogen peroxide. This balance, together with sequestering of metal ions, is thought to be important in preventing the formation of highly toxic hydroxyl radical via the metal-dependent Haber-Weiss or Fenton reactions (Mittler, 2002).

The Haber-Weiss reaction generates •OH (Hydroxyl Radicals) from $H_2O_2$ and superoxide.

### *First Step*

$Fe^{3+} + \bullet O_2^- \rightarrow Fe^{2+} + O_2$

The second step is the Fenton Reaction:

$Fe^{2+} + H_2O_2 \rightarrow Fe^{3+} + OH^- + \bullet OH$

### *Net reaction*

$\bullet O^{2-} + H_2O_2 \rightarrow \bullet OH + OH- + O_2$ (Michalek, 2006).

Plants have developed various mechanisms to cope with the oxidative stresses caused by unfavorable environments, such as modulating the expression of stress tolerance genes and synthesizing compatible solutes (Ahmad et al., 2010). Antioxidant enzymes such as SOD, APX, catalase and GR help in the regeneration of redox ascorbate and glutathione metabolites (Asada 1996; Foyer and Nector 2000). The coordinated functions of antioxidant enzymes that can be best described by different ROI-scavenging and ROI-producing mechanisms have been revealed in transgenic plants with suppressed production of ROI-detoxifying mechanism. For example, it

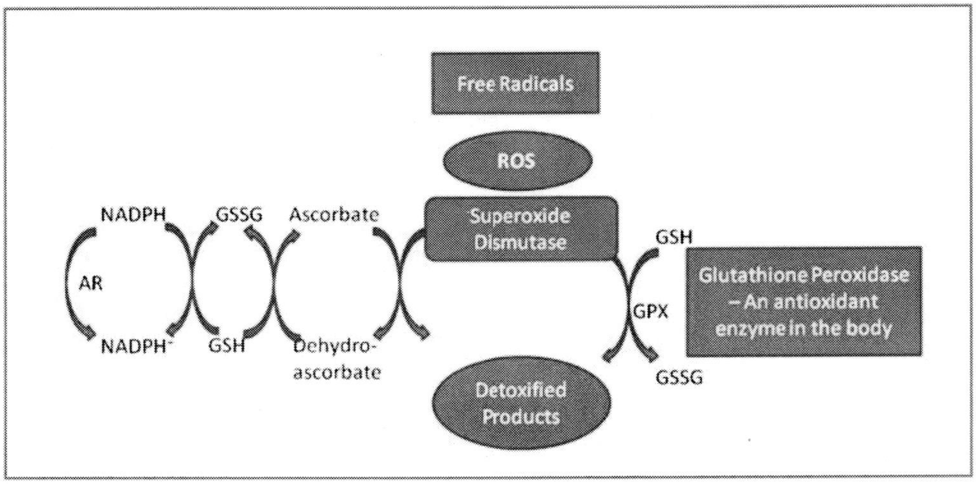


**Fig. 5.** Mechanism of antioxidant enzymes in combating oxidative stress.

is seen that plants with suppressed APX production induce SOD, CAT and GR to compensate for the loss of APX, whereas plants with suppressed CAT production induce APX, GPX and mitochondrial AOX (Willekins et al. 1997). CAT and APX are not completely redundant because they do not compensate for the lack of each other, as observed through the studies demonstrating the sensitivity of plants.

To start with the role of antioxidant enzyme in plants, one should consider the important aspects of the defense mechanisms which are mainly based on chelation and subcellular compartmentalization. Chelation of heavy metals during its uptake by plants is a ubiquitous detoxification strategy described in a wide variety of plants. Phytochelatins (PCs), a principal class of heavy metal chelator known in plants, belongs to the family of Cys-rich peptides. PCs are family of metal complexing peptides that have a general structure (γ-Glu Cys)n -Gly where n = 2-11 and are rapidly induced in plants by heavy metal treatment. PCs are synthesized non-translationally from reduced glutathione (GSH) in a trans-peptidation reaction catalyzed by the enzyme phytochelatin synthase (PCS). Therefore, availability of glutathione is very essential for PCs synthesis in plants at least during their exposure to heavy metals. Now coming to Glutathione (GSH), which is a key member of antioxidant enzyme family, a tripetide that contains an unusual peptide linkage between the amine group of cysteine (which is attached by normal peptide linkage to a glycine) and the carboxyl group of the glutamate side-chain. It is an antioxidant, preventing damage to important cellular components caused by reactive oxygen species such as free radicals and peroxides (Pompella et al. 2003).

The affinity of heavy metals such as Cd to bind to GSH, forming metal-thiolate compounds, proved the fact that GSH (Fig. 6) involved in the synthesis of phytochelating, which could detoxify Cd ions (Smeets et al. 2005). Potentially toxic heavy metal ions are firstly chelated by GSH and then transferred to PCs for eventual sequestration. GSHs act as a first line of defense against metal toxicity by

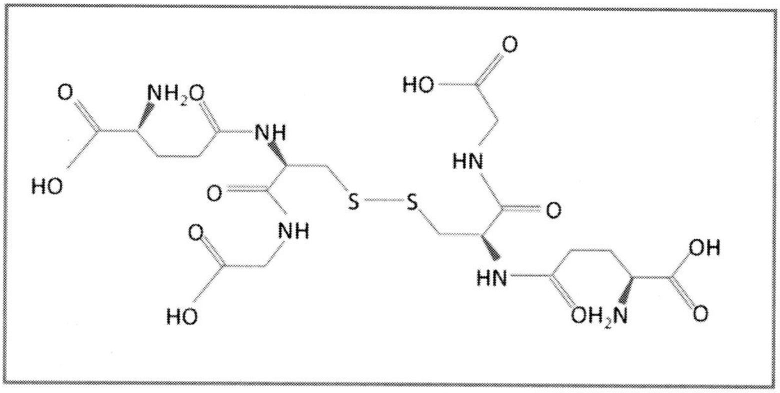

**Fig. 6.** Structure of Glutathione (Adapted from Bhaduri and Fulekar, 2011).

complexing metals before the induced synthesis of PCs arrives at effective levels (Freedman et al. 1989). The main function of Glutathione (GSH) is to directly reduce most active oxygen species. It is a major water-soluble antioxidant in plants cells having low-molecular weight thiol tripeptide, involved in cellular defense against the toxic action of xenobiotics, oxyradicals as well as metal cations (Meister and Anderson 1983). Owing to its redox active thiol group, GSH is involved in the redox regulation of cell cycle (Sánchez-Fernandez et al. 1997) and has often been considered to play an important role in defense of plants and other organisms against oxidative stress. Various free radicals and oxidants are able to oxidize GSH to GSSG (Noctor et al. 1998). This reaction maintains a proper GSH/GSSG concentration ratio in cells (Rendon et al. 1995). It is also able to modify metal toxicity by chelating metal ions in cells and plays a key role in protecting macromolecules from damage by free radicals by trapping them in an aqueous phase (Freedman et al. 1989). Glutathione reductase contains a highly conserved disulphide bridge between Cys76 and Cys81 (Lee et al., 1998), which may undergo cleavage by heavy metals. Heavy metal-induced loss in glutathione reductase has frequently been observed in pea by Zn, Cu and Fe (Bielawski and Joy, 1986), in sunflower by Fe, Cu and Cd (Gallego et al., 1996), and in *Lemna minor* by Cu (Teisseire and Guy, 2000).

The ascorbate-glutathione cycle seems to be a mechanism of great importance in controlling the cellular redox status especially after application of heavy metals such as copper, zinc and cadmium (Smeets and Cuypers 2005). Both, ascorbic acid that controls the concentration of oxygen and its derivatives (OH (AsA), and glutathione (GSH) in the oxidized and reduced forms are among the most important non-enzymatic cellular antioxidant defense compounds by quenching ROS and are important in controlling the metal.

Another enzyme in the family of enzymes is "Peroxidases" (Gaspar et al., 1991). Peroxidases, which are widely accepted as 'stress enzymes', typically catalyze a reaction of the form:

ROOR′ + electron donor ($2e^-$) + $2H^+$ → ROH + R′OH

Peroxidases may contain a heme cofactor in their active sites, or redox-active cysteine or selenocysteine residues. Changes in peroxidase activity have been associated with a wide array of physiological processes involved with auxin function and cell wall synthesis. The association with auxin and lignifications has made peroxidase analysis informative in the study of plant response to external stimuli (Cakmak and Marschner 1992). Peroxidases are heme containing proteins that utilize $H_2O_2$ in the oxidation of various organic and inorganic substrates (Asada 1994). Peroxidases utilizing guaiacol as electron donor *in vitro* are guaiacol peroxidases and participate in developmental processes, lignification, ethylene biosynthesis, defense, wound healing, etc.

Peroxidase, which participates in lignin biosynthesis might built up a physical barrier against poisoning of heavy metals (Rai et al. 2004). The other group of peroxidases scavenge $H_2O_2$ in cell and utilize glutathione, Cyt c, pyridine nucleotide and ascorbate as electron donors *in vitro*. Guaiacol peroxidases are glycoproteins, located in cytosol, vacuole, cell wall and in extracellular space, while the other group is non glycosylated and localized in chloroplasts and cytosol (Asada 1992). Peroxidase activity and photosynthetic pigments are sensitive indicators of heavy metal stress and can be used to anticipate events on the organism level (Wu et. al., 2003; Mac Farlane and Burchett 2001).

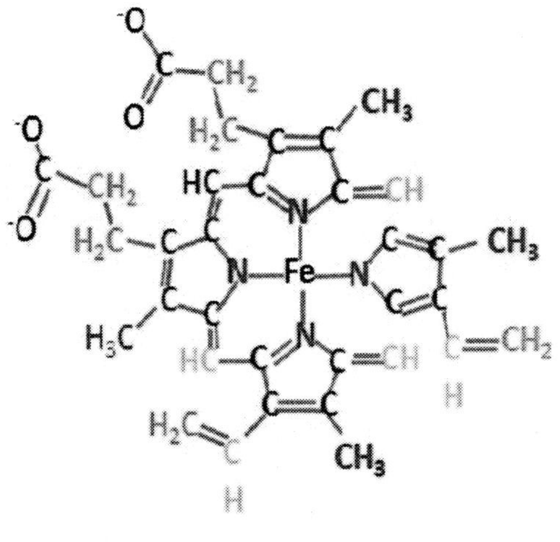


**Fig. 7.** Structure of catalase.

CAT is a universally present oxidoreductase and an important heme-containing enzyme that catalyses the dismutation of $H_2O_2$ to $H_2O$ and oxygen and is localized in the peroxisomes. Most catalases exist as tetramers of 60 or 75 kDa, each subunit containing an active site heme group buried deep within the structure, but which is accessible from the surface through hydrophobic channels. The very rigid, stable structure of catalases is resistant to unfolding, which makes them uniquely stable enzymes that are more resistant to pH, thermal denaturation and proteolysis than most other enzymes. Their stability and resistance to proteolysis is an evolutionary advantage, especially since they are produced during the stationary phase of cell growth when levels of proteases are high and there is a rapid rate of protein turnover. CAT is found in most plant and animal cells that function as an oxidative catalyst and an indispensable enzyme required for ROS detoxification in plants. Catalase decomposes $H_2O_2$ to water and molecular oxygen, and it is one of the key enzymes involved in the removal of toxic peroxides (Lin and Kao 2000). Catalases are involved in scavenging $H_2O_2$ generated during the photo-respiration and b-oxidation of fatty acids (Morita et al. 1994).

While the complete mechanism of catalase is not currently known, the reaction is believed to occur in two stages:

$$H_2O_2 + Fe(III)\text{-}E \rightarrow H_2O + O{=}Fe(IV)\text{-}E(.+)H_2O_2 + O{=}Fe(IV)\text{-}E(.+) \rightarrow H_2O + Fe(III)\text{-}E + O_2$$

Here Fe()-E represents the iron center of the heme group attached to the enzyme. Fe(IV)-E(.+) is a mesomeric form of Fe(V)-E, meaning that iron is not completely oxidized to +V but receives some "supporting electron" from the heme ligand. This heme has to be drawn then as radical cation (.+).(Boon and Marshey, 2007).

Ascorbate peroxidase is one of the important peroxidases of ubiquitous occurrence in plants. It is regarded as a universal housekeeping protein in the cytosol and chloroplasts of plant cells. Ascorbate peroxidases use ascorbate as a substrate and are believed to scavenge excess of $H_2O_2$ formed in plant cells under both normal and stress conditions (Asada 1992). The product of ascorbate oxidation by ascorbate peroxidase is an ascorbate-free radical, which is reduced back to dehydroascorbate by the enzyme monohydroascorbate reductase with NADPH as an electron donor (Asada et al 1996). Increase in ascorbate peroxidase activity in response to air pollutants, specially $O_3$, has been demonstrated in several species such as wheat (Bender et al, 1994), spinach (Tanaka et al. 1985), pumpkin (Ranieri et al. 1994), and *Picea abies* (Sehmer et al. 1998). The different affinities of APX (mM range) and CAT (mM range) for $H_2O_2$ suggest that they belong to two different classes of $H_2O_2$-scavenging enzymes: APX might be responsible for the fine modulation of ROIs for signaling, whereas CAT might be responsible for the removal of excess ROIs during stress (Mittler, 2002).

Superoxide dismutase (SOD) is an enzyme composed of metal-containing proteins that convert superoxide radicals into less toxic agents. It constitutes the main enzymatic mechanism for clearing superoxide radicals from the plants. It represents a group of multimeric metalloenzymes catalyzing the disproportionation of superoxide free radicals, generated by univalent reduction of molecular oxygen to $H_2O_2$ and $O_2$ in different cellular compartments (Fridovich 1989).

The reaction is as follows:

$$M^{(n+1)+}\text{-}SOD + O_2^- \rightarrow M^{n+}\text{-}SOD + O_2$$

$$Mn^{+}\text{-SOD} + O_2^{-} + 2H^{+} \rightarrow M^{(n+1)+}\text{-SOD} + H_2O_2.$$

Here, M = Cu (n = 1), Mn (n = 2), Fe (n = 2), Ni(n = 2)

In this reaction the oxidation state of metal ion oscillates between n, n+1.

The hydrogen peroxide molecule, which is still a danger to cells, is then further processed to nontoxic by-products. In an aqueous environment, the enzyme catalase degrades the hydrogen peroxide as follows:

$$2H_2O_2 \rightarrow 2H_2O + O_2$$

$H_2O_2$ is relatively stable and not very reactive electrically neutral ROS, but is very dangerous because it can pass through cellular membranes and reaches cell compartments far from the site of its formation.

At least three types of SODs with several isoforms are present in plants. These are: (i) chloroplastic or cytosolic Cu–Zn SOD; the cytosolic Cu–Zn SOD is referred to as Cu–Zn SOD I, and chloroplastic one is referred to as Cu–Zn SOD II; (ii) Mitochondrial Mn SOD; and (iii) chloroplastic Fe SOD. The amino acid sequence of Fe–SOD and Mn–SOD (apoproteins are similar, whereas Cu–Zn SOD is different). The genes for different types of SODs are also identified. For example, *Arabidopsis* contains multiple SOD genes encoding at least three Cu–Zn SODs, three Fe–SODs and one Mn–SOD (Srivastava, 1999) (Fig. 8).

Ascorbate is a major antioxidant and radical scavenger in plants. Monodehydroascorbate reductase (MDAR) and dehydroascorbate reductase (DHAR) are two enzymes of the ascorbate-glutathione cycle that maintain ascorbate in its reduced state. These enzymes metabolites such as monodehydro ascorbate

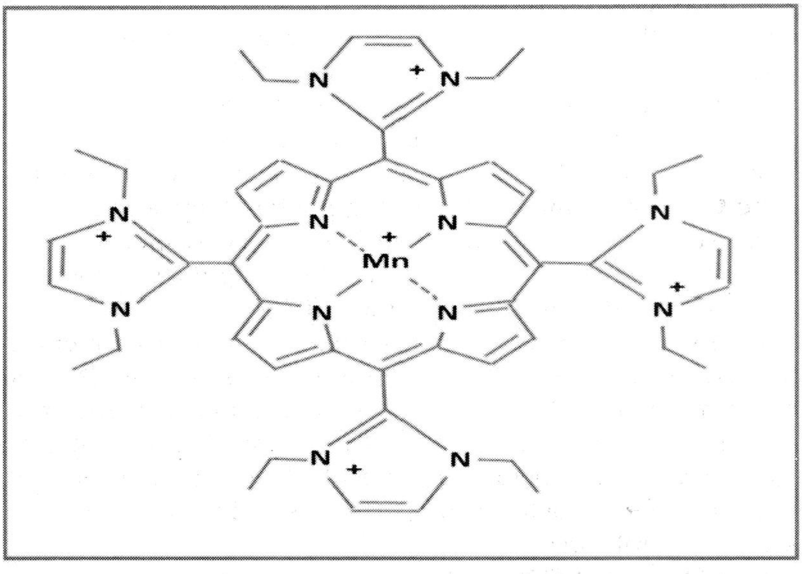


**Fig. 8.** Structure of Mn-SOD (Adapted from Bhaduri and Fulekar, 2011).

reductase (MDAR), dehydroascorbate reductase (DHAR) and glutathione reductase (GR) also play a significant role in scavenging $H_2O_2$ mainly in chloroplasts and in maintaining the redox status of the cell (Foyer et al. 1997) of nitrogen fixing root nodules, where ROS are produced under unstressed conditions also. It may constitute up to 1% of the total protein (Dalton 1995).

## TOLERANCE MECHANISM OF PLANTS DUE TO HEAVY METAL STRESS

There are many biochemical and physiological responses implicated when living organisms encounter any abiotic or biotic stress. Similarly during the process of Phytoremediation of toxic heavy metals, it is seen many a times that primary response of plants is the generation of reactive oxygen species (ROS) upon exposure to high levels of heavy metals. Various metals either generate ROS directly through Haber-Weiss reactions or overproduction of ROS and consequently oxidative stress in plants could be the outcome of heavy metal toxicity (Wojtaszek 1997). The indirect mechanisms include the interaction of plants with the antioxidant system disrupting the electron transport chain (Qadir et al., 2004).

The ability of plants to increase antioxidative protection to combat negative consequences of heavy metal stress appears to be limited since many studies showed that exposure to elevated concentrations of redox reactive metals resulted in decreased and not increased activities of antioxidative enzymes. It is also observed in many studies that metals do not interfere directly with cellular oxygen metabolism; then the question arises as to the reasons of the observed oxidative stress. Exposure to heavy metals also provoked pronounced responses of antioxidative systems, but the direction of responses is dependent on the plant species and tissue analysed, the metal used for the treatment, and the intensity of the stress (Table 1). On the other hand, some common reaction patterns can be found in most cases irrespective of the plants species or metals used. When the plants are exposed to heavy metals it initially resulted in a severe depletion of GSH, for example, Cd: *Rauvolfia serpentina*: Grill et al., 1987; pine: Schützendübel et al., 2001; carrot: di Toppi et al., 1999; tobacco: Vögeli-Lange and Wagner, 1996; Cu: *Silene cucubalus*: de Vos et al., 1992; Cu or Cd: *Arabidopsis*: Xiang and Oliver, 1998; Ni and Zn: pigeonpea: Rao and Sresty, 2000; Fe, Cu or Cd: sunflower leaves: Gallego et al., 1996) to name a few. Metal phytotoxicity occurs when metals move from soil to plant roots and are further transported to various sites in the shoots. The most pronounced effect of heavy metals when accumulated in plant tissues is the development and growth inhibition, which is inseparably connected with cell division (Kumar and Rai 2007). It can be concluded from studies that when plants are subjected to heavy metal stress, it is possible that heavy metal stress reduces the capacity of plants to assimilate carbon, which in turn triggers an increase in photosynthetic electron flux to molecular oxygen, resulting in the increased production of superoxide, hydrogen peroxide, and hydroxyl radical. Since these reactive oxygen species are damaging to lipids, proteins and pigments, they are rapidly scavenged by antioxidant enzymes. There are evidences that increased

**Table 1** Examples antioxidant enzymes interaction with plants during phytoremediation of heavy metals. (Adapted from Bhaduri and Fulekar, 2011)

| *Metals* | *Plants* | *Enzymes* |
|---|---|---|
| Cu | *Alhagi camelorum* Fisch | PC synthesis induced by metals is accompanied by a rapid depletion of total GSH in plants (Boojar and Tavakoli,2010) |
| | *Hydrilla verticillata* (L.f.) | Decrease in GSH and an elevation in PC levels (Boojar and Tavakoli 2010) |
| | *Carthamus tinctorius* | Enhanced production of SOD, POD and CAT (Ahmed et al 2010) |
| Ni | *Alyssum murale* | The significant induction of SOD activity catalase activity and the significant reduction in GR activity displayed a typical antioxidative enzyme response pattern. (Babaoglu et al, 2009) |
| Cd | *Maize seedlings* | Enhanced the activities of APX and GPX with variation in CAT activity (Malekzadeh et al., 2007) |
| Hg | *Abelmoschus esculentus* L. | Increased in the activity of SOD, APX and GR, but CAT activity decreased with the increased treatment. (Hameed et al, 2011) |
| | *Phaseolus aureus* | Increase in the activity of APX (Shaw, 1995) |
| Cd, Ni, and Pb | *Pteris vittata* L. | Catalase activity in the plant increased with the increased concentrations of the metal, but CAT activity decreased at higher concentration as compared to lower concentration or control in case of Zn and Ni (Fayiga et al 2004) |
| Zn | *Jatropa* seedlings | POD activity in the cotyledons and hypocotyls with the increasing zinc concentrations gradually increased the CAT activity in the cotyledons and hypocotyls increased with the increasing zinc concentration (Luo et al. 2010) |
| | *E. crassipes* | CAT activity increased in response to increasing concentrations of metals like Ag, Cd, Cr, Pb, Cu, but showed decrease in activity with increase in concentration in Zn and Hg (Odjegba and Fasidi, 2007) |
| Cu, Pb | Sainfoin, (*Onobrychis viciifolia*) | Activity and function of three enzymes SOD, CAT and GPX in the leaves increased. (Beladi, 2011) |
| Zn | Maize (*Zea mays* L.) | SOD and POD activity increased but at the same concentration of Zn CAT activity decreased with increase in metal concentrations |

levels of these scavenging enzymes may play a role in limiting the degree of photo damage to plants (Hodges et al. 1997; Collen and Davison 1999; Rao and Sresty 2000; Schutzendubel et al. 2001). However, the mechanisms involved in those processes are still not completely understood. In-depth studies in the past few decades have shown that metals like iron, copper, cadmium, mercury, nickel, lead and arsenic possess the ability to generate reactive radicals, resulting in cellular damage like depletion of enzyme activities, damage to lipid bilayer and DNA

(Stohs and Bagchi 1995). These reactive radical species include a wide variety of oxygen-, carbon-, sulfur- and nitrogen- radicals, originating not only from superoxide radical, hydrogen peroxide, and lipid peroxides, but also from chelates of amino acids, peptides, and proteins forming complexes with the toxic metals. Recently it has been reported that antioxidant prevents certain types of damage in living cells, appearing to allow some kinds of plants to thrive on metal-enriched soils that typically kill other plants, according to recent research in Purdue University, USA. This finding provides an important new insight into the development of plants that could be used to help clean polluted sites. The work also answers a fundamental question for researchers studying how certain types of plants can tolerate levels of metals in their tissues that are toxic to most other plants. Therefore, it is crucial that plants should maintain the activities of these enzymes in order to accommodate these oxidative stresses. The main enzyme determining the resistance to the oxidative stresses might be dependent on the plant species and the metal toxicity.

It is seen that within the cortex, the metals are transported in the apoplastic space according to their concentration gradient and also accumulate in the cell walls (Arduini et al 1996). The toxic effects of heavy metals are exerted at the plasma membrane and within the cell in two different uptake routes that have been reported across the membrane: (a) passive uptake, only driven by the concentration gradient across the membrane and (b) inducible substrate-specific and energy-dependent uptake (Nies, 1999; Williams et al., 2000). A common transmembrane transporter was found for Cd, Cu, and Ni. The uptake of these compounds was competitively inhibited by K, Ca, and Mg (Clarkson and Lüttge, 1989). Thus the tolerance may involve the protection of plasma membrane integrity against heavy metal damage that would produce increased leakage of solutes from cells.

There are good number of studies that report uptake of metals and defence mechanism of plants due to heavy metals stress. The present paper focusses particularly on role of enzymes in plants during such stress.

## Cadmium

The variation in response of antioxidant enzymes to Cd stress could be due to the variability of plant species in producing free radicals (Mazhoudi et al. 1997). Cadmium causes oxidative stress probably through indirect mechanisms such as interaction with the antioxidative defence, disruption of the electron transport chain or induction of lipid peroxidation. This finding of induction of lipid peroxidation was further strengthened by the study where the activation of lipoxygenase, an enzyme that stimulates lipid peroxidation, has been observed after cadmium exposure (Smeets et al. 2005). When coming to the network of antioxidant enzymes, it is seen in several studies that roots and leaves of Cd-exposed *Phaseolus vulgaris* as well as suspension cultures of tobacco (*Nicotiana tabacum*) cells contained elevated APX activities after Cd exposure (Chaoui et al. 1997). In *Phaseolus aureus*, Cd induced elevated guaiacol peroxidase (POD) but decreased CAT activities. (Shaw 1995). The increase of APX and GPX in different plants has been reported with Cd (Ashraf et al. 2003; Karataglis et al. 1991; Rama Devi and Prasad 1998).

### Lead

Pollution due to increased content of heavy metals in soil and water bodies is becoming concern for the present era. Lead is one of the most abundant heavy metals and non-essential metal polluting the soil environment and water and is considered to be one of the biggest sources of pollutants. This metal is very harmful for plants mainly because it is readily absorbed by plants mainly through the root system and thereafter exerts its toxicity symptoms.

Its increasing levels in soil environment inhibit germination of seeds and exerts a wide range of adverse effects on growth and metabolism of plants (Moustakas et al. 1994).

The toxic symptoms resulting due to Pb phytotoxicity include stunted growth, chlorosis, blackening of the root systems, ultimately damaging the water and nutritional status of plants. Several studies with this metal have reported the pivotal role of the antioxidant enzyme system that helps in combating the toxic effect incurred due to this metal in plants system. A study by Verma and Dubey in the year 1993 clearly depict in rice plants an enhancement in the activity of guaiacol peroxidase during Pb induced stress, suggesting that this enzyme serves as an intrinsic defense tool to resist Pb-induced oxidative damage in rice plants. Glutathione action in stress management is through its direct participation in $H_2O_2$ reduction catalysed by the enzyme glutathione peroxidase. Glutathione peroxidase activity and sequences encoding glutathione peroxidase-like genes have been demonstrated in several species including *Nicotiana sylvestris, Citrus sinensis, Arabidopsis thaliana, Avena fatua,* and *Brassica compestris,* indicating that glutathione peroxidase is present in plants (Eshdat 1997). Not only the role of guaicol peroxidase shows elevated results with Pb toxicity, but enzymes like SOD and GR also appear to play a pivotal role in combating oxidative stress in plants (Verma and Dubey 2003). Most of the results suggest that plants possess antioxidant enzymes, which operate either unspecifically (SOD and APX) or depending on the nature of the contamination (CAT, GPX, GRD). Positive relationship between increased POD and CAT enzyme activity and amounts of heavy metals such as Cu, Pb and Zn in plant tissue (Mazhoudi et al. 1997; Mocquot et al. 1996) suggest the pivotal role of these enzymes in plants defense to combat heavy metal stress. These enzymes remove superoxide radicals, which are harmful to cell membranes. Over expression of genes encoding these enzymes in several transgenic plant species conferring protection against free radicals has also been demonstrated (Allen 1995). For example, it is reported that Pb toxicity resulted in oxidative stress in rice plants and the enzymes peroxidases, SOD and GR appear to play a pivotal role in combating oxidative stress in plants. Unlike iron, Pb is not an oxido-reducing metal, so the oxidative stress induced by Pb in growing rice seedlings appears to be indirect effect of Pb toxicity leading to production of ROS with simultaneous increase of antioxidant enzymes (Verma and Dubey 2003).

## Chromium

Chromium is one of the most important heavy metal, and it is considered to cause a serious environmental pollution. Hexavalent chromium is highly toxic than trivalent form. The toxicity of chromium to plants depends upon its valency state. Modern industrial use of chromium has led to its ubiquitous distribution in our environment, and at higher concentrations, chromium shows negative effects on germination, growth, biochemical changes and enzyme activities (Shanker et al. 2007). Like copper and iron, chromium is also a redox metal and its redox behaviour exceeds that of other metals like Co, Fe, Zn, Ni, etc. Cr was thought to be a non-redox metal that could not participate in fenton reactions, however, other studies have shown that Cr can indeed participate in Fenton reactions, proving its redox character (Shi and Dalal, 1989). Cr reactivity can be considered from its interaction with glutathione, NADH and $H_2O_2$, forming OH- radicals in cell-free systems (Shi and Dalal, 1989; Aiyar et al., 1991). Production of $H_2O_2$, $OH^-$ and $O_2^-$ under Cr stress has been demonstrated in many plants, generating oxidative stress leading to damage of DNA, proteins and pigments as well as initiating lipid peroxidation (Panda, and Choudhury, 2005). The redox behavior can thus be attributed to the direct involvement of chromium in inducing oxidative stress in plants. Chromium can affect antioxidant metabolism in plants. Antioxidant enzymes like SOD, CAT, POX and GR are found to be susceptible to chromium resulting in a decline in their catalytic activities. Unlike other heavy metals, Cr detoxification using phytoremediation has been studied little (Panda, and Choudhury, 2005).

It must be noted that Cr is a non essential element and toxic metal for plant and hence plant may not possess any specific mechanism for its transport. In most of the studies conducted, a gradual decrease in CAT activity was observed in plants, while in mosses both alleviation and decline was observed (Panda and Choudhury, 2005). During Cr uptake by plants, APX was found to be more efficient in destroying $H_2O_2$ than was catalase under both speciation of Cr. The reason for this could be that unlike CAT, which is present only in the peroxisome and has low substrate affinities since it requires simultaneous access of two molecules of $H_2O_2$, APX is present throughout the cell and has higher substrate affinity in the presence of ascorbic acid as a reductant (Shanker et al., 2003).

## Copper

Copper (Cu) is a micronutrient, therefore it is required only in small quantity, and as any other nonessential metal it is also detrimental to plants after certain threshold. It also plays an important role in $CO_2$ assimilation and ATP synthesis. Exposure of plants to excess Cu generates oxidative stress and ROS (Stadtman and Oliver, 1991). The protective mechanism against these stress is seen in many plants, e.g., *C. tinctorius* showed the stimulation of SOD activity along with CAT, which seemed to play a protective role against membrane damage, as Cu is particularly toxic to membranes (Mazhoudi et al. 1997). Thus, it can be inferred that production of SOD, POD and CAT may serve as useful biomarkers for Cu tolerance in plants.

## Zinc

Zinc (Zn) is an essential micronutrient required by the plants for normal growth and development. Like other heavy metals, excess Zn invariably shows marked alterations in electron transport, membrane permeability and uptake and translocation of nutrient elements (Wang et al. 2009b) resulting in inhibition of various plant metabolic functions, retarded growth, and senescence. Although it is a micronutrient and essential nutrient for living organisms, zinc toxicity in plants limits the growth of both root and shoot (Choi et al., 1996; Ebbs and Kochian, 1997) causing chlorosis in the younger leaves, which can extend to older leaves after prolonged exposure to high soil Zn levels (Ebbs and Kochian, 1997) like any other non essential metal like Cd or Hg. The chlorosis may arise partly from an induced iron (Fe) deficiency as hydrated $Zn^{+2}$ and $Fe^{+2}$ ions have similar radii (Marschner, 1986). Excess Zn can also give rise to manganese (Mn) and copper (Cu) deficiencies in plant shoots. Such deficiencies have been ascribed to a hindered transfer of these micronutrients from root to shoot. The increase in GR activity in plants growing in soils with Zn addition may be related to the maintenance of intracellular levels of reduced glutathione, which is required for phytochelatins biosynthesis (Gomes-Junior et al., 2006) to combat oxidative stress in plants. In fact, Zn ions have been shown to induce the synthesis of these peptides in plants (Gasic and Korban, 2007). GR can also be involved in $H_2O_2$ removal by the activation of ascorbate-glutathione cycle (Asada, 1999), in which APX also plays a role converting $H_2O_2$ to water. APX can be found in different cells compartments, such as the cytosol and plastids, possibly participating in the fine modulation of ROS for signaling. In contrast, CAT might be responsible for the removal of $H_2O_2$ when accumulated Zn in plants tissues exerts stress conditions (Foyer and Noctor, 2005), particularly in the peroxisomes. A decline in CAT activity was reported in many plants grown under excess Zn (Andrade et al. 2009). The decrease in CAT activity observed in plant supplemented with excess Zn might be due to inhibition of enzyme synthesis or a change in the assembly of enzyme subunits (Radiæ et al. 2010). However, in *Brassica juncea* grown under excess Zn, increased CAT activity was reported (Prasad et al. 1999). These inconsistent results regarding CAT activity might be due to differences in the plant organs studied, the plant growth conditions, the durations and concentrations of the metals utilized and the plant species.

## CONCLUSION

Antioxidant enzymes play a major role during the stress induced by the heavy metal or during uptake of heavy metal in the technique termed as phytoremediation. Antioxidant enzymes are not limited to combat heavy metal stress, but a variety of environmental stresses like soil salinity, drought, extremes of temperature are known to cause oxidative damage to plants either directly or indirectly by triggering an increased production of reactive oxygen species (ROS) (Verma and Dubey, 2003), and antioxidant enzymes plays a pivotal role in tolerance mechanisms of plants. The tolerance mechanisms induced by the spontaneous functioning of various antioxidant enzymes help in slowing down the oxidation of biomolecules and block the process of oxidative chain reactions. This role of antioxidant enzymes enables the plant to survive in stress condition, which is reached when plants accumulate excessive concentrated metals in their organs. It can be inferred from the various reported studies that antioxidative enzymes protect the plants to some extent against oxidative damage, but the direction of response is combination of various parameters vis plant species, the metal used for the treatment, or the type of stress conditions. It can be concluded from the various studies to date that antioxidant enzyme activities of plants under heavy metal stress during phytoremediation have evolved as an efficient defense system by which the ROS is scavenged by superoxide dismutase, catalase, peroxidase, and glutathione reductase (Joseph and Jini 2010).

It is a well-known fact that although phytoremediation is the use of living green plants to clean the contaminated water and soil. Specially selected or engineered plants are used in the process with combination of several factors to optimize the tolerance of a particular plant to a specific metal (http://arabidopsis info./students/dom/main page html). Therefore, we need to extend our understanding for factors influencing the tolerance mechanism or enzymes inhibition or induction under various stressful environments by carrying out studies with various plants and varying with different parameters vis temperature, time, pH and other parameters and different concentrations of different metals. These would facilitate us to develop strategies for useful plants to be well adapted to environmental stress through manipulation of antioxidant system and further increase the productivity of the plant under stress conditions also. For example, genetic modification offers a new hope for phytoremediation as GM approaches can be used to over express the enzymes involved in the existing plant metabolic pathways, or introduce new pathways into plants (Gallegos et al., 2002). Over expression of genes encoding these enzymes can be used for designing transgenic plants, which enable phytoremediation above the toxic threshold of a particular metal. Understanding of antioxidant enzyme response to heavy metal stress could also provide vital clue to optimizing efficient ways of phytoremediation and selection of the most suitable plants for the same.

## REFERENCES

Cakmak I. and Marschner H. (1992). Magnesium deficiency and high light intensity enhances activities of super oxide dimutase, ascorbate peroxidase and glutathione reductase in bean leaves. Plant Physiol: 98: 1222–1227.

Gallegos O.R.S., Oliveira A.V. Sliveira L.M. and Silveira E.R. (2002). Organic carbonate from Calotropics procera leaves, Fitoterapia. 73(3): 263–265.

Gill S.S. and Tuteja N. (2010). Reactive oxygen species and antioxidants machinery in abiotic stress tolerance in crop plants. Physiology and Biochemistry 48: 909–930.

Streit B. Strumm W. (1933). Chemical properties of metals and the process of bioaccumulation in terrestrial plants as Biomonitors in: Indications for Heavy Metals in Terresthial Environment, VCH Weinheim, 381.

# 12

# Reuse and Recycling of Treated Distillery Spent Wash: A Novel Approach

**VASANTHY, M*., THAMARAI, C**., SWABNA, A. AND A. KARTHIGA**

*Department of Environmental Biotechnology, School of Environmental Sciences, Bharathidasan University, Tiruchirapalli-620024*
**Correspondence Author: E-mail: drvasanthy02@gmail.com*
***Department of Biotechnology, Mother Teresa University for Women, Kodaikanal*

## ABSTRACT

The present investigation deals with the treatment of spent wash diluted with sugar effluent (named as sugar wash) using chemicals and biological material. As the initial concentration of physico-chemical parameters such as colour, TS, TDS, BOD, COD, chloride, nitrate and sulphate was higher, it is clear that before the treatment of the effluent, it should be diluted. Hence, it was diluted with sugar effluent, therefore, the resultant diluted effluent was called sugar wash (50%). The removal of colour, TDS and COD from the spent wash using chemical agents such as coagulating agent(CaO), oxidizing agent ($H_2O_2$) and adsorbent(CAC) was studied both individually and in combination. Complete colour was obtained from 100 ml sugar wash when treated with 7.5 g of CaO followed by different dosages of hydrogen peroxide and commercial activated carbon. It removed about 85.3% of TDS and 88% of COD when the sugar wash was treated with 7.5 g of CaO, 3.5 ml of $H_2O_2$ and 5 g of CAC. It was achieved by the coagulating effect of CaO and oxidizing property of hydrogen peroxide and adsorption capability of CAC. The optimum dosage of CaO alone is 7.5 g for the highest removal of (45.2%) TDS and (47.9%) COD. Further 7.5 g of CaO combined with 3.5 ml of $H_2O_2$ removed 90% of colour, 70.7% of TDS and 77% of COD. When 10 g of CAC was added to 7.5 g of CaO, it removed about 100% of colour, 71.1% of TDS and 79.1% COD, whereas the combined treatment of 7.5 g of CaO, 3.5 ml of $H_2O_2$ and 5 g of CAC was found to be the optimum dosage of the coagulating agent, oxidizing agent and adsorbent for the removal of colour (100%), TDS (85.3%) and COD (88%). The treatment of sugar wash with chemicals could not result in higher reduction of TDS and COD. But sugar wash treated with about 7.5 g of CaO, 5 g of nirmali seeds and 5 g of CAC showed the highest reduction of colour (100%), TDS (91%) and (96%) COD. The treated effluent was further neutralized and was used to irrigate the soil to grow horticultural plant, ***Tagetes erecta*** **L.** From the results it is clear that the treated sugar wash would not reduce the flowering capacity of ***Tagetes erecta*** **L.** which has economic importance. No remarkable change was noticed in the physico-chemical characterization of soil irrigated with treated effluent especially with respect to macro (N, P and K) and micronutrients (Fe, Cu, Mn and Zn). The solid waste generated during the treatment was subjected to compost and the manural value was also checked. The amount of N, P and K in the compost was appreciable along with its C: N ratio which happens to be 10:1. This proves the possibility of utilizing the sludge positively for the preparation of compost, which possess manural value too.

From the analysis, it may be concluded that the combined treatment of sugar wash with chemical and biological material (CaO 7.5 g + 5 g of Nirmali seeds +CAC 5 g) has shown the highest removal of colour, TDS and COD. And the horticultural plant growth has proved the possibility of reuse of treated sugar wash. Therefore, the reuse of the distillery spent wash, sugar industry effluent and the treated sludge is possible in the days to come.

**Keywords:** Sugar Wash, CaO, CAC, Nirmali Seeds, *Tagetes erecta*.

## INTRODUCTION

Today, the demand for energy is increasing at an alarming rate, which necessitates much work in the concerned field. With the tremendous urbanization and population explosion, the fuel consumption has increased to a maximum, thereby posing a threat to the non-renewable resources. Currently much importance is being given to select suitable renewable energy resources. One such resource happens to be the product of distillery i.e., ethanol. With an increase in the demand for ethanol, the problem of utilization of distillery effluent also enlarges into a major problem. The current condition also makes us feel the importance of water to a greater extent. Any method or idea that facilitates the reuse of any effluent will be highly appreciated. Of the various industries, which uses higher amount of water for the processes, distillery plays a major and an important role. These industries concentrate on the production of valuable ethanol accompanied with the discharge of spent wash, which has importance.

The Central Pollution Control Board (CPCB) has listed distilleries as one of the seventeen most polluting industries. The distillery wastewater is known as spent wash. For every litre of alcohol produced, molasses based distilleries have been reported to generate 8-15L of wastewater characterized by high BOD (45-60000mg/l), high COD (80-160000mg/l) and dark colour (CPCB, 2003).Spent wash is characterized by dark brown colour, high temperature, low pH, high percentage of dissolved organic and inorganic matter which may be present as reducing sugars (Bhavik et al., 2010). The spent wash is acidic (pH 3.94-4.30), dark brown liquid with high BOD (45000–100000 mg l-1) and COD (90000–210000 mg l-1), and emits obnoxious odour. Although it does not contain toxic substances, its discharge without any treatment brings about immediate discolouration and depletion of dissolved oxygen in the receiving water streams, posing serious threat to the aquatic flora and fauna (Murthy and Chaudhari, 2009).

The high chemical oxygen demand (COD) and biochemical oxygen demand (BOD) of the distillery spent wash are due to the presence of a number of organic compounds, such as polysaccharides, reduced sugars, lignin, proteins, melanoidin, waxes, etc. It contains dark brown recalcitrant pigment called melanoidin (nearly 2%), which is formed in Maillard reactions between amino acids and carbohydrates. These Maillard compounds have antioxidant properties that render them toxic to aquatic micro- and macro-organisms. These colored components lead to a reduction in sunlight penetration in rivers, lakes or lagoons, which in turn decreases both photosynthetic activity and dissolved oxygen concentrations causing serious harm to aquatic biota (Kaushik G. and Thakur I.S. 2009).

Inthorn et al. (2001) have stated that conventional biological process such as activated sludge treatment is inefficient to decolorize melanoidin-containing wastewater. An inexpensive and effective technology for decolorizing distillery wastewater has not yet been developed for practical applications.Asthana et al. (2001) have observed that even after anaerobic treatment, the effluent does not meet the stringent standards laid down by Central Pollution Control Board, India, in terms of very high levels of BOD, COD, total solids etc. These observations indicate the necessity for suitable alternative method which would facilitate the treatment of distillery wastewater.

***Tagetes erecta*** **L.** was selected for the present investigation due to the following reasons. The flowers of ***Tagetes erecta*** **L.** are commonly used for garland making in many states of India. It has gained popularity amongst the gardeners and flower dealers on account of its easy culture methodology and wide adaptability. Due to their free flowering nature, short duration needed to produce marketable flowers, long shelf life and attractive colours of the flowers, the plant cultivation has become popularized. Soule (1993) further reported that, is being used as a beverage in South Carolina and Southern U.S.A. Shweta et al. (2006) have reported the same concept further justifying the sensation of Indian floriculturists to grow ***Tagetes erecta*** **L.** due to their economic importance too.

Hence, as a lot of research is essential to arrive at a suitable methodology for the treatment of spent wash, the present work was focused on decolourisation and reduction of TDS and COD by using chemicals, biological agent and microbial treatment. Among the treatments, the higher percent removal of colour (100%), COD (96%) and TDS (91%) using chemical combined with biological material and adsorbent was observed. Such treated effluent was used to grow the horticultural plant, ***Tagetes erecta*** **L.**

## MATERIALS AND METHODS

### Sampling of Effluent

Spent wash and sugar effluent were collected from sugar cum distillery plant located at Trichy, Tamilnadu. All the chemicals used were of Analytical grade. All glass double distilled water was used throughout the investigation.

### Physico-chemical Characterization of Spent Wash and Sugar Effluent

The physico-chemical characteristics of spent wash, sugar effluent such as Colour, Odour, pH, temperature, total solids (TS), total dissolved solids (TDS), total suspended solids (TSS), sodium, potassium, sulphate, chloride, nitrate, phosphate, dissolved oxygen (DO), biological oxygen demand (BOD) and chemical oxygen demand (COD) were determined using standard analytical methods (APHA, 1989). Due to higher pollution potential of spent wash, it was diluted with sugar effluent.

### Dilution of Spent Wash with Sugar Effluent

Most of the distilleries are located along with sugar industries. The colour, BOD and COD of sugar effluent was not so high when compared to spent wash. Hence, distillery spent wash was diluted with sugar effluent to obtain 50% dilution. The combination of 50% sugar effluent and 50% spent wash was named as sugar wash.

### Physico-chemical Characterization of Sugar Wash

The various characteristics of sugar wash such as colour, odour, temperature, pH, TS, TDS, TSS, DO, BOD, COD, nitrate, chloride, sodium, potassium, phosphate and sulphate were estimated as per APHA (1989).

### Batch Type Experiments

### Treatment of Sugar Wash

Sugar wash was subjected to treatment with chemical and biological material for the reduction of colour, TDS and COD.

#### *Effect of treatment with individual chemicals on sugar wash*

The oxidizing agent (hydrogen peroxide ($H_2O_2$)), coagulating agent (calcium oxide (CaO)), and adsorbent (commercial activated carbon - CAC) were selected and used for colour, TDS and COD removal. The CAC was obtained from Nice Chemicals Pvt. Ltd., Cochin. Whatman - No-1 paper was utilized for all the filtration.

#### *Effect of Individual chemicals (CaO, $H_2O_2$ and CAC) on the colour, TDS and COD removal from the sugar wash*

Three sets of 6 conical flasks were used and 100ml of sugar wash was dispensed into each flask. To 2 sets of 6 flasks, 6, 6.5, 7, 7.5, 8 and 8.5 g of CaO and CAC were added individually to the appropriate conical flasks. To the remaining set of 6 flasks 2, 4, 6, 8, 10 and 12 ml of $H_2O_2$ was added. They were allowed for 30 minutes for chemical reaction. The flasks were taken and filtered through the whatman No.1 filter paper. The colour, TDS and COD were estimated and recorded. The experiments performed were tabulated as follows (Table 1).

**Table 1** Dosage of chemicals (CaO, $H_2O_2$ and CAC) used for the treatment of sugar wash

| *S. No.* | *No. of Flasks Taken* | *Nature of the Agent* | *Chemicals Used* | *Dosage* |
|---|---|---|---|---|
| 1. | 6 | Coagulating agent | CaO(g) | 6, 6.5, 7, 7.5, 8, 8.5 |
| 2. | 6 | Adsorbent | CAC (g) | 6, 6.5, 7, 7.5, 8, 8.5 |
| 3. | 6 | Oxidising agent | $H_2O_2$ (ml) | 2, 4, 6, 8, 10, 12 |

### *Effect of coagulating agent and oxidizing agent on the colour, TDS and COD removal from the sugar wash*

Three sets of 6 conical flasks were used, and to each flask 100ml of sugar wash was added. To all the flasks, 7.5g of calcium oxide was added and left for half an hour and was filtered. After the filtration, the first set was subjected to treatment with 2.5, 3, 3.5, 4, 4.5 and 5 ml of $H_2O_2$ and was allowed for 30 minutes after which colour, TDS and COD removal were estimated following the filtration (Table 2).

### *Effect of coagulating agent and adsorbent on the colour, TDS and COD removal from the sugar wash*

To the second set of CaO treated effluent, 6, 7, 8, 9, 10 and 11 g of CAC was added and allowed for 30 minutes in static condition. After the filtration the percentage of colour, TDS and COD removal was estimated and recorded (Table 2).

### *Coagulating agent, oxidizing agent and adsorbent used for the removal of colour, TDS and COD from the sugar wash*

The third set was treated with 3.5ml of $H_2O_2$ followed by 1, 2, 3, 4, 5 and 6 g of CAC. The content was left for 30 minutes and the above said parameters were analysed. The treatment performed is presented in Table 2.

**Table 2** Dosage of combined chemicals used for the treatment of sugar wash

| *S. No.* | *No. of Flasks Taken* | *Nature of the Agents* | *Chemicals Used* | *Dosage* |
|---|---|---|---|---|
| 1. | 6 | Coagulating agent (g)+ Oxidising agent (ml) | CaO+ $H_2O_2$ | 7.5 + 2.5, 7.5 + 3, 7.5 + 3.5, 7.5 + 4, 7.5+ 4.5, 7.5 + 5. |
| 2. | 6 | Coagulating agent (g)+ Adsorbent (g) | CaO+ CAC | 7.5 + 6, 7.5+ 7, 7.5 + 8, 7.5 + 9, 7.5+10, 7.5+11. |
| 3. | 6 | Coagulating agent (g)+ Oxidizing agent (ml)+ Adsorbent (g) | CaO+ $H_2O_2$ +CAC | 7.5+3.5+1, 7.5+3.5+2, 7.5+3.5+3, 7.5+3.5+4, 7.5+3.5+5, 7.5+3.5+6 |

### Effect of Treatment with Biological Material on the Colour, TDS and COD Removal from the Sugar Wash

Nirmali seeds (***Strychnos Potatorum* L.**) obtained from the Tamilnadu Agricultural University, Coimbatore, were crushed and sieved through BSS No.18. About 100ml of sugar wash was added to 6 conical flasks. To the flasks 1, 2, 3, 4, 5 and 6 g of ***Strychnos potatorum* L.** was added and was allowed for 30 minutes. The sugar wash was subjected to colour, TDS and COD analysis.

#### *Combined effect of treatment with chemical and biological agent on the colour, TDS and COD removal from the sugar wash*

About 100 ml of sugar wash was taken in 6 conical flasks and was treated with 7.5 g of CaO followed by 1, 2, 3, 4, 5 and 6 g of ***Strychnos potatorum* L.** It was left for half an hour and filtered. The filtrate was subjected to treatment with 5 g of CAC. After the filtration, the above-mentioned parameters were estimated in the filtrate. The treatment was performed as in Table 3.

**Table 3** Dosage of chemical agents and biological materials used for the treatment of sugar wash

| *S. No.* | *No. of Flasks Taken* | *Nature of the Agent* | *Chemicals/biological Material Used* | *Amount of the Agent Used (g)* |
|---|---|---|---|---|
| 1. | 6 | Coagulating | *Strychnos Potatorum* L. | 1, 2, 3, 4, 5 and 6 |
| 2. | 6 | Coagulating + Coagulating and reducing + adsorbent | *CaO+Strychnos Potatorum* L.+CAC | 7.5+1+5<br>7.5+2+5<br>7.5+3+5<br>7.5+4+5<br>7.5+5+5<br>7.5+6+5 |

### Biochemical Characteristics of the Biological Materials

Biochemical characters of nirmali seed were analysed. The parameters such as water solubility, swelling factors and mucilage property of nirmali seeds were estimated (Kokate, 1993).

Further the moisture content, amount of protein, carbohydrate, fibre, fat and ash present in the biological material were also estimated (Horwitz, 2000). Biochemical components and minerals were estimated using gas chromatography-mass spectroscopy and AAS, respectively.

#### *Decolourising power of nirmali seeds (ISI, 1977)*

About 100mg of the nirmali seed was taken in a 25 ml conical flask. Methylene blue solution (0.15%) was added from a burette and shaken for 5 minutes. The methylene blue solution was added with continuous shaking till the blue colour persisted at least for five minutes.

$$\text{Decolourising power (mg of methylene blue Adsorbed by g of adsorbent)} = \frac{1.5 \times V}{A}$$

V = Volume of methylene blue solution consumed (ml).
A = Weight of the adsorbent (g)

### Characterization of the Adsorbent (ISI, 1977)

Apparent density, moisture content, ash content, decolorizing power of CAC, pH, surface area (acetic area adsorption method–Shoemaker and Garland, 1967) water soluble matter, acid-soluble matter were estimated.

### Effect of Treated Sugar Wash on the growth Parameters of Marigold (*Tagetes erecta* L.) under Experimental Conditions

The seeds of ***Tagetes erecta*** **L.** were procured from Tamil Nadu Agricultural University (TNAU), Coimbatore.

#### *Sterilization of seeds*

The selected healthy seeds were divided into lots of 20 each. Prior to treatment with sugar wash, the seeds were surface sterilized with 0.1% mercuric chloride for 1-2 minutes followed by repeated rinsing with distilled water.

#### *Effect of treated sugar wash on the seed germination, growth parameters and flowering capacity of Marigold (Tagetes erecta L.)*

The experiments were conducted under uniform condition of light and equal amount of soil was placed in three containers. About 20 seeds were evenly placed in each container. One was maintained as control and tap water was used for irrigation. The second one was irrigated with the treated effluent regularly (DI-Direct irrigation). The third set was alternatively flooded (ADI-Alternative days irrigation) with tap water and treated effluent. The percentage of seed germination was recorded daily at 11.00 am for a time period of 10 days.The length of the root and shoot of the 90-days-old seedlings was measured and the mean length was calculated.

After harvesting, the fresh weight was taken immediately. Then the plants were dried in the oven at 80°C for 72 hours following which they were weighed and recorded as dry weight. The flowering capacity of the plant was noticed after 90 days. The number of flowers were counted and recorded as the yield.

### Physico-chemical Characterisation of Soil before and after Irrigation with Treated Sugar Wash

Before and after the experiment, the soil from control and the soil irrigated with treated effluent were subjected for analysis. The pH, EC, available nitrogen, potassium, phosphorus, Fe, Zn, Mn and Cu of soil were estimated (Muthuvel and Udayasoorian, 1999).

### Preparation of used Adsorbents and Coagulants Sludge Material for Composting

The sludge material obtained after the treatment of sugar wash was subjected to composting. Two sets of sludge materials were prepared.

**Set (A)**—About two parts of cow dung, 1 part of garden waste, 2 parts of sludge were mixed to get 2:1:2 ratio.

**Set (B)**—About two parts of cowdung, 2 parts of garden waste, 1 part of sludge were mixed to get 2:2:1 ratio.

#### *Composting process of sludge material*

Prepared sets of A and B sludge materials were subjected to aerobic degradation in a container. Water was sprinkled on the surface to maintain the required moisture level. Thorough mixing of the materials was done regularly until the composting process got complete. (Vasanthy et al. 2005)

#### *Studies on the manural value of the compost*

After 60 days, the compost was taken from the pit and subjected to analysis. The organic carbon, available nitrogen, potassium, phosphorus, total organic matter and C/N were estimated as per the standard method (Muthuvel and Udayasoorian, 1999).

## RESULTS AND DISCUSSION

### Physico-Chemical Characterization of Spent Wash and Sugar Effluent

The colour and odour of the spent wash were dark brown and objectionable respectively. The pH of the sample was of acidic range and the temperature was about 90°C. The DO of the spent wash was nil. Hence the BOD and COD of spent wash were found to be 60000 mg/L and 126000 mg/L respectively. The TDS of spent wash was 90000 mg/L. The amounts of chloride, nitrate, sulphate and potassium were 7000, 2500, 7000 and 12000 mg/L respectively. The colour may be due to the presence of melanoidin pigments present in the spent wash. These melanoidin pigments have been reported as natural polymers, highly resistant to biodegradation (Inthorn et al. 2001). Anthrocyanine and tannin are formed due to the non-coloured sugars like caramels (Saxena and Rai, 2000, Kasturibai and Ganga, 1996). The dark brown melanoidin pigment was found to be resistant to reduction even after subjecting the spent wash to secondary waste water treatment (Patil and Kapadnis, 1995).

Most malodour of an inorganic nature has been reported to arise from the anaerobic decomposition of compounds containing nitrogen and sulphur, such as ammonia, hydrogen sulphide, mercaptans, amines, aldehydes, ketones, indole and skatole (Hwang et al. 1994).

The distillery spent wash is hot and highly acidic due to the presence of organic acids too. Binkly and Wolfrom (1953) have reported that 10% of acids such as aconitic, citric, malic, glutamic, succinic, acetic, lactic acid are produced during the fermentation of molasses. This may be the reason for acidic pH. They have also reported that higher solids, BOD and COD could be attributed to 40-50% of total reducing sugars and unfermentable sugars like pentoses and polysaccharides, sterols, wax, proteins along with organic acids, nitrogenous compounds and minerals too (Bardiya et al. 1995). Similarly Deshananda et al. (1993) reported that Indian spent wash contains large percentage of potassium in comparison to spent wash from other countries. This may be due to the higher accumulation of potassium by the sugarcane grown in the soil with higher K content originating from the minerals.

The total solids, total dissolved solids of the spent wash were so high, which may be the reason for higher BOD and COD.

**Table 4** Physico-chemical characteristics of the effluents used for the present study

| *S. No.* | *Parameters* | *Spent Wash** | *Sugar Effluent** | *Sugar Wash (50%)** |
|---|---|---|---|---|
| 1. | Colour | Dark brown | Pale yellow | Dark brown |
| 2. | Odour | Objectionable | Sweet smell | Objectionable |
| 3. | Temperature | 90°C | 45°C | 32°C |
| 4. | pH | 4.5 | 6.4 | 5.2 |
| 5. | TS | 100000 | 3600 | 52600 |
| 6. | TDS | 90000 | 3000 | 52060 |
| 7. | TSS | 10000 | 600 | 580 |
| 8. | DO | – | 0.9 | – |
| 9. | BOD | 60000 | 5050 | 32060 |
| 10. | COD | 126000 | 6165 | 63500 |
| 11. | Chloride | 7000 | 95 | 3600 |
| 12. | Nitrate | 2500 | 100 | 1650 |
| 13. | Phosphate | 250 | 180 | 250 |
| 14. | Potassium | 12000 | 750 | 6700 |
| 15. | Sodium | 200 | 500 | 325 |
| 16. | Sulphate | 7000 | 170 | 3600 |

*All the values are tabulated in mg/L, except pH and temperature.

The colloidal nature of caramels and melanoidin were also found to increase the amount of total solids, BOD and COD (Saxena and Rai 2000). Large quantities of inorganic constituents contribute to the high total solids content of about 100000 mg/L, chloride 7000 mg/L, about 7000 mg/L of sulphate and 2500 mg/L of nitrates.

Similar results of low pH, dark brown colour, objectionable odour, higher amount of solids, BOD, COD, potassium and nitrate have been reported by Handa and Seth (1990), Kasturibai and Ganga (1996), Srivastava and Pathak (1998), Vasanthy et al. (2004a, 2006a, 2006b). The higher values of TDS, BOD, COD and chloride proves the toxicity of the effluent, and this effluent cannot be discharged into inland surface water, public sewers or as such for irrigational purpose because the values are not within the tolerance limit. Hence, the spent wash must be subjected to treatment prior to disposal and it needs dilution before treatment.

The characterization of the sugar effluent was also done (Table 4). The colour and odour of the sugar effluent was pale yellow and sweet smell respectively. The pH of the sugar effluent was 6.4. The total solids and total dissolved solids of sugar effluent were 3,600 mg/L and 3,000 mg/L respectively. The DO of sugar effluent was 0.9 mg/L. The BOD and COD were 5050 mg/L and 6165 mg/L respectively.

### Dilution of Spent Wash with Sugar Effluent

Many distilleries are located along with sugar industries. The colour, higher BOD and COD of spent wash render them very difficult to be treated. Hence, the dilution of spent wash is utmost necessary. The physico-chemical characteristics such as colour, TDS, BOD and COD of sugar effluent were not so high when compared to spent wash. Reduction of rainfall and ground water resources has resulted in water crisis problem. Hence, sugar effluent was thought to be suitable diluent for the dilution of spent wash.

### Physico-chemical Characterization of Sugar Wash

The colour and odour of the sugar wash were very dark brown and objectionable, respectively. The pH of the sugar wash was almost neutral. The DO was nil. The BOD and COD of sugar wash was found to be 32060 mg/L and 63500 mg/L, respectively (Table 4).

## BATCH TYPE EXPERIMENTS

### Effect of Individual Chemicals on Sugar Wash

The effect of the coagulating agent, oxidizing agent and adsorbent on colour, TDS and COD removal from the sugar wash was determined (Tables 5, 6 and 7).

#### *Effect of coagulating agent on the colour, TDS and COD removal from the sugar wash*

About 60, 65 and 68% of colour removal took place, when the diluted spent wash was treated with 6, 6.5 and 7 g of CaO respectively. All other treatments showed about 70% of colour reduction. About 45.2% of TDS and 47.9% COD were removed when the sugar wash was treated with 7.5 g of CaO. This suggests that could be the optimum dosage of CaO for the treatment of sugar wash. With further increase in the dosage of CaO, TDS and COD removal showed a very slight reduction.

**Table 5** Effect of coagulating agent (CaO) on the colour, TDS and COD removal from the sugar wash

| *Sl. No.* | *Dosage of Calcium Oxide (g)* | *Colour Removal %* | *TDS Reduction %* | *COD Reduction %* |
|---|---|---|---|---|
| 1. | 6 | 60 | 16.1 | 20 |
| 2. | 6.5 | 65 | 23.3 | 32.7 |
| 3. | 7 | 68 | 32.5 | 41.8 |
| 4. | 7.5 | 70 | 45.2 | 47.9 |
| 5. | 8 | 70 | 45 | 47.5 |
| 6. | 8.5 | 70 | 45.1 | 47.2 |

The colour causing substances such as melanoidin should have been degraded when the sugar wash was treated with different dosages of CaO. Colour removal increased with increasing dosage of CaO upto 7.5 g/100 ml sugar wash after which there was no significant colour reduction. Treatment with 8, 8.5 g of CaO resulted in 70% of colour reduction. This removal must have been achieved by the hydroxyl radicals donated by CaO. Similarly Inthorn et al. (2001) have reported that the hydroxyl radicals generated are capable of degrading the contaminants, especially chromophore or chromogen through the process of oxidation. Similarly Vasanthy et al. (2006b) have reported that 0.75 g of CaO is the optimum dosage for the highest removal of TDS and COD from the 5% sugar wash.

### *Effect of oxidizing agent on the colour, TDS and COD removal from the sugar wash*

There was no colour removal in sugar wash when treated with 2 ml and 4 ml of $H_2O_2$. Only 10 % of colour removal was noticed in the spent wash when treated with 10 ml and 12 ml of hydrogen peroxide. Also the treatment resulted in very poor reduction of TDS and COD (Fig. 5). It was clear that $H_2O_2$ could not oxidize the impurities from the sugar wash individually. But hydrogen peroxide combined with ferrous iron has shown higher reduction of colour, TDS and COD (Inthorn et al. 2001). Similarly Vasanthy et al. (2006a) have reported that hydrogen peroxide combined with CaO was capable of removing colour, TDS and COD.

**Table 6** Effect of oxidising agent ($H_2O_2$) on the colour, TDS and COD removal from the sugar wash

| *Sl. No.* | *Dosage of Hydrogen Peroxide (ml)* | *Colour Removal %* | *TDS Reduction %* | *COD Reduction %* |
|---|---|---|---|---|
| 1. | 2 | No change | 0.1 | 0.15 |
| 2. | 4 | No change | 0.2 | 0.45 |
| 3. | 6 | 2 | 1.2 | 2.4 |
| 4. | 8 | 4 | 1.8 | 3.9 |
| 5. | 10 | 10 | 3.9 | 5.6 |
| 6. | 12 | 10 | 3.8 | 5.6 |

### *Effect of adsorbent (CAC) on the colour, TDS and COD removal from the sugar wash*

Not much colour, TDS and COD removal were noticed from the sugar wash when treated with commercial activated carbon. Only 15% of colour removal and 12% removal of TDS and COD took place when the sugar wash was treated with 8 and 8.5 g of CAC without any agitation. Similar results have been reported by Vijayakumari (2006) and Kalpana (2006). It is clear that the process of adsorption alone could not result in higher removal of colour, TDS and COD from the sugar wash.

**Table 7** Effect of adsorbent (CAC) on the colour, TDS and COD removal from the sugar wash

| *S. No.* | *Dosage of Commercial Activated Carbon (g)* | *Colour Removal %* | *TDS Removal %* | *COD Removal %* |
|---|---|---|---|---|
| 1. | 6 | No change | 0.11 | 0.15 |
| 2. | 6.5 | 1 | 0.14 | 0.78 |
| 3. | 7 | 5 | 2 | 3.9 |
| 4. | 7.5 | 8 | 3.9 | 4.7 |
| 5. | 8 | 15 | 11.8 | 12.2 |
| 6. | 8.5 | 15 | 11.9 | 12 |

## Effect of Combined Chemical Treatment on Sugar Wash

### *Effect of coagulating agent and oxidizing agent on the colour, TDS and COD removal from the sugar wash*

Higher percentage of colour reduction of about 90% was noticed when the spent wash was treated with 7.5 g of calcium oxide and 3.5, 4, 4.5 and 5 ml of hydrogen peroxide, and the other treatments also showed appreciable reduction of colour i.e., 85%. Sugar wash treated with 7.5 g of CaO, and 3.5 ml of $H_2O_2$ showed the highest reduction of 70.7% of TDS and 77% of COD. It must be the optimum dosage for the TDS and COD removal. When the dosages of calcium oxide and hydrogen peroxide were increased from the optimum dose, the removal capacity of agents decreased slightly. The combined dosage of CaO and $H_2O_2$ resulted in appreciable colour, TDS and COD removal. It must have been achieved by the hydroxyl radicals donated by CaO combined with $H_2O_2$, which could have produced perhydroxyl anions ($HOO^-$), which oxidizes the impurities. Similar results have been reported by Patil and Kapadnis (1995). During oxidation, the impurities might have got converted into $CO_2$, nitrogen and water. Similar view has been stated as the reason by Berchmans and Vijayavalli (1989) for the higher reduction of colour, TDS and COD.

**Table 8** Effect of coagulating agent and oxidizing agent on the colour, TDS and COD removal from the sugar wash

| *S. No.* | *Calcium Oxide(g)+ Hydrogen Peroxide(ml)* | *Colour Removal %* | *TDS Removal %* | *COD Removal %* |
|---|---|---|---|---|
| | *Dosage* | | | |
| 1. | 7.5+2.5 | 85 | 65 | 70.8 |
| 2. | 7.5+3 | 85 | 66.9 | 74.3 |
| 3. | 7.5+3.5 | 90 | 70.7 | 77 |
| 4. | 7.5+4 | 90 | 70 | 76.6 |
| 5. | 7.5+4.5 | 90 | 69 | 76 |
| 6. | 7.5+5 | 90 | 68.2 | 76.2 |

### *Effect of coagulating agent and adsorbent on the colour, TDS and COD removal from the sugar wash*

The combined effect of CaO and CAC showed 100% colour reduction when the sugar wash was treated with 7.5 g of CaO followed by 11 and 10 g of CAC. Other treatments showed about 70–95% of significant colour reduction. The highest reduction of TDS i.e., about 71.1% and 79.1% of COD were noticed from the sugar wash, when treated with 7.5 g of CaO and 10 g of CAC. This must have been possible due to the combined effect of OH capable of oxidizing the impurities and due to the adsorption capability of CAC. Similar results have been reported by Vasanthy et al. (2006b), Ramteke et al. (1989). The authors reported that activated carbon is the most efficient adsorbent for colour causing organics, non-colour causing organics and COD removal.

**Table 9** Effect of coagulating agent and adsorbent on the colour, TDS and COD removal from the sugar wash

| *S. No.* | *Calcium Oxide(g)+ CAC(g)* | *Colour Removal %* | *TDS Removal %* | *COD Removal %* |
|---|---|---|---|---|
| | *Dosage* | | | |
| 1. | 7.5+6 | 70 | 66.9 | 74.3 |
| 2. | 7.5+7 | 80 | 67.3 | 75.2 |
| 3. | 7.5+8 | 90 | 68 | 76.3 |
| 4. | 7.5+9 | 95 | 70.8 | 76.9 |
| 5. | 7.5+10 | 100 | 71.1 | 79.1 |
| 6. | 7.5+11 | 100 | 69.8 | 79.1 |

### *Effect of coagulating agent, oxidizing agent and adsorbent on the colour, TDS and COD removal from the sugar wash*

Sugar wash treated with different dosages of CaO, $H_2O_2$ and CAC has shown 100% removal of colour, 85.3% of TDS and 88% of COD when treated with 7.5 g of CaO, 3.5 ml of $H_2O_2$ and 5 g of CAC. The optimum dosage is found to be 7.5 g of CaO, 3.5 ml of $H_2O_2$ and 5 g of CAC for 100 ml of sugar wash, as this dosage has resulted in maximum removal of colour, TDS and COD. Further increase or decrease in the dosage of agents from the optimum dose has not shown much difference with respect to TDS and COD removal. The mechanism may be as follows. The solution becomes alkaline due to $OH^-$ ions formed on reaction with CaO. In the presence of $OH^-$ ions, the $H_2O_2$ reacts with these hydroxyl ions, which gives mainly per hydroxyl anions ($HOO^-$). It may be the reason for the highest level of oxidation of organic matter, thereby resulting in the highest reduction of TDS and COD. Similar results have been reported by Patil and Kapadnis (1995), Ramteke et al. (1989) along with Vasanthy et al. (2006b). Berchmans and Vijayavalli (1989) have stated that the final products of oxidation of organic matter could be carbon dioxide, water and nitrogen.

**Table 10** Effect of coagulating agent, oxidizing agent and adsorbent on the colour, TDS and COD removal from the sugar wash

| *S. No.* | *Calcium Oxide(g)+ Hydrogen Peroxide(ml) + CAC(g)* <br> *Dosage* | *Colour Removal %* | *TDS Removal %* | *COD Removal %* |
|---|---|---|---|---|
| 1. | 7.5+3.5+1 | 100 | 78.8 | 80.9 |
| 2. | 7.5+3.5+2 | 100 | 79.7 | 81.9 |
| 3. | 7.5+3.5+3 | 100 | 80.7 | 84.2 |
| 4. | 7.5+3.5+4 | 100 | 82.4 | 85.1 |
| 5. | 7.5+3.5+5 | 100 | 85.3 | 88.0 |
| 6. | 7.5+3.5+6 | 100 | 85 | 88 |

## Effect of Treatment with Biological Material on Sugar Wash

The use of biocoagulants is an important traditional practice employed in making drinking water as well as wastewater cleaner. People of Chattisgarh and Tamil Nadu are use nirmali seeds for water purification traditionally. Hence, nirmali seeds were collected and used for the treatment of sugar wash.

### *Effect of treatment with biological material on the colour, TDS and COD removal from the sugar wash*

No colour removal was noticed when the sugar wash was treated with nirmali seeds alone. It may be due to the fact that nirmali seeds do not posses oxidizing power for degrading the colour causing substances. About 25.4% and 27.6% of TDS and COD have been removed from the sugar wash when treated with 6 g of nirmali seeds. It might have been achieved by the antioxidant property of nirmali seeds. This property has been reported by Anitha (2006), Vasanthy et al. (2006c). Nirmali seeds contain an antioxidant like palmitic acid, which may be the reason for the reduction of TDS and COD (www. Dr. Duke's phytochemical and ethnobotanical databases.com).

### *Combined effect of treatment with chemical agents and biological material on the colour, TDS and COD removal from the sugar wash*

The combined treatment has shown the highest removal of colour and remarkable reduction of TDS and COD. About 91% of TDS and 96% of COD have been removed by the combined treatment with 7.5 g of CaO followed by 5 g of nirmali seeds and 5 g of CAC without any agitation. The highest removal of TDS and COD was accompanied by the oxidation of impurities by hydroxyl radicals donated by CaO, followed by the coagulating and anti-oxidising property of nirmali seeds, followed by the adsorbing capability of activated carbon. The coagulating property of Nirmali seed has been reported by Rangwala et al. (1997) (www.natural coagulant.com). From the phytochemical analysis of the seed, it is clear that it has an important antioxidant compound such as n-hexadecanoic acid (palmitic acid ). (www.Dr.Duke's phytochemical and ethnobotanical databases.com). This antioxidant must have reduced the oxygen demanding species like $NO^-_3$, $NO^-_2$, $PO^{3-}_3$, $SO^{2-}_4$ etc. Nitrate removal property of nirmali seeds has also been reported by Anitha (2006). Thereby, it must have removed the chemical species, which results in the highest removal of TDS and COD.

It is thus clear that the chemical and biological treatments when combined together, might result in better treatment of spent wash diluted with sugar effluent called sugar wash. Such treatment would make the recycling of spent wash possible. The sugar wash subjected to treatment with 7.5 g of CaO, 5 g of nirmali seeds and 5 g of CAC was neutralized and was used for irrigating the plants of Marigold.

**Table 11** Combined effect of chemical and biological material on the colour, TDS and COD removal from the sugar wash

| *S. No.* | *Agent* | *Dosage (g)* | *Colour Removal %* | *TDS Reduction %* | *COD Reduction %* |
|---|---|---|---|---|---|
| 1. | CaO +***Strychnos potatorum*** L.+CAC | 7.5+1+5 | 100 | 76 | 80.9 |
| 2. | CaO + ***Strychnos potatorum*** L.+CAC | 7.5+2+5 | 100 | 80.77 | 84 |
| 3. | CaO + ***Strychnos potatorum*** L.+CAC | 7.5+3+5 | 100 | 84 | 88 |
| 4. | CaO + ***Strychnos potatorum*** L.+CAC | 7.5+4+5 | 100 | 86.72 | 90 |
| 5. | CaO + ***Strychnos potatorum*** L.+CAC | 7.5+5+5 | 100 | 91 | 96 |
| 6. | CaO + ***Strychnos potatorum*** L.+CAC | 7.5+6+5 | 100 | 91 | 96 |

### *Biochemical characteristics of biological material*

The water solubility, swelling factors, mucilage property and moisture content of nirmali seeds were 9.74%, 13%, 29% and 11.76% respectively. About 8.9% of protein was found to be the constituent of the nirmali seeds along with 66.8% of carbohydrate and 10.9% of fibre.

About 47 phytocomponents, which include carbohydrates, acids, proteins, alcohol and vitamins, were observed (Table 12). The presence of sucrose, D-Glucose, vitamin D3, palmitic acid, oleic acid, octadecanoic acid was of higher significance and these activity has been reported. The other components such as glycerin, furanmethanol, acetylmannose etc., are also found, but their activities have not been reported (Table 12)

The coagulating property of nirmali seeds must be due to the presence of coagulating protein such as 4,5-diamino-2-hydroxypyrimidine, 4-amino-2,6-dihydroxy-5-nitrosopyrimidine and 1 hexyl-1-2-nitrocyclohexane. Similarly, Ghobremichael (2004) has purified and isolated the coagulant protein from the seeds of ***Moringa oleifera*** (www.Natural coagulant.com).

### *Decolourisation power of Nirmali seeds*

The decolourisation power was not noticed.

**Table 12** Phytochemical components of known activity present in Nirmali seeds

| *Sl. No.* | *Component Name* | *Activity* |
|---|---|---|
| 1. | 1,5-Anhydro-d-mannitol | Anti-inflammatory |
| 2. | Sucrose | Antioxidant, Preservative, Antiophthalmic, Sweetener |
| 3. | Vitamin D3 | Anti-tumor |
| 4. | D-Glucose | Antivaricose, Hyperglycemic Antiketonic |
| 5. | Erucic acid | Anti-tumor, Lubricant |
| 6. | n-Hexadecanoic acid (Palmitic acid) | Antioxidant Hypocholesterolemic<br>Nematicide, Pesticide<br>Lubricant, Antiandrogenic<br>Flavor, Hemolytic |
| 7. | Oleic acid | Flavor, Cancer preventive<br>Anti inflammatory, Hypocholesterolemic<br>Dermatitigenic, Perfumery |
| 8. | 9-Hexadecanoic acid (palmitoleic acid) | Hypocholesterolemic |
| 9. | 6-Octadecanoic acid, (Z)-(cis- Oleic acid) | Flavor, Cancer preventive,<br>Anti inflammatory, Hypocholesterolemic<br>Dermatitigenic, Perfumery |
| 10. | Hexadecanoic acid, Z-11-(cis-Palmitoleic acid) | Hypocholesterolemic |

(www.Dr.Duke's phytochemical and ethnobotanical data bases.com)

### Characteristics of the Adsorbent-CAC

The pH of CAC was 7.1. The surface area of CAC was 320 $m^2/g$, and its corresponding decolourising power was 62.

## Effect of Treated Sugar Wash on the Seed Germination, Growth Parameters and Flowering Capacity of Marigold (*Tagetes erecta* L.)

The seed germination percentage of plants irrigated with treated effluent daily (DI) decreased when compared with plants irrigated on alternative days and control. It may be due to the higher osmotic pressure caused by the salt concentration in the effluent (Subramani et al. 1995a). Similar results have been reported by Vasanthy et al. (2004b). They have stated that the germination percentage of the seed irrigated with the diluted treated effluent has shown a gradual reduction with an increase in the effluent concentration.

The root and shoot length of the seedlings irrigated with treated effluent daily and on alternative days has not shown much variation when compared with control. The values were not statistically significant. Similar results have been reported by Vasanthy (1996) while growing maize irrigated with carbon treated aqueous dye solution.

No remarkable change has been noticed in the fresh and dry weight of the *Tagetes erecta* L. About 13 flowers were seen after a time period of 75 days. No significant reduction has been noticed with control, alternative irrigated plant (ADI) and daily irrigated (DI) plant.

Hence, it is clear that 50% spent wash diluted with sugar effluent termed sugar wash after being treated with CaO, nirmali seeds and CAC and neutralized could be used for irrigating the horticultural plant, Marigold, positively.

**Table 13** Effect of treated sugar wash on the seed germination, growth parameters and flowering capacity of *Tagetes erecta* L.

| *S. No.* | *Parameters* | *Control* | *Alternative day Irrigation (ADI)* | *Direct Irrigation (DI)* |
|---|---|---|---|---|
| 1. | Seed Germination | 70 % | 67% | 50% |
| 2. | Length of root(cm) | 35 ± 2 | 36 ± 2 | 33 ± 2 |
| 3. | Length of Shoot(cm) | 152 ± 2 | 150 ± 2 | 145 ± 2 |
| 4. | Root / Shoot ratio | 0.23 ± 0.02 | 0.24 ± 0.02 | 0.23 ± 0.02 |
| 5. | Fresh Weight (g) | 170 | 169 | 165 |
| 6. | Dry weight | 40 | 39.5 | 38 |
| 7. | Yield (No. of Flowers) | 15 | 14 | 13 |

All values are average of three determination

## Physico-chemical Characterization of Soil before and After Irrigation with Treated Sugar Wash

The pH is one of the most important factor in soil quality management. The pH influences many physical activities of microorganisms. The pH of all the soil samples was around 8.5. Similar results have been reported by Nemade and Shrivastava (1996a).

There was no remarkable difference in terms of macronutrients N, P and K have been seen in control soil, soil irrigated on alternative day (ADI) and daily irrigated soil (DI). Similar results have been reported by Elavarasi (2005) for the growth of Marigold and by Elayaraja (2005) for the plant ***Gomphrena globusa*** **L.**

**Table 14** Physico-chemical characterization of soil before and after irrigation with treated sugar wash

| *S. No.* | *Parameters* | *Values* | | | |
|---|---|---|---|---|---|
| | | *Initial Soil* | *Control* | *Alternative Day Irrigation(ADI)* | *Direct Irrigation(DI)* |
| 1. | Texture | Sandy loamy | Sandy loamy | Sandy loamy | Sandy loamy |
| 2. | pH | 8.26 | 8.50 | 8.51 | 8.53 |
| 3. | EC (mmhos) | 0.16 | 0.17 | 0.18 | 0.20 |
| 4. | N (Kg/ha) | 81.1 | 81.2 | 81.3 | 81.5 |
| 5. | P (Kg/ha) | 2.1 | 2 | 2 | 2 |
| 6. | K (Kg/ha) | 3.50 | 3.55 | 3.50 | 3.52 |
| 7. | Fe (ppm) | 3.084 | 3.270 | 3.100 | 3.150 |
| 8. | Zn (ppm) | 2.77 | 2.761 | 2.760 | 2.750 |
| 9. | Mn (ppm) | 2.506 | 2.504 | 2.500 | 2.510 |
| 10. | Cu (ppm) | 0.750 | 0.742 | 0.735 | 0.740 |

The result of this study indicates that the soil irrigated with treated sugar wash has shown very slight change in the levels of nutrients from the control. Hence, the treated effluent would not affect the growth of plant. Thus, it proves the possibility of reuse of treated spent wash for irrigation.

## Studies on the Manurial Value of the Compost

The organic carbon has got reduced in its content by 75% in the compost. When compared with the raw waste, the reduction may be due to the oxidation of organic matter. Similar results have been given by Mali (2002) and Gomathi et al. (2004).

The percentage of nitrogen has increased in the compost than in the initial waste. This may be attributed to an increase in the microbial population, which could perform the process of nitrogen fixation. Similar results have been cited by Kale et al. (1992).

The reduction of the C/N ratio is an indication of completion of composting process. If C/N ratio is about 10:1, it denotes the completion of composting process and the manural value is high. Similar result has been reported by Mali (2002).

**Table 15** Manural value of the compost

| *S. No.* | *Parameters* | *SETS* | | |
|---|---|---|---|---|
| | | *Initial* | *A* | *B* |
| 1. | Organic carbon (%) | 47.02 | 16.50 | 24.6 |
| 2. | Organic matter (%) | 55.07 | 28.44 | 27.22 |
| 3. | Total nitrogen (%) | 1.89 | 1.86 | 2.44 |
| 4. | Total phosphorus (%) | 1.00 | 1.26 | 1.26 |
| 5. | Total potassium (%) | 0.90 | 0.96 | 0.60 |
| 6. | C/N Ratio | 25:1 | 19:1 | 10:1 |

The total phosphorus and potassium content of compost was higher when compared to the initial raw waste. The increase in phosphorous might be due to the humic substances, which would increased phosphorous availability as they have very high cation exchange capacity. Similar results have been reported by Gomathi et al. (2004), Vasanthy et al. (2005), Mali (2002) and Kalaiselvi and Ramasamy (1996).

Among the treatment, the combined effect of chemical and biological treatment (CaO 7.5 g + 5 g of Nirmali seeds +CAC5 g) has shown the highest removal of colour, TDS and COD. And the horticultural plant growth and yield has proved the possibility of reuse of sugar wash.

## REFERENCES


Anitha, A. (2006). Effective nitrate removal using nirmali seeds. M.Sc, Dissertation. Bharathidasan University. Trichy.

APHA. American Public Health Association. (1989). Standard methods for the examinations of water and wastewater. American Public Health Association. 17th edition. Washington DC. Newyork.

Asthana, A.K., Misra, S.K., Chandra, R. and Gunu, R. (2001). Treatment of colour and biochemical oxygen demand of anaerobically treated distillery effluent by aerobic bacterial strains. *IJEP.*, **21(12):** 1070-1072.

Bardiya, M.C., Hashia, R. and Chandna, S. (1995). Performance of hybrid reactor for anaerobic digestion of distillery effluent. *Journal IAEM,* **22:** 237-239.

Bhavik. K.A., Mohana, S. and Datta, M. (2008).Anaerobic treatment of distillery spent wash—A study on upflow anaerobic fixed film bioreactor, *Bioresource Technology* **99:** 4621-4626.

Binkly, W.W. and Wolfrom, M.L. (1953). Composition of cane juice and cane final molasses.Ad. in carbohydrate chemistry. Academic Press Inc. Publishers New York. USA. *Adv. Carbohydr. Chem.* **8:** 291-314.

CPCB (2003). Central Pollution Control Board, Annual Report. New Delhi, India.

Deshananda, C.V., Kaul, S.N. and Nandy, T. (1993). Pollution control methods applicable to distilleries. Pollution control handbook, National Safety Council. Delhi. 291-306.

Elavarasi, V. (2005). Positive utilization of lime treated chrome liquor for the growth of ornamental plant (*Tagetes erecta (L)*) M.Sc., Dissertation. Bharathidasan University, Trichy.

Elayaraja, N. (2005).Utilization of treated and diluted tannery wastewater for the growth of horticultural plant. M.Sc., Dissertation Bharathidasan University. Trichy.

Ghobremichael, K.A. (2004). Moringa seed and pumice as alternative natural materials for drinking water treatment. KTH land and water resources engineering. Royal Institute of Technology. Ph.D thesis. Sweden.

Gomathi, M., Subash, A. and Parvatham, R.P. (2004). Bioremediation of tannery solid waste by vermicomposting using *Eudrilus eugeniae* and *Lampito mauritii*. *J. Industrial Pollution Control.* **20(2):** 331-338.

Handa, B.K. and Seth, R. (1990). Waste management in distillery industry. *J. IAEM.,* **17:** 44-54.

Horwitz, W. (2000). Association of official analytical chemistry methods. 17th edition., Maryland-USA.

Hwang, Y., Matsuo, T., Hanaki, K. and Suzuki, N. (1994). Removal of odorous compounds in wastewater by using activated carbon, ozonation and aerated biofilter. *Wat. Res.* **28(11):** 2309-2319.

Inthorn, D., Silapanuntakul, S. and Chanchitprecha, C. (2001). Improvement of colour and COD removal in molasses wastewater by using advanced oxidation processes. *Poll. Res.*, **20(4):** 691-696.

ISI. (1977). Methods for sampling and test of activated carbon used for decoloursing vegetable oils and sugar solution. Burearu of Indian Standards. New Delhi. 877.

Kalaiselvi, T. and Ramasamy, K. (1996). Compost maturity—can it be evaluvated? *Madras Agric J.*, **83(10):** 609-618.

Kale, R.D., Mallesh, B.C., Bano, K. and Bagyaraj, D.J. (1992).Influence of vermicompost application on the available macronutrients and subsequent microbial populations in the paddy field. *Soil biology and bio-chemistry*, **24:** 1307-1320.

Kalpana, S. (2006). Removal of colour, odour, TDS and COD from the 10% diluted spent wash by using coagulating agents and adsorbents. M.Sc., Dissertation Bharathidasan University. Trichirappalli.

Kasturibai, R. and Ganga, N. (1996). Treatment strategies for the decolorisation of distillery effluent. *J. Industrial Pollution Control.*, **12(1):** 1-8.

Kaushik, G. and Thakur, I.S. (2009). Isolation and characterization of distillery spent wash color reducing bacteria and process optimization by Taguchi approach. Int. Biodeterior. *Biodegrad*, **63:** 420-426.

Kokate, C.K. (1993). Practical Pharmacognosy. Kakatiya university, Warrangal, Andhra Pradesh .pp. 66-67.

Mali, D.S. (2002). Aerobic composting of spent wash. *J. Ecotoxicol. Environ. Monit.*, **12(2):** 101-104.

Murthy, Z.V.P. and Chaudhari. L.B. (2009). Treatment of distillery spent wash by combined UF and RO processes, Global NEST Journal, 11, No 2, pp. 235-240.

Muthuvel, P. and Udayasoorian, C. (1999). Soil, plant, water and agrochemical analysis. Tamil Nadu Agricultural University, Coimbatore.

Patil, N.B. and Kapadnis, B.P. (1995). Decolourisation of melanoidin pigment from distillery spentwash. *Indian J. Environ. Hlth.*, **37(2):** 84-87.

Rangwala, S.C., Rangwala, K.S. and Rangwala, P.S. (1997). Water supply and sanitary engineering (IIedition). Charotar publishing house.Anand. 181.

Saxena, M. and Rai, J.P.N. (2000). Colour removal from anaerobically treated distillery waste. *Indian J. Environ. and Ecoplan.*, **3(1):** 9-12.

Shoemaker, D.P. and Garland, C.W. (1967). Experiments in physical chemistry. II edition, McGraw Hill Kogakusha Ltd., Tokyo. 258-261.

Shweta, Sharma, D. and Sonal, Kumar, K. (2006). Effect of earthworm exudate on growth and yield of *Tagetes erecta L.* (Family: compositae). *J. Curr. Sci.*, **9(11):** 381-386.

Soule, J.A. (1993). *Tagetes minuta*—A potential new herb from south America. *In:* Janick, J. and Simon, J.E. (Eds.). New crops. Wiley. Newyork. 649-654.

Srivastava, A. and Pathak, A.N.(1998). Modern technologies for distillery effluent treatment. *J. Scientific and Industrial Research*, **57:** 388-392.

Subramani, A., Sundramoorthy, P. and Lakshmanachary, A.S. (1995). Effect of biologically treated distillery effluent on seed germination and seedling growth of Green gram *(Vigna radiata(L.)* Wilczek VAR Co.2). *Poll Res.*, **14(1):** 37-41.

Vasanthy, M. and Thamaraiselvi, C. (2006). Bio remediation of distillery spent wash using indigenous microbial strains and cashew nut hull carbon. *Ecotoxicol. Environ. Monit.*, **16(4):** 301-305.

Vasanthy, M., Thamaraiselvi, C. and Sangeetha, M. (2004). Studies on colour and odour removal from distillery spent wash. Proceeding of the 3rd national seminar on inland water resources and environment. Dept. of Environmental sciences, Kerala university, on 2nd–4th Feb, Trivandrum. 152-154.

Vasanthy, M., Thamaraiselvi, C. and Sangeetha, M. (2006). Effective removal of TDS and COD by using coagulants and adsorbents. *Nature and Biosphere,* **11(1&2):** 5-8.

Vasanthy, M., Thamaraiselvi, C. and Mangalasundari, R. (2006d). Effective treatment of spent wash diluted with sugar effluent by using chemical agents. *Nature Environment and Pollution Technology,* **5(3):** 361-365.

Vasanthy, M., Thamaraiselvi, C. and Sangeetha, M. (2006). Effective treatment of distillery spent wash by using calcium oxide, hydrogen peroxide and commercial activated carbon. *Nature Environment and Pollution Technology,* **5(1):** 67-70.

Vasanthy, M., Thamaraiselvi, C. and Ramadoss, A. (2005). Utilization of garden waste for the production of vermicompost. *J. Ecotoxicol Environ Monit.,* **15(3):** 207-211.

Vasanthy, M., Thamaraiselvi, C. and Sathya, R. (2004). Effect of untreated distillery effluent on growth and biochemical characters of *Trigonella foenum. J. Eco. Env. & Cons.,* **10(3):** 239-242.

Vasanthy, M., Thamaraiselvi, C. and Velmurugan, R. (2004). Physico-chemical characterisation of untreated and anaerobically treated distillery effluent. *J. Industrial Pollution Control,* **20(1):** 125-130.

Vasanthy, M., Thamaraiselvi, C. and Anitha, A . (2006). Effective denitrification of drinking water using (*Strychnos potatorum L.*) plant material.Proceedings of National seminar on recent trends in Biosciences. Vivekanandha college of Arts and sciences for women.Elayampalayam.Tiruchengode. 23 th and 24 th Feb. pp. 11.

Vasanthy, M. (1996). Removal of reactive dyes from aqueous solutions and dyeing industrial effluent by cashewnut hull carbon. Ph.D thesis Bharathidasan University, Coimbatore.

Vijayakumari, R. (2006). Effective treatment of diluted spent wash by using selective coagulating agent, oxidising agent and adsorbents.M.Sc dissertation. Bharathidasan University. Trichy.

# 13

# Uptake of Chromium and Inorganic Nutrients from Aqueous Solutions by Two Aquatic Plants

LEENA APPAVOO, ACKMEZ MUDHOO* AND ARVINDA KUMAR RAGEN

*Department of Chemical and Environmental Engineering, University of Mauritius, Réduit, Mauritius*
*E-mail: ackmezchem@yahoo.co.uk*

**ABSTRACT**

Water hyacinth (*Eichhornia crassipes*) and water lettuce (*Pistia stratiotes*) were grown in batch systems and water quality was monitored for 20 days. Water hyacinth was a better macrophyte than water lettuce for the phytoremediation of similar wastewaters. The water hyacinth system had a higher removal of 79.7%, 50.6% and 47.3% in synthetic wastewaters against 73.4%, 39.0% and 35.2% for the water lettuce systems containing 6.4, 16.4, 26.4 mg/L $NO_3$-N, respectively. The average removal for $NO_3$-N from the water hyacinth systems was 83.6%, 57.5% and 62.0 % in systems containing 6.1, 26.1 and 46.1 mg/L of $NO_3$-N against 75.4%, 49.8% and 46.4% for the water lettuce systems, respectively. The highest removal for phosphate was 64% for the water hyacinth system and 52% for the water lettuce system under the same operating conditions. The highest increase in biomass for both macrophytes was for those systems containing the ratio of N: P of 4.6:1.The water hyacinth systems also had a higher % removal of chromium as compared to the water lettuce systems. The concentration of chlorophyll of the leaves had decreased in all systems. The highest reduction of 75.3% was for the water lettuce system in the highest concentration of chromium against 37.2% for the water hyacinth system. The aquatic plants also accumulated more of the metal in their roots as supported by their high BCF factor.

**Keywords:** *Eichhornia Crassipes*, *Pistia Stratiotes*, Nitrate, Phosphate, Chromium, Macrophyte, Synthetic Wastewaters, Phytoremediation.

## INTRODUCTION

Water pollution is one of the most serious problems of today's civilization. Water pollution can be defined as the contamination of water bodies and it occurs when pollutants are discharged directly or indirectly without adequate treatment. Both human and natural activities can change the physical, chemical and biological characteristics of water. Water quality is affected by changes in nutrients, sedimentation, temperature, pH, heavy metals, non-metallic toxins, organics, pesticides and biological factors, among others. Continuous discharge of wastewater in water bodies has several environmental and health impacts. Water contamination weakens and ultimately destroys natural ecosystems that support human health, food production and biodiversity. It is therefore necessary that economical and simple processes for wastewater treatment should be adopted for minimizing the discharge of pollutants in water bodies (Lavrova and Koumanova, 2007).

Human activities can accelerate the rate at which nutrients enter ecosystems. Among the various contaminants in the environment, nitrates play a significant role in ground and surface water contamination due to their toxicity and widespread presence. The main sources of nitrate pollution are related to the use of nitrogen fertilizers (Fahrner, 2002). Several tertiary treatment methods have been developed and implemented to treat nitrate either in wastewater or drinking water. These include biological denitrification, ion-exchange and reverse osmosis. Phosphorus, on the other hand, occurs in natural water and wastewater mainly as inorganic phosphates. Both domestic and industrial activities are the major sources of phosphorus discharge into wastewater. About 30-50% of the phosphorus present in domestic wastewater comes from human waste and the remaining from synthetic detergents (Bashan and Bashan, 2004). The industrial sources of phosphorus generally originate from fertilizer manufacturing, metal finishing, flour processing, dairy, commercial laundry and slaughterhouse industry waste. Together with phosphorus, nitrate causes eutrophication. Removal of phosphorus from wastewater can be achieved by chemical and biological means or a combination of both (Della Rocca et al., 2007). Similarly, heavy metals are one of the most persistent inorganic pollutants in wastewater. Unlike other pollutants, they cannot be degraded. Their presence in wastewater is due to the discharge from industrial dwellings and groundwater infiltration. Conventional primary and secondary unit processes at municipal wastewater treatment plants are inadequate for efficient removal of heavy metals. Advanced processes include chemical precipitation, electrolysis, reverse osmosis and ion exchange, and they are used for pretreatment of known sources of heavy metals in industrial wastewater (Akpor and Muchie, 2010).

Constructed wetlands are planned systems designed and constructed to employ wetland vegetation to assist in treating wastewaters in a more controlled environment than occurs in a natural wetland. Constructed wetlands used for wastewater treatment can be categorized as Free Water Surface (FWS), Subsurface

Flow (SSF) and Floating Aquatic Macrophytes Systems (FAMS). The concept of using plants to clean and restore soil and wastewater is called phytoremediation. The field of phytoremediation is one more area of endeavor closely tied to the principles of constructed wetlands for water renovation and the broader principles of sustainability. It utilizes a variety of plants, biological processes and the physical characteristics of plants to aid in-site remediation. Phytoremediation has been employed for over 300 years. Several floating aquatic macrophytes are used to clean-up wastewater through phytoremediation and are gaining popularity since the macrophytes are productive, easily available and stocked without deterioration (Debusk and Reddy, 1987). Phytoremediation is a promising clean-up solution for a wide variety of pollutants, and it is a process whereby plants remove, transfer, stabilize or destroy contaminants in the soil, groundwater and wastewater. Phytoremediation is well suited for use at very large field sites where other methods of remediation are not effective or practicable, at site with low concentrations of contaminants and where only "polishing treatment" is required over a long period of time and in conjunction with other technologies. This study presents the results of an investigation on the use of two floating aquatic macrophytes systems to remove nutrient load (phosphate and nitrate) and chromium subjected to varying concentration of synthetic wastewaters. The objectives of this work study were to assess the efficiency of removal of nitrate, phosphate and chromium in batch systems containing the aquatic plants, namely water hyacinth (*Eichhornia crassipes*) and water lettuce (*Pistia stratiotes*), and to monitor pH, temperature and conductivity at intervals of 6 or 7 days, to monitor variation in biomass in terms of the weight of the plants, to determine changes in chlorophyll level to analyze any reduction in chlorophyll due to heavy metal uptake, to monitor the Bioconcentration Factor (BCF) for the plants in terms of total chromium, and to model the data obtained as a first and zero order degradation process. BCF is an index of plants to accumulate the metals with respect to the metal concentration in the substrate and has been extensively used to evaluate parameters linked to sublethal effects of heavy metals in plants (Homyok et al., 2004).

## EXPERIMENTAL METHODOLOGY

### Plant Selection and Preparation

Two aquatic floating macrophytes were selected. These were water hyacinth (*Eichhornia crassipes*) and water lettuce (*Pistia stratiotes*) as they were locally available. Young plants of both species were uprooted from their natural habitats. Water lettuces were collected from Grand River South East and water hyacinths from St-Pierre. The plants were washed with tap water to remove mud and other foreign matter. Care was taken not to destroy the root structure and the associated bacteria that are important for microbial processes. In order to acclimatize the plants to their new environmental conditions, they were placed in containers filled with the synthetic wastewaters loaded with $NO_3^-$, $PO_4^{3-}$ and total chromium. The withering parts of the macrophytes were removed as far as possible before being introduced into the treatment container, and they were allowed to drain before measuring the total fresh weight using electronic balances.

### Experimental Set Up and Analytical Procedures

The experiments were conducted on a batch mode. Each batch consisted of plastic container filled with 4 or 5 L of synthetic wastewater of a certain concentration. One plant was placed in the container and the removal of pollutants was assessed over a period of 20 days. Initially, 5 batches for each plant (10 batches in all) were prepared at varying concentration of $NO_3^-$ from 5.5 mg/L to 105.5 mg/L. The synthetic wastewater was prepared using silver nitrate salt. It is to be noted that one batch was used as a control, in which only water was filled. However, it was observed after 3 days that all the plants in the batches containing $AgNO_3$ had died. Consequently, the experiment was restarted using potassium nitrate salt as a source of $NO_3^-$, with reduced concentration of $NO_3^-$.

Three batches were then prepared at 3 varying concentrations of nitrate, i.e., 6.4, 16.4 and 26.4 mg/L. Each plant was treated under the same conditions. The experiment was run for a period of 20 days. Samples were taken at an interval of 2 or 3 days and analyzed for $NO_3^-$ concentration by Hach DR/2500 Spectrophotometer method 359 N. pH, temperature and conductivity were also monitored at intervals of 6 or 7 days. An Oakion Model 510 pH meter was used to monitor the pH and temperature of the synthetic wastewater. The conductivity of the wastewater was read from a Hanna Model 9932 conductivity meter. It was found that nitrogen fixing plants are unable to fix nitrogen if there is no sufficient supply of phosphate. This was investigated in a series of 3 batches, for each plant, containing both $NO_3^-$ and $PO_4^{3-}$ species. The concentrations of $NO_3^-$ and $PO_4^{3-}$ were 6.1, 26.1 and 46.1 mg/L and 5.0, 10.0 and 10.0 mg/L, respectively. This resulted in an increasing ratio of $NO_3^-/ PO_4^{3-}$ of 1.2, 2.6 and 4.6, respectively in the batches.

Samples of wastewater from the container were collected at intervals of 2/3 days and analyzed for both $NO_3^-$ concentration and $PO_4^{3-}$ concentration as per the Hach method Spectrophotometer 535 P. pH, temperature and conductivity were also analyzed at intervals of 6 or 7 days. To investigate chromium removal, a series of 3 batches, for each plant, was set up at varying concentration of chromium of 10, 20 and 30 mg/L.

Samples of wastewater from the containers were collected at intervals of 2 or 3 days and analyzed for chromium by using the Atomic Absorption Spectrometer (nitrous oxide/ acetylene flame). pH, temperature and conductivity were also recorded at intervals of 6/7 days. The leaves of the plants were sampled once every week and analyzed for chlorophyll content according to the method described by Arnon (1949). 1 g of leaf sample was cut into small pieces and homogenized in a pre-cooled mortar and pestle using 80% (v/v) acetone. Mature leaves were chosen because, in general, leaves produced earlier in the season accumulated more metals than those produced later (Ma, 2005). The extract was transferred in test tubes of 10 mL and centrifuged at 3000 rpm for 15 minutes. The clear solution was decanted off and then made up to 25 mL with 80% (v/v) acetone. 10 mL of clear solution was then transferred and the absorbance was read at 645 nm and 663 nm by the use of DR/2500 Spectrophotometer against a blank sample containing 80% acetone. The chlorophyll concentration was calculated as described by Bunluesin (2004): total chlorophyll (mg/L) = (8.02 × A663) + (20.2 × A645) where A663 = absorbance at 663 nm, and A645 = absorbance at 645 nm.

It should be noted that in cases where the residue was not white, i.e., not all the chlorophyll was removed, the leaves were crushed again and a two-step centrifuge was carried out to ensure that all the chlorophyll was extracted. Two-step centrifuges were mainly used for the water hyacinth leaves since it contained more chlorophyll. The plants were recovered from the synthetic solutions containing chromium at the end of the experiment, and were separated into aerial parts (leaves and shoots) and roots and analyzed for bioconcentration factor. The separated parts were oven dried at 105 $^0$C for 24 hr. For the water lettuce systems, some leaves were harvested before they dissolved in the synthetic wastewater. They were dried and kept in air tight plastic bags before being merged with the other leaves at the end of 20 days. The dried plant parts were then ashed in a furnace for 4 hr. The ash was then transferred to a 1000 mL round bottom flask containing 15 mL concentrated HCl (which had been diluted to a 1:1 ratio). The ash-HCl mixture was heated for 20 minutes in a water bath until the ash dissolved completely. The resulting solution was cooled, filtered to a 100 mL volumetric flask and made up to the volume with distilled water prior to heavy metal analysis using AAS (Jayaweera et al., 2006; Lu et al., 2004; Misra et al., 2010). BCF was determined as expressed below by Bunluesin (2004).

$$BCF = \frac{[M_{tissue}]}{[M_{initial}]}$$

where $[M_{tissue}]$ is the metal concentration in the plant tissue (whole plant) (mg/kg) and $[M_{tissue}]$ is the initial concentration of metal in substrate (mg/L).

## Labeling of Experimental Batches

The experimental set-ups were labelled in the order in which they were carried out in the research work. The first experiments analyzed the efficiency of nitrate removal and these have been denoted using the subscript 1. The batches for both plants were labelled in order of increasing concentration of $NO_2^-$, $NO_3^-$ and $PO_4^{3-}$ and chromium (Table 1).

**Table 1** Labeling protocol of the batches for experimental runs

| *Experimental Runs* | *Concentration of the Species (mg/L)* | | | *Water Hyacinth* | *Water Lettuce* |
|---|---|---|---|---|---|
| | $NO_3^-$ | $PO_4^{3-}$ | *Cr* | | |
| Efficiency of $NO_3^-$ removal by the plants by using $AgNO_3$ | 5.5 | – | – | $WH_{A1}$ | $WL_{A1}$ |
| | 30.5 | – | – | $WH_{B1}$ | $WL_{B1}$ |
| | 55.5 | – | – | $WH_{C1}$ | $WL_{C1}$ |
| | 80.5 | – | – | $WH_{D1}$ | $WL_{D1}$ |
| | 105.5 | – | – | $WH_{E1}$ | $WL_{E1}$ |
| Efficiency of $NO_3^-$ removal by the plants by using $KNO_3$ | 6.4 | – | – | $WH_{A2}$ | $WL_{A2}$ |
| | 16.4 | – | – | $WH_{B2}$ | $WL_{B2}$ |
| | 26.4 | – | – | $WH_{C2}$ | $WL_{C2}$ |
| Efficiency of $NO_3^-$ and $PO_4^{3-}$ removal by the plants | 6.1 | 5.0 | – | $WH_{A3}$ | $WL_{A3}$ |
| | 26.1 | 10.0 | – | $WH_{B3}$ | $WL_{B3}$ |
| | 46.1 | 10.0 | – | $WH_{C3}$ | $WL_{C3}$ |
| Efficiency of Cr removal by the plants | – | – | 10 | $WH_{A4}$ | $WL_{A4}$ |
| | – | – | 20 | $WH_{B4}$ | $WL_{B4}$ |
| | – | – | 30 | $WH_{C4}$ | $WL_{C4}$ |

## RESULTS AND DISCUSSIONS

### Effect of using Silver Nitrate Salt on Plant Growth

Initially when $AgNO_3$ was used as a salt to prepare the synthetic wastewaters, it was observed that after 3 days, the plants were toxified as depicted in Figure 1. The weight of the water hyacinths was monitored after 3 days and results are given in Table 2. The adverse effects of $AgNO_3$ salt were indicated by the loss in biomass. The weight of the water lettuces was not noted since the roots had become detached from the leaves and had also started to decompose. Table 2 shows the % reduction in biomass for the water hyacinth in batches containing silver nitrate salt.

**Table 2** Weight of the water hyacinth before and after the experiment

| *Batch Systems* | *Fresh Weight Before Experiment (g)* | *Weight After 3 Days (g)* | *% Reduction in Biomass* |
|---|---|---|---|
| $WH_{A1}$ | 147.4 | 140.5 | 4.7 |
| $WH_{B1}$ | 140.4 | 113.2 | 19.3 |
| $WH_{C1}$ | 190.4 | 149.4 | 21.5 |
| $WH_{D1}$ | 182.0 | 117.5 | 35.4 |
| $WH_{E1}$ | 190.8 | 111.3 | 40.7 |

The % reduction in biomass for the water hyacinths increased with increasing concentration of nitrate. The maximum % reduction was 40.7% when the plant was exposed to 105.5 mg/L of $NO_3^-$. The water hyacinth system having a load of

6.4 mg/L of $NO_3^-$ showed the least reduction of 4.7% in biomass. The average reduction in biomass was 24.3 % for the water hyacinth systems. The % reduction in biomass for the water hyacinth systems increased with increasing concentration of nitrate. The highest reduction of 40.7% was observed for the water hyacinth in a synthetic solution containing the highest concentration of silver nitrate salt. Another important observation was that the water hyacinth acting as the control also showed a reduction in biomass. This could be attributed to a lack of optimum amount of sunlight or drainage of minerals in the wastewater during the adaptation phase of the plant to its new environment. The experiment was restarted using potassium nitrate with reduced nitrate concentration in the synthetic wastewater.

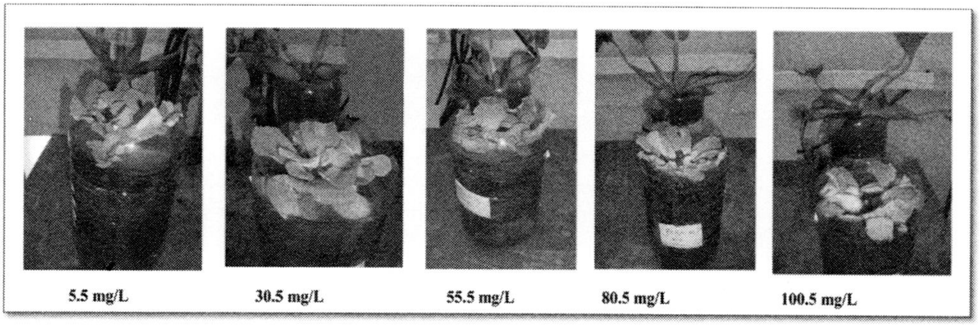


**Fig. 1.** Effects of silver nitrate on plants.

## Uptake of Nitrate by using Potassium Nitrate Salt

### *Variation in pH*

Table 3 shows the variation in pH at intervals of 6 or 7 days in synthetic wastewaters containing nitrate.

**Table 3** Variation in pH for the synthetic wastewaters containing nitrate

| *Batch Systems* | *pH* | | | |
|---|---|---|---|---|
| | *Number of Operating Days* | | | |
| | *1* | *6* | *13* | *20* |
| $WH_{A2}$ | 6.17 | 7.13 | 7.62 | 7.05 |
| $WH_{B2}$ | 6.52 | 7.80 | 7.77 | 7.17 |
| $WH_{C2}$ | 6.78 | 7.82 | 7.80 | 7.24 |
| $WL_{A2}$ | 6.19 | 7.54 | 7.79 | 7.24 |
| $WL_{B2}$ | 6.54 | 7.90 | 7.84 | 7.44 |
| $WL_{C2}$ | 6.70 | 7.94 | 7.87 | 7.48 |

$NO_3$-N is removed in aquatic macrophyte systems by denitrification and plant uptake (Chew, 2006). As shown in Table 3, the pH for the water hyacinth and the water lettuce systems varied between 6.2-7.9 and 6.2-8.0 respectively. Hence, pH condition in the water columns was found to be favorable for a denitrification environment. pH measured was always between 6 and 8, the range cited as being optimal for denitrification (Fahrner, 2002). Figures 2 and 3 depict the aquatic macrophyte systems initially and at the end of 20 days. The biomass had increased in all the systems at the end of the experiment. The water lettuce systems achieved 100% coverage by the growth of numerous daughter plants. New shoots and leaves had emerged from the water hyacinth systems.

**Fig. 2.** Three batches containing water hyacinth initially and after 20 days.

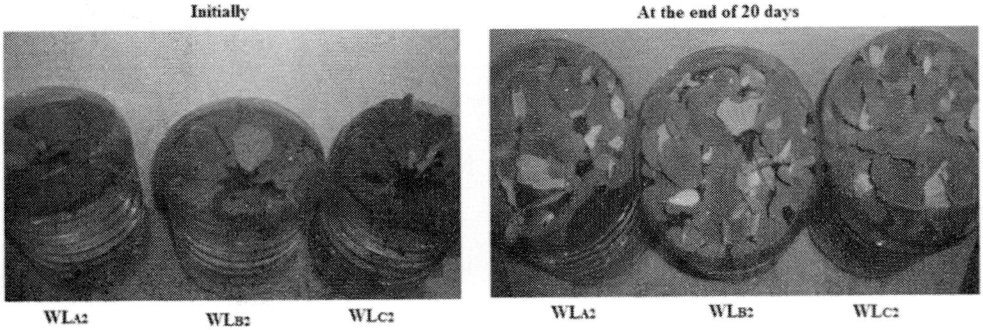


**Fig. 3.** Three batches containing water lettuce initially and after 20 days.

### *Variation in temperature*

Table 4 shows the variation in temperature at intervals of 6 or 7 days in synthetic wastewaters containing nitrate.

**Table 4** Variation in temperature for the synthetic wastewaters containing nitrate

| *Batch Systems* | *Temperature/°C* | | | |
|---|---|---|---|---|
| | *Number of Operating Days* | | | |
| | *1* | *6* | *13* | *20* |
| $WH_{A2}$ | 24.4 | 24.2 | 24.3 | 24.5 |
| $WH_{B2}$ | 24.3 | 24.3 | 24.4 | 24.5 |
| $WH_{C2}$ | 24.5 | 24.3 | 24.3 | 24.4 |
| $WL_{A2}$ | 24.4 | 24.3 | 24.3 | 24.4 |
| $WL_{B2}$ | 24.3 | 24.4 | 24.1 | 24.5 |
| $WL_{C2}$ | 24.4 | 24.3 | 24.2 | 24.4 |

Over the period of monitoring, the changes in temperature were small for both the water hyacinth and the water lettuce systems. The temperature for the water hyacinth systems varied between 24.2°C to 24.5°C, while that for the water lettuce systems were between 24.1°C and 24.5°C as shown in Table 4. Hence, the temperature for both macrophyte systems throughout the whole duration of the experiment was constant. A constant temperature indicated that the rate of removal of $NO_3$-N also remained constant for all the systems (Ma, 2005).

### *Variation in conductivity*

Figure 4 illustrates the variation in conductivity of the batches with time. Conductivity is an indication of the amount of ionic species dissolved in solution. The general trend observed was a decrease in conductivity with time for all the batches. The decrease in conductivity was more pronounced for the water hyacinth systems than the water lettuce systems, since the decrease in the concentration of $NO_3$-N was more than the water lettuce systems under the same operating conditions as shown in Figure 4.

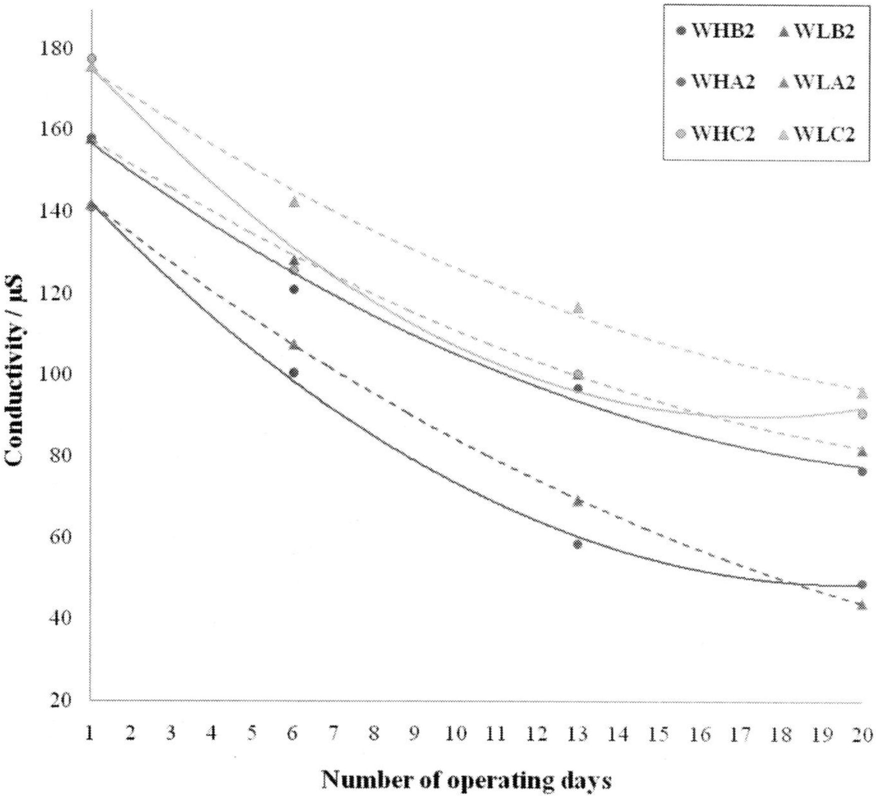


**Fig. 4.** Variation in conductivity for synthetic wastewater containing nitrate.

### *Variation in the concentration of $NO_3$-N with time*

Figure 5 illustrates the variation in the concentration of $NO_3$-N with time in all the batches containing the aquatic plants. As shown in Figure 5, the concentration of $NO_3$-N decreased in all the systems containing the aquatic macrophytes. The decrease in the concentration of $NO_3$-N was more in the water hyacinth systems than the water lettuce systems for the same operating conditions. It can also be observed that at a low concentration of 6.4 mg/L of $NO_3^-$, the decrease in the concentration varied linearly with time as compared to other systems.

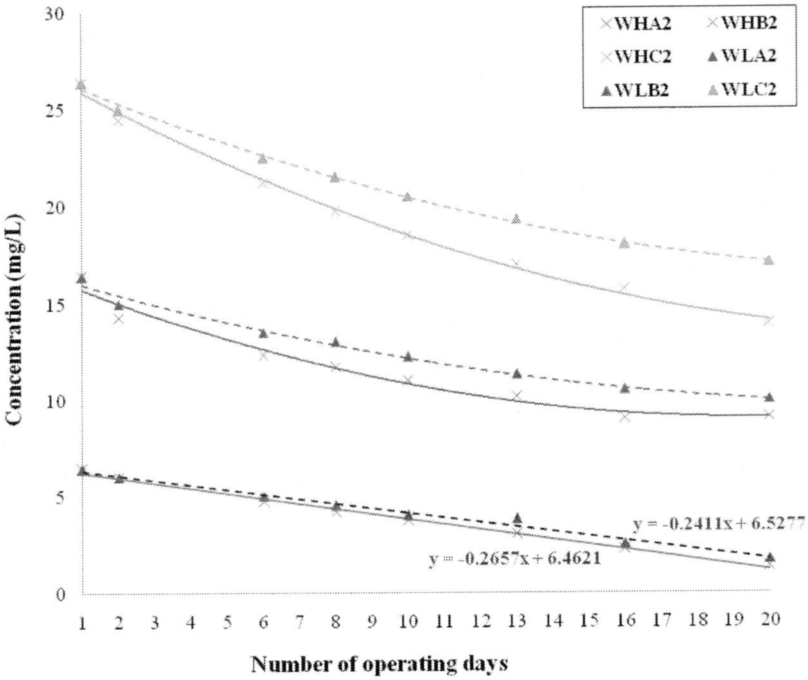


**Fig. 5.** Variation in the concentration $NO_3$-N with time.

Table 5 shows how the % removal of $NO_3$-N varied as the concentration of nitrate was increased in the synthetic wastewaters.

**Table 5** Concentration and % removal of $NO_3$-N in the synthetic wastewater

| *Batch Systems* | *[$NO_3^-$] (mg/L)* | *[$NO_3$-N]$_{t=20}$ (mg/L)* | *[$NO_3$-N] Removed (mg/L)* | *% Removal of $NO_3$-N* |
|---|---|---|---|---|
| $WH_{A2}$ | 6.4 | 1.3 | 5.1 | 79.7 |
| $WH_{B2}$ | 16.4 | 8.1 | 8.3 | 50.6 |
| $WH_{C2}$ | 26.4 | 13.9 | 12.5 | 47.3 |
| $WL_{A2}$ | 6.4 | 1.7 | 4.7 | 73.4 |
| $WL_{B2}$ | 16.4 | 10.0 | 6.4 | 39.0 |
| $WL_{C2}$ | 26.4 | 17.1 | 9.3 | 35.2 |

As shown in Table 5, the concentration of $NO_3$-N had decreased in all the systems after a period of 20 days, but the % removal of $NO_3$-N differed. The water hyacinth system containing 6.4 mg/L of $NO_3^-$ had a higher removal of 79.7% $NO_3$-N against 73.4% for the water lettuce system. As the nitrate load was increased in the wastewater, the concentration of $NO_3$-N removed increased, but the % removal decreased in all the batches. In general, denitrification primarily occurs in the underlying sediments with little or no denitrification occuring in the overlying waters. The dense cover of floating water hyacinths depleted dissolved $O_2$ of the underlying water, thus creating anaerobic conditions (Sooknah, 2000). These conditions favoured the denitrification process, thus maximizing the removal of $NO_3$-N. This explained why the water hyacinth systems had a higher removal of 6.3%, 11.6% and 12.1% than the water lettuce systems for 6.4, 16.4, 26.4 mg/L of nitrate in the wastwater, respectively. Although in this study, no direct evidence for denitrification was available, the dense coverage of the macrophytes suggested the possibility of this process. However, a larger proportion of nitrate removal was due to plant uptake than denitrification.

### *Changes in biomass*

Figure 6 illustrates the % increase in biomass in the batches containing varying concentration of nitrate.

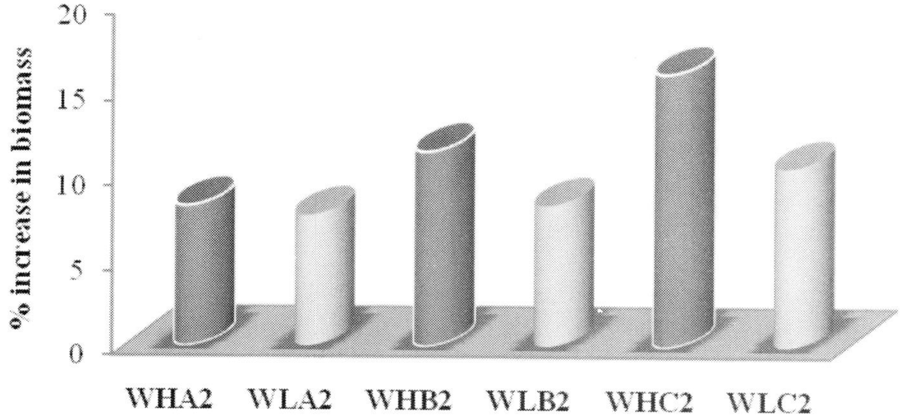


**Fig. 6.** Increase in biomass for synthetic wastewater containing nitrate.

Figure 6, the % increase in biomass increased as the load of nitrate was increased in the wastewater. The highest increase in biomass for the water hyacinth system was 16.1% in a synthetic wastewater containing 26.4 mg/L $NO_3^-$ against 10.6% for the water lettuce system under the same operating conditions. The average increase in biomass for the water hyacinth systems was 12%, and it was 8.9% for the water lettuce systems.

## Uptake of Nitrate and Phosphate in Batch Systems Containing the Macrophytes

### *Variation in pH*

Table 6 shows the variation in pH at intervals of 6 or 7 days in wastewaters containing nitrate and phosphate.

**Table 6** Variation in pH for the synthetic wastewaters containing the nutrients

| *Batch Systems* | *pH* | | | |
|---|---|---|---|---|
| | *Number of Operating Days* | | | |
| | *0* | *7* | *14* | *20* |
| $WH_{A3}$ | 7.24 | 7.10 | 7.61 | 7.15 |
| $WH_{B3}$ | 7.11 | 6.80 | 7.54 | 7.19 |
| $WH_{C3}$ | 7.23 | 7.12 | 7.48 | 7.28 |
| $WL_{A3}$ | 7.26 | 7.40 | 7.84 | 7.55 |
| $WL_{B3}$ | 7.12 | 7.45 | 7.61 | 7.53 |
| $WL_{C3}$ | 7.25 | 7.23 | 7.37 | 7.46 |

As shown in Table 6, the pH for the water hyacinth and water lettuce systems varied between 6.8-7.61 and 7.12-7.84, respectively, which was optimum for the denitrification process, as pH for all the batches were between 6 and 8 throughout the experiment (Fahrner, 2002). Figures 7 and 8 depict the batches containing the macrophytes at regular intervals and up to a period of 20 days. The biomass had increased in all the systems at the end of the experiment.

### *Variation in temperature*

Table 7 shows the variation in temperature at intervals of 6 or 7 days in wastewaters containing nitrate and phosphate.

**Table 7** Variation in temperature for synthetic wastewater containing the nutrients

| *Batch Systems* | *Temperature (°C)* | | | |
|---|---|---|---|---|
| | *Number of Operating Days* | | | |
| | *0* | *7* | *14* | *20* |
| $WH_{A3}$ | 24.5 | 24.4 | 24.5 | 24.3 |
| $WH_{B3}$ | 24.6 | 24.5 | 24.6 | 24.4 |
| $WH_{C3}$ | 24.4 | 24.4 | 24.5 | 24.3 |
| $WL_{A3}$ | 24.4 | 24.3 | 24.4 | 24.4 |
| $WL_{B3}$ | 24.5 | 24.2 | 24.5 | 24.4 |
| $WL_{C3}$ | 24.3 | 24.3 | 24.3 | 24.4 |

There were no significant changes in temperature as shown in Table 7, indicating that the rate of removal of $NO_3$-N and $PO_4^{3}$-P remained constant for all the batches. The temperature for the water hyacinth systems varied between 24.3°C-24.6°C while for the water lettuce systems varied between 24.2°C-24.5°C.

**Fig. 7.** Water lettuce systems in synthetic wastewater containing both nitrate and phosphate.

**Fig. 8.** Water hyacinth systems in synthetic wastewater containing both nitrate and phosphate.

*Variation in conductivity*

Figure 9 illustrates the variation in conductivity with time in batches containing varying concentration of nutrients. The general trend observed was a decrease in conductivity with time for all the batches. The decrease in conductivity was more for the water hyacinth systems than the water lettuce systems for the same operating conditions.

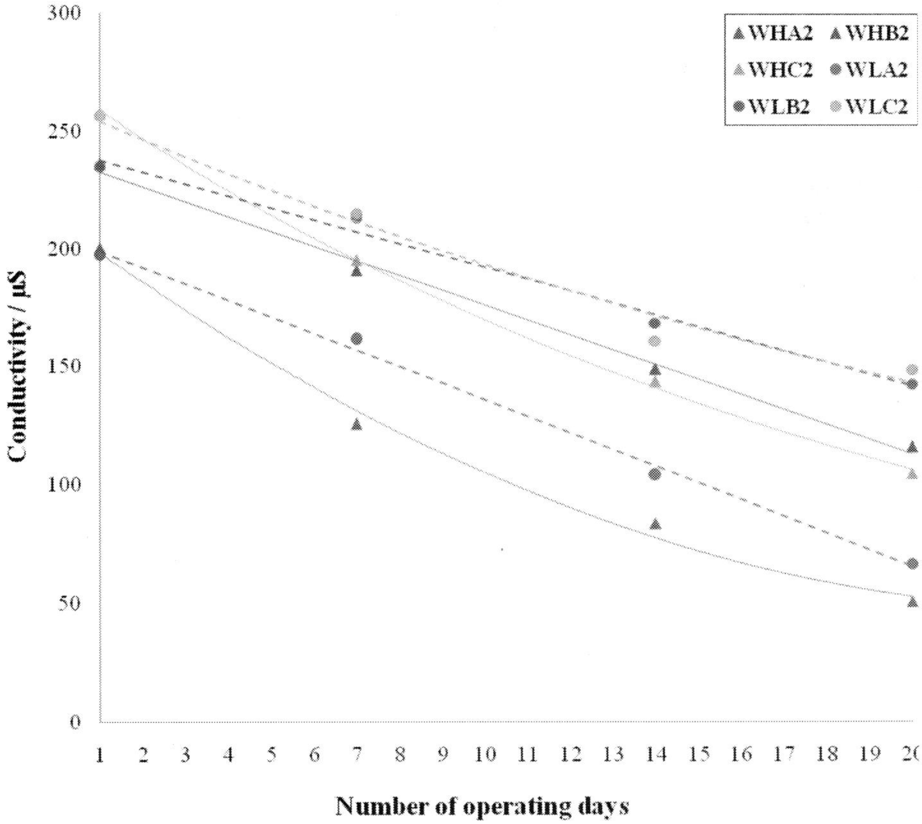


**Fig. 9.** Variation in conductivity for systems containing $NO_3^-$ and $PO_4^{3-}$.

## *Changes in biomass*

Figure 10 illustrates the % increase in biomass in the batches containing varying concentration of nutrients.

As shown in Figure 10, the increase in biomass was greater for the systems containing water hyacinth than for those containing the water lettuce for the same operating conditions. The biomass increased with increasing concentration of nutrients in all the batches. The highest increase of 21.5 % in biomass was in the water hyacinth system containing 46.1 mg/L $NO_3^-$ and 10 mg/L $PO_4^{3-}$ in the synthetic wastewater against 14.6% for the water lettuce system under the same

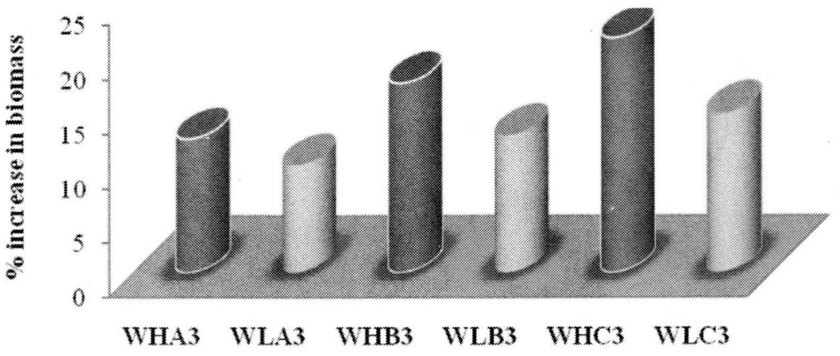


**Fig. 10.** % increase in biomass for systems containing $NO_3^-$ and $PO_4^{3-}$.

operation conditions. The average increase in biomass for the water hyacinth systems was 17.1%, and it was 12.4% for the water lettuce systems.

### *Variation in $NO_3$-N for macrophyte systems containing $NO_3^-$ and $PO_4^{3-}$*

Figure 11 illustrates the variation in the concentration of $NO_3$-N with time as the concentration of nutrients increased in the wastewater.

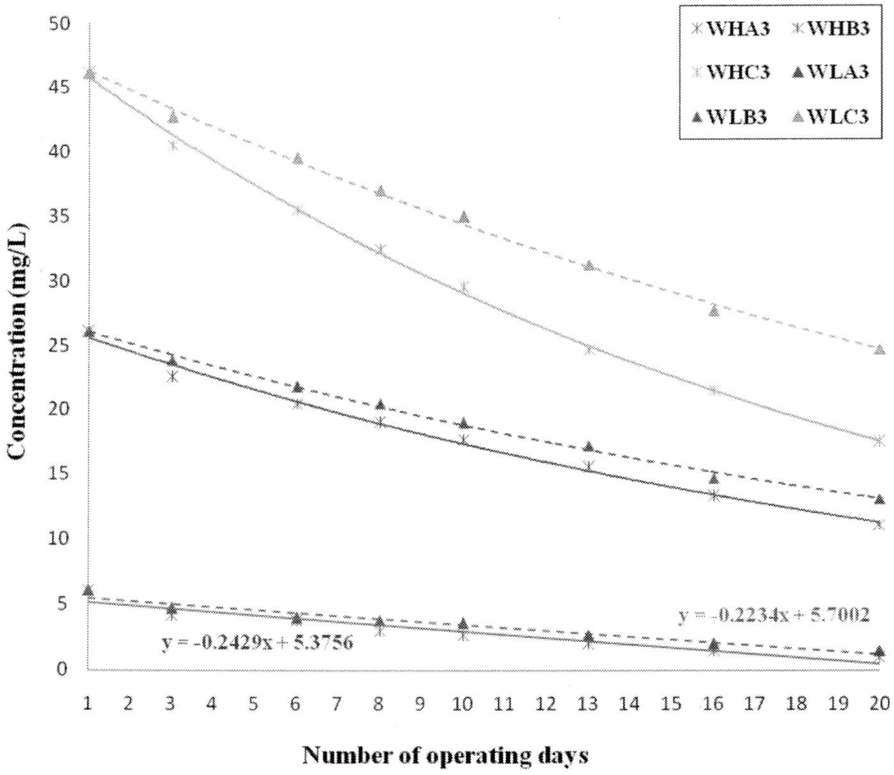


**Fig. 11.** Concentration of $NO_3$-N with time for systems containing $NO_3^-$ and $PO_4^{3-}$.

As shown in Figure 11, the concentration of $NO_3$-N decreased in all the systems. The decrease in the concentration of $NO_3$-N was more for the water hyacinth systems as compared to the water lettuce systems for the same operating conditions. At low concentration of 6.1 mg/L $NO_3^-$, the decrease in concentration was linear with time as compared to other systems. The % removal of $NO_3$-N for the batches containing varying concentration of the nutrients is given in Table 8.

**Table 8** % removal of $NO_3$-N for systems containing both nutrients

| *Batch Systems* | *$[NO_3^-]$ (mg/L)* | *$[NO_3\text{-N}]_{t=20}$ (mg/L)* | *$[NO_3$-N] Removed (mg/L)* | *% Removal of $NO_3$-N* |
|---|---|---|---|---|
| $WH_{A3}$ | 6.1 | 1.0 | 5.1 | 83.6 |
| $WH_{B3}$ | 26.1 | 11.1 | 15.0 | 57.5 |
| $WH_{C3}$ | 46.1 | 17.5 | 28.6 | 62.0 |
| $WL_{A3}$ | 6.1 | 1.5 | 4.6 | 75.4 |
| $WL_{B3}$ | 26.1 | 13.1 | 13.0 | 49.8 |
| $WL_{C3}$ | 46.1 | 24.7 | 21.4 | 46.4 |

As shown in Table 8, the concentration of $NO_3$-N had decreased in all the systems after 20 days. The % removal of $NO_3$-N varied from 75.4% to 46.4% as the concentration was increased from 6.1 to 46.1 mg/L $NO_3^-$ in the water lettuce systems. The water hyacinth system containing the highest concentration of 46.1 mg/L $NO_3$ nitrate had an efficiency removal of 62% against 46.4% for the water lettuce systems under the same operating conditions. The systems containing water hyacinth had a higher % removal of $NO_3$-N as compared to the systems containing water lettuce for the same operating conditions. This was due to the fact that the dense cover of floating water hyacinths depleted dissolved $O_2$ of the underlying water, creating anaerobic conditions (Sooknah, 2000).

### *Variation in $PO_4^{3}$-P for systems containing $NO_3^-$ and $PO_4^{3-}$*

Figure 12 illustrates the variation in the concentration of $PO_4^{3}$-P with time. The concentration of $PO_4^{3}$-P decreased in all the batches containing the water hyacinth. For the water lettuce system containing the lowest concentration of 5 mg/L of $PO_4^{3-}$ and 6.1 mg/L of $NO_3^-$, the concentration of $PO_4^{3}$-P decreased exponentially with time. As the concentration of phosphate in the water lettuce systems was increased to 10 mg/L, the concentration of $PO_4^{3}$-P decreased to reach a minimum. For the water lettuce system $WL_{C3,}$ the concentration of $PO_4^{3}$-P decreased to reach a minimum of 6.9 mg/L $PO_4^{3}$-P on day 13. The concentration was then found to increase in the system. A similar pattern was observed for the system $WL_{B3}$.

Table 9 shows the % removal of $PO_4^{3}$-P as the concentration of nutrients was increased in the batches.

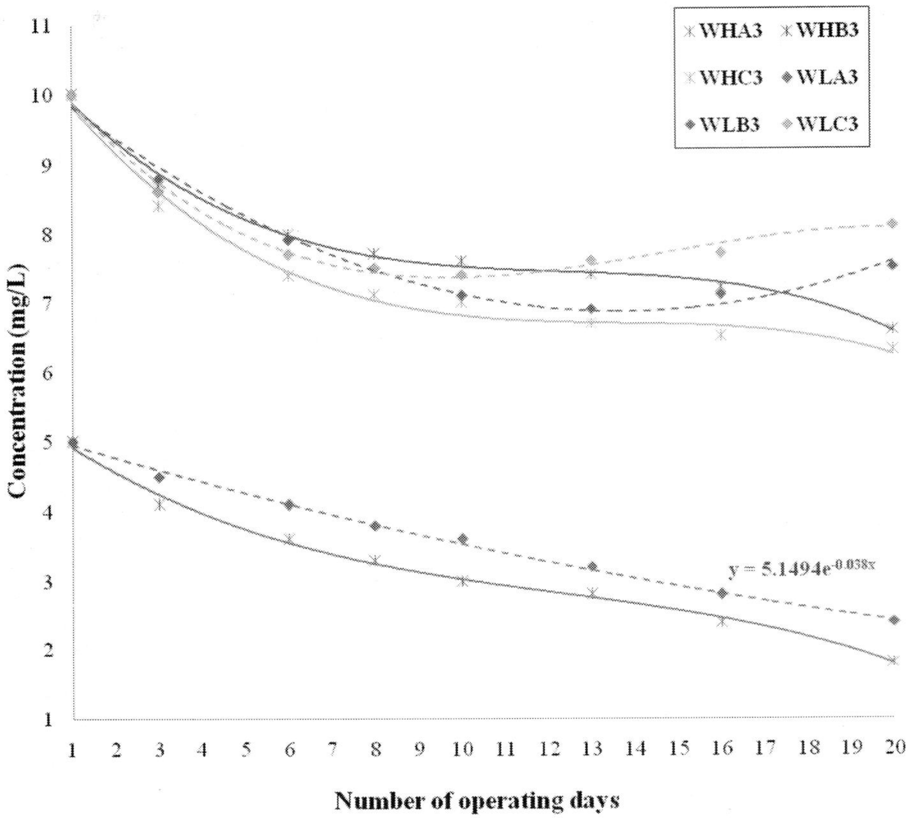


**Fig. 12.** Variation in the concentration of $PO_4^{3-}$-P with time.

**Table 9** % increase in $PO_4^{3-}$-P for the wastewaters containing $PO_4^{3-}$ and $NO_3^-$

| *Batch Systems* | *$[PO_4^{3-}]$ (mg/L)* | *$[PO_4^{3-}\text{-P}]_{t=20}$ (mg/L)* | *$[PO_4^{3-}\text{-P}]$ Removed (mg/L)* | *% Removal $PO_4^{3-}$-P* |
|---|---|---|---|---|
| $WH_{A3}$ | 5.0 | 1.8 | 3.2 | 64 |
| $WH_{B3}$ | 10.0 | 6.6 | 3.4 | 34 |
| $WH_{C3}$ | 10.0 | 6.3 | 3.7 | 37 |
| $WL_{A3}$ | 5.0 | 2.4 | 2.6 | 52 |
| $WL_{B3}$ | 10.0 | 7.5 | 2.5 | 25 |
| $WL_{C3}$ | 10.0 | 8.1 | 1.9 | 19 |

As shown in Table 9, the concentration of $PO_4^{3-}$-P had decreased for all batches after 20 days. The % removal of $PO_4^{3-}$-P decreased in the water lettuce systems as the concentration of nutrients was increased. The water hyacinth system containing 10 mg/L $PO_4^{3-}$-P and 46.1 mg/L $NO_3$-N had a removal of 37 % of $PO_4^{3-}$-P against 19% for the water lettuce system operated under the same conditions. In aquatic system, phosphate is removed by plant uptake, microbial assimilation, precipitation with cations such as calcium, magnesium, iron etc. through immobilization in the debris plant tissue and adsorption by the underlying sediment (Debusk and Reddy, 1987). Hence, it can be deduced that initially the concentration of $PO_4^{3-}$-P decreased in all the batches due to precipitation of the phosphate with the cations present in tap water. For the water hyacinth systems, the concentration of $PO_4^{3-}$-P decreased with time. Water hyacinth had a dense root structure, and since detritus was observed as deposits at the bottoms of all containers as a result of natural die-off while draining the synthetic wastewater, a fraction of the removal of $PO_4^{3-}$-P was due to immobilization in the debris of plant tissue and adsorption by the underlying sediment (Debusk and Reddy, 1985).

On the other hand, the root system of water lettuce was less dense than those of the water hyacinth. Hence the removal of $PO_4^{3-}$-P from the water lettuce systems was mainly by plant uptake and precipitation with the cations present in tap water. Initially and up to day 13, the concentration of $PO_4^{3-}$-P decreased in all the water lettuce systems. After day 13 for system $WL_{C3}$ and day 16 for $WL_{B3,}$ the concentration of $PO_4^{3-}$-P increased in both the systems. This could be due to the fact that detritus of the plants started decomposing and releasing $PO_4^{3-}$-P into the synthetic wastewater. A point in time was reached for the water lettuce systems, $WL_{C3}$ and $WL_{B3}$ whereby the rate of uptake of $PO_4^{3-}$-P was less than the rate at which the phosphate was being released by decomposition of the plant debris. The conductivity for both water hyacinth and water lettuce systems decreased throughout the experiment since the total concentration of nutrients decreased with time. The decrease in conductivity for the water hyacinth systems was more than the water lettuce systems as they removed the highest % of nutrients for the same operating conditions.

Table 10 shows the % removal of $PO_4^{3-}$-P and $NO_3$-N and the % increase in biomass as the concentration of nutrients was increased in the batches. The highest removal of $NO_3$-N for the systems $WH_{C3}$ and $WL_{C3}$ was 28.6 mg/L and 21.4 mg/L, respectively, as shown in Table 10. Indeed, both systems have $NO_3^-$ to $PO_4^{3-}$ ratio of 4.6:1 which has been considered as the optimum ratio for this study, and this could be further demonstrated by the highest increase in biomass for both systems.

**Table 10** Removal of the nutrients from the synthetic wastewater

| *Batch Systems* | *$PO_4^{3-}$ (mg/L)* | *[$PO_4^{3-}$-P] Removed (mg/L)* | *$NO_3^-$ mg/L)* | *[$NO_3$-N] Removed (mg/L)* | *$NO_3^-/PO_4^{3-}$* | *% Increase in Biomass* |
|---|---|---|---|---|---|---|
| $WH_{A3}$ | 5.0 | 3.2 | 6.1 | 5.1 | 1.22:1 | 12.3 |
| $WH_{B3}$ | 10.0 | 3.4 | 26.1 | 15.0 | 2.61:1 | 17.4 |
| $WH_{C3}$ | 10.0 | 3.7 | 46.1 | 28.6 | 4.61:1 | 21.5 |
| $WL_{A3}$ | 5.0 | 2.6 | 6.1 | 4.6 | 1.22:1 | 9.86 |
| $WL_{B3}$ | 10.0 | 2.5 | 26.1 | 13.0 | 2.61:1 | 12.6 |
| $WL_{C3}$ | 10.0 | 1.9 | 46.1 | 21.4 | 4.61:1 | 14.6 |

### Comparison between the Systems Containing $NO_3^-$ only and those Containing both $PO_4^{3-}$ and $NO_3^-$

Table 11 shows the % removal of $PO_4^{3-}$-P and $NO_3$-N and the % increase in biomass in wastewater containing nitrate only, and for those containing both nitrate and phosphate.

**Table 11** Changes in biomass and nutrients in the synthetic wastewaters

| *Batch Systems* | *% Increase in Biomass* | *Initial Concentration (mg/L)* | | *$NO_3$-N Removed (mg/L)* | *% Removal of $NO_3$-N* | *$PO_4^{3-}$- P Removed (mg/L)* | *% Removal of $PO_4^{3-}$- P* |
|---|---|---|---|---|---|---|---|
| | | *$NO_3^-$* | *$PO_4^{3-}$* | | | | |
| $WH_{A2}$ | 8.28 | 6.4 | – | 5.1 | 79.7 | – | – |
| $WH_{A3}$ | 12.3 | 6.1 | 5 | 5.1 | 83.6 | 3.2 | 64 |
| $WL_{A2}$ | 7.74 | 6.4 | – | 4.7 | 73.4 | – | – |
| $WL_{A3}$ | 9.86 | 6.1 | 5 | 4.6 | 75.4 | 2.6 | 52 |
| $WH_{C2}$ | 16.1 | 26.4 | – | 12.5 | 47.3 | – | – |
| $WH_{B3}$ | 17.4 | 26.1 | 10 | 15.0 | 57.5 | 3.4 | 34 |
| $WL_{C2}$ | 10.6 | 26.4 | – | 9.3 | 35.2 | – | – |
| $WL_{B3}$ | 12.6 | 26.1 | 10 | 13.0 | 49.8 | 2.5 | 25 |

From Table 11, the role of $PO_4^{3-}$- P in the removal of $NO_3$-N can be analyzed since the systems had the same operating conditions. Upon comparison of $WH_{A2}$ with $WH_{A3}$ for the water hyacinth systems and $WL_{A2}$ and $WL_{A3}$ for the water lettuce systems, it can be observed that the amount of $NO_3$-N removed was the same, i.e., 5.1 mg/L for the water hyacinth systems $WH_{A2}$ and $WH_{A3}$. This indicated that a ratio of $NO_3^-$: $PO_4^{3-}$ of 1.22:1 has no effect on the uptake of nitrate from the systems as compared to systems having the same concentration of nitrate but no phosphate.

On comparing $WH_{C2}$ with $WH_{B3}$ for the water hyacinth systems and $WL_{C2}$ and $WL_{B3}$ for the water lettuce systems, it can be observed that the removal of $NO_3^-$N differed by 10.2% and 14.6% respectively, i.e., the systems containing the phosphate showed much higher % removal of $NO_3$-N for the same operating conditions. The % increase in biomass was also greater for systems containing phosphate. The presence of phosphate caused an increase in biomass since it is an important constituent of cell membrane and also because it increased the uptake of nitrate by the aquatic plants.

### Uptake of Chromium by the Aquatic Macrophyte Systems

#### *Variation in pH*

Table 12 shows the variation in pH at intervals of 6 or 7 days in wastewaters containing varying concentration of chromium.

**Table 12** Variation in pH for the synthetic wastewaters containing chromium

| *Batch Systems* | *pH* | | | |
|---|---|---|---|---|
| | *Number of Operating Days* | | | |
| | *1* | *8* | *14* | *20* |
| $WH_{A4}$ | 6.88 | 8.17 | 7.84 | 7.66 |
| $WH_{B4}$ | 6.75 | 8.09 | 7.74 | 7.85 |
| $WH_{C4}$ | 6.47 | 7.48 | 7.70 | 7.73 |
| $WL_{A4}$ | 6.88 | 7.55 | 7.74 | 8.82 |
| $WL_{B4}$ | 6.75 | 7.39 | 7.58 | 7.32 |
| $WL_{C4}$ | 6.47 | 7.24 | 7.44 | 8.20 |

In general, external factors, such as temperature and pH not only influence growth but also affect metal uptake (Bunluesin, 2004). For the water hyacinth systems, the pH varied between 6.47-8.09, while that for the water lettuce systems it was between 6.47-8.82 as shown in Table 12. Figures 13 and 14 depict the macrophytes in synthetic wastewater containing varying concentration of chromium at regular intervals.

### *Variation in temperature*

Table 13 shows the variation in temperature at intervals of 6 or 7 days in wastewaters containing varying concentration of chromium.

**Table 13** Variation in temperature for synthetic wastewaters containing chromium

| *Batch Systems* | *Temperature (°C)* | | | |
|---|---|---|---|---|
| | *Number of Operating Days* | | | |
| | *1* | *8* | *14* | *20* |
| $WH_{A4}$ | 25.4 | 25.3 | 25.4 | 25.4 |
| $WH_{B4}$ | 25.5 | 25.2 | 25.3 | 25.2 |
| $WH_{C4}$ | 25.5 | 25.2 | 25.4 | 25.3 |
| $WL_{A4}$ | 25.4 | 25.3 | 25.8 | 25.8 |
| $WL_{B4}$ | 25.5 | 25.4 | 25.8 | 25.9 |
| $WL_{C4}$ | 25.5 | 25.5 | 25.7 | 25.8 |

Temperature, which has a direct affect on the rate of uptake, remained more or less constant since the temperature for the water hyacinth systems varied between 25.2°C to 24.5°C and those for water lettuce systems were between 25.3°C to 25.9°C.

### *Variation in conductivity*

Figure 15 illustrates the variation in conductivity with time for varying concentration of chromium. The general trend observed was a decrease in conductivity with time for all the batches. The decrease in conductivity was more for the water hyacinth systems than for the water lettuce systems for the same operating conditions.

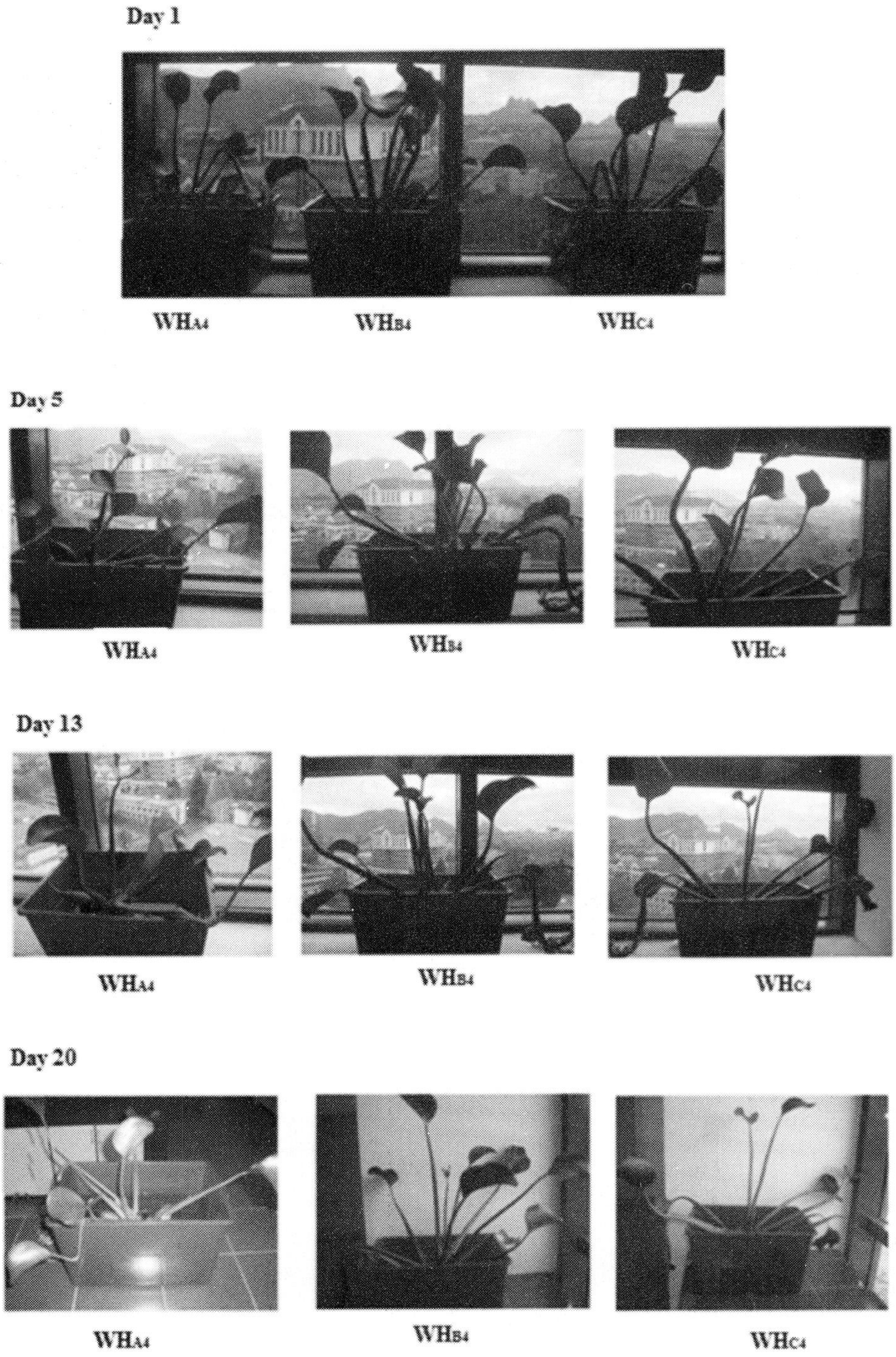


**Fig. 13.** Water hyacinth systems in varying concentration of chromium.

**Fig. 14.** Water lettuce systems in varying concentration of chromium.

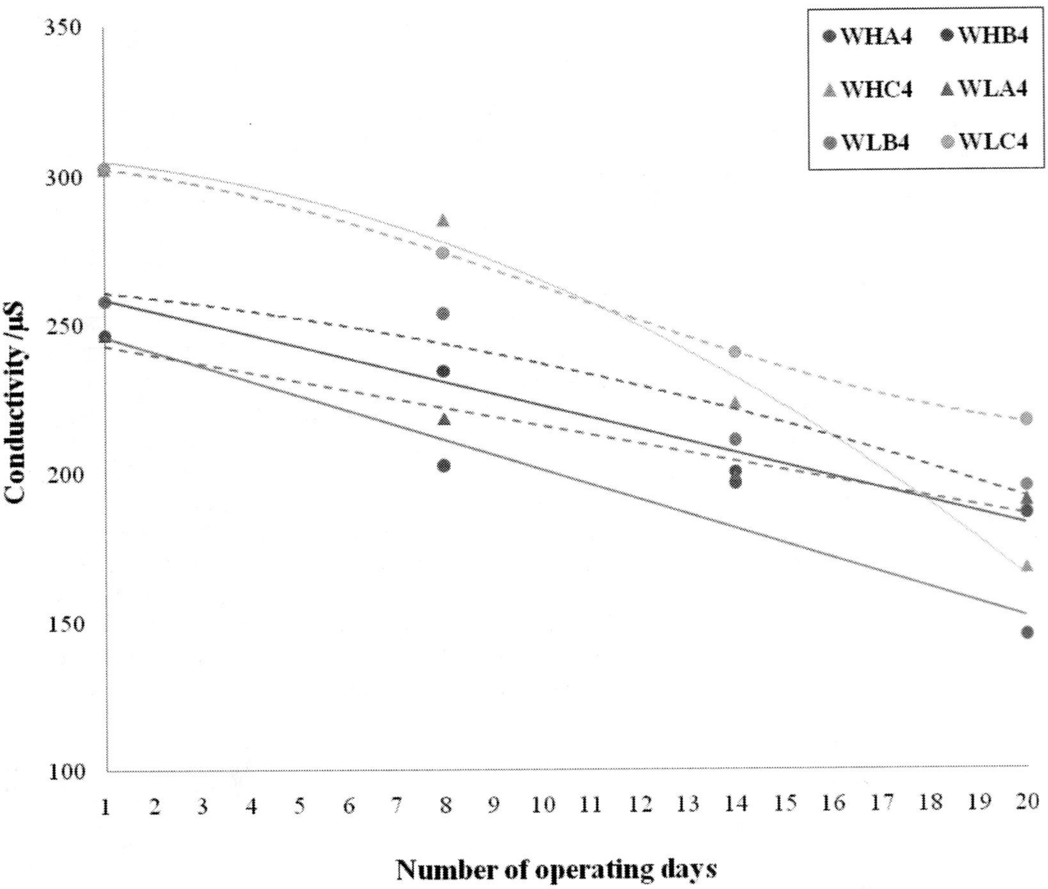


**Fig. 15.** Variation of conductivity for systems containing chromium.

### *Variation in the concentration of chromium*

Figure 16 illustrates the variation in the concentration of chromium with time for the batches containing the aquatic plants.

As shown in Figure 16, the concentration of chromium decreased in all the systems. The decrease in the concentration of chromium was more for the water hyacinth systems than the water lettuce systems under the same operating conditions. At low concentration of 10 mg/L of chromium, the concentration decreased linearly with time for both macrophyte systems $WH_{A4}$ and $WL_{A4}$ as compared to other systems.

Table 14 shows the % removal of chromium with increasing concentration of the species.

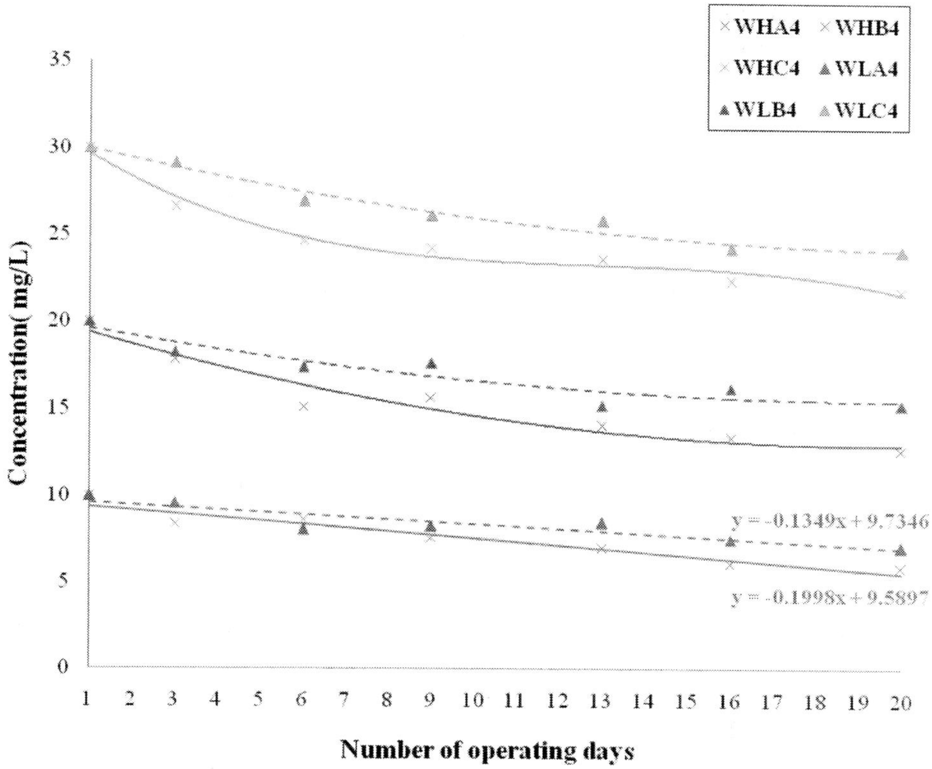


**Fig. 16.** Variation in the concentration of chromium with time.

**Table 14** % removal of chromium from the synthetic wastewaters

| *Batch Systems* | *[Cr] initially (mg/L)* | *$[Cr]_{t=20}$ (mg/L)* | *[Cr] Removed (mg/L)* | *% Cr Removal* |
|---|---|---|---|---|
| $WH_{A4}$ | 10 | 5.876 | 4.124 | 41.2 |
| $WH_{B4}$ | 20 | 12.724 | 7.276 | 36.9 |
| $WH_{C4}$ | 30 | 21.815 | 8.185 | 27.3 |
| $WL_{A4}$ | 10 | 7.078 | 2.922 | 29.2 |
| $WL_{B4}$ | 20 | 15.317 | 4.683 | 23.4 |
| $WL_{C4}$ | 30 | 24.142 | 5.858 | 19.5 |

As shown in Table 14, the concentration of chromium decreased in all the systems after 20 days. As the load of chromium was increased in the wastewater, the concentration of chromium increased, but the % removal decreased in all the batches. The highest removal was 41.2% for the water hyacinth system containing 10 mg/L Cr against 29.2% for the water lettuce system at the same operating conditions. The decrease in conductivity was more for the water hyacinth systems as compared to the water lettuce systems under the same operating condition. This was due to the fact that the water hyacinth systems had a removal of 12%,

16.2% and 7.8% more than the water lettuce systems in synthetic wastewaters containing 10, 20 and 30 mg/L Cr, respectively. Water hyacinth has an excellent potential for phytoremediation of chromium in wastewater (Hansen et al., 1998). Water hyacinth roots harbour bacteria of the genera *Pseudomanas* and *Bacillus* in significant numbers. These bacteria have the ability to reduce $Cr^{6+}$ to $Cr^{3+}$ (Jayaaweera et al., 2006).

Water hyacinth can also accumulate chromium in high concentration in its tissues, especially in roots (Hansen et al., 1998). Warrier and Saroja (2008) found that water hyacinth dosed with Cr (V1) accumulated Cr (III) in the roots and stem tissues, and after the Cr (VI) was reduced to Cr (III) in the fine lateral roots, the Cr (III) was translocated to leaf tissues.

### *Variation in chlorophyll level with time*

Figure 17 illustrates the variation in the concentration of chlorophyll in mature leaves as the load of chromium was increased in the water hyacinth systems.

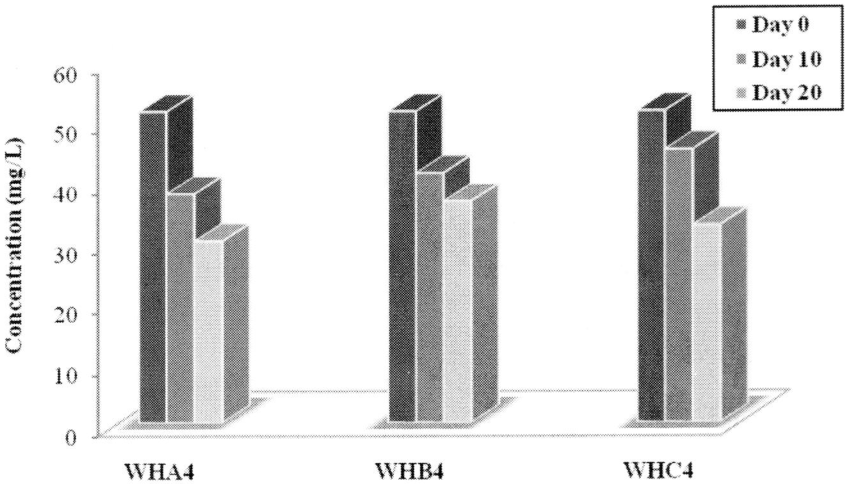


**Fig. 17.** Variation in chlorophyll for the water hyacinth systems in chromium.

Table 15 shows the % reduction in chlorophyll for the leaves of water hyacinth systems at intervals of 10 days.

**Table 15** Reduction in chlorophyll for the water hyacinth systems

| *Batch Systems* | *% Reduction in Chlorophyll* | |
|---|---|---|
| | *Number of Operating Days* | |
| | *10* | *20* |
| $WH_{A4}$ | 26.5 | 41.8 |
| $WH_{B4}$ | 19.8 | 29.0 |
| $WH_{C4}$ | 22.2 | 37.0 |

As shown in Figure 17, the chlorophyll level in the leaves of the water hyacinth systems decreased with time. The highest reduction in the chlorophyll level was 41.8% at the end of 20 days for the systems containing the lowest concentration of 10 mg/L chromium in the synthetic wastewater. Figure 18 illustrates the variation in the concentration of chlorophyll in mature leaves as the load of chromium was increased in the water lettuce systems.

Table 16 shows the % reduction in chlorophyll for the leaves of the water hyacinth systems at intervals of 6 or 7 days.

**Table 16** % reduction in chlorophyll for the water lettuce systems

| *Batch Systems* | *% Reduction in Chlorophyll* | | |
|---|---|---|---|
| | *Number of Operating Days* | | |
| | *7* | *13* | *20* |
| $WL_{A4}$ | 30.0 | 52.3 | 63.8 |
| $WL_{B4}$ | 47.8 | 53.4 | 68.4 |
| $WL_{C4}$ | 51.2 | 63.7 | 75.3 |

The concentration of chlorophyll in the mature leaves of water lettuce systems decreases with time and with increasing load of chromium. The highest reduction in chlorophyll was 75.3% at the end of the experiment for the water lettuce in 30 mg/L of Cr as shown in Table 16. In aquatic plants growing in heavy metal contaminated environment, the metabolically active sites for chlorophylll have been reported to be substituted by heavy metals (Bunluesin, 2004). From Figures 17 and 18, it is observed that exposure of macrophytes to synthetic wastewater containing chromium resulted in a decrease in the concentration of chlorophyll in

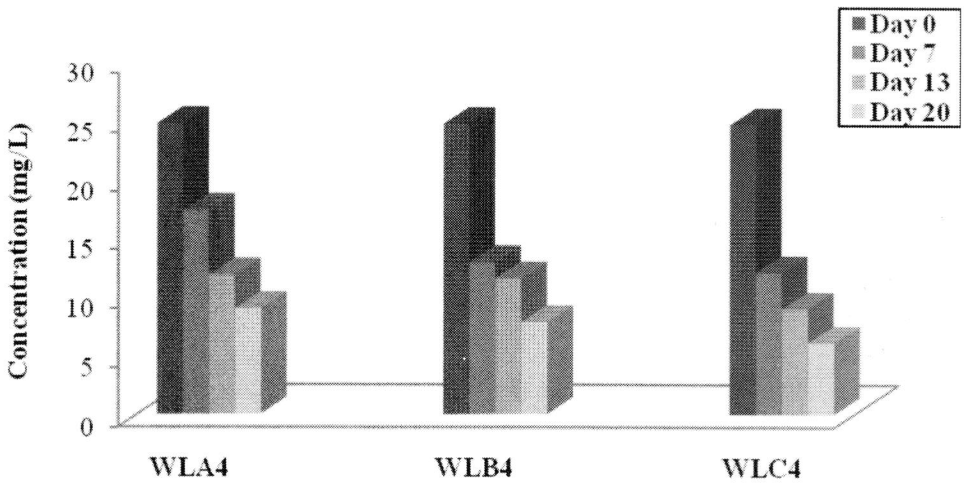


**Fig. 18.** Variation in chlorophyll for the water lettuce systems in chromium.

the mature leaves. In both plant species, there was a significant decrease in total chlorophyll when the load of chromium in the synthetic wastewater was increased. At the end of the experiment, the highest reduction of 75.3% in chlorophyll level was for the water lettuce system $WL_{C4}$. This indicated that chromium had a more phytotoxic effect on water lettuce than water hyacinth, and if the water lettuces were left for a longer period in the chromium solution, they could have been completely toxified. Hence, chromium inhibited photosynthesis in both macrophytes at physiological levels since chorophyll content is an indicator of the photosynthesis activity of a plant and it decreased with time.

### Bioconcentration Factor (BCF)

In general, plants exhibit variable degrees of heavy metal accumulation in their physiological parts. Metal bioaccumulation depends upon numerous biotic and abiotic factors such as temperature, pH and dissolved ions in water (Ma, 2005).

The BCF factor for the water hyacinth systems varied in the order of roots > shoots > leaves except for the water hyacinth $WH_{A4}$. For the water hyacinth system in 30 mg/L of Cr, the BCF factor was 7.93, 0.43 and 0.10 in roots, shoots and leaves, respectively. The leaves and shoots of the water hyacinth system $WH_{A4}$ had relatively shorter shoots as compared to other systems. Some of the leaves and part of the shoots were submerged in the synthetic wastewater. Hansen et al. (1998) observed that some chromium precipitated on the aerial parts of water hyacinth when in contact with the solution in which they were placed. This explained why the BCF factor for leaves and shoots were greater than roots for the system $WH_{A4}$.

The BCF varied in the order of roots > leaves for all the systems. For the roots, the BCF factor increased as the load of chromium was increased in the synthetic wastewater. The BCF factor was 2.92, 0.64 and 1.55 in the leaves of the water lettuce systems in 10, 20 and 30 mg/L of chromium.

The BCF factor for roots was greater than the BCF factor for the aerial parts. This confirmed the ability of roots to accumulate metals and may be treated as the most important storage organ (Chiang, 2009). Since higher biomass has a higher ability for the uptake of metals (Bunluesin, 2004), the BCF factor was used to analyze the translocation of metal in different parts of each macrophyte. BCF factor in this study will not be used to compare the accumulation of chromium in different parts of one macrophyte with respect to the other. This is due to the fact that BCF factor does not take into consideration the time factor. Table 17 shows the % reduction in chlorophyll and the variation in the mass of chromium that accumulated in the plant parts.

**Table 17** Relationship between reduction in chlorophyll and amount of Cr in leaves

| *Batch Systems* | *Mass of Cr in Leaves (mg)* | *% Reduction in Chlorophyll After the Experiment* |
|---|---|---|
| $WH_{A4}$ | 0.841 | 41.8 |
| $WH_{B4}$ | 0.045 | 29.0 |
| $WH_{C4}$ | 0.049 | 37.0 |
| $WL_{A4}$ | 0.994 | 63.8 |
| $WL_{B4}$ | 0.787 | 68.4 |
| $WL_{C4}$ | 3.219 | 75.3 |

The leaves of the water lettuce systems accumulated more chromium than those of the water hyacinth systems, and hence had a reduction in chlorophyll of 22%, 39.4% and 38.3 % more in synthetic wastewaters containing 10, 20 and 30 mg/L of Cr, respectively. This could be due to the fact that chromium precipitated on the leaves of the water lettuce since the mature leaves were in direct contact with the synthetic solution (Hansen et al., 1998), or that the translocation of chromium from roots to leaves was higher for water lettuce systems since the latter are directly translocated to leaves without having to pass through shoots as compared to water hyacinths.

## Total Mass of Chromium Accumulated in Aquatic Plants

The water hyacinth systems had accumulated the highest amount of chromium after 20 days, which is in accordance with the fact that water hyacinth has a higher % removal of chromium from the synthetic wastewater as compared to water lettuce systems under same operating conditions. The water hyacinths in 10, 20, 30 mg/L chromium had accumulated 11.186, 17.391 and 23.362 mg of Cr against 5.439, 7.851 and 14.039 mg of Cr in the water lettuces, respectively.

## Modeling

In order to understand the removal mechanism, the data obtained were modeled as a zero or first order degradation process.

### *Zero order degradation process*

For a zero order degradation process, a plot of the concentration of the species as a function of time will be linear with slope (-k). At low concentration of nitrate and chromium, the concentration of the species decreased linearly with time. The equations, $R^2$ and K values of all the degradation processes which followed a zero order degradation process are given in Table 18.

**Table 18** Zero order degradation process with equations, $R^2$ and k values

| *Batch Systems* | *Concentration (mg/L)* | | | *Equations* | $R^2$ | *k* |
|---|---|---|---|---|---|---|
| | $NO_3^-$ | $PO_4^{3-}$ | *Cr* | | | |
| $WH_{A2}$ | 6.4 | – | – | y = –0.2657 x + 6.4621 | 0.9942 | –0.2657 |
| $WH_{B2}$ | 6.4 | – | – | y = –0.2411 x + 6.5277 | 0.9868 | –0.2411 |
| $WH_{A3}$ | 6.1 | 5.0 | – | y = –0.2429 x + 5.3756 ($NO_3^-$) | 0.9118 | –0.2429 |
| $WL_{A3}$ | 6.1 | 5.0 | – | y = –0.2234 x + 5.7002 ($NO_3^-$) | 0.9511 | –0.2234 |
| $WH_{A4}$ | – | – | 10 | y = –0.1998 x + 9.5897 | 0.9223 | –0.1998 |
| $WL_{A4}$ | – | – | 10 | y = –0.1349 x + 9.7346 | 0.8048 | –0.1349 |

As shown in Table 18, for low concentrations of 6.1 and 6.4 mg/L nitrate and 10 mg/L chromium in the synthetic wastewaters, the degradation process was zero order with respect to the species.

### *First order degradation process*

If a degradation process is first order, the slope of a plot of $[\ln(C_{AO}/C_A)]$ as a function of time is linear with slope k. The following equation indicated a first order degradation process:

$$ln \frac{C_{AO}}{C_A} = kt$$

whereby $C_{AO}$ is the concentration of the species initially and $C_A$ is the concentration of the species at time *t*. Figure 19 illustrates the variation in the value of $ln(C_{AO}/C_A)$ with time for the removal of nitrate in the synthetic wastewater containing nitrate and phosphate.

As shown in Figure 19, the values of $ln(C_{AO}/C_A)$ increase linearly with time for the removal of nitrate in synthetic wastewater containing 26.1 and 46.1 mg/L $NO_3^-$ and 10 mg/L $PO_4^{3-}$. Hence, the degradation processes are first order with respect to the concentration of the species. Table 19 shows the equations, $R^2$ and k values of all the degradation processes which are first order.

As shown in Figure 19, the degradation processes are first order with respect to nitrate at high concentration of 26.1 and 46.1 mg/L of the species in the synthetic wastewater containing 10 mg/L $PO_4^{3-}$. The removal of phosphate for the water lettuce system containing 6.1 mg/L $NO_3^-$ and 5 mg/L $PO_4^{3-}$ was also a first order

**Table 19** First order degradation process with $R^2$ and k values

| *Batch Systems* | *Concentration (mg/L)* | | | *Equations* | $R^2$ | *k* |
|---|---|---|---|---|---|---|
| | $NO_3^-$ | $PO_4^{3-}$ | *Cr* | | | |
| $WH_{B3}$ | 26.1 | 10.0 | – | $y = 26.697e^{-0.043x}$ ($NO_3^-$) | 0.9943 | 0.0432 |
| $WH_{C3}$ | 46.1 | 10.0 | – | $y = 48.095\ e^{-0.05x}$ ($NO_3^-$) | 0.9986 | 0.0504 |
| $WL_{B3}$ | 26.1 | 10.0 | – | $y = 27.054\ e^{-0.036x}$ ($NO_3^-$) | 0.9954 | 0.0363 |
| $WL_{C3}$ | 46.1 | 10.0 | – | $y = 47.095\ e^{-0.033x}$ ($NO_3^-$) | 0.9969 | 0.0330 |
| $WL_{A3}$ | 6.1 | 5.0 | – | $y = 5.1494\ e^{-0.038x}$ ($PO_4^{3-}$) | 0.9971 | 0.0377 |

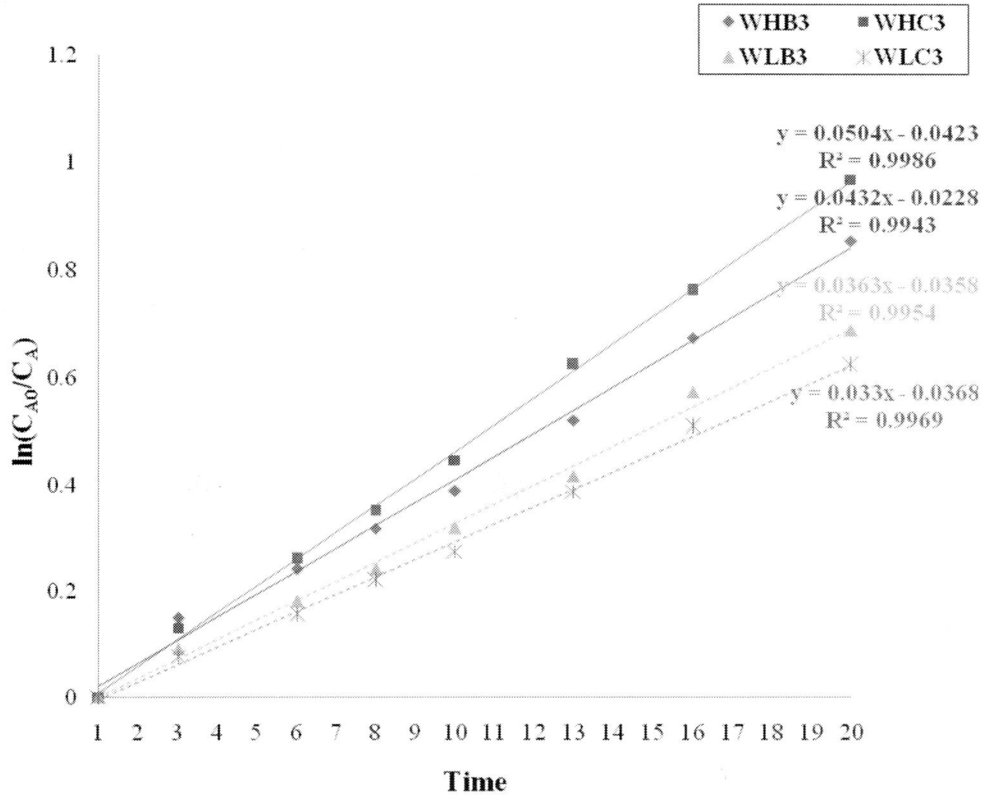


**Fig. 19.** Graph of *ln* $(C_{A0}/C_A)$ with time for nitrate removal in synthetic wastewaters containing both nutrients.

degradation process with respect to phosphate.

## CONCLUSION

Phytoremediation of synthetic wastewater containing nitrate, phosphate and chromium in batch systems using water lettuce and water hyacinth was evaluated. Samples were collected and analyzed for water quality parameter. Data from the study showed that water hyacinth systems had a higher removal of 79.7%, 50.6% and 47.3% in systems containing 6.4, 16.4, 26.4 mg/L $NO_3$-N against 73.4%, 39.0% and 35.2% for the water lettuce systems, respectively. The water hyacinth systems also had a higher biomass production as compared to water lettuce systems under the same operating conditions.

For the phytoremediation of synthetic wastewater containing both phosphate and nitrate, water quality in all the systems was improved, as evidenced by the significant decrease in nutrient concentration. Water hyacinth had a higher %

removal of $NO_3$-N as well as $PO_4^{3}$-P. The removal of $NO_3$-N for the water hyacinth systems was 83.6%, 57.5% and 62.0% for the systems containing 6.1, 26.1 and 46.1 mg/L of $NO_3$-N in the synthetic solution against 75.4%, 49.8% and 46.4% for the water lettuce systems, respectively. The highest removal of phosphate was 64% for the water hyacinth system against 52% for the water lettuce systems. The highest increase in biomass for both macrophytes was for those systems containing the ratio of $NO_3^-$: $PO_4^{3-}$ of 4.6:1, which is in accordance with literature review. The presence of phosphate in the synthetic wastewater increased the % removal of $NO_3$-N, resulting in an increase in biomass. The water hyacinth systems had a higher increase in biomass, making them suitable for animal feed or methane production.

It may be concluded that both macrophytes improved water quality through the uptake of nutrients and chromium from the synthetic wastewaters. However, water hyacinth had higher removal in all the systems, which makes it a better macrophyte as compared to water lettuce for the phytoremediation of similar wastewaters.

## REFERENCES

Akpor, O.B. and Muchie, M. (2010). Remediation of heavy metals in drinking water and wastewater systems: processes and applications. *International J. Physical Sciences*, **5:** 1807-1817.

Arnon, D.I. (1949). Copper enzymes in isolated chloroplasts. Polyphenoloxidase in *Beta Vulgaris*. *Plant Physiology*, **24:** 1-15.

Bashan, L.E. and Bashan, Y. (2004). Recent advances in removing phosphorus from wastewater and its future use as fertilizer. *Water Research*, **38:** 4222-4246.

Della Rocca, C., Belgiorno, V. and Meric, S. (2007). Overview of *in-situ* applicable nitrate removal processes. *Desalination*, **204:** 46-62.

Bunluesin, S. (2004). Phytoremediation of cadmium contaminated wetland using aquatic macrophytes. Thesis (PhD), University of Mahidol.

Chiang, B. (2009). Effect of land on river water quality, Johor Strait coastal water and Green Mussel. Thesis (Undergraduate), University of Malaysia.

Chew, A.L. (2006). Nutrient removal from leachate using horizontal subsurface constructed wetlands. Thesis (Master), University of Malaysia.

Debusk, T.A., Peterson, J.E. and Reddy, K.R. (1995). Use of aquatic and terrestrial plants for removing phosphorus from dairy wastewater. *J. Ecotechnology*, **5:** 371-390.

Debusk, T.A. and Reddy, K.R. (1987). State-of-the-art utilization of aquatic plants in water pollution control. *Water Science and Technology*, **19:** 61-79.

Debusk, W.F. and Reddy, K.R. (1985). Nutrient removal potential of selected aquatic macrophytes. *J. Environmental Quality*, **14:** 459-461.

Fahrner, S. (2002). Groundwater nitrate removal using a bioremediation trench. Thesis (Honours), University of Western Australia.

Hansen, O., Lytle, C.M. and Lytle, F.W. et al. (1998). Reduction of Cr (VI) to Cr (III) by wetland plants: Potential for *in-situ* heavy metal detoxification. *Environmental Science and Technology*, **32:** 3087-3093.

Homyok, K., Kruatrachue, M. and Lu, X. et al. (2004). Removal of cadmium and zinc by water hyacinth, *Eichhornia crassipes*. *Science Asia*, **30:** 93-103.

Jayaweera, M.W., Kasige, A.P. and Kularatne, R.K.A. et al. (2006). Removal mechanism and efficiency of selected heavy metal in constructed wetlands. *Sustainable Development of Water Resources Water Supply and Environmental Sanitation*, 112-116.

Lavrova, S. and Koumanova, B. (2007). Polishing of aerobically treated wastewater in a constructed wetland system. *J. University of Chemical Technology and Metallurgy*, **42:** 195-200.

Ma, Y. (2005). Monitoring of heavy metals in the Butterfly River using *Typha capensis* and *Phragmites australis*. Thesis (Master), University of Western Cape.

Misra, A., Srivastava, A.K. and Srivastava, N.K. (2010). Cr-acquisition and reactive oxygen species role in growth, photosynthesis, photosynthesis pigments, and biochemical changes in essential oils(s). *International Research J. Plant Science*, **1:** 14-20.

Warrier, R.R. and Saroja, S. (2008). Histochemical studies on water hyacinth with particular reference to water pollution. *International J. Integrative Biology*, **3:** 96-99.

Sooknah, R. (2000). A review of the mechanism of pollutant removal in water hyacinth systems. *Research J. Science and Technology*, **6:** 50-57.

# 14

# Smog Pollution (PAN and Ozone): A Case Study of Delhi

SANYOGITA, AMIT PRAKASH, PRADIP BAURI*, UJJWAL KUMAR**, KRISHAN KUMAR, AND V.K. JAIN***

*Jawaharlal Nehru University, New Delhi-110067*
**Presently at Central University of Gujarat, Gandhinagar*
***Presently at KNMI - Royal Netherlands Meteorological Institute, Netherlands*
****Corresponding Author: E-mail: vkj0400@mail.jnu.ac.in*

**ABSTRACT**

The modern-day typical urban atmosphere is marred by tremendous quantity and variety of air pollutants due to a large number of their emission sources. When these pollutants enter into the atmosphere, they undergo a variety of physico-chemical processes depending upon the weather conditions. Under such conditions, a hazy atmosphere is created which not only possesses the characteristics of reduced visibility, but also carries a number of undesirable properties detrimental to human health, plants, animals and materials. In such an atmosphere, the presence of $NO_x$, $O_3$, PAN and aerosol constitutes the condition of smog in urban areas. The monitoring study, carried out in Delhi at several locations, reveals the presence of high concentrations of $O_3$, PAN and NOx along with PM10 (RSPM). PAN, an important constituent of smog, is found to be strongly correlated with $O_3$, NOx and RSPM in winter season. In winter, smog was observed mostly during the afternoon period (3 to 7 pm) at most of the locations. In summer season, PAN concentration was extremely low due to its thermochemical instability at high temperatures (>25°C). The high concentrations of $O_3$ and PAN observed in the ambient atmosphere of Delhi are likely to have serious implications for human health and vegetation.

**Keywords:** Smog, Pollution, PAN, Ozone, Human Health.

## INTRODUCTION

In ambient environment, smog is formed when an inversion or stable atmospheric condition is established and large amount of air pollutants are trapped within this stable atmospheric condition. Under such conditions, a hazy atmosphere is created due to formation of smog. The word "smog" is a portmanteau of smoke and fog. However, in modern terms, smog is a kind of air pollution or more appropriately a by-product of air pollution. Industrial sources such as coal-based power plants emit huge amount of sulphur compounds, particulates, carbon monoxide etc. Traffic emissions include carbon monoxide, oxides of nitrogen, volatile organic compounds, particulate matters etc. When these pollutants enter into the atmosphere, they undergo a variety of physical and chemical processes depending upon the weather conditions. As a result, many different kinds of secondary air pollutants such as ozone, PAN, aldehydes, etc. are generated which further worsen the quality of air. Moreover, when a "*temperature inversion*" occurs and air becomes thermally stable blocking the vertical flow of air, entire pollutants are trapped within a limited height and thus a condition of smog is created.

Near the ground, air temperature normally decreases with height. When the rate of decrease is rapid, there is a pronounced tendency toward vertical air mixing. On the other hand, when air temperature increases with height, vertical air motions are suppressed. This temperature structure is called a "temperature inversion" because it is "inverted" from the normal condition of temperature decreasing with height.

An inversion may take place at the surface or in the upper air. A surface or radiation inversion usually occurs on clear nights with low wind speed. In this situation, the ground cools the surface air by nature of its own cooling due to long wave radiation to the outer atmosphere. The surface air becomes cooler than the upper layers, and vertical air flow is halted (Hot air rises, cool air sinks.). A parcel of warm air trying to rise finds the air above it hotter than itself. Thus, it will not rise because the temperature gradient is inverted. This type of inversion is common and is broken up as the sun once again heats the ground next morning.

## THE SMOG STUDIES

The characteristics of smog can best be understood by studying the characteristics of the smog episodes that occurred so far in the different parts of the world, which have been great cause of concern and also extensively studied. We hereby present the characteristics of Donora smog, London smog and Los Angeles smog, the differences and similarities among them, and how their study can help us to understand and characterize the smog phenomena in Delhi.

### Donora Smog

Donora, lying near the bottom of a steep valley, is about 500 feet below the surrounding terrain. In the fall of 1948, a high pressure area moved over western

Pennsylvania on October 26, 1948, and remained fairly stationary for the next 5 days. During the period October 26 to 31, 1948, an inversion capped the valley. Pilot reports and weather balloons showed that the cap, at least part of the time, was less than 1000 feet above the town. Thus, smoke fed to the atmosphere was largely confined within the valley walls and the inversion top. The air near the ground was very humid. Fog formed in the night, and in some low lying areas of western Pennsylvania, it persisted during the day. At Donora, the visibility cut by smoke and fog, ranged from about 0.6 to 1.5 miles.

### London Smog

The London smog also referred to as *Great Smog* or *Big Smoke*, befell London starting on December 5th, 1952, and lasted until December 9th, 1952. The *Great Smog of 1952* darkened the streets of London and killed approximately 4,000 people in the short time of 4 days (a further 8,000 died from its effects in the following weeks and months).

Early in December 1952, a cold fog descended upon London. Because of the cold, Londoners began to burn more coal than usual. The resulting air pollution was trapped by the inversion layer formed by the dense mass of cold air. Concentrations of pollutants, coal smoke in particular, built up dramatically. The problem was made worse by use of low-quality, high-sulphur coal for home heating in London in order to permit export of higher-quality coal, because of the country's tenuous postwar economic situation (Camps, 1976). The "fog", or smog was so thick that driving became difficult or impossible. It entered indoors easily, and concerts and screenings of films were cancelled as the audience could not see the stage or screen.

### Los Angeles Smog

Being in a low basin surrounded by mountains, Los Angeles is notorious for its smog. Los Angeles city was often engulfed by a brownish-yellow haze. Initially, the cause of this haze was very little known, so early efforts were not very effective. It was not until Dr. Haagen-Smit of the California Institute of Technology showed that the Los Angeles smog was the result of atmospheric reactions between sunlight, oxides of nitrogen, and olefinic organics (Haagen-Smit, 1952; Haagen-Smit and Fox, 1956), that more effective control measures were implemented.

The complexity of Los Angeles smog can be studied using diurnal variation of major pollutants, ozone, and its precursors, oxides of nitrogen and reactive hydrocarbons. A major source of precursor pollutants is the automobile. During morning rush hour traffic, automobiles emit nitric oxide (NO) and unburned hydrocarbons. Mixing in the atmosphere is typically limited in the cool morning hours and the emission result in a rapid rise in NO and hydrocarbon concentrations in the near-surface atmosphere. Then NO is oxidized to nitogen dioxide ($NO_2$) beginning the complex series of chemical reactions leading to the formation of ozone and photochemical smog.

### Comparison between "London smog" and "Los Angeles Smog"

Table 1 shows the comparison that makes Los Angeles smog different from London smog.

**Table 1** Comparison between "London smog" and "Los Angeles smog"

| *Condition* | *London* | *Los Angeles* |
|---|---|---|
| Fuel | Coal and hydrocarbon | Hydrocarbons |
| Seasons during year | Dec-Jan | Aug-Sep |
| Peak-Time | Early Morning | Mid-day |
| Temperature | 30 to 40°F | 75 to 90°F |
| | (-1.1 to 4.4°C) | (23.9 to 32.2°C) |
| Humidity | High with fog | Low-relatively clear sky |
| Wind Speed | Calm | < 5mph |
| Inversion | Surface (Radiation) | Overhead (Subsidence) |
| Visibility | Severely reduced (<1000 ft) | Reduced (1/2 to 1 mile) |
| Principal Constituents | Sulfur compounds, CO, particulates | $O_3$, Nitrogen Oxides, CO, PAN |
| Effects on human | Bronchial Irritation | Eye and Skin irritation |
| Chemical Effects | Reducing | Oxidizing |

They are similar in that they result from community air pollution, and major source is the combustion of fuels. In London, it is coal and different hydrocarbons; in Los Angeles, it is primarily hydrocarbons. The peak time in London is early morning. In Los Angeles, the peak time is mid-day. In London, temp is 30 to 40°F; in Los Angeles 75 to 90°F. The humidity is high with fog in London. Generally, in Los Angeles, pollution occurs on a relatively clear day with low humidity. The inversions in London are at the surface; in Los Angeles, inversions are overhead. Visibility is severely reduced in London but only partially reduced in Los Angeles. The effects in London are to produce bronchial irritation, whereas in Los Angeles the effects are to produce eye and skin irritations. In Los Angeles, the smog is primarily produced through photochemical oxidation of hydrocarbons by the ozone and nitrogen oxides that are in the atmosphere. The product of this photochemical reaction is an organic type molecule that causes plant damage and reduced visibility, and irritates skin and eyes very badly. The London type chokes us. We get a feeling of being in midst of a big smoke, because that is primarily what it is—smoke and fumes mixed with moisture in the air. Undesirable effects in Los Angeles are due to photochemical reactions irradiated by solar radiation. Due to oxidant formation such as PAN, photochemical smog produces undesirable effects like eye and skin irritation, plant damage, and reduced visibility. The sources of organic compounds and oxides of nitrogen largely are the internal combustion engine.

## CLASSICAL SMOG AND PHOTOCHEMICAL SMOG

Based on the above studies, the smog is now broadly classified into two categories: (i) Classical Smog; and (ii) Photochemical Smog.

**(i) Classical Smog:** The mixture of particulates with gaseous oxides of sulphur present in atmosphere is called classical smog. The principal source of classical smog is the combustion of industrial and household fuels (coal and petroleum). The major constituents of classical smog are particulates and $SO_2$. The sources are coal, biomass (wood, dungcakes, crop-residues) burning.

**(ii) Photochemical Smog:** Photochemical smog occurs due to the action of sunlight on air pollutants, generating photochemical reactions. The principal constituents of the photochemical smog are particulates, $SO_2$, $O_3$, PAN, aldehydes, VOCs, unburnt hydrocarbons and fog particles. The sources are coal and biomass burning + vehicular and industrial emissions.

Under certain conditions, fog + all constituents of photochemical smog will be commonly known as smog. In Delhi, average ambient temperature during November-February is 22.7°C (Ref: of the year 2007-08), while during March-June it is 28.59°C (Ref: of the year 2008). Since July-September witnesses rainy season, the probability of occurrence of smog during this period is very low. Further, the main source of air pollution in Delhi is traffic emissions. Thus, the weather condition and the sources of pollution in Delhi most likely favor the formation of *photochemical smog*.

## CHEMICAL CONSTITUENTS OF SMOG

We, hereby, focus on photochemical smog which is most likely to occur in the ambient urban environment of Delhi. In this chapter, we will also discuss another important constituent of Delhi's smog, suspended particulate matter (SPM), which is highly prevalent in the Delhi's atmosphere and consistently exceeds the national ambient air quality standard (www.cpcb.nic.in).

In a typical polluted urban atmosphere, when oxides of nitrogen, various hydrocarbons, and sunlight come together, they initiate a complex set of reactions that produce a number of secondary pollutants known as photochemical oxidants [such as ozone ($O_3$) and peroxyacetyl nitrate (PAN)]. In the very simplest terms, we can express the formation of photochemical smog as

$$\text{Hydrocarbons} + NO_x + \text{Sunlight} \rightarrow \text{Photochemical smog} \quad (1)$$

This reaction gives us only the simplest overview. We add a few details below to have meaningful picture of the formation of photochemical smog. The traffic sources emit nitric oxide (NO) in the atmosphere due to combustion of fuel. The nitric oxide thus emitted can oxidize to $NO_2$:

$$2NO + O_2 \rightarrow 2NO_2 \tag{2}$$

If sunlight is available, $NO_2$ can photolyze, and the freed atomic oxygen can then help to form ozone, as the following pair of reaction suggests:

$$NO_2 + hn \rightarrow NO + O \tag{3}$$

$$O + O_2 + M \rightarrow O_3 + M \tag{4}$$

where hn represents a photon (1 < 0.38 mm), and M represents a molecule (usually $O_2$ or $N_2$ since they are most abundant in air) whose presence is necessary to absorb excess energy from the reaction. Without M, the ozone ($O_3$) would have too much energy to be stable, and it would dissociate back to O and $O_2$.

$O_3$ can then convert NO back to $NO_2$:

$$O_3 + NO \rightarrow NO_2 + O_2 \tag{5}$$

This set of reactions creates a cycle. However, the chemistry of photochemical smog is exceedingly complex. A multitude of organic chemicals are introduced to the atmosphere when fuel burns or volatiles evaporate, and many more are produced in the atmosphere as a result of chemical reactions. A very important radical that plays a crucial role in the atmospheric chemistry is OH radical. The OH radical is formed when atomic oxygen reacts with water:

$$O + H_2O \rightarrow 2OH \tag{6}$$

The OH radical is extremely reactive, and its atmospheric concentration is so low that it has been difficult to detect. Nevertheless, it plays a key role in many reactions, including the oxidations of $NO_2$ to nitric acid and CO to $CO_2$.

$$OH + NO_2 \rightarrow HNO_3 \tag{7}$$

$$OH + CO \rightarrow CO_2 + H \tag{8}$$

The following three reactions provide one explanation for the way that hydrocarbons can enhance the rate of conversion of NO to $NO_2$ and hence increase $O_3$ concentrations. Letting RH be a hydrocarbon, we have

$$RH + OH \rightarrow R + H_2O \tag{9}$$

$$R + O \rightarrow RO_2 \tag{10}$$

$$RO_2 + NO \rightarrow RO + NO_2 \tag{11}$$

These reactions are not complete, though they do show how hydrocarbons can help convert NO to $NO_2$, and thus increase $O_3$. Similar reactions of OH with other volatile organic compounds (VOCs) occur that enhance the formation of $O_3$. In the above set of three reactions, it appears that OH is destroyed in the process. Unless there is some way to rejuvenate that OH, these reactions could not continue for long. The following pair of reactions shows not only one way that OH is regenerated, but also, in the process, how another NO is converted to $NO_2$. As well, an aldehyde can be formed from this process:

$$RO + O_2 \rightarrow HO_2 + RCHO \tag{12}$$

$$HO_2 + NO \rightarrow NO_2 + OH \tag{13}$$

The net effect of reactions (9) – (11) is that one hydrocarbon molecule converts two molecules of NO to $NO_2$ and produces an aldehyde, RCHO. The removal of NO slows the rate at which $O_3$ is removed, while the addition of $NO_2$ increases the rate at which it is produced, thus leading to higher levels of $O_3$ in the air.

*Formation of PAN*: The above relationship can be also extended to show the formation of another eye-irritants, peroxyacetyl nitrate (PAN). Acetaldehyde ($CH_3CHO$), formed from ethane emissions and by following the reaction set (9)-(11), reacts with OH as follows

$$CH_3CHO + O_2 + OH \rightarrow CH_3C(O)O_2 + H_2O \tag{14}$$

The resulting acetylperoxy radical can react with $NO_2$ to create PAN:

$$CH_3C(O)O_2 + NO_2 \rightarrow \underset{\text{(PAN)}}{CH_3C(O)O_2NO_2} \tag{15}$$

However, the two important constituents of photochemical smog, $O_3$ and PAN, deserve special treatment pertinent to the present project work. The following two sub-sections are devoted to describe their properties and behaviour in the air.

### Ozone

Ozone ($O_3$), one of the key trace gases in the atmosphere, is most abundant tropospheric oxidant. The residence time (life cycle) of tropospheric $O_3$ varies according to season and altitudes between a few days (5–8) and a few weeks (3–15). Tropospheric $O_3$ is not only a chemically active component, but also an important greenhouse gas. Tropospheric $O_3$ changes contribute roughly $0.35 W/m^2$ (about 8–15%) to the total radiative forcing associated with the greenhouse gas increase since pre-industrial times (e.g., Marenco et al., 1994; Portmann et al., 1997; Brasseur et al., 1998).

Ozone is responsible for many of the undesirable properties of photochemical smog, from chest constriction and irritation of the mucous membrane in people to cracking of rubber products and damage to vegetation. Breathing ozone, even at relatively low abundances, is correlated with pulmonary damage and asthma attacks (e.g., Peden, 2001; Desqueyroux et al., 2002; Mortimer et al., 2002). Negative effects of tropospheric ozone in plants and agriculture include reduced crop and grain yields (EPA, 1996; Finnan et al., 1997; Agrawal et al., 2003). In addition to reducing plant growth, exposure to elevated $O_3$ can also alter plant tissue chemistry (Findlay et al., 1996) and reduce allocation of carbon to roots and root exudates (Coleman et al., 1995; Andersen, 2003).

## PAN

As discussed earlier in this section, Peroxyacetyl nitrate (PAN) is formed photochemically in air polluted with nitrogen oxides and volatile organic compounds (VOCs). Formation rates are greatest in warm, sunny conditions, but PAN is thermally unstable, decomposing to form the peroxyacetyl radical (PA) and nitrogen dioxide ($NO_2$). In the absence of gases such as nitric oxide (NO), which rapidly remove the PA radical, an equilibrium between PAN, PA and $NO_2$ occurs (Orlando et al., 1992). PAN decomposition is slow at low temperatures (<10°C); as PAN is itself not easily photolysed (Talukdar et al., 1995), it may be transported for very long distances in winter, or in the upper troposphere, and act as a source of 'reactive' nitrogen in regions remote from pollution (Penkett and Brice, 1986). Although often associated with smog and urban pollution, PAN is a ubiquitous component of the atmosphere in regions influenced by anthropogenic pollutants (Moxim et al., 1996; Spicer et al., 1983). The chemistry and biological effects of PAN were reviewed in 1994 (Kleindienst, 1994).

PAN shows a very strong cytotoxicity, comparable with exposure to much higher concentrations of ozone (Aufderheide, 1998). PAN is a potent lachrymator and acts as a strong eye-irritant (Gaffney and Marley, 1993; Ramanathan et al, 1985; Gaffney et al., 1984). The results of Sun and Huang (1995) show that 4-5 ppbv of PAN is sufficient to injure the native cultivars of lettuce (a type of vegetable) and other sensitive species in Taiwan. PAN effects on plant species (of many varieties) include silvering or bronzing symptoms on the lower leaf surface, distributed in transverse bands on young leaves.

## SUSPENDED PARTICULATE MATTER (SPM)

Suspended particulate matter (SPM) is a major contributor to atmospheric visibility impairment, a symptom of smog. The concentration of SPM is a valuable index of air pollution as the increased level of SPM in the atmosphere cause severe respiratory problems. The ability of human respiratory system to defend itself against SPM is, to a large extent, determined by the size of the particles. Large particles that enter the respiratory system can be trapped by the hairs, and lining of the nose. Once captured, they can be driven out by a cough or sneeze. Smaller particles that make into the tracheobronchial system can be captured by mucous, worked back to throat by tiny hair like cilia, and removed by swallowing and spitting. Particles larger than 10 μm are quite effectively removed in the upper respiratory system by these defense mechanisms. Smaller particles (< 10 μm), however, are often traverse the many turns and bends in the upper respiratory system without being captured on mucous lining.

Therefore, particles of size less than 10 μm ($PM_{10}$) have been the focus of the study by pollution control agencies as well as researchers. Further, chemical compositions of the particulate are extremely complex in a city's air. More than 20 metallic compounds have been found by chemical and spectroscopic analysis of the inorganic fraction. The concentration of these chemical elements depends on the nature of the area and type of pollution sources present. Some of the other studies reported 30-35 heavy metals in atmospheric particulate matter (Schroder et al. 1987).

Suspended particulate matter is a mixed primary and secondary pollutant. Coarse particles (= 2 μm) are mainly primary pollutants, with dominant sources that include fugitive dust from roads, construction, and agricultural emissions. Wind erosion is another important source, especially during dry seasons. Industrial emissions are significant sources of both coarse and fine particles (= 2 μm). A major transportation source of fine particulate matter is diesel soot. Secondary processes contribute to fine particulate mass, including oxidation of hydrocarbons into low vapor-pressure organics, and the formation of aerosol salts such as ammonium nitrate and ammonium sulphate.

In Delhi, concentration of SPM has been consistently exceeding the CPCB national ambient air quality standard (www.cpcb.nic.in). The SPM levels in Delhi are varying between 255 and 728 $\mu g/m^3$, but never within the limits of annual ambient air quality standard. NEERI reported SPM and RSPM ($PM_{10}$) annual average concentration in the range of 358 to 562 $\mu g/m^3$ and 142 to 236 $\mu g/m^3$, respectively. The concentration of trace metals was also reported high in Delhi by various researchers. The levels of trace metals like Ni, Cd, Cr, Pb were found comparatively higher than the European and American cities. The level of Fe in the residential area of Delhi has been reported to be about 15 $\mu g/m^3$, which is twice the levels in Taiwan, and several times higher than that in European cities.

Industries, vehicular traffic, and domestic burning are the major contributors to pollution in Delhi. Vehicular pollution contributes 67% of the total air pollution load (approximately 3000Mt per day) in Delhi. Another 25% of air pollution is generated by industries and coal-based thermal power plants. The three power plants in Delhi generate 6000 Mt of fly ash per day (White paper on pollution in Delhi, http://envfor.nic.in/divisions/cpoll/delpolln.html). There are 3 coal based and 2 gas based thermal power stations in Delhi, namely Badarpur (705 MW coal based), Indraprastha (247.5 MW coal based), Rajghat power house (135 MW coal gas), Gas turbine (282 MW gas based) and Pragati power plant (330 MW, gas based), respectively.

## SMOG DURING FOGGY WINTER

We have already discussed in section 2 how London deadly smog was a result of dense 'fog' and 'smoke from fossil-fuel burning (especially, coal)'. We, hereby, discuss the formation and composition of fog, especially with reference to Delhi.

Fog is formed due to the condensation of water vapour on condensation nuclei that are always present in natural air. Whenever the air is cooled, by any means whatever, below its dew point, a portion of the water vapour present separates out on such dust particles or other condensation nuclei as may happen to be present. If this process happens to occur at the surface of the Earth, it is called fog, no matter how shallow or how deep it may be.

Fog plays an important role as processors of atmospheric aerosols and soluble gases. It affects the fate of some atmospheric species via both chemical and physical processes. They play the role in new particle production e.g., uptake and oxidation of $SO_{2(g)}$ to $SO_{4\,(aq)}^{2-}$ in drops, and particle removal (e.g., via particle scavenging and removal by fog deposition). The drop surface also provides an important reaction site. For example, the presence of a wet surface layer on atmospheric aerosol particles has been associated with the production of $HONO_{(g)}$ (Notholt, et. el., 1992). Fog also interacts with the incoming short wave radiations as well as long wave radiations. Hence, it reduces visibility which disrupts air, rail and road traffic. The transportation system comes to standstill during foggy conditions, which have a very serious implication for Indian economy. Foggy conditions have been classified according to the visibility status in the atmosphere.

Early results from CHEF (Schemenauer, 1986; Schemenaur and Winston, 1988) and other projects (Mohnen and Kaleck, 1989; Saxena et al., 1989) indicated that fogs covering upland areas were both more acidic and had higher concentrations of major ions than did precipitation at the same sites

## WINTER SMOG IN DELHI: A CASE STUDY

In Delhi, normally December to February is winter season. It is these months in Delhi when fog is frequently reported. In ambient environment, when an inversion or stable atmospheric condition is established and large amount of air pollutants are trapped within this stable atmospheric condition, a haze-type atmosphere is created, smog is formed, which not only possesses the characteristics of reduced visibility but also carries a number of undesirable properties detrimental to human health, plants, animals and materials. Under certain conditions, fog + all secondary pollutants along with SPM etc. is commonly known as smog. When smog occurs due to the action of sunlight on air pollutants, generating photochemical reactions, it is called photochemical smog. The principal constituents of the photochemical smog are particulates, $SO_2$, $O_3$, PAN, aldehydes, VOCs, unburnt hydrocarbons and fog particles. The sources are coal and biomass burning + vehicular and industrial emissions. In Delhi, average ambient temperature during November-February is 22.7°C (Ref: of the year 2007-08) while during March-June it is 28.59°C (Ref: of the year 2008). Further, the main source of air pollution in Delhi is traffic emissions.

The surrounding areas of Delhi witness heavy activities of biomass burning. Thus, the weather condition and the sources of pollution in Delhi most likely favor the formation of *photochemical smog.*

## Sampling Details

In the present study, the characteristic constituents of smog such as peroxyacetyl nitrate (PAN), $NO_2$, $NO_x$ and Ozone ($O_3$) along with daily average concentration of RSPM ($PM_{10}$) and meteorological parameters were monitored during January to May, 2008. The monitoring/sampling for the required parameters has been carried out at 7 selected sites within Delhi. The sites include residential areas (JNU, Dwarka and Ashok Vihar), heavy traffic zones (Delhi secretariat and Laxmi Nagar), and industrial sites (Okhla and Wazirpur). At three sites, namely at JNU, Dwarka and Ashok Vihar, sampling/monitoring were carried out twice during both winter and summer months. At other sites, namely, Delhi Secretariat, Laxmi Nagar, Okhla and Wazirpur, monitoring/sampling were carried out for one season.

## CPCB & WHO Standards for Selected Parameters

To study the exceedances, CPCB National Ambient Air Quality standards for 24 hour average $NO_2$ (80 $\mu g/m^3 \sim$ 39 ppb) and 24 hour average PM10 (100 $\mu g/m^3$ for residential and 150 $\mu g/m^3$ for industrial area) while WHO standards for 8 hour average $O_3$ (100 $\mu g/m^3 \sim$ 46 ppb) and PAN (5 ppb) have been used.

## RESULTS AND DISCUSSIONS

Figures 1 to 6 depict the variations in the concentrations of major constituents of smog at different sampling locations in Delhi viz. JNU, Dwarka, Delhi Secretariat, Ashok Vihar, Okhla, and Wazirpur, respectively. Figure 7 presents the RSPM concentrations at different sampling locations.

### (a) JNU (Jawaharlal Nehru University)

In JNU, during winter months, the 24 hour average PM10 concentration always exceeded up to two to three times of the standards set by the CPCB (Figure 7). PAN and $NO_2$ concentration exceeded up to 4.6% and 2%, respectively, of time during the total sampling duration. Measurements indicate that $NO_2$ and $NO_x$ typically exceed 40 ppb during five days of study. PAN was observed to attain a maximum of 10 ppb value during five days of study. Levels of PAN showed a diurnal variation with PANs maximum during evening hours or afternoon and falling to less than 0.1 ppb during night hours. The $O_3$ concentration reached up to a maximum of 30 ppb at JNU during the five days of study. The high level of $O_3$ is commensurate with the high level of PAN in the air. The elevated average level of SPM is accompanied by elevation of average $NO_x$ concentrations. This also indicates their similar source, i.e., traffic.

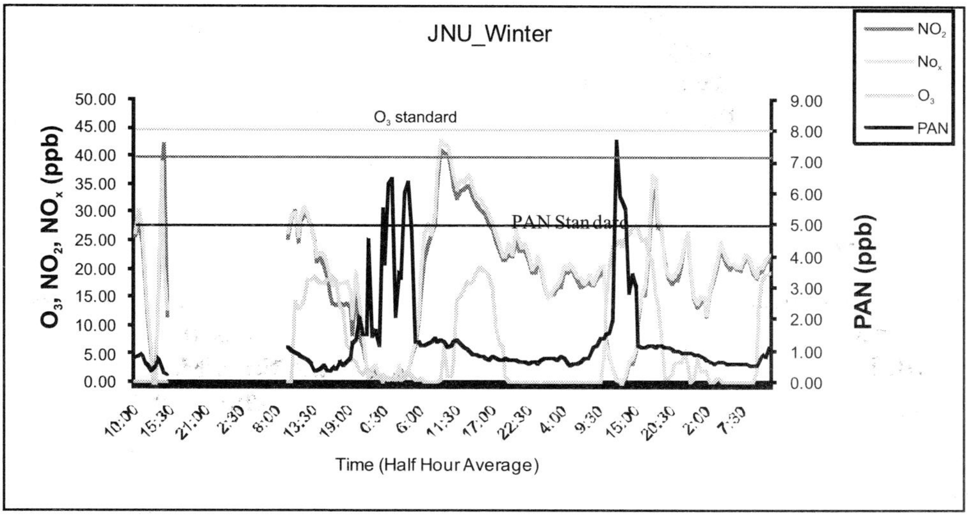


**Fig. 1.** The variation of half hour average concentrations of $NO_2$, NOx, $O_3$ and PAN at JNU during 17/1/2008 to 21/1/2008. $O_3$ continuous measurements were taken since 10:53 AM, 18/1/2008 till 11:58 AM, 21/1/2008.

## (b) Dwarka

At Dwarka also, PM10 concentration always exceeded CPCB standard reaching up to two to three times of the standards set. PAN was observed to exceed 0.7% of the times over total sampling duration. The observed maximum $NO_2$, and $NO_x$

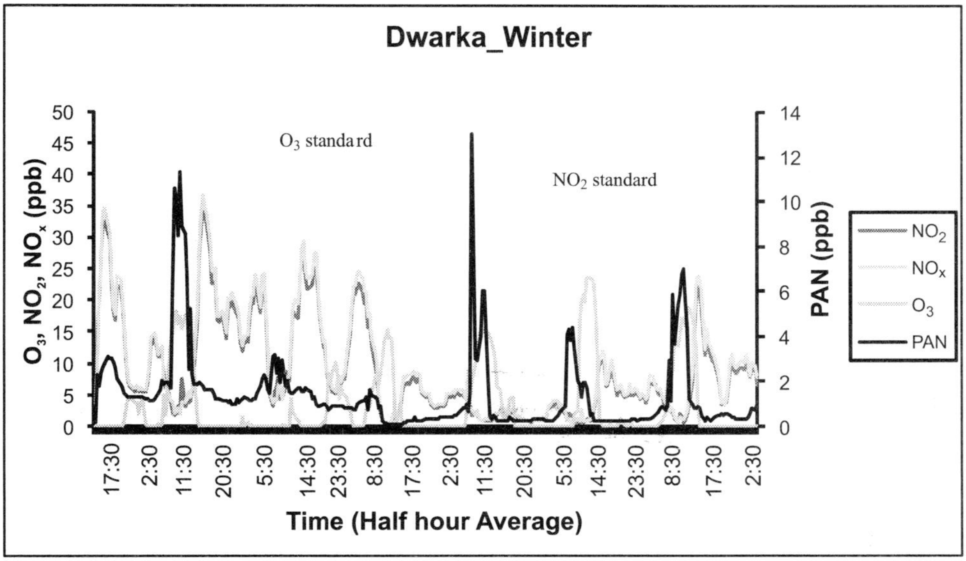


**Fig. 2.** The time series of half hour average concentrations of PAN, $NO_2$, $NO_x$, $O_3$ variations at **Dwarka** during 22/1/2008 to 29/1/2008. Continuous measurement for PAN, $NO_2$, $NO_x$ were taken since 5:51 PM, 22/1/2008 till 8:05 AM, 29/1/2008. $O_3$ Measurements taken were continuous since 16:36, 22/1/2008 till 18:21, 28/1/2008 except brief interruptions (due to power-failure or other technical reasons).

concentrations were 35.0 ppb and 37.7 ppb. $O_3$ concentration was observed to reach up to a maximum of 24.4 ppb with strong diurnal variation. The daily maximum PAN and the daily maximum $NO_2$ are well correlated (0.5) indicating that higher level of $NO_2$ induces the higher level PAN in a day. The daily maximum PAN and daily maximum $O_3$ are also moderately correlated indicating that elevation of PAN in a day is accompanied by increased level of $O_3$ in a day.

### (c) Delhi Secretariat

At Delhi Secretariat, during winter, 24 hour average PM10 concentrations were always observed to much higher almost two to three times than that of CPCB standard. PAN concentration exceeded only 0.7% of the time. Maximum $NO_2$ and $NO_x$ were recorded as 36.8 ppb and 44.6 ppb in the afternoon of 6/2/2008. The PAN concentration level was observed to reach to the maximum of 7.799 ppb in the evening hours. The maximum $O_3$ recorded at Delhi Secretariat is 50 ppb. PAN is found to be have strong association with NOx and $O_3$, indicating that elevation of PAN in a day is accompanied by increased level of $O_3$ in a day. The mean SPM concentration is strongly correlated (0.96) with mean PAN concentrations indicating the contribution of PAN in the aerosol formation.

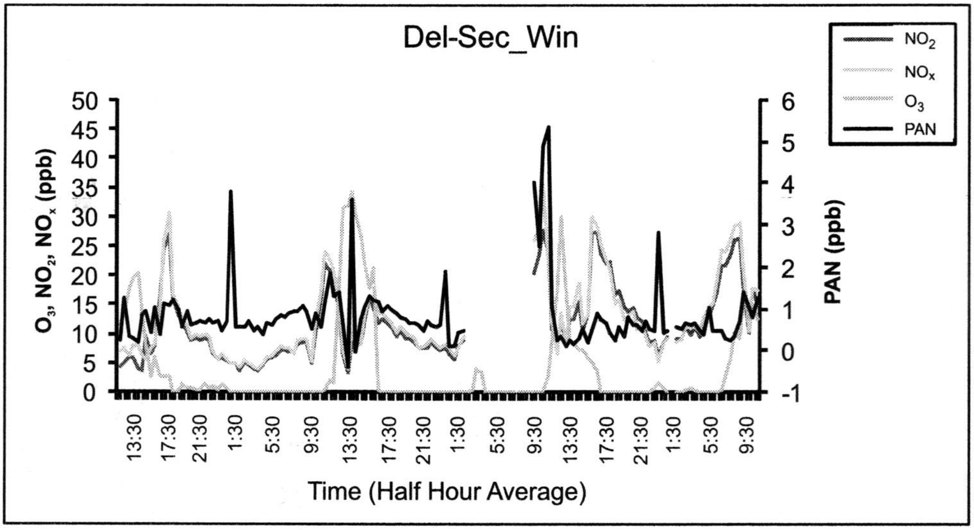


**Fig. 3.** The time series of half hour average concentrations of PAN, $NO_2$, $NO_x$, $O_3$ variations at **Delhi Secretariat** during 04/02/2008 to 07/02/2008.

### (d) Ashok Vihar

In winter season at Ashok vihar, 24 hour average PM10 concentration was always found to be almost three or more times higher than that of CPCB residential PM10 standard. $O_3$ concentration exceedance was 41% of the total sampling time.

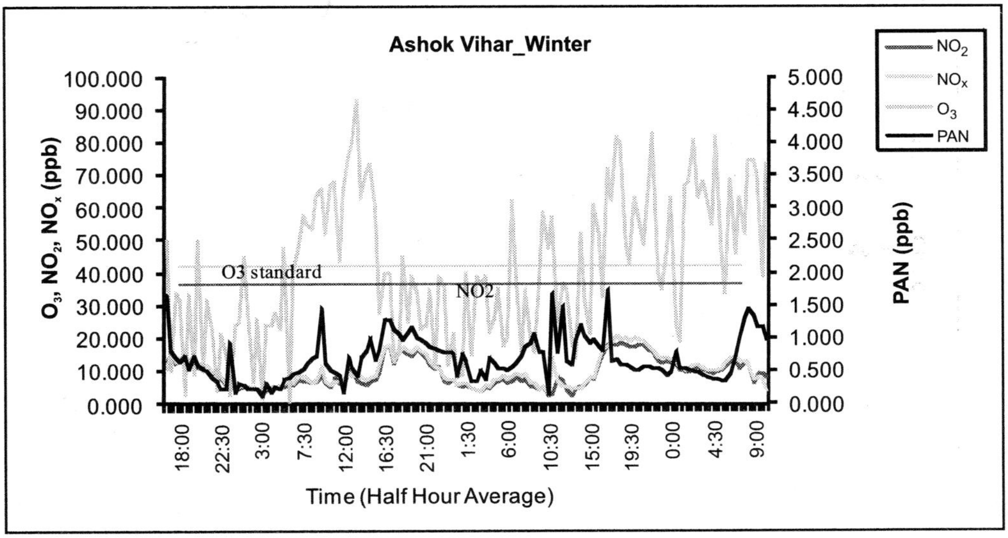


**Fig. 4.** The time series of half hour average concentrations of PAN, $NO_2$, $NO_x$, $O_3$ variations at Ashok Vihar during 08/02/2008 to 11/02/2008.

Measurements of $NO_2$ and $NO_x$ indicate that their concentrations reach up to a maximum of 25 ppb during five days of study. The $NO_2$ and $NO_x$ concentration show a very strong diurnal variation with NOx maximum during early afternoon and falling to less than 5 ppb during night. Results indicated that there exists strong correlation between daily maximum PAN and daily maximum $NO_2$ or $NO_x$. That is, if there is an increase in $NO_x$ concentration, there will certainly be an increase in PAN concentration in a day.

### (e) Okhla

At **Okhla**, PM10 concentration was always found to be more than double than that of CPCB industrial standard for PM10 and more than three times than that of residential standard. PAN exceeded only 0.85% of the total time, while $NO_2$ exceeded only 1.3% of the time. PAN was observed to be varying from 0.00 to 6.39 ppb with an average 0.74±0.51 ppb. Average $NO_2$ concentration was 34.86, while $O_3$ ranges from 0 to 35 ppb, which has not exceeded the prescribed limits. Here, PAN and $NO_2$ were positively correlated (0.68). Perhaps the temperature at the last week of February and first week of March was not sufficient enough to break the PAN. $O_3$ and PAN correlation (0.82) was also observed to be high. It means during those period both high ozone and high PAN situation prevailed. The correlation between PAN and $NO_2$ concentration was high (0.80), i.e., high $NO_2$ is associated with high formation of PAN, i.e., PAN decomposition phenomena is too slow to be significant as noted earlier. Also, PAN and PM10 correlation (0.66) was significant.

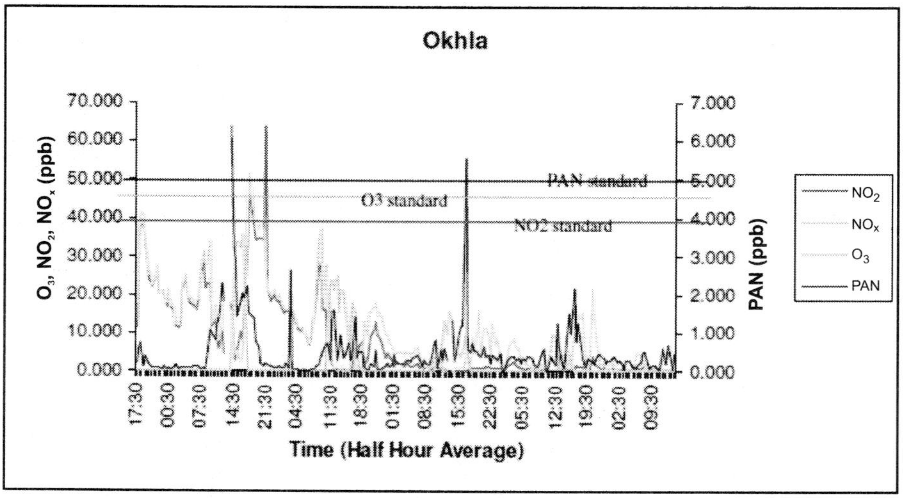


**Fig. 5.** The time series of half hour average concentrations of PAN, $NO_2$, $NO_x$, $O_3$ variations at **Okhla** during 28/2/2008 to 04/03/2008.

## (f) Wazirpur

At **Wazirpur** in summer, PM10 concentration has always exceeded the industrial standard by more than double. $NO_2$ exceeded 7.5% of the total time, while $O_3$ exceeded 10% of the total time. PAN concentration variation was very low ranging from 0.26 to 0.71 ppb. $NO_2$ concentration was observed to reach to a maximum of 95.5 ppb. $O_3$ concentration was observed to reach as high as 95 ppb. $O_3$ and PAN were found to be positively correlated (0.57). PM10 and $NO_2$ concentration correlation was again observed to be high (0.81), indicating similar emission sources. The daily maximum PAN is positively associated (0.62) with daily maximum $NO_2$. PAN is also positively correlated (0.46) with $O_3$ at Wazirpur.

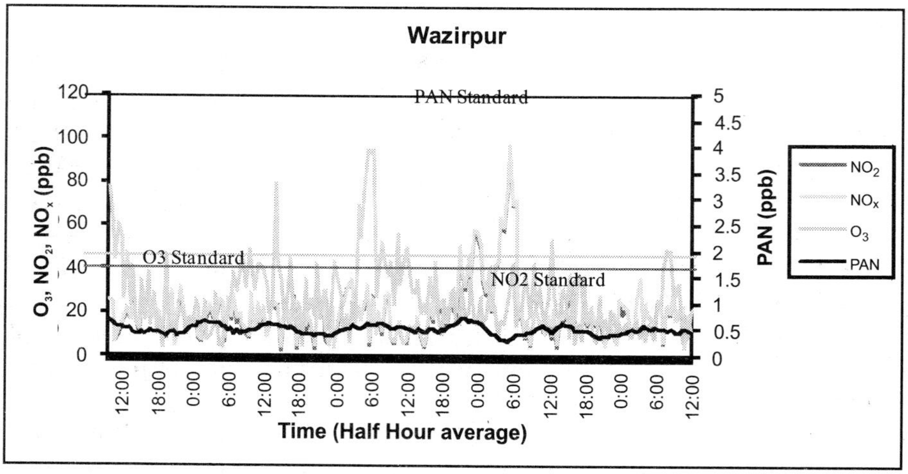


**Fig. 6.** The time series of half hour average concentrations of PAN, $NO_2$, $NO_x$, $O_3$ variations at **Wazirpur** during 26/5/2008 to 30/05/2008.

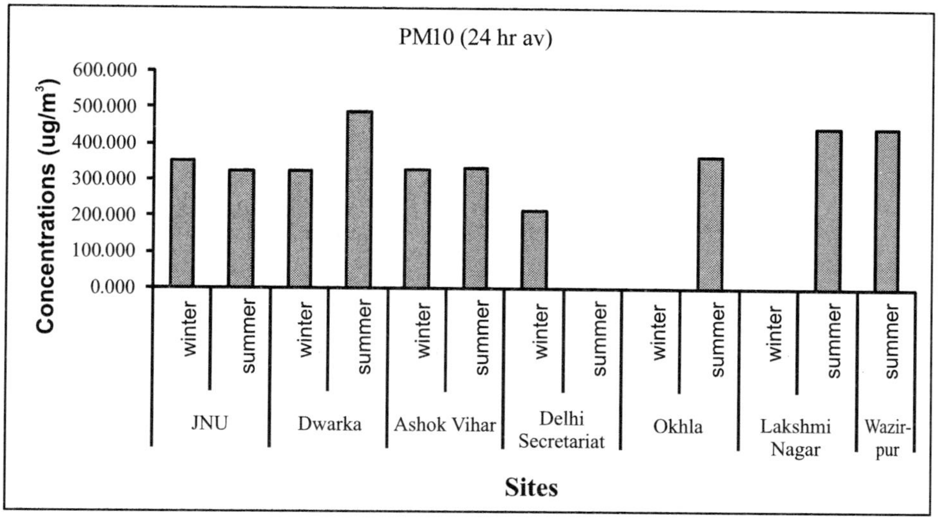


**Fig. 7.** RSPM (PM10) Concentrations at different monitoring location in Delhi during summer and winter months.

## Exceedances Table

Table 2 shows the percentage of time that each pollutant exceeded the prescribed standard at a particular site.

**Table 2** Exceedance of pollutants from their prescribed standards

| *Sites* | *PM10 Percentage of Time it Exceeded the Standard* | *PAN Percentage of Time it Exceeded the Standard* | *$NO_2$ Percentage of Time it Exceeded the Standard* | *$O_3$ Percentage of Time it Exceeded the Standard* |
|---|---|---|---|---|
| JNU (Winter) | 100% | 4.6% | 2% | 0% |
| Dwarka (Winter) | 100% | 5.6% | 0% | 0% |
| Delhi Secretariat (Winter) | 100% | 0.7% | 0% | 0% |
| Ashok Vihar (Winter) | 100% | 0% | 0% | 41% |
| Okhla (Spring) | 100% | 0.85% | 1.3% | 0% |
| Wazirpur (Summer) | 100% | 0% | 79% | 6% |

## Meteorological Conditions Responsible for Smog Formation

During the total sampling events carried out at 6 different locations in Delhi, foggy conditions were observed only during winter season. But fog need not necessarily constitute the condition of smog. The simultaneous presence of High $O_3$ and High PAN concentrations in presence of sunlight constitutes the condition of smog. Such conditions are observed during late afternoon periods mostly around 3 to 7 pm at JNU and Dwarka and between 2 and 5 pm at Delhi Sec. except at Ashok Vihar where it occurs during before noon period. The fact is also evident from Figures 1, 2, 3 and 4 respectively, for JNU, Dwarka, Delhi Sec.

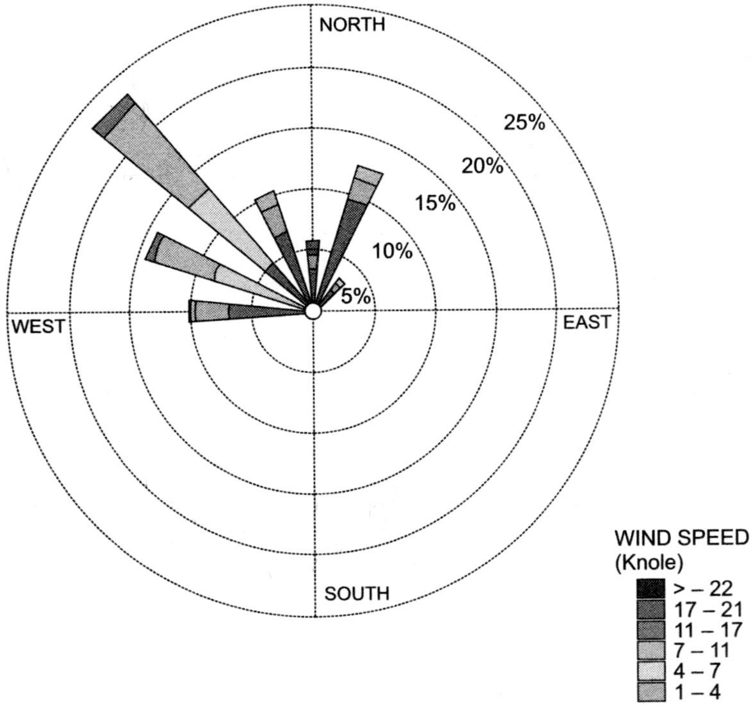


**Fig. 8.** The typical wind rose plot at Dwarka during winter season.

and Ashok Vihar. The prevailing atmospheric condition during the smog formation is observed as mostly hazy and smoky. These conditions were observed at all the above mentioned four sites where sampling was carried out during winter. At the same sites during summer seasons, almost negligible or no PAN concentrations were observed. This may be because of the thermo-chemical instability of PAN at high temperature. The prevailing day time temperature during summer season in Delhi averaged around 35°C. At the temperature more than 20-25°C PAN decomposes and hence during summer PAN concentrations is almost negligible. Therefore, the smog formation is ruled out in summer. Table 3 shows the meteorological conditions responsible for smog formation. Figure 8 depicts the typical wind rose plot at the Dwarka during winter season. As evident from the windrose plots, during winter season, westerly and north westerly wind predominates over Delhi, which is usually dry and cold. Whenever moisture introduced into the atmosphere due to western disturbances or any other reasons, it leads to fog formation. This fog when associated with high PAN and ozone concentrations creates smog like situations.

**Table 3** The meteorological conditions during smog events

| Site | Time of Smog | Temp | | RH | | WS | |
|---|---|---|---|---|---|---|---|
| | | Min | Max | Min | Max | Min | Max |
| JNU | 3-7 PM | 12.00 | 23 | 32 | 72 | 3.5 | 16.1 |
| Dwarka | 3-7 PM | 11.00 | 19 | 16 | 58 | 1.9 | 3.1 |
| Delhi Sec. | 2-5 pm | 15 | 20 | 23 | 77 | 4.6 | 13.8 |
| Ashok Vihar | 10-12 am | 11 | 36 | 22 | 67 | 4.6 | 18.4 |

Temperature: ~10°C to ~20°C; RH: 15:70; and Wind Speed: 3 to 18 m/s

## CONCLUSION

RSPM (PM10) load in Delhi's air is very high. At no site, the 24 hour average PM10 concentration was found to be lower than CPCB prescribed standard on any occasion. Thus RSPM levels are quite alarming from the point of view of their impact on health. Comprehensive measures are needed to control RSPM load in Delhi. In our study of PAN, $NO_x$ and $O_3$, it is obvious that $O_3$ concentration is higher during summer season while PAN concentration is quite low in summer as PAN dissociates at high temperatures. As the cold season approaches, PAN concentration starts increasing and becomes significant. In the smoggy conditions during summer, $O_3$ is the most abundant constituent of smog while during mild winter (15°C to 25°C), PAN is found to be a significant constituent. However, both are secondary pollutants and their existence can be attributed to mainly $NO_x$ emissions. Although, VOCs may also play an important role in determining the concentration of $O_3$ at times (Kumar et al, 2008) it is the NOx levels which govern the $O_3$ levels in the ambient atmosphere. In light of the findings of the present study, it is imperative to devise measures to control $NO_x$ emissions to reduce the menace of smog in Delhi.

## REFERENCES


Agrawal, M., Singh, B., Rajput, M., Marshall, F. and Bell, J.N.B. (2003). Effect of air pollution on periurban agriculture: a case study. *Environmental Pollution,* **126(3):** 323-339.

Andersen, C.P. (2003). Source-sink balance and carbon allocation below ground in plants exposed to ozone. *New Phytologist,* **157:** 213-228.

Aufderheide M., Ritter D. and Knebel J.W. (1998). Cytotoxic activity of pan (peroxyacetyl nitrate), an environmental pollutant, on rat lung cells. Toxicology Letters, 95, (Supplement 1), 240-240.

Brasseur, G.P., Hauglustaine, D.A., Walters, S., Rasch, P.J., Muller, J.-F., Granier, C. and Tie, X.X. (1998). MOZART, a global chemical transport model for ozone and related chemical tracers, Part 1. Model description. *J. Geophysical Research,* **103:** 28265-28289.

Camps, Francis E. (Ed.) (1976). Gradwohl's Legal Medicine, 3rd edition. *Bristol: John Wright & Sons Ltd,* ISBN 0 7236 0310, p. 236.

Coleman, M.D., Dickson, R.E., Isebrands, J.G. and Karnosky, D.F. (1995). Carbon allocation and partitioning in aspen clones varying in sensitivity to tropospheric ozone. *Tree Physiology,* **15:** 593-604.

Desqueyroux, H., Pujet, J.C., Prosper, M., Squinazi, F. and Momas, I. (2002). Short-term effects of low-level air pollution on respiratory health of adults suffering from moderate to severe asthma. *Environmental Research,* **89:** 29-37.

EPA, (1996). Air quality criteria for ozone and related photochemical oxidants. United States Environmental Protection Agency (EPA), pp. 1.1-1.33.

Findlay, S., Carreiro, M., Krischik, V. and Jones, C.G. (1996). Effects of damage to living plants on leaf litter quality. *Ecological Applications,* **6:** 269-275.

Finnan, J.M., Burke, J.I. and Jones, M.B. (1997). An evaluation of indices that describe the impact of ozone on the yield of spring wheat (Triticum aestivum L.). *Atmospheric Environment,* **31:** 2685-2693.

Gaffney, J. S., Fajer, R. and Senum, G.I. (1984). An improved procedure for high purity gaseous peroxyacyl nitrate production: Use of heavy lipid solvents. *Atmospheric environment,* **18:** 215-218.

Gaffney, J.S. and Marley, N.A. (1993). Measurement of peroxyacetyl nitrate at a remote site in the Southwestern united states: tropospheric Implications. *Environmental Science & Technology,* **27:** 1905-1910.

Haagen-Smit, A.J. (1952). Chemistry and Physiology of Los Angeles Smog. *Industrial and Engineering Chemistry,* **44(6):** 1342-1346.

Haagen-Smit, A.J. and Fox, M.M. (1956). Ozone formation in photochemical oxidation of organic substances. *Industrial and Engineering Chemistry,* **48(9):** 1484-1487.

Kleindienst, T.E. (1994). Recent developments in the chemistry and biology of peroxyacetyl nitrate. *Res. Chem. Intermed,* **20:** 335-384.

Kumar, U., Prakash, A. and Jain, V.K. (2008). Photochemical Modelling Approach to Investigate $O_3$ Sensitivity to NOx and VOCs in the urban atmosphere of Delhi. *Aerosol and Air Quality Research,* **8(2):** 147-159.

Marenco, A., Gouget, H., Nedelec, P., Pages, J.P. and Karcher, F. (1994). Evidence of a long term increase in tropospheric ozone from Pic du Midi data series: consequences: positive radiative forcing. *J. Geophysical Research,* **99:** 16617-16632.

Mortimer, K.M., Neas, L.M., Dockery, D.W., Redline, S. and Tager, I.B. (2002). The effect of air pollution on inner-city children with asthma. *Eur. Respiratory J.,* **19:** 699-705.

Moxim W.J., Levy H. and Kasibhatla P.S. (1996). Simulated global tropospheric PAN—its transport and impact on NOx. *J. Geophys Res.,* **101:** 12621-12638.

Orlando, J.J., Tyndall, G.S. and Calvert, J.G. (1992). Thermal decomposition pathways for peroxyacetyl nitrate (PAN): Implications for atmospheric methyl nitrate levels. *Atmospheric Environment,* **26:** 3111-3118.

Peden, D.B. (2001). Air pollution in asthma: effect of pollutants on airway inflammation. *Annals Allergy Asthma & Immunology,* **87:** 12-17.

Penkett, S.A. and Brice K.A. (1986). The spring maximum in photo-oxidants in the northern hemisphere troposphere. *Nature,* **7:** 319-655.

Portmann, R.W., Solomon, S., Fishman, J., Olson, J.R., Kiehl, J.R. and Briegleb, B. (1997). Radiative forcing of the earth's climate system due to tropical tropospheric ozone production. *J. Geophysical Research,* **102:** 9409-9417.

Ramanathan, V., Cicerone, R.J., Singh, H.B. and Kiehl, J.T. (1985). Trace gas trends and their potential role in climate change. *J. Geophysical Research,* **90:** 5547-5566.

Spicer, C.W., Holdren, M.W. and Keigley, G.W. (1983). The ubiquity of peroxyacetyl nitrate in the continental boundary-layer. *Atmospheric Environment,* **17:** 1055-1058.

Sun, E. and Huang, M. (2005). Detection of peroxyacetyl nitrate at phytotoxic level and its effects on vegetation in Taiwan. *Atmospheric Environment,* **29(21):** 2899-2904.

Talukdar, R.K., Burkholder, J.B., Schmoltner, A.M., Roberts, J.M., Wilson, R.R. and Ravishankara, A.R. (1995). Investigation of the loss processes for peroxyacetyl nitrate in the atmosphere-UV photolysis and reaction with OH. *J. Geophysical Research,* **100:** 14163-14173.

# 15

# Metagenomics: A Novel Technique and Approaches for Uncultured Microorganisms

**M.H. Fulekar and Bhawana Pathak**

*School of Environment and Sustainable Development Central University of Gujarat Sector-30, Gandhinagar-382030*
*Corresponding Author: E-mail: mhfulekar@yahoo.com*

**ABSTRACT**

Metagenomics has emerged as a powerful tool that can be used to analyze microbial communities regardless of the ability of member organism to be cultured in the laboratory. Metagenomics is based on the genomic analysis of microbial DNA that is extracted directly from communities in environmental samples. Metagenomics has opened new avenues of research by enabling unprecedented analysis of genome heterogeneity and evolution in environmental context and providing access to far more microbial diversity. The fundamentals of metagenomics, metagenomic analysis-approaches, including isolation of metagenomic DNA from water as well as soil samples have been discussed in order to identify microbial diversity. Global Metagenomic initiative has also been highlighted.

**Keywords:** Metagenomics, DNA, Genome, Microbial Diversity, Water, Soil-environmental Samples.

## INTRODUCTION

*"Metagenomics give scientists access to millions of microbes that have not previously been studied"* (National Research Council Report). Microbiology has experienced a transformation during the last 25 years that has altered microbiologist's view of

microorganisms. The realization that most microorganisms cannot be grown readily in pure culture force microbiologists to question their belief that the microbial world has been conquered (Handelsman Jo, 2004). A revolution in microbiological thought- demonstrate that the uncultured microbial world far outsized the cultured world and that this unseen world could be studied (Handelsman Jo, 2004). The vast majority of the microbial world has been inaccessible to science because merely a miniscule fraction most scientists estimate less than 1% of the estimated millions of microbial species on earth can be cultured (National Research Council Report).The outcome of the reorganization of uncultured microorganisms are worthy of examination. One of these outcomes, metagenomics is further shaping microbiology. Metagenomics has already opened new avenues of research by enabling unprecedented analysis of genome heterogeneity and evolution in environmental context and providing access to far more microbial diversity than has been viewed in the petri-dish. This will explore the origin of metagenomics and examine its recent application in microbial ecology and biotechnology (Handelsman Jo, 2004).

Metagenomics has emerged as a powerful tool that can be used to analyse microbial communities regardless of the ability of member organisms to be cultured in the laboratory (Logan). Metagenomics is based on the genomic analysis of microbial DNA that is extracted directly from communities in environmental samples. This technology-genomics on a huge scale-enables a survey of different micro-organisms present in a specific environment such as water or soil to be carried out. By integrating the information gleared with information about biological functions within the community, the structure of microbial communities can potentially be probed (Logan). Metagenomics could also unlock the massive uncultured microbial diversity present in the environment to provide new molecules for therapeutic and biotechnological application.

A broad range of fields: including chemistry, genetics, microbiology, biochemistry, pathology, ecology, evolution, soil and atmospheric sciences, geology, oceanography, statistics, computer sciences, database development, mathematics, engineering and others have application to metagenomics (National Research Council Report).

## METAGENOMICS—CONCEPT

Metagenomics is an emerging field in which the power of genomic analysis (positive analysis of the entire DNA in an organism) is applied to entire communities of microbes, bypassing the need to isolate and culture individual microbial species (National Research Council Report). Metagenomics includes a wide variety of novel techniques and approaches and it is likely that other new methods will arise as the field progress. In metagenomic process:

- Researchers often start metagenomic work by taking a sample of particular environment (such as soil, seawater or the human mouth) and extracting the DNA from all the microbes in the sample.

- This genetic information is then studied to reveal the important characteristics of microbial community.
- In sequence based metagenomics, researchers study the entire genetic sequence-pattern of the four different nucleotide bases (A, T, G and C) in the DNA strands found in the samples.
- In function-based metagenomics, researchers screen the extracted genes for various functions such as vitamins or antibiotic production. New antibiotics have already been discovered using this approach.

## METAGENOMICS ANALYSIS—APPROACHES

The word metagenomics was coined to capture the notion of analysis of a community genomics, environmental genomics and population genomics (Handelsman Jo,2004). The idea of cloning DNA directly from environmental samples was first proposed by Pace in 1991,the first such cloning in a phage vector was reported. The next advance was the construction of a metagenomic library with DNA derived from a mixture of organisms enriched on dried grasses in the laboratory. Clones expressing cellulolytic activity were found in these libraries. The work of DeLong's group defined the field when they reported libraries constructed from prokaryotes in seawater. They identified a 40-kb clone that contained a 16srRNA gene indicating that the clone was derived from an archaeon that had never been cultured. Construction of libraries with DNA extracted from soil lagged due to difficulties associated with maintaining the integrity of DNA during its extraction and purification from a soil matrix, but eventually produced analysis analogous to those from sea water. Metagenomic analysis involves isolating DNA from an environmental sample, cloning the DNA into a suitable vector, transforming the cloning into a host bacterium and screening the resulting transformants. The clone can be screened for phylogenetic markers as "anchors", such as 16srRNA and recA or for other conserved genes by hybridization or multiplex PCR or for expression of specific traits such as enzyme activity or antibiotic production (Courtoisetal, 2003; Diaz-Torres et al., 2003; Gillespie et al., 2002; Knietsch et al., 2003; Lorenz et al., 2002; Lorenz & Schleper, 2002; MacNeil et al., 2001; Majernik et al., 2001; Rondon et al., 2000; Schloss et al., 2003), or they can be sequenced randomly. Each approach has strengths and limitations; together these approaches have enriched our understanding of the uncultured world, providing insight into a group of prokaryotes that are otherwise entirely unknown.

## SEQUENCE BASED ANALYSIS

Sequence-based analysis can involve complete sequencing of clones containing phylogenetic anchors that indicate the taxonomic group, which is the probable source of DNA fragment. Alternatively random sequencing can be conducted and once a gene of interest is identified, phylogenetic anchors can be brought in the flanking DNA to provide a link of phylogeny with the functional gene. Sequence analysis guided by the identification of phylogenetic markers is a powerful approach first proposed by the DeLong group, which produced the first genomic sequence linked to a 16srRNA gene of an uncultured archaeon (Stein et al., 1996). Subsequently, they identified an insert from seawater bacteria containing a 16srRNA gene that activated with the ã-proteobacteria. The sequence of flanking DNA revealed a bacteriorhodospin-like gene. Its gene products was shown to be an authentic photoreceptor leading to the insight that bacteriorhodopsin genes are not limited to archae but are in fact abundant among the proteobacteria of the ocean (Bejaetal 2001; Beja et. al, 2000). A promising application of phylogenetic anchor-guided sequencing is to collect and sequence many genomic fragments from a taxon. In more complex environments and taxon, reassembly of a genome may not be feasible, but inference about the physiology and ecology of the members of the group can be gleaned from sequence data. This approach has been initiated with clones from diverse soils carrying 16S rRNA genes that affiliate with the Acidobacteria phylum, which is abundant in soil and highly diverse (Barns et al., 1999; Buckley et al., 2003; Furlong et al., 2002; Janssen et al., 2002; Saitetal 2002) and about which little is known (Quaiser et al., 2003; Liles et al., 2003). Complete sequencing of the estimated ~500kb of acidobacterium DNA in metagenomic library may provide insight into the subgroups of bacteria in this phylum that have never been cultured.

## METAGENOMICS—FUNDAMENTAL

A powerful yet challenging approach to metagenomic analysis is to identify clones that express a function. Success requires faithful transcription and translation of the gene of interest and secretion of the gene product, if the screen or assay requires it to be extra cellular. Functional analysis has identified novel antibiotics (Courtois et al., 2003; MacNeil et al., 2001; Gillespie et al., 2002; Venter et al., 2004; Wang et al., 2000) antibiotic resistance genes (Diaz-Torres et al., 2003; Riesenfeld), $Na^{+}(Li^{+})/H^{+}$ transporters (Majernik et al., 2001) and degradative enzymes (Healy et al.,1995; Henne et al., 1999; Henne et al., 2000). The power of the approach is it does not require that the genes of interest be recognisable by sequence analysis, making it the only approach to metagenomics that has the potential to identify entirely new class of genes for new or known functions. The significant limitation is that many genes, perhaps most, will not be expressed in any particular host bacterial selected for cloning. In fact, there is an inherent contradiction in this approach—genes are cloned fromexotic organisms to discover new motifs in biology and yet these genes are required to be expressed in *E. coli* or another demonstrated bacterium in order to be detected. The diversity of the organisms whose DNA has been successfully expressed in *E. coli* (Rondon et al., 2000; Black et al., 1995; Chávez et al., 1995; Ding et al., 1993; Ferreyra et al., 1993; Rondon et al., 1999; Rondon & Raffel et al., 1999), but heterologous expression remains a barrier to extracting the maximum information from functional metagenomics analysis.

## DNA ISOLATION KIT

DNA isolation kit is designed to isolate randomly-sheared, high molecular weight metagenomic DNA free of humic and folic acid, directly from unculturable or difficult to culture microbial species present in environmental water or soil samples. The DNA isolating kit is approximately 40kb in size and is readily used for immediate end repair reaction and subsequent cloning into fosmid vector. The given protocol streamlines standard fosmid library construction by eliminating the traditional mechanical shearing and size-selection steps. Additionally, the DNA can be used as a substrate for PCR or sequencing.

### Isolation of Metagenomic DNA from Water Samples

1. To remove debris from the collected water sample, pour the water through Miracloth filtration material (Calbiochem), sterile cheesecloth or centrifuge at 1,000 × g for 5 minutes and collect the water in a sterile container.
2. Filter the water (i.e., 100 ml) through a 0.45 μm membrane to trap microbes onto the filter, using the appropriate filtration apparatus (e.g., Millipore, catalog no. XX1004700).
3. Using forceps and scissors pre-soaked in 70% ethanol, remove the membrane from the filter apparatus, cut the membrane in half, and place each half (rounded side down) along the side (near the bottom) of a 50 ml sterile conical tube. The upper surface of the filter needs to face the center (not wall) of the tube. Do not allow the filter membrane to dry out.
4. Prepare the filter wash buffer by adding 2 μl of Tween-20 to 1 ml of filter wash buffer immediately before use. Add 1 ml of filter wash buffer containing 0.2% Tween-20 to the filter pieces in the tube to wash off the microbes trapped on the membrane.
5. Vortex the tube at a low speed setting to re-wet the filter pieces, and then increase the setting to the highest speed for ~2 minutes with intermittent breaks.
6. Transfer the cell suspension to a clean microcentrifuge tube, and then centrifuge the tube at 14,000 × g for 2 minutes to pellet the cells. Discard the supernatant.
7. Re-suspend the cell pellet in 300 ml TE Buffer, and then add 2 μl of Ready-Lys lysozyme Solution and 1 μl of RNase A to the cell suspension. Mix by vortexing.
8. Incubate at 37°C for 30 minutes.
9. Add 300 μl of Meta-Lysis Solution (2X) and 1 ml of Proteinase K to the tube. Mix by vortexing.
10. Briefly pulse centrifuge the tube to ensure that all of the solution is at the bottom of the tube.
11. Incubate at 65°C for 15 minutes.
12. Cool to room temperature, and then place on ice for 3-5 minutes.
13. Add 350 μl of MPC protein precipitation reagent to the tube and mix by vortexing vigorously for 10 seconds.

14. Pellet the debris by centrifugation for 10 minutes at 20,000 x g or at maximum speed in a microcentrifuge at 4°C.
15. Transfer the supernatant to a clean 1.7 ml microcentrifuge tube and discard the pellet.
16. Add 570 μl of isopropanol to the supernatant. Mix by inverting the tube several times.
17. Pellet the DNA by centrifugation for 10 minutes at 20,000 × g or at maximum speed in a microcentrifuge at 4°C.
18. Use a pipet tip to remove the isopropanol without dislodging the DNA pellet. Briefly pulse centrifuge the sample and remove any residual liquid with a pipet tip, without disturbing the pellet.
19. Add 500 μl of 70% ethanol to the pellet without disturbing the pellet. Then centrifuge for 5 minutes at 20,000 × g or at maximum speed, in a microcentrifuge at 4°C.
20. Use a pipet tip to remove the ethanol without dislodging the DNA pellet. Briefly pulse centrifuge the sample and remove any residual liquid with a pipet tip, without disturbing the pellet.
21. Air-dry the pellet for 8 minutes at room temperature. **Note:** Do not over dry the pellet.
22. Re-suspend the DNA pellet in 50 μl of TE Buffer.
23. Validate the size and concentration of the isolated DNA by comparison to the Fosmid control DNA (40 kb; 100 ng/μl) provided in the kit via gel electrophoresis on a 1% agarose gel. Use 2 μl of the isolated DNA preparation for this analysis.

The isolated DNA is ready for PCR or end-repair and subsequent cloning into a vector for construction of a fosmid library.

## Isolation of Metagenomic DNA from Soil Samples

### *Extraction*

Before starting, add Tween 20 to the extraction buffer (1 ìl Tween 20 per milliliter of extraction buffer) to achieve a final concentration of 0.1% Tween 20. Each extraction will use 50 ml of extraction buffer with Tween 20.

1. Add 1 g of wet soil (or 200-300 mg of compost) to a 50-ml screw-cap conical tube and add 10 ml of extraction buffer with Tween 20.
2. Mix by vortexing at maximum speed for 1 minute to disperse and dissociate the soil particles.
3. Centrifuge the soil suspension at 1,600 × g for 4 minutes in a tabletop centrifuge. Pour the supernatant into a new 50-ml tube.
4. Add 20 ml of remaining extraction buffer with Tween 20 (per extraction) to the soil pellet and mix by vortexing at maximum speed for 1 minute.

5. Centrifuge the soil suspension at 900 × g for 3 minutes in a tabletop centrifuge. **Important:** Do not exceed 900 × g. Combine the supernatant with the previously collected supernatant.
6. Reextract the soil pellet again by repeating steps 4 and 5 (above) and then proceed to step 7 (below).
7. Briefly centrifuge the pooled supernatant at 900 × g for 2 minutes in a tabletop centrifuge and transfer the supernatant to a fresh 50-ml tube.

### Filtration, Lysis, and Protein Precipitation

1. Pour the collected supernatant (50 ml) through four layers of Miracloth filtration material (Calbiochem).
2. Prefilter the sample through 1.2 μm filter membrane provided using an appropriate filterware apparatus (e.g., Millipore, catalog no. XX1004700). Collect the filtrate.
3. Pass the collected filtrate through 0.45 μm filters membrane provided using an appropriate filterware apparatus in order to trap the microbial mass on the filter. Retain the filter membrane.
4. Using forceps and scissors presoaked in 70% ethanol, remove the membrane from the filter apparatus, cut the membrane in half, and place each half (rounded side down) along the side (near the bottom) of a 50 ml sterile conical tube. The upper surface of the filter needs to face the centre (not wall) of the tube. Do not allow the filter membrane to dry out.
5. Prepare the filter wash buffer by adding 1.5 μl of Tween 20 to 1.5 ml of filter wash buffer immediately before use. Add 1.5 ml of filter wash buffer containing 0.1% Tween 20 to the filter pieces in the tube.
6. Vortex the tube at the low speed setting to rewet the filter pieces, and then increase the setting to the highest speed for ~2 minutes to wash off the microbes trapped on the filter membrane.
7. Transfer the cell suspension to a clean microcentrifuge tube, and then centrifuge the tube at 14,000 × g for 2 minute to pellet the cells. Discard the supernatant.
8. Resuspend the cell pellet in 300 μl of TE buffer, and then add 2 μl of ready-lyse lysozyme solution and 1 μl of RNase A to the cell suspension. Mix and centrifuge briefly to bring all of the solution to the bottom of the tube.
9. Incubate at 37°C for 30 minutes.
10. Add 300 μl of Meta-Lysis Solution (2X) and 1 μl of Proteinase K to the tube. Mix by vortexing.
11. Briefly pulse-centrifuge the tube to ensure that all of the solution is at the bottom of the tube.
12. Incubate at 65°C for 15 minutes.
13. Cool to room temperature, and then place on ice for 3-5 minutes.
14. Add 350 μl of MPC protein precipitation reagent to the tube and mix by vortexing vigorously for 10 seconds.

15. Pellet the debris by centrifugation for 10 minutes at 20,000 × g or at maximum speed in a microcentrifuge at 4°C.
16. Transfer the supernatant to a clean 1.7-ml microcentrifuge tube and discard the pellet
17. Add 570 μl of isopropanol to the supernatant. Mix by inverting the tube several times.
18. Pellet the DNA by centrifugation for 10 minutes at 20,000 × g or at maximum speed in a microcentrifuge at 4°C.
19. Use a pipet tip to remove the isopropanol without dislodging the DNA pellet. Briefly pulse-centrifuge the sample and remove any residual liquid with a pipet tip, without disturbing the pellet.
20. Add 500 μl of 70% ethanol to the pellet without disturbing the pellet. Then centrifuge for 5 minutes at 20,000 × g or at maximum speed in a microcentrifuge at 4°C.
21. Use a pipet tip to remove the ethanol without dislodging the DNA pellet. Briefly pulse-centrifuge the sample and remove any residual liquid with a pipet tip, without disturbing the pellet.
22. Air-dry the pellet for 8 minutes at room temperature. **Note:** Do not over-dry the DNA pellet.
23. Resuspend the DNA pellet in 40 μl of TE buffer.
24. Validate the size and concentration of the isolated DNA by comparison to the fosmid control DNA (40 kb; 100 ng/μl) provided in the kit, via gel electrophoresis on a 1% agarose gel. Use 5 μl of the isolated DNA preparation for this analysis.

The isolated DNA is ready for PCR or end-repair and subsequent cloning into vector for construction of a fosmid library.

## CLONING SYSTEM—WORKS

1. Ligate the DNA of interest into the lineralized and dephosphorylated CopyControl Cloning- Research vector supplied with the respective kit.
2. Package the ligated DNA into the lambda phage and infect EP1300-T1$^{R}$*E.coli* and select on LB-Chloramphenicol plates. Under these conditions, the *trf*A gene is not expressed and the clones are maintained at single copy.
3. Pick individual CopyControl clones from the plate and grow in culture.
4. Add the Copy Control Fosmid Autoinduction Solution (included) or Copy Control Induction Solution (available separately) to induce expression of the *trf*A gene product and subsequent amplification of the clones to high-copy number.
5. Purify plasmid DNA for sequencing, fingerprinting, subcloning, or other applications.

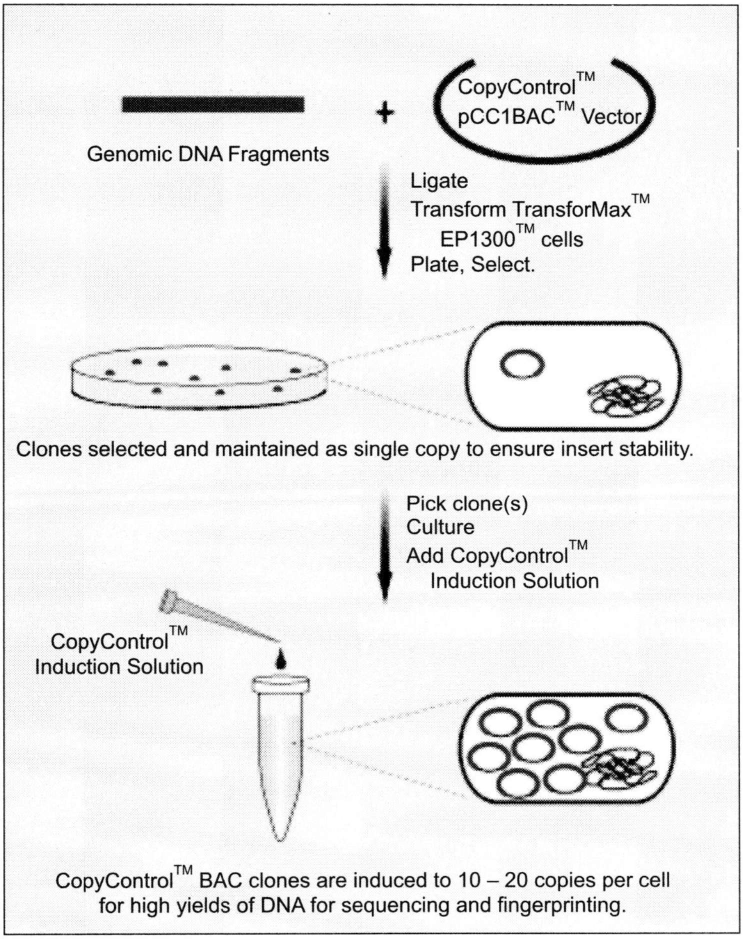


**Fig. 1.** Overview of the copy control™ system.

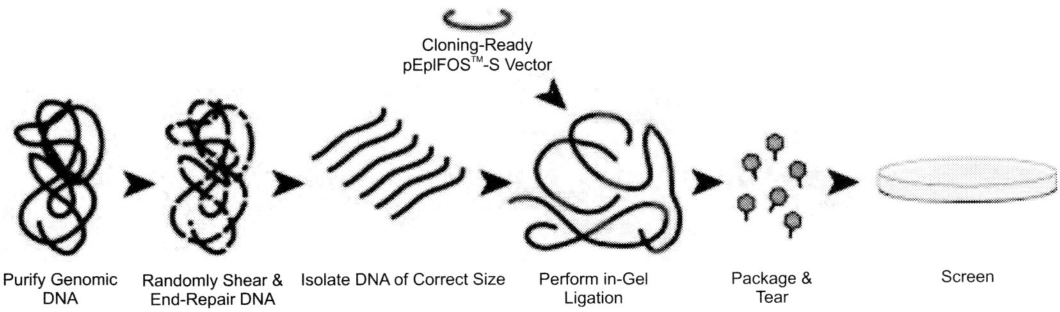


**Fig. 2.** Production of a copycontrol™fosmid library and subsequent induction of clone to high copy number.

## FOSMID LIBRARY PRODUCTION PROCESS

The steps involved (protocols for steps 2-8 are included in this manual):

1. Purify DNA from the desired source (the kits do not supply materials for this step).
2. Shear the DNA to approximately 40-kb fragments.
3. End-repair the sheared DNA to blunt, 5′-phosphorylated ends.
4. Isolate the desired size range of end-repaired DNA by LMP agarose gel electrophoresis.
5. Purify the blunt-ended DNA from the LMP agarose gel.
6. Ligate the blunt-ended DNA to the cloning –Ready CopyControl pCC2FOS Vector or pCC2FOS vector.
7. Package the ligated DNA and plate on EPI300-T1$^{R}$ plating cells. Grow clones overnight.
8. Pick CopyControl fosmid clones of interest and induce them to high-copy number using the copy-control fosmid autoinduction Solution.
9. Purify DNA for sequencing, fingerprinting, subcloning or other applications.

## GLOBAL METAGENOMICS INITIATIVE

In metagenomics, scientists access a community's genome without relying on pure cultures. Metagenomics transcends the limitations of classical genomics and microbiology.

- Metagenomic data should be stored in databases that use common standards and are freely assessable to all.
- Databases should include specialised tools that enable different scientists to analyse the data in different ways.
- A global metagenomic initiative should be established that includes a small number of large scale, comprehensive projects, a large number of medium sized projects and many small scale projects.
- Large scale projects will explore a microbial community in great depth, analysing a habitat with attention to potential for discovery, variation, commonalities and detailed investigation. Possible candidate communities include:
  - A community in a natural environment(such as sea water or soil)
  - A host associated community (such as the microbes in the human gut or mouth) and
  - A "managed" environment community (such as sewage treatment facility or bioremediation site)
- Medium-sized projects will encompass the greater diversity of microbial habitat and should include a multi-disciplinary approach.
- Small scale projects will be initiated by individual scientists to examine a slice of a community, a particular function in multiple communities, or a specific technical advance.

## CONCLUSION

Metagenomics has changed the microbiologists' concept of a genome and accelerated the rate of gene discovery. Metagenomics has potential applications in biotechnology. Realizing the potential for discovery from metagenomics is dependent on the advancement of methods that are central to library construction and analysis. Microbiology has long relied on diverse methods for analysis and metagenomics can provide the tools to balance the abundance of knowledge attained from culturing with an understanding of the uncultured majority of microbial life. Metagenomics will expand and continue to enrich our understanding of microorganisms.

## REFERENCES

Barns, S.M., Takala, S.L. and Kuske, C.R. (1999). Wide distribution and diversity of members of the bacterial kingdom Acidobacterium in the environment. *Appl. Environ. Microbiol.* **65:** 1731-1737.

Beja, O., Aravind, L., Koonin, E.V., Suzuki, M.T., Hadd, A., Nguyen, L.P., Jovanovich, S.B., Gates, C.M., Feldman, R.A., Spudich, J.L., Spudich, E.N. and DeLong, E.F. (2000). Bacterial rhodopsin: evidence for a new type of phototrophy in the sea. *Science* **289:** 1902-1906.

Beja, O., Spudich, E.N., Spudich, J.L., Leclerc, M. and DeLong, E.F. (2001). Proteorhodopsin phototrophy in the ocean. *Nature* **411:** 786-789.

Black, C., Fyfe, J.A.M. and Davies, J.K. (1995). A promoter associated with the neisserial repeat can be used to transcribe the *uvrB*gene from *Neisseria gonorrhoeae. J. Bacteriol.* **177:** 1952-1958.

Buckley, D.H. and Schmidt, T.M. (2003). Diversity and dynamics of microbial communities in soils from agro-ecosystems. *Environ. Microbiol.* **5:** 441-452.

Chávez, S., Reyes, J.C., Cahuvat, F., Florencio, F.J. and Candau, P. (1995). The NADP-glutamate dehydrogenase of the cyanobacterium *Synechocystis* 6803: cloning, transcriptional analysis and disruption of the *gdhA*gene. *Plant Mol. Biol.* **28:** 173-188.

Courtois, S., Cappellano, C.M., Ball, M., Francou, F.X., Normand, P., Helynck, G., Martinez, A., Kolvek, S.J., Hopke, J., Osburne, M.S., August, P.R., Nalin, R., Guerineau, M., Jeannin, P., Simonet, P. and Pernodet, J.L. (2003). Recombinant environmental libraries provide access to microbial diversity for drug discovery from natural products. *Appl. Environ. Microbiol,* **69:** 49-55.

Diaz-Torres, M.L., McNab, R., Spratt, D.A., Villedieu, A., Hunt, N., Wilson, M. and Mullany, P. (2003). Novel tetracycline resistance determinant from the oral metagenome. Antimicrob. *Agents Chemother,* **47:** 1430-1432.

Ding, M.A.D.B.Y. (1993). Cloning and analysis of the *leuB*gene of *Leptospira interrogans* serovarpomona. *J. Gen. Microbiol.* **139:** 1093-1103.

Ferreyra, R.G., Soncini, F.C. and Viale, A.M. (1993). Cloning, characterization, and functional expression in *Escherichia coli* of chaperonin (*groESL*) genes from the prototrophic sulfur bacterium *Chromatium vinosum. J. Bacteriol.* **175:** 1514-1523.

Furlong, M.A., Singleton, D.R., Coleman, D.C. and Whitman, W.B. (2002). Molecular and culture-based analyses of prokaryotic communities from an agricultural soil and the burrows and casts of the earthworm *Lumbricus rubellus. Appl. Environ. Microbiol.* **68:** 1265-1279.

Gillespie, D.E., Brady, S.F., Bettermann, A.D., Cianciotto, N.P., Liles, M.R., Rondon, M.R., Clardy, J., Goodman, R.M. and Handelsman, J. (2002). Isolation of antibiotics turbomycin A and B from a metagenomic library of soil microbial DNA. *Appl. Environ. Microbiol*, **68:** 4301-4306.

Gillespie, D.E., Brady, S.F., Bettermann, A.D., Cianciotto, N.P., Liles, M.R., Rondon, M.R., Clardy, J., Goodman, R.M. and Handelsman, J. (2002). Isolation of antibiotics turbomycin A and B from a metagenomic library of soil microbial DNA. *Appl. Environ. Microbiol.* **68:** 4301-4306.

Handelsman, Jo, (2004). Metagenomics: Application of Genomics to Uncultured Microorganisms, *Microbiology and Molecular Biology Reviews*, p. 669-685.

Healy, F.G., Ray, R.M., Aldrich, H.C., Wilkie, A.C., I.L. O. and Shanmugam, K.T. (1995). Direct isolation of functional genes encoding cellulases from the microbial consortia in a thermophilic, anaerobic digester maintained on lignocellulose. *Appl. Microbiol. Biotechnol.* **43:** 667-674.

Henne, A., Daniel, R., Schmitz, R.A. and Gottschalk, G. (1999). Construction of environmental DNA libraries in *Escherichia coli* and screening for the presence of genes conferring utilization of 4-hydroxybutyrate. *Appl. Environ. Microbiol.* **65:** 3901-3907

Henne, A., Schmitz, R.A., Bomeke, M., Gottschalk, G. and Daniel, R. (2000). Screening of environmental DNA libraries for the presence of genes conferring lipolytic activity on *Escherichia coli. Appl. Environ. Microbiol.* **66:** 3113-3116.

Janssen, P.H., Yates, P.S., Grinton, B.E., Taylor, P.M. and Sait, M. (2002). Improved culturability of soil bacteria and isolation in pure culture of novel members of the divisions acidobacteria, actinobacteria, proteobacteria, and verrucomicrobia. *Appl. Environ. Microbiol.* **68:** 2391-2396.

Knietsch, A., Bowien, S., Whited, G., Gottschalk, G. and Daniel, R. (2003). Identification and characterization of coenzyme B12-dependent glycerol dehydratase- and dioldehydratase-encoding genes from metagenomic DNA libraries derived from enrichment cultures. *Appl. Environ. Microbiol.* **69:** 3048-3060.

Liles, M.R., Manske, B.F., Bintrim, S.B., Handelsman, J. and Goodman, R.M. (2003). A census of rRNA genes and linked genomic sequences within a soil metagenomic library. *Appl. Environ. Microbiol.* **69:** 2684-2691.

Logan N.A., Lappin-Scott H.M., Oyston P.C.F., Society of General Microbiology Symposium 66, Prokaryotic Diversity: Mechanism and Significance.

Lorenz, P. and Schleper, C. (2002). Metagenome—a challenging source of enzyme discovery. *J. Mol. Catal. B Enzym.* **19:** 13-19.

Lorenz, P., Liebeton, K., Niehaus, F. and Eck, J. (2002). Screening for novel enzymes for biocatalytic processes: accessing the metagenome as a resource of novel functional sequence space. *Curr. Opin. Biotechnol*, **13:** 572-577.

MacNeil, I.A., Tiong, C.L., Minor, C., August, P.R., Grossman, T.H., Loiacono, K.A., Lynch, B.A., Phillips, T., Narula, S., Sundaramoorthi, R., Tyler, A., Aldredge, T., Long, H., Gilman, M., Holt, D. and Osburne, M.S. (2001). Expression and isolation of antimicrobial small molecules from soil DNAlibraries. *J. Mol. Microbiol. Biotechnol*, **3:** 301-308.

Majernik, A., Gottschalk, G. and Daniel, R. (2001). Screening of environmental DNA libraries for the presence of genes conferring $Na^+(Li^+)/H^+$antiporter activity on *Escherichia coli*: characterization of the recovered genes and the corresponding gene products. *J. Bacteriol*, **183:** 6645-6653.

National Research Council Report: The New Science of Metagenomics: Revealing the Secrets of our Microbial Planet.

Quaiser, A., Ochsenreiter, T., Lanz, C., Schuster, S.C., Treusch, A.H., Eck, J. and Schleper, C. (2003). Acidobacteria form a coherent but highly diverse group within the bacterial domain: evidence from environmental genomics. *Mol. Microbiol.* **50:** 563-575.

Riesenfeld, C.S., Goodman, R.M. and Handelsman, J. Uncultured soil bacteria are a reservoir of new antibiotic resistance genes. *Environ. Microbiol.* **6:** 981-989.

Rondon, M.R., August, P.R., Bettermann, A.D., Brady, S.F., Grossman, T.H., Liles, M.R., Loiacono, K.A., Lynch, B.A., MacNeil, I.A., Minor, C., Tiong, C.L., Gilman, M., Osburne, M.S., Clardy, J., Handelsman, J. and Goodman, R.M. (2000). Cloning the soil metagenome: a strategy for accessing the genetic and functional diversity of uncultured microorganisms. *Appl. Environ. Microbiol,* **66:** 2541-2547.

Rondon, M.R., Goodman, R.M. and Handelsman, J. (1999). The Earth's bounty: accessing soil microbial diversity. *Trends Biotechnol.* **17:** 403-409.

Rondon, M.R., Raffel, S.J., Goodman, R.M. and Handelsman, J. (1999). Toward functional genomics in bacteria: analysis of gene expression in *Escherichia coli* from a bacterial artificial chromosome library of *Bacillus cereus. Proc. Natl. Acad. Sci. USA* **96:** 6451-6455.

Sait, M., Hugenholtz, P. and Janssen, P.H. (2002). Cultivation of globally distributed soil bacteria from phylogenetic lineages previously only detected in cultivation-independent surveys. *Environ. Microbiol.* **4:** 654-666.

Schloss, P.D. and Handelsman, J. (2003). Biotechnological prospects from metagenomics. *Curr. Opin. Biotechnol,* **14:** 303-310.

Stein, J.L., Marsh, T.L., Wu, K.Y., Shizuya, H. and DeLong, E.F. (1996). Characterization of uncultivated prokaryotes: isolation and analysis of a 40-kilobase-pair genome fragment from a planktonic marine archaeon. *J. Bacteriol.***178:** 591-599.

Venter, J.C., Remington, K., Heidelberg, J.F., Halpern, A.L., Rusch, D., Eisen, J.A., Wu, D., Paulsen, I., Nelson, K.E., Nelson, W., Fouts, D.E., Levy, S., Knap, A.H., Lomas, M.W., Nealson, K., White, O., Peterson, J., Hoffman, J., Parsons, R., Baden-Tillson, H., Pfannkoch, C., Rogers, Y.H. and Smith, H.O. (2004). Environmental genome shotgun sequencing of the Sargasso Sea. *Science* **304:** 66-74.

Wang, G.Y., Graziani, E., Waters, B., Pan, W., Li, X., McDermott, J., Meurer, G., Saxena, G., Andersen, R.J. and Davies, J. (2000). Novel natural products from soil DNA libraries in a streptomycete host. *Org. Lett.* **2:** 2401-2404.

# 16

# Nodulated Native Legumes in an Arid Environment of Indian Thar Desert

DHEEREN PANWAR, NISHA TAK AND H.S. GEHLOT

*BNF & Stress Biology Laboratory, Department of Botany, J.N. Vyas University, Jodhpur-342001, Rajasthan, India*

**ABSTRACT**

Fraction of earth's surface occupies arid and semi-arid areas and covers a large up to 45% (Schimel 2010). Desert floras contain significant amount of legume component. Legumes play an important role in providing 33% of global human protein in human and animal diet. Legumes have been gathered for food, fodder, medicine and shelter throughout the globe. The present review focuses on native nodulating legumes from arid and semi arid regions of W. Rajasthan. The arid and semi arid regions of Western Rajasthan possess several economically important native legumes providing food, feed, shelter, medicine and improve soil fertility and add to economy of the rural people. However there is no in-depth investigation on nodulating native legumes and associated indigenous rhizobia in arid regions of Indian Thar Desert. The present investigation is an initiative to explore the nodulating native legumes reported in the "Flora of the Indian Desert" (Bhandari 1990). This is probably the first record of documentation/digitalization of large number of species of native legumes from the arid regions of Indian Thar desert at one place.

## INTRODUCTION

Fraction of earth's surface occupies arid and semi-arid areas and covers a large up to 45% (Schimel 2010). Desert floras contain significant amount of legume component. Extensive literature is available on nitrogen fixation studies in temperate and sub-tropical as compared to extreme environmental conditions such as drought, salinity, high and low temperature, low nutrients and toxicity. The recent hypothesis based on large number of evidences suggests that legumes are evolved in dry areas bounding the Tethys Sea (Schrire et al., 2005). According to annual precipitation or months of dry season, the arid and semi arid areas can be divided into hyper-arid 12 months of dry season), arid (9-11 months), and semi arid (7-8 month). Again all these arid and semi-arid regions can be further divided into high and low altitude areas (Sprent and Gehlot, 2010). Legumes grow in a diverse array of habitats including savannas, tropical and subtropical, dry forest, low land tropical rain forest, mid-alevation subtropical, desert grass land and wet land.

Indian Thar desert also known as Sahara-Thar desert is at low altitude ((1500m a.s.l.). North West part of Rajasthan with an approximate area of 25,000 square kilometer lies between latitude 25° 2' to 28° 10' N and longitude 69° 30° to 74°E. It includes the arid and semi arid area having scattered saline tracts in the districts of Jodhpur, Jalore, Bikaner, Barmer and Jaisalmer (Bhandari, 1990). The soil of arid region is alkaline throughout the W Rajasthan. The climate of this region is characterized by extreme of temperature (-1°C in winter to 48°-49°C in summer and even beyond 50°C at remote places in Jaisalmer, and Churu districts). Severe drought coupled with high wind velocity, low relative humidity, high irradiation, salt tracts, poor nutrients and low precipitations are the major constraints of the region. Almost 61.9% of the hot and arid regions of India lie in Western Rajasthan. Erratic rainfall and recurrent famine is common phenomenon in this region. The annual precipitation varies from 50 mm to 400 mm. Pearl millet, Sesame, Guar, Mung and Moth beans are common rain-fed crop in this region. Major portion of export of guar gum is from W Rajasthan. The leguminosae family is the most dominating among dicotyledonous plants in terms of both number of genera and species in W Rajasthan.

Legumes play an important role in providing 33% of global human protein in human and animal diet. Legumes have been gathered for food, fodder, medicine and shelter throughout the globe. Legume products contribute significantly to the world's economy through their multipurpose uses as food, feed and drink, pharmaceuticals and medicines, bio diesel fuel, building and construction, textile, furniture and crafts, paper and pulp, chemical and fertilizers, waste recycling, horticulture, pest control and ecotourism. Leguminosae (nom. alt. Fabaceae) is the third largest family of the flowering plants with approximately 678 genera and 18000 species (Polhill & Raven, 1981). The number of species has raised from about 1,8000 in 1994 to about 19,325 in 2001(Legumes of the World, Lewis et al., 2005). The legumes species are distributed throughout the world in almost all habitats. They also encompass a wide variety of life forms as tiny herbs, vines, lianas, shrubs to gigantic trees in the forest.

However, quite a good number of wild legumes are also found in the arid regions of the Thar Desert as herbs and shrubs (small trees) representing many important tribes. *Tephrosia* and *Indigofera* genus is well represented in the region with more than 5 species. Legumes and their role in increasing soil fertility is not new, however occurrence of nodulation and nitrogen fixation in extreme environment is important (Zaharan, 1999, 2001 and Zakhia et al 2006). Studies on native legumes become low priorities due to the cumbersome field work and decreasing interest in taxonomy. Loss of natural habitat by expansion of human activity in this region makes it difficult to locate plants and collection of their germplasm. However, with the global climate change and food security, such studies are important.

The present review focuses on native nodulating legumes from arid and semi arid regions of W. Rajasthan. The arid and semi arid regions of Western Rajasthan possess several economically important native legumes providing food, feed, shelter, medicine and improve soil fertility and add to economy of the rural people. However there is no in-depth investigation on nodulating native legumes and associated indigenous rhizobia in arid regions of Indian Thar Desert. Today there have been only a handful published scientific studies in our country on our native legumes as well as indigenous rhizobia. Much of the information on biological nitrogen fixation had been carried out in Europe, USA, and Australia. Even African countries (Kenya, Sudan, Morocco, Ethiopia, Cape Town etc.) have now information about nodulating legumes and associated rhizobia of their region (Sylla et al., 2002; Benata et al., 2008; Zahran, 2001; Benhizia et al., 2004; Zakhia et al., 2006; Endalkachew et al., 2004). India is lagging behind in characterizing indigenous rhizobia associated with native legumes of extreme habitats. There is only one extensive survey that had been carried out in India by Pokhriyal et al. (1993 and 2003) and in University of Delhi by C.R. Babu group (Jha P.K.1988, and Subramaniam et al., 1994) for recording nodulation status in wild legumes and was only restricted to sub-Himalayan regions. No effort has been made on isolation and characterization of rhizobial diversity associated with native legumes. Even in other countries most of the laboratories are working on a few agricultural and industrial crops like soybean, *Cicer,* etc.

Studies on native legumes become low priorities due to the cumbersome field work and decreasing interest in taxonomy. The present investigation is an initiative to explore the nodulating native legumes reported in the "Flora of the Indian Desert" (Bhandari 1990). Loss of natural habitat by expansion of human activity in this region makes it difficult to locate plants and collection of their germplasm.

This is probably the first record of documentation/digitalization of large number of species of native legumes from the arid regions of Indian Thar desert at one place.

## SOIL AND CLIMATIC CONDITIONS IN THE ARID ENVIRONMENT OF INDIAN THAR DESERT

The soil of arid region is classified as "desert soils" and "grey brown soils" (Ray Choudhury et al., 1963). The soil is light-textured, having low organic carbon and available phosphorus. The available phosphorus for plant in arid soil is only 0.7 to 1.8 % (Tarafdar et al., 2006). The P solubilization is much more by the microbial activity than the plants. Soil is characterized as alkaline from having pH in the range of 7.5 to 8.9.

Most of the soil in arid and semiarid regions has $CaCO_3$ in the soil profile. If calcareous soils have an excess of exchangeable Na and high pH, caused by the presence of $NaHCO_3$ and $NaCO_3$, the soils are termed alkali soils. These soils have a poor physical structure, which together with Na toxicity adversely affect plant growth. The elements commonly observed in toxic concentrations include Na, Mo, Bo, and sometimes Se. Roy et al (1978) divided the soils into "desert soils" and "sand dune soils". Dhir (1977) divided Indian Desert soils into seven major groups. Poor soil texture of arid and semi-arid regions contains organic contents at most of the places in arid regions from 0.02 to 0.04% whereas available N and P are 0.002-0.004 % and 0.7-1.2% respectively (Yadav et al, 2011).

The climate of arid and semi arid region is mainly characterized by low annual precipitation. Precipitation exceeds evapotranspiration. The major part of the annual rainfall is received during southwest monsoon between July to September. The total rainfall in one rainy season (July - September) is 91 to 96% of the total Rajasthan and whole of the Gujarat (Ramakrishna et al., 1992). Jodhpur received 250 to 350 mm of total rainfall annually, whereas Jaisalmer and Barmer has average rainfall below 180 mm and at certain places it is less than 100 mm annually. Dust storms are frequent in the month of May to June exceeding wind speed of 25-30 $kmh^{-1}$ in the Western part of this region.

The mean diurnal variation is 15-17°C. The mean summer temperature (May-June) varies between 40°C to 43°C, but sometimes mercury touches 46°C to 48°C. In winter, temperature varies between 4°C to 12°C. Strong North winds prevail in the month of December and January for a week or so causing chilling.

Underground water is at very low level and unsuitable for irrigating crops due to high salt content. Increasing desertification and the shortage of fuel wood are some of the problems arising in this region. The area under permanent pastures and long fallows has been decreasing rapidly, whereas the population of cattle's has been increasing. Farming lands have become fragile due to overexploitation of the natural resources by the increasing human populations. Agro forestry is the solution to this problem, particularly by growing the legume trees in the farms.

The arid and semi arid regions of Indian Thar Desert possess eight genera among the top twenty largest legumes genera of the world having more than 165 species per genera. These are *Acacia* (c. 1450 spp.), *Indigofera* (c. 700 spp.), *Crotalaria* (c. 690 spp.), *Mimosa* (c. 500 spp.), *Tephrosia* (c. 350 spp.), *Chamaecrista* (c. 330 spp.), and *Rhynchosia* (c. 250 spp.). The great biological diversity in *Acacia, Prosopis, Indigofera* and *Tephrosia* is reflected in their wide distribution, ecological aptitude and particularly value in the tolerance of some species to extreme of drought, salinity and high temperature. *Prosopis cineraria* is a multipurpose tree of an immense value in arid and semi arid regions. Species of *Indigofera* (baker) are used as source of feed, dye (indigo) and medicine. *Crotalaria burhia* and species of *Tephrosia* are good soil binder and contribute in improving soil fertility through BNF. Abdel-Ghaffar, 1989 and Zahran, 1991, reported that naturally occurring forage legumes (annuals and perennials) in Egyptian desert soil are well nodulated, and their roots nodules are active in fixing nitrogen. These legumes may be found in desert or cultivated lands as wild plants. Jha et al (1995) suggested the suitability of rhizobium-inoculated wild herb legumes for providing vegetation cover and improving soil fertility in unreclaimed lands.

The seminal text by Allen and Allen (1981) is the first large documentation of nodulating legumes of the world. Since then nodulation has been reported in many species of legumes of the world. The major sources of nodulation database are: International Legumes Database & Information Service (ILDIS) U.K., rhizobia nodulation data in GRIN (Germplasm Resources Information Network) online database USDA, Beltsvillel, Maryland; Nodulation in legumes (Sprent, 2001) and The Bean Bag (Joseph H Krikbride).

Well-characterized nodule forming and nitrogen fixing rhizobia associated with legumes forms less than 2% of the known nodulating legumes.

## LEGUMES OF THE WORLD

The leguminosae (or Fabaceae) comprise 650 genera and 18,000 species and is the largest family of flowering plants after the Compositae and Orchidaceae (Polhill et al. 1981). Polhill (1994) recognized 671 genera, whereas recently published book "legumes of the world" (Lewis et al. 2005) described 727 genera with c. 19,327 species. The family is currently divided in three subfamilies and 36 tribes. According to Lewis et al. 2005, Caesalpiniodeae comprises four tribes and c. 3270 species; Mimosoideae with four tribes and c. 3000 species; and Papilionoideae with 28 tribes and c. 13,800 species. Nearly one-third of the total species (>18,000) are contained in six genera *Acacia, Astragalus, Cassia, Crotalaria, Indigofera,* and *Mimosa,* all of which are highly characteristics of open and distributed places habitats that have expanded greatly from the Miocene onward (Polhill et al. 1981), of the top twenty largest legume genera recognized by Lewis et al. 2005 in "Legumes of the world", twelve are found in Rajasthan and eight are found in arid and semi arid part of Western Rajasthan. These are *Acacia, Indigofera, Crotalaria, Mimosa, Tephrosia, Chamaecrista, Senna, & Rhynchosia.* Recently Simon et al. (2011) reported an elegant account on evolutionary history of *Mimosa* towards a phylogeny of the sensitive plant where he described the *Mimosa* in the old and new world.

In the last 30 years, legumes systematic has benefited from major advances in our understanding of the morphology and classification of the family (Polhill and Raven 1981; Polhill 1994), as well as from more than 10 years of intensive molecular phylogenetic research by several group of researches around the world.

Legumes have been used by humans for millennia and are as important as grasses in global term. Grain and forage legumes are grown on approximately 180 million hectare (12 to 15%) of the earth's arable surface and account for 27% of the world's primary crop production with grain legumes alone contributing 33% of the dietary protein nitrogen need of humans (Graham & Vance 2003).

Forage legumes provide the protein, fiber and energy that have underpinned dairy and meat production for centuries. In temperate regions, Alfa-Alfa (*Medicago sativa*) is an important forage crop for cattle.

## LEGUMES OF INDIA AND THAR DESERT (W. RAJASTHAN)

As compared to more than 19,000 species of legumes reported in "Legumes of the World" (Lewis et al., 2005), 179 genera and 1152 species are reported from India (Sanjappa 1992) including 35 subspecies and 103 varieties. Twenty-three percent of total leguminous taxa reported from India are endemic. Rajasthan possesses 64 genera and 225 species of legumes (Shetty and Singh 1987), whereas 38 genera and 88 species of native legumes have been reported in "Flora of the Indian Desert" (Bhandari, 1990) from W. Rajasthan. Some arid legumes such as *Cymopsis tetragonoloba* (Guar) have been commercialized for industrial uses and source of major export as guar gum. Guar gum industries in Jodhpur are advocating, advertising and prioritizing for growing guar in the region. Alfa-Alfa, also locally known as 'Rizka', is a source of major nutritious fodder for cattle and is widely grown throughout North India, also in Nagure near Jodhpur. Guar as well as Rizka are cash crops for the farmers in this region. Both these legumes increase soil fertility, so do not require any organic fertilizers. Native Rhizobia supports growth and biomass of these legumes crops in this region. In arid and semi arid regions of Rajasthan, leaves and pods of *Acacia* species (*A. nilotica, A. tortilis (now radiana), A. jacquemontii*), *Prosopis* species (*P. cineraria, P. juliflora*), *Mimosa hamata, Crotalaria burhia, Chamaecrista pumila* are used as feed and fodder for cattles, goats and sheeps. *Prosopis cineraria* is a multipurpose tree of this region (Sandison and Harris, 1991). *Acacia* and *Prosopis* have wide distribution and ecological amplitude in the tolerance of drought and salinity. The pods and browse, used as animal fodder, wood of both genera is essential source of fuel in arid and semi arid areas, and timber is also valued for a wide range of purpose. Gums and tannins from *Acacia* and *Prosopis* have immense value in industries (Fagg and Stewart 1994).

The plant characteristics, taxonomic characters including vegetative, flower and pod characters have been studied in the field with the support of flora of the Indian desert (Bhandari, 1990), and results are presented in Table 1. Digital pictures of important parts of plants helpful in identification purposes have been presented in Table 1. Information regarding the use of legumes and their products as food, feed, medicines and others was gathered from the literature as well as native villagers of the region, and presented in Table 2.

## NODULATION IN NATIVE LEGUMES

It has been more than 27 years since the classic volume 'The leguminosae', a source book of characteristics, uses and nodulation was published (Allen & Allen 1981). The Bean Bag maintain nodulation record and reports of nodulation in legumes not mentioned in Allen & Allen 1981 (Published and maintained by Joseph H Kirkbride Jr. from Herbarium, Royal Botanic Garden, Kew, U.K.). Nodulation in legumes by J. I. Sprent (2001) is another source book that gives latest information about the various aspects of nodulation and associated rhizobia. Rhizobial nodulation data in GRIN (Germplasm Resource Information Network online database) of USDA, USA is linked to ILDIS (International Legumes Database and Information Service, UK) and provide instant facilitation of searching/knowing nodulation status of any legumes listed in ILDIS.

In Allen & Allen (1981), only 15% legumes have been examined for nodulation. The nodulation frequency varies among the genera belonging to sub-families. Sprent reported nodulation frequency of 5% at genera level and 21% at species level in Caesalpiniodeae, 88% genera and 97% species in Mimosoideae and 93% genera and 99% species in Papilionoideae (Sprent 2001). In India, a massive survey was conducted earlier at Forest Research Institute (FRI), Dehra Dun, for the identification of leguminous $N_2$ fixing herb, shrub, climber and tree species distributed in the Himalayan region to assess their nodulation and nitrogen fixation behavior (Pokhriyal et al. 1993). Among legumes, maximum 93% nodulated species were recorded in Papilionoideae followed by 4% in Mimosoideae and 3% in Caesalpiniodeae. Amongst them 54%, 20%, 10% and 8% were herb, shrub, climber and tree species respectively (Pokhriyal et al. 1993). Survey conducted from variable altitude in Garhwal Himalaya (300-3900 m) examined more than 250 legumes species, among which maximum 15 species were noted in genus *Desmodium*, followed by 9 species in *Indigofera*, 8 species in *Crotalaria*, 5 species in *Vigna*, 4 species in *Alysicarpus* and 3 species each in *Astragalus* and *Flemingia*.

An extensive survey of nodulation status of legumes viz. chickpea, pigeon pea, moong bean, soybean, and groundnut with native rhizobia during 1967-72 (Sundra et al. 1969; Subba Rao et al. 1972) and in 1977-80 (Rewari 1979) under All India coordinated Pulse Improvement Program has belied this assumption since except for groundnut, most legumes nodulated poorly at more than 50% of the places surveyed. There was a deficiency of specific *Rhizobium* even in traditional legume growing areas. A detailed eco-serological survey of chickpea in 13 major soils of India revealed three broad serological groups of which Sero-group I was widely distributed. Sero-group II was limited to grey and brown soil types and Sero-group III which recognized among strains of American origin did not occur in any Indian soil. Field trials conducted in India showed that nearly 50% of nitrogenous fertilizer can be saved through rhizobial inoculations with considerable increase in the yield depending on the legumes, soil and agro-climatic condition (Rewari & Tilak 1988 and Tilak 1993).

A survey of nodulated legumes (native as well as exotic) in arid regions of Indian Thar desert around Jodhpur was conducted during 2006-2008 during monsoon and post-monsoon period (Dheeren 2009). The positively nodulated legumes in field as well as pots have been presented in Table 3.

## Caesalpiniodeae

The only genera whose most species nodulate is *Chamaecrista* within the tribe cassieae of subfamily Caesalpiniodeae; other well-confirmed nodulation genus is

**Table 1** Description of wild legumes of W Rajasthan

| *Name of plant and habitat* | *Flower characters* | *Fruit characters* | *Seed characters* | *Nodule characters* |
|---|---|---|---|---|
| ***Acacia nilotica*** (Linn.) Del subsp. indica (Benth.)<br><br>**Habitat:**<br>A moderate size tree, evergreen with slender, terete, pubescent branches when young. | **Flowers**:<br>Inflorescence globose head 6-15 mm in diameter. Head yellow.<br> | **Pods:**<br>Pods 7-15 x 1.2-1.5 cm, moniliform, compressed constricted at sutures between seeds<br> | **Seeds**:<br>Seeds 8-12, black brown, smooth, sub orbicular, compressed.<br> | **Nodules**:<br>Elongated- Cylindrical Red- brown.<br> |
| ***Acacia jacquemontii*** Benth<br><br>**Habitat:**<br>A much branched, 3-4 m high, bushy shrub. Branches stiff, smooth and reddish brown. | **Flowers**:<br>Inflorescence globose head, 12-16 mm in diameter. Head yellow.<br> | **Pods**:<br>Pods 3-6 x 1.2-1.5 cm, ovate-oblong, rounded at base, flat, straight and glabrous.<br>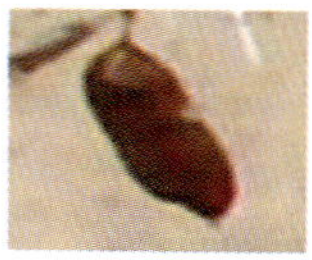 | **Seeds**:<br>Seeds 5-8, compre-ssed disc shape, light brown.<br> | **Nodules**:<br>Elongated or Oblong elliptical. Pale brown or red- brown.<br> |
| ***Acacia senegal*** (Linn.) Willd.<br><br>**Habitat:**<br>A small, prickly tree, branches fluxous, glabrous-grey. | **Flowers**:<br>Flowers fragrant, spike's 5-10 cm longs. Flowers yellowish white.<br> | **Pods**:<br>Pods 7-8 x 2.5 cm, linear oblong, thin, flat, and attenuated at both ends.<br> | **Seeds**:<br>Seeds 5-6, sub orbicular 8-12 mm in diameter, impressed on sides, brown.<br> | **Nodules**:<br>Irregular Pale brown.<br> |
| ***Alysicarpus vaginalis***(Linn.) DC.<br><br>**Habitat:**<br>Prostrate, procumbent or diffuse annual herb, grow on waste field | **Flowers**:<br>Flowers in axillary or terminal, 4-12-flowered racemes, along an elongated axis. Corolla light orange<br> | **Pods**:<br>Pod 20-30 x 2.2 mm, slightly moniliform, reticulately veined, joint 4-8, and becomes black at maturity.<br> | **Seeds**:<br>Elongated yellowish-brown in color. 2.5-3 x 1.5 mm<br>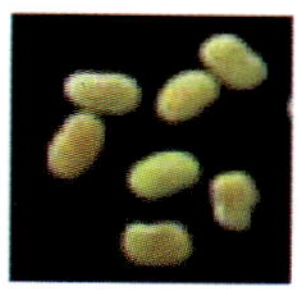 | **Nodules**:<br>Globular Off white with pinkish appearance.<br>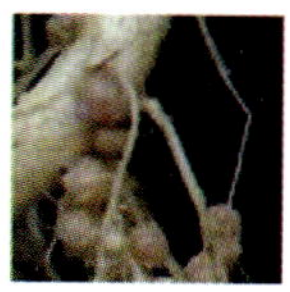 |
| ***Chamaecrista pumila*** (Lam.) K. Larsen<br><br>**Habitat:**<br>A diffuse or sub erect annual herb, 20-30 cm long branches. | **Flowers**:<br>Flowers solitary axillary, corolla yellow.<br> | **Pods**:<br>3 x 0.4-0.5 mm long, with oblique partition between the seeds.<br> | **Seeds**:<br>Seeds 6-12, obovoid-cuneate, compressed, and chocolate brown.<br> | **Nodules**:<br>Oblong Red- brown.<br> |

**Table 1** Contd.

| *Name of plant and habitat* | *Flower characters* | *Fruit characters* | *Seed characters* | *Nodule characters* |
|---|---|---|---|---|
| ***Crotalaria burhia*** Buch.-Ham.<br><br>**Habitat:**<br>A rigid under shrub, Much branched. | **Flowers:**<br>Flowers in 7-8 cm long, elongated, terminal raceme, corolla yellow with reddish streaks.<br> | **Pods:**<br>Pods 5 x 4 mm densely pilose, obliquely-ovoid, beaked, brownish.<br>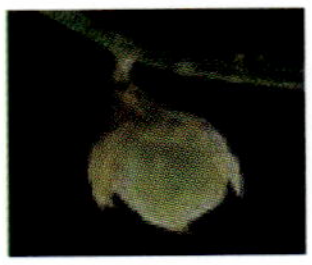 | **Seeds:**<br>One, rarely 2 per pod, obliquely semi lunar, grayish-brown, notched at hilum, 2.5 x 1.8-2. 2 mm, compressed.<br> | **Nodules:**<br>Elongated branched Dirty yellow or yellowish brown.<br>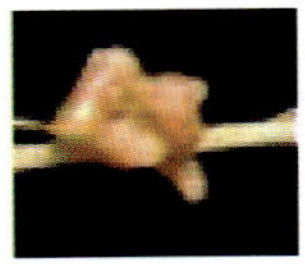 |
| ***Crotalaria medicaginea*** Lamk var. ***luxurians*** (Benth.)<br><br>**Habitat:**<br>An erect, much branched herb, 1-2 feet high. | **Flowers:**<br>Flowers in terminal and leaf-opposed raceme, 5-15 flowered raceme. Corolla lemon yellow in color.<br> | **Pods:**<br>Obliquely sub-globose, 4 mm long, curved beak at tip, 1-3 seeded, cream in color.<br> | **Seeds:**<br>Compressed, sub-orbicular, notched at hilum, 2 x 2 mm long.<br> | **Nodules:**<br>Elongated or branched Dirty yellow or yellowish brown.<br>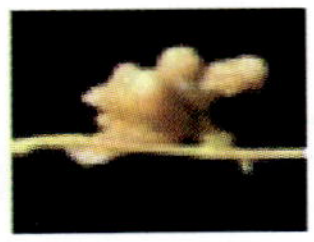 |
| ***Indigofera cordifolia*** Heyne ex Roth.<br><br>**Habitat:**<br>A prostrate or sub erect annual herb, enormously grown on barren area or waist fields, and rocky slopes. | **Flowers:**<br>Flowers in dense, axillary, 4-8 flowered head, corolla bright pink.<br>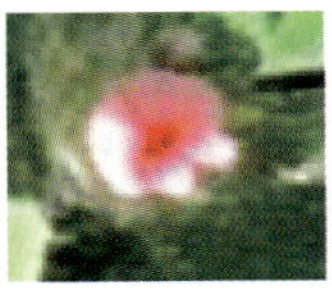 | **Pods:**<br>Pods cylindrical, 4.5 mm long, oblong, straight or moniliform, densely pubescent, 1-2 seeded.<br> | **Seeds:**<br>Pale-brown, globular or irregular ovoid, truncate at one end, 1 mm in diameter.<br> | **Nodules:**<br>Elongated Red-brown with cream tip.<br>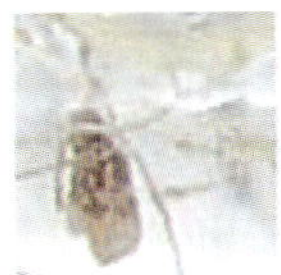 |
| ***Indigofera hochstetteri*** Backer<br><br>**Habitat:**<br>An erect annual herb, growsin open fields, some time found in cultivated fields | **Flowers:**<br>Inflorescence in 12-15 flowered axillary raceme, corolla light orange in color.<br> | **Pods:**<br>Occur in bunch of 12-15 falcate, papery, rigid, and distinctly partitioned between seeds, 20 x 3 mm.<br> | **Seeds:**<br>Seeds 2 mm wide, compressed more or less quadrangular and dirty yellow in color.<br>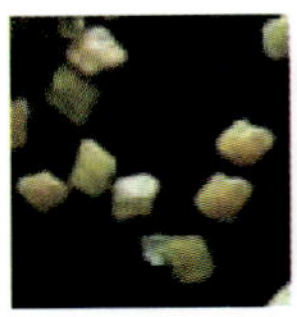 | **Nodules:**<br>Elongated Pale-brown.<br>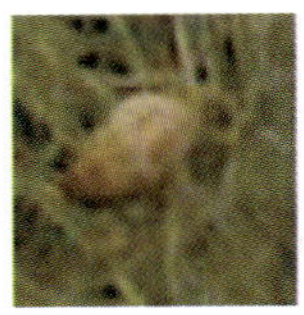 |
| ***Indigofera linifolia*** (Linn.f.) Retz.<br><br>**Habitat:**<br>A prostrate or sub-erect annual herb, grown in sandy area or open fields. | **Flowers:**<br>Flowers in dense, sessile or on shortly peduncled, axillary raceme up to 1-1.5 cm long, 8-12 flowered. Corolla bright pink<br>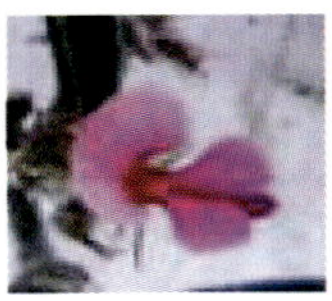 | **Pods:**<br>Globose, 2 mm in diameter, compactly arranged on linear axis.<br> | **Seeds:**<br>Seeds 1-2 mm in diameter, globular, brown.<br> | **Nodules:**<br>Elongated Brown.<br> |

**Table 1** Contd.

| *Name of plant and habitat* | *Flower characters* | *Fruit characters* | *Seed characters* | *Nodule charcaters* |
|---|---|---|---|---|
| ***Indigofera linnaei***Ali<br><br>**Habitat:**<br>A prostrate crown forming annual or biennial herb, grown in barren area or open fields. | **Flowers**:<br>Flowers in compact, axillary heads or racemes, corolla light red.<br> | **Pods**:<br>Pods 3-5 x 1.8 mm, cylindric, 2-seeded with a slight depression between seeds.<br> | **Seeds**:<br>Seeds ovoid-globose more or less truncate at one side, pale brown 1-2 mm in diameter.<br> | **Nodules**:<br>Elongated Red- brown or dark brown.<br> |
| ***Indigofera oblongifolia*** Forsk.<br><br>**Habitat:**<br>An erect, ashy-grey, twiggy, much branched shrub, 4-5 feet high. | **Flowers**:<br>Flowers small in many flowered, axillary raceme, corolla purplish pink in color.<br>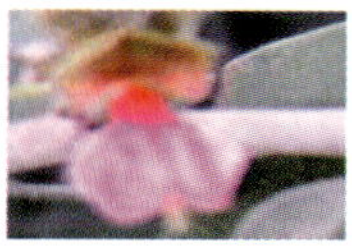 | **Pods**:<br>Compactly arranged on an axis, usually 25-30 in number, at maturity purplish brown colored. Slightly moniliform between seed 20 x 0.2 mm.<br> | **Seeds**:<br>Seeds oblong-globose, yellow, smooth tetra gonous (1.5-2 mm).<br> | **Nodules**:<br>Oblong elongated Light red- brown.<br>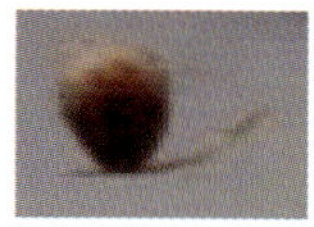 |
| ***Indigofera sessiliflora***DC<br><br>**Habitat:**<br>An annual herb. Stem prostrate, procumbent or rarely sub-erect. | **Flowers**:<br>12-20-floweres in dense, sessile, axillary, villous head. Corolla pale-rose.<br> | **Pods**:<br>Pods up to 1 cm long, linear, cylindric, torulose, straight, acute at apex.<br> | **Seeds**:<br>Seeds 1 mm in diameter. Quadrangular globose, pale-brown.<br>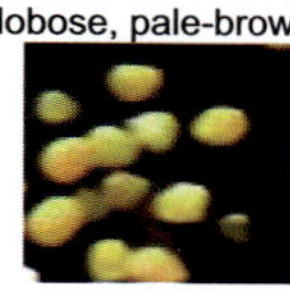 | **Nodules**:<br>Elongated Light red-brown.<br> |
| ***Indigofera tinctoria***Linn.<br><br>**Habitat:**<br>An extensively branched, 2-3 feet high shrub. Usually cultivated also grows in open fields. | **Flowers**:<br>Flowers in 4-6 cm long, many flowered, lax, axillary raceme. Corolla light orange in color.<br> | **Pods**:<br>Pods 3-4 cm long, linear, straight or more usually slightly deflexed and curved, 8-12 seeded. Golden brown at maturity<br> | **Seeds**:<br>Seeds 1.8-2.2 mm wide, brown with few darker blotches, more or less oblong-quadrate.<br>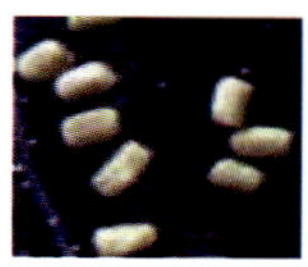 | **Nodules**:<br>Oblong or elongated.<br>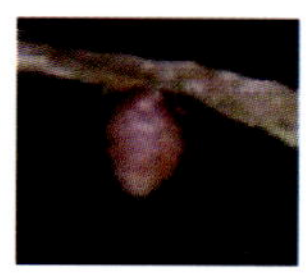 |
| ***Melilotus indica*** (Linn.) All.<br><br>**Habitat:**<br>An annual erect herb grows as weed of cultivated fields or open fields. | **Flowers**:<br>Flowers in 4-5 cm long raceme; Calyx 1-1.8 mm long, pale yellow.<br> | **Pods**:<br>Pods ellipsoid, compressed globular, red spot at apex.<br> | **Seeds**:<br>Seeds 1 per pod oblong, ellipsoid, 2 mm long, brown.<br>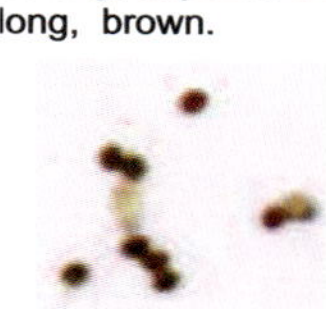 | **Nodules**:<br>Cylindrical Brown with yellow tip.<br> |

**Table 1** Contd.

| *Name of plant and habitat* | *Flower characters* | *Fruit characters* | *Seed characters* | *Nodule characters* |
|---|---|---|---|---|
| ***Mimosa hamata*** Willd.<br><br>**Habitat:**<br>A much branched, armed shrub, branches downy, with numerous straw-colored, curved or straight prickles. | **Flowers**:<br>Flowers tetra-merous in globose head, peduncles axillary. Head pink in color.<br> | **Pods**:<br>Pods 5-7 x 1.5 cm, falcate, consisting of 4-8, one-seeded joints, prickly sutures.<br> | **Seeds**:<br>Seeds 6 x 4 mm, chestnut-brown.<br> | **Nodules**:<br>Branched brownish<br> |
| ***Prosopis cineraria***(Linn.) Druce.<br><br>**Habitat:**<br>A large, much branched shrub or small tree. Branches slender, glabrous, armed with somewhat compressed. | **Flowers**:<br>Flowers in axillary spikes which are 7-11 cm long solitary or terminal panicles.<br>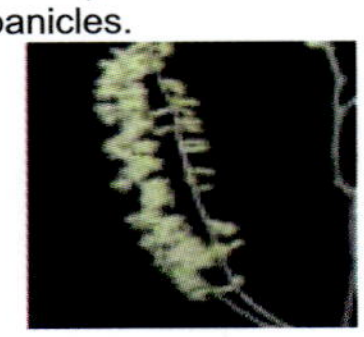 | **Pods**:<br>Pods 10-20 cm long, rigid, straight, cylindric, torulose, glabrous, 10-15 seeded.<br> | **Seeds**:<br>Seeds embedded in brown pulp, 3-8 mm long, dull-brown.<br> | **Nodules**:<br>Irregular Pinkish white.<br> |
| ***Rhynchosia minima***(Linn.) DC. var. ***laxiflora***(Camb.)<br><br>**Habitat:**<br>A climber, not very common, grows in open fields. | **Flowers**:<br>Flowers in 6-12 flowered lax axillary raceme, corolla yellow.<br> | **Pods**:<br>16-20 x 5.5 mm, 2 seeded turgid but compressed between seed.<br> | **Seeds**:<br>3.5-4 x 2.5-3 mm, turgid, pale olive green, mottled with grey and black, ovate in shape.<br> | **Nodules**:<br>At young stage globular but later on become elongated. Yellowish cream.<br> |
| ***Sesbania bispinosa***(Jacq.) W.F.Wight.<br><br>**Habitat:**<br>A sparingly branched annual, up to 2.5 m high. Stem semi woody, branches and leaf rachis more or less prickly | **Flowers**:<br>Flowers in lax, axillary, 3-6 flowered, drooping raceme. Corolla yellow, usually mottled with red dots on the back.<br> | **Pods**:<br>Pods 15-27 cm, slightly falcate, 20-35 seeded, beaked, beak 1-2 cm long.<br> | **Seeds**:<br>Seeds oblong, 3-3.5 mm long, olive green with shiny appearance.<br> | **Nodules**:<br>Globular or callus type. Dark brown.<br>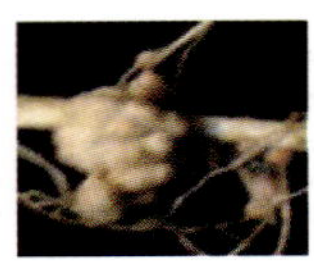 |
| ***Tephrosia leptostachya*** (DC.)<br><br>**Habitat:**<br>A straggling or diffuse, annual herb, woody at base. Stem slender, wiry, pubescent, reddish brown. | **Flowers**:<br>Flowers few on 6-7 cm long peduncle, bright pink.<br> | **Pods**:<br>Pods usually in pairs, 40-50 x 3-3.5 mm, slightly curved at the tip. Slightly pubescent, 7-9 seeded.<br> | **Seeds**:<br>Seeds 3-3.5 mm long, cylindrical loosely packed in the pod, pale-brown with dark vertical lines along with width.<br> | **Nodules**:<br>At young stage globular but later on become elongated or branched. Yellowish white.<br>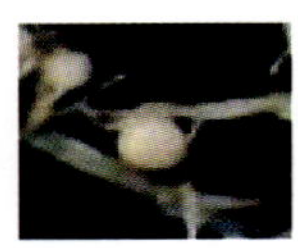 |

**Table 1** Contd.

| *Name of plant and habitat* | *Flower characters* | *Fruit characters* | *Seed characters* | *Nodule characters* |
|---|---|---|---|---|
| ***Tephrosia purpurea*** (Linn.) Pers.<br><br>**Habitat:**<br>A much branched, erect, perennial herb, 3-4 feet in height. Stem more or less hairy. | **Flowers:**<br>Flower purplish pink, on terminal or leaf opposed peduncles, laxly 6-25 flowered raceme.<br> | **Pods:**<br>Pods 3-4 x 0.4-0.5 cm, slightly recurved, glabrous or softly pubescent, 5-6 seeded.<br> | **Seeds:**<br>Seeds 5-6 mm long, pale brown mottled with black spot.<br> | **Nodules:**<br>At young stage globular but later on become elongated or mucunoid. Yellowish brown.<br>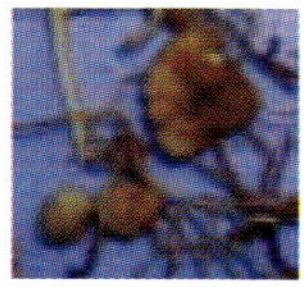 |
| ***Tephrosia subtriflora*** Hochst. ex Baker.<br><br>**Habitat:**<br>A perennial much branched herb, form a woody base. Branches slender, more or less angular. | **Flowers:**<br>Flowers in axillary, 1-3 flowered fascicles, corolla light purple<br> | **Pods:**<br>Pods 3 x 0.3 cm, turgid, 1-3 in axillary fascicles linear, highly curved, shortly obliquely-triangular at apex, 6-8 seed.<br> | **Seeds:**<br>Seeds slightly ranniform, 2.2 x 1.7 mm, pale brown mottled with darker lines.<br> | **Nodules:**<br>At young stage globular but later on become elongated or branched. Yellowish cream.<br>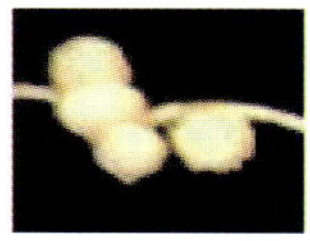 |
| ***Tephrosia uniflora*** Pers. Subsp ***petrosa*** (Blatt. & Hallb.)<br><br>**Habitat:**<br>A profusely branched perennial. Branches erect, diffuse **or** procumbent. | **Flowers:**<br>Flowers axillary, solitary or often geminate. Corolla bright pink.<br> | **Pods:**<br>Pods 3-4.5 x 0.25-0.3 cm, linear, curved, silvery white hair, slightly red from one side 7-9 seed.<br> | **Seeds:**<br>3.5 x 2.5 mm, subcylindric, somewhat compressed at one end.<br>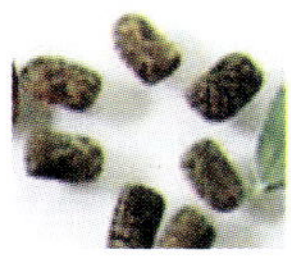 | **Nodules:**<br>At young stage globular but later on become elongated or branched. Light pink.<br> |
| ***Tephrosia villosa*** (Linn.) Pres.<br><br>**Habitat:**<br>Annual gregarious, erect bushy herb. Stem angular, densely hairy with white or grayish hair. | **Flowers:**<br>Flowers in lax raceme. Corolla purple.<br> | **Pods:**<br>Pods 28-30 x 5-6 cm falcate curved upwards, densely, persistently velvety hairy.<br> | **Seeds:**<br>3-5 x 2.5 mm blackish brown, some what compressed at both ends.<br> | **Nodules:**<br>At young stage globular but later on become elongated or branched. Pinkish white.<br> |
| ***Tephrosia wallichii*** Graham<br><br>**Habitat:**<br>An erect much branched 3-4 feet high under-shrub. | **Flowers:**<br>Inflorescence up to 10 cm long, with many flowers together at each node. Corolla bright pink.<br> | **Pods:**<br>Pods spreading on all sides of the inflorescence, 4-4.5 x 0.45 cm, 6-7 . seeded<br>Pale brown at maturity.<br> | **Seeds:**<br>4-4.5 x 2.5-2.75 mm, reniform, loosely packed in the pod, greenish pale with brown. Spot<br> | **Nodules:**<br>At initial stage globular but later on become much branched. Pinkish white.<br> |

**Table 2** Economic importance and uses of some native legumes of Thar desert

| *Species* | *General Uses* | *Industrial Uses* | *Medicinal Uses* |
|---|---|---|---|
| *Acacia jacquemontii* | Fodder & forage. Gun & seed are edible. Soil binding and improvement | Oil, gum, tannin & timber. Beverage obtained from bark. The bark of the roots is used in distillation of spirits (also used as source of fermenting microbes for making local wine/ crude form in villages). Wood burns intensively and is used by gold, silver & iron smiths. | Vertebrate poisons: mammals Two novel cassane diterpenoids, with potential anti-microbial use |
| *Alysicarpus vaginalis* | Fodder, forage & pasture soil improver; green manure & cover crop. | Seed is a source of industrial gum. | Root decoction used for irregular menstruation, uterine bleeding & leprosy. Plant is antipyretic and expectorant; used in bone setting, pain in joints & coughs. Leaf infusion is anticancer & purgative. |
| *Crotalaria burhia* | Fodder for camels. | Branches are very strong & used for cordage, coarse gunny cloth & canvas. | Branches & leaves are used in preparation of indigenous medicines that have cooling effect in burns & swellings; useful in gout & hydrophobia. |
| *Indigofera cordifolia* | Very good fodder for cattle & goats. Green manure. Edible seeds | Blue color dye (indigo) obtained from leaves. Seed mixed with others such as millet used for making bread. Poisonous when eaten unmixed. | Whole plant as tonic A remedy for hepatitis and nervousness, purgative |
| *Indigofera linnaei* | Fodder and manure. Leaves are edible. | Piscicide/molluscicide. Seeds contain saponins. | Antifertility, digestive disorders, intestinal infections |
| *Indigofera sessiliflora* | Fodder | None known | **None known** |
| *Mimosa hamata* | Fodder, forage & manure. Ornamental | None known | Urinary complaints, glandular swellings, blood-purifier, Diarrhoea, inflammation & lung diseases. Five grams of seeds pounded and boiled in buffalo milk is taken as tonic against weakness, but must not be taken in excess. |
| *Prosopis cineraria* | Leaves as green manure and fodder. Pods as cattle feed. | Bark is used for tanning. Wood is hard but of inferior | Useful in gynaecological problems, cough, dysentery, giddiness, haemorrhage, |

**Table 2** Contd.

| *Species* | *General Uses* | *Industrial Uses* | *Medicinal Uses* |
|---|---|---|---|
| | Hedge plant. Gum, leaf, pod & seed are edible. Bark is sweet and edible; eaten by poor during famine. | quality & hence used for rough wood work and fuel in villages. | leucoderma, diarrhoea, piles, skin diseases, rheumatism & dyspnoea. Removal of skin hairs. Root is antidysenteric. |
| *Rhynchosia aurea* | Green manure. | Not known | Not known |
| *Rhynchosia minima* | Fodder & green manure | Fungicide, pesticide, piscicide, molluscicide & viricide. | Plant is analgesic, apoptotic, antioxidant, antiseptic, antiviral, cancer preventive, diuretic, bronchodilator & nephrotoxic. Leaves used in asthma & piles. Also used as abortifacient. |
| *Tephrosia purpurea* | Fodder, forage & manure.Soil binder; & cover crop. Roots & seeds are edible. | Dye of fast shades of blue color is obtained from leaves. Whole plant extracts used as insecticides and piscicides. Leaves, roots & seeds contain saponins. | Antidiabetic, anti-inflammatory, antioxidant, antidote, blood purifier & cancer preventive. Plant has antiviral, antifungal, antibacterial activity. Useful in digestive disorders, fever, impotency, liver complaints, urinary diseases, dental problems, etc. |
| *Tephrosia villosa* | Green manure. | Piscicide/molluscicide. Leaves & seeds contain saponins. | Shoot extracts used for colds and coughs, digestive disorders & lung diseases. Roots used in hypoglycemia & also useful in scrotal pain. Leaf juice is given in diabetes & dropsy. |
| *Tephrosia uniflora* | | Insecticidal & piscicidal. Leaves & seeds contain saponins. | Leaves are boiled in water & eaten; shoot extracts considered good remedy for syphilis, bleeding piles, boils, bronchitis & pimples. Roots are smoked for relief from asthma. Seed oil is specific against eczema, itching & scabies. |

**Table 3** Nodulated native and exotic legumes in arid and semi arid regions of Western Rajasthan. Taxonomy as in Bhandari, 1990 and Kumar & Sane (2003).

| *Name of Plant (Native Species)* | *Sub-Family* | *Habit* | *Nodulation* |
|---|---|---|---|
| *Cassia fistula* L. | Caesalpinioideae | Shrub | - |
| *Chamaecrista pumila* (Lam.) K. Larsen | Caesalpinioideae | Herb | + |
| *Senna auriculata* (L.) Roxb. | Caesalpinioideae | Shrub | - |
| *Senna italica* Mill. | Caesalpinioideae | Under shrub | - |
| *Acacia jacquemontii* Benth. * | Mimosoideae | Shrub | + |
| *Acacia leucophloea*(Roxb.) Willd | Mimosoideae | Tree | + |
| *Acacia nilotica* (Linn.) Del ubs.. *Indica* (Benth.) *Acacia nilotica* (Linn.) Del ubs.. | Mimosoideae | Tree | + |
| Cupressiformis (J.L. Stewart) | Mimosoideae | Tree | + |
| *Acacia senegal*(Linn.) | Mimosoideae | Tree | + |
| *Albizia lebbeck* (Linn.) Willd. | Mimosoideae | Tree | + |
| *Dichrostachys cinerea* (Linn.) Wt. & Arn. | Mimosoideae | Shrub | + |
| *Leucaena leucocephala (Lamk.)* | Mimosoideae | Shrub | + |
| *Mimosa hamata* Willd. | Mimosoideae | Shrub | + |
| *Mimosa himalayana* Gamble | Mimosoideae | Shrub | + |
| *Prosopis cineraria* (Linn.) Druce. | Mimosoideae | Tree | + |
| *Alysicarpus monilifer* (Linn.) DC. | Papilionoideae | Herb | + |
| *Alysicarpus vaginalis* (Linn.) DC. | Papilionoideae | Herb | + |
| *Crotalaria burhia* Buch.-Ham. | Papilionoideae | Under shrub | + |
| *Crotalaria medicaginea* Lamk var. *luxurians* (Benth.) | Papilionoideae | Herb | + |
| *Cyomopsis tetragonoloba* (Linn.) Taub. | Papilionoideae | Herb | + |
| *Dalbergia sissoo* Roxb. | Papilionoideae | Tree | + |
| *Indigofera cordifolia* Heyne ex Roth. | Papilionoideae | Herb | + |
| *Indigofera hochstetteri* Backer | Papilionoideae | Herb | + |
| *Indigofera linifolia* (Linn.f.) Retz. | Papilionoideae | Herb | + |

**Table 3** Contd.

| *Name of Plant (Native species)* | *Sub-Family* | *Habit* | *Nodulation* |
|---|---|---|---|
| *Indigofera linnaei* Ali | Papilionoideae | Herb | + |
| *Indigofera oblongifolia* Forsk. | Papilionoideae | Shrub | + |
| *Indigofera sessiliflora* DC. | Papilionoideae | Herb A | + |
| *Indigofera tinctoria* Linn. | Papilionoideae | Under-shrub | + |
| *Melilotus indica* (Linn.) All. | Papilionoideae | Herb | + |
| *Rhynchosia minima* (Linn.) DC. var. *laxiflora* (Camb.) | Papilionoideae | Climber | + |
| *Rhynchosia aurea* (Will.) DC. * | Papilionoideae | Climber | + |
| *Taverniera cuneifolia* (Roth) Arn. | Papilionoideae | Under shrub | + |
| *Trigonella corniculata* (Linn.) Syst. | Papilionoideae | Herb | + |
| *Tephrosia falciformis* * Ramaswami<br>*Tephrosia purpurea* (L.) | Papilionoideae | Under-shrub | + |
| Pers. sub sp. *purpurea* (L.) Pers<br>*Tephrosia purpurea* (L.) | Papilionoideae | Herb | + |
| Pers. sub sp *leptostachya* DC. * | Papilionoideae | Herb | + |
| *Tephrosia strigosa* (Dalz.) Santapaw & Maheshwari. | Papilionoideae | Herb A | + |
| *Tephrosia subtriflora* Hochst. ex Baker.<br>*Tephrosia uniflora* Pers. | Papilionoideae | Herb A | + |
| Subsp. *petrosa* (Blatt. & Hallb.) | Papilionoideae | Herb P | + |
| *Tephrosia villosa* (Linn.) Pres. | Papilionoideae | Herb | + |
| *Tephrosia wallichii* Graham | Papilionoideae | Herb | + |
| *Vigna trilobata* | Papilionoideae | Herb | + |
| **(Exotic species)** | | | |
| *Acacia radiana* (Forssk) Hayne | Mimosoideae | Tree | + |
| *Mimosa pudica* L. | Mimosoideae | Tree | + |
| *Prosopis juliflora* (Sw.) DC.<br>*Colophospermum mopane* (Benth.) Leonard** | Papilionoideae | Tree | - |

(+) where nodulation was found, (-) where no nodules were found. (*) new record of nodulation
** exotic species

*Dicorymia,* for which there is some evidence of nodulation (Sprent 2001). Faria et al. (1987) described for the first time persistent infection thread in the genus *Chamaecrista* sp. In advanced legumes, bacteria are released from the infection thread before $N_2$ fixation can begin, whereas in *Chamaecrista* sp. it appears that bacteria are retained within the thread during the $N_2$ fixation process (Naisbitt et al 1992). In other three tribes of Caesalpiniodeae, the genus either shows negative nodulation report or nodulation and isolation of rhizobia found unsuccessful and hence are considered erroneous. There is negative nodulation report in *Colophospermum* (J. Leonard), a monotypic shrub/tree of tropical Africa with two negative reports on nodulation. *Colophospermum mopane* belongs to Detarieae tribe of Caesalpiniodeae, recently found to possess functionally diazotrophic diversity (Burbano 2008). In arid regions of Indian Thar Desert, around Jodhpur, one species of *Chamaecrista* (*C. pumila* Lam. erlier *Cassia pumila* Lam.) is widely distributed. It is an annual herb and drought tolerant.

### The Following Genera and Species from Caesalpiniaceae are Reported in Western Rajasthan (Bhandari, 1990)

The main genera are *Bauhinia, Cassia, Delonix, Parkinsonia,* and *Tamarindus*

The species are:

*Bauhinia racemosa* (Lamk) (introduced)
*Cassia auriculata* Linn.
*C. italica* (Mill.)
*C. obtusifolia* Linn.
*C. occidentalis* Linn.
*C. pumila* (Lam) [Now *Chamaecrista pumila* (Lam) K. Larsen]
*C. sophera* Linn.
Delonix elata
*Parkinsonia aculeata* Linn. (introduced)
*Tamarindus indica* Linn. (introduced)

Nodulation recorede in *Chamaecrista pumila* growing around Jodhpur. Nodulated *C. pumila* was also recorded in NEHU-Shilong Campus

### Mimosoideae

The Mimosoideae has approximately 78 genera and 3000 species, but not all genera within the tribes can nodulate. There are well-confirmed good nodulation records from tribe Acacieae, Ingeae and Mimoseae. The monotypic tribe Mimozygantheae has recently been reported to nodulate (Faria & Lima 1998). *Acacia* is the second genus from leguminosae family having more than 1400 species (Lewis et al. 2005). More than 90% species surveyed possess nodules, mostly indeterminate elongated less branched. Faria et al. (1999) made a broad study of cross inoculation between strains isolated from trees in the Ingeae (Species of *Albizia, Calliandra, Enterolobium, Inga, Pithecellobium*) and other legumes. Faria et al. studied 16 isolates from arboreal species of the mimoseae and found that these nodulated tree species are from the

Acacieae, Ingeae and Phaseoleae like other members of the mimoseae. *Dichrostachys* genus contains 14 species, among which six species are known to nodulate, including *D. cinerea* which is also found in India. *Mimosa* L. is a large genus of about 480 species, ranging from trees through shrubs to herbs and occurring in tropical and warm areas. More than fifty species of *Mimosa* have been reported as nodulated by range of rhizobia (Allen & Allen 1981; Sprent 2001). Six species of *Mimosa* found in North and South India two have been reported native to arid regions of Rajasthan is *M. hamata* and *M. himalayana*. However *M.himalayana* was not seen in arid regions of Rajasthan by the author (*M. himalayana* is found only in Eastern part of Rajasthan and way to Madhya Pradesh from Rajasthan). *Prosopis* L. genera consist of forty four spp. of shrubs or trees from various warm regions of the world, but specially Central America (Sprent 2001). *Prosopis cineraria* is native to arid regions of India and Pakistan, while *P. juliflora* is exotic spreading in almost each state of India and drought and salinity tolerant. The other native Acacia species found in arid and semi arid regions are *Acacia senegal, Acacia nilotica* found throughout arid and semi arid regions of India, whereas *A. jacquemontii*, locally called as "Banwali or Boo banwali" is native to arid and semi arid regions of Rajasthan, Gujarat, Saurashtra and Delhi. *Acacia radiana (earlier A.toltilis), Acacia leucophloea and Leucaena leucocephala* are some exotic acacias growing in arid and semi arid regions of W Rajasthan.

### The Following Genera and Species from Mimosoideae are Reported in Western Rajasthan (Bhandari, 1990)

The main genera are *Leucaena, Albizia, Mimosa, Pithecellobium, Dichrostachys, Prosopis* and *Acacia.*

The main species of genus *Acacia* found in Western Rajasthan are *A. pennata, A. senegal, A. nilotica* (*A. nilotica* subsp. *cupressiformis* and *Acacia nilotica* subsp. *indica*).
*Albizia lebbeck* (Pan tropical Africa)

*Dichrostachys cinerea* (Tropical and subtropical Africa, Arabia, North Australia)
*Leucaena latisiliqua* (Tropical American)
*Mimosa hamata*
*Pithecellobium dulce* (native of Mexico)
*Prosopis cineraria* (Native to India and Pakistan)
*P. chilensis* (introduced)
*P. glandulosa* (not indigenous of India)
Exotic species of Acacias growing in W Rajasthan are—
Acacia radiana (earlier A. toltilis)
Acacia leucophloea
*Leucaena leucocephala.*

### Papilionoideae

Papilionoideae is the largest of the legume sub families and is the one containing the highest proportion of nodulated species. All known types of nodules are found in this sub family. The nodulated genera of various tribes found in arid and semi arid regions are *Abrus* from *Abreae, Cicer* from *Cicereae, Crotalaria* from crotalarieae, *Aeschynomene* and *Dalbergia* from Dalbergieae, *Alysicarpus* from Desmodieae, *Alhagi* from Galegeae, *Taverniera* from Heydisareae, *Indigofera* from Indigofereae, *Clitoria, Dolichos, Mucuna, Puraria,* and *Rhynchosia* from Phaseoleae, *Psoralea* from Psoraleeae, *Sesbania* from Robinieae, *Medicago, Melilotus* and *Trigonella* from Trifolieae, *Tephrosia* from Millettieae (Sprent 2001, ILDIS).

### Following are the Major Genera and Species from Papilionoideae Reported in Western Rajasthan (Bhandari, 1990)

The major genera are *Abrus, Aeschynomene, Alhagi, Alysicarpus, Atylosia, Butea, Clitoria, Crotalaria, Cymopsis, Dalbergia, Derris, Goniogyna, Indigofera, Lathyrus, Medicago, Melilotus, Microtyloma, Psoralia, Rhynchosia, Sesbania, Taverniera, Tephrosia, Trigonella, Vicia, Vigna* and *Zornia.*

The species are:

*Abrus precatorius* is common in rocky habitats, climbing on *Euphorbia caducifolia.*

*Aeschynomene indica*

*Alhagi maurorum* (*Alhagi pseudoalhagi*)

*Alysicarpus heterophyllous, A. monilifer, A. longifolius, A. glumaceus, A. styracifolius, A. procumbens* and *A. vaginalis*

*Atylosia scarabaeoides* (Linn.)

*Butea monosperma* (Lamk)

*Clitoria ternatea* (Linn.)

*Crotalaria burhia* (Buch-am.), growing in extremely dry area is sand dunes stabilizer. Branches of the plant are very strong and used in the preparation of ropes and cordage (strong and not prone to decay)

*Crotalaria medicaginea* (Lamk.)

*Crotalaria retusa* (Linn.)

*Cyamopsis tetragonoloba* (Linn.) The young pods are consumed as vegetable, whole plant is used as green manure. Boiled grains are given to cattle, source of guar gum (India is number one guar gum exporter). In Jodhpur there are guar gum industries.

*Dalbergia sissoo* (Roxb.)

*Derris indica* (Lamk)

*Goniogyna hirta* (Willd.)

*Indigofera hochstetteri* (Baker in Fl. Trop. Afr. 2:101. 1871), *I. astragalina* (DC Prodr.), *I. sessiliflora* (DC. Prodr.), *I. linifolia* (Linn. F.), *I. cordifolia* (Heyne), *I. linnaei* (Ali in Bot.), *I. argentea* (Burm.), *I. tinctoria* (Linn), *I. caerulea* (Roxb.), *I. oblongifolia* (Forsk.).

*Lathyrus aphaca* (Linn.), *L. sativus* (Linn.)

*Macrotyloma uniflorum* (Lamk.)+

*Medicago laciniata* (Linn.)

*Medicago sativa* (Linn.)

*Melilotus alba* (Medik.), *M. indica* (Linn.)

*Psoralea odorata* (Blatt. & Halb.) Found in North West Rajasthan. A sweet smelling perennial herb is very similar *Psoralea plicata*. Lack nodulation report (ILDIS).

*Rhynchosia aurea* (Willd.)
*Rhynchosia minima* (Linn.) var. *laxiflora* (Camb.),
*Rhynchosia schimperi* (Hochst.) lack nodulation report (ILDIS),
*Rhynchosia pulverulenta* (Stocks in Hook.).
*Sesbania bispinosa* (Jacq.)
*Sesbania sesban* (Roth.) var. *picta*
*Taverniera cuneifolia* (Roth.)
*Tehrosia falciformis* (Ramaswami)
*T. laptostachya* (DC. Prodr.)
*T. purpurea* (Linn.)
*T. strigosa* (Daiz.)
*T. subtriflora* (Hochst.)
*T. uniflora* Pers. subsp. *petrosa* (Blatt & Hallb)
*T. villosa* (Linn.)
*T. wallichii* (Graham in Wall.)
*Trigonella corniculata* (Linn.)
*Trigonella foenum-graecum* (Linn)
*T. hamosa* Linn. subsp. *uncata* (Boiss. & Noe)
*T. monantha* C.A.Meyer. subsp. *incsia* (Benth.)
*T. occulta* Delie Fl.
*Vicia sativa* (Linn.) var. *angustifolia* (Roth.)
*Vigna aconitifolia* (Jacq.)
*Vigna mungo* (Linn.)
*Vigna radiata* (Linn.)
*Vigna trilobata* (Linn.) wild moth, medicinal value
*Zornia gibbosa* Span.

## FUTURE PERSPECTIVE

Native legumes growing in arid and semi arid regions of Indian Thar Desert are important source of food, feed, medicines and shelter for the local inhabitants. These legumes should be emphasized at National level for conservation and improvement in existing germplasm. There are only few legume trees of arid and semi arid regions of the Thar desert such as *Prosopis cineraria, Acacia Senegal* and *Acacia nilotica*. These plants should be planted in the area of shifting sand dunes as well as on degraded lands to protect soil erosion and to improve soil fertility. *Rhynchosia* species (*R. minima, R. aurea* and *R. schimperi*) belong to important tribe Phaseolae to which important grain legumes belong. Species of *Rhynchosia* can be improved through breeding. Some inhibitors and anti-nutritional factors are present in the species of *Tephrosia*, which can be removed using modern techniques to make it more palatable for cattle feed. *Tephrosia falciformis* is a medicinal plant but is in the list of threatened plants due to loss of habitat and expansion of urban area. The list of such plants may increase in the future if adequate measures of conservation are not taken.

*Crotalaria burhia* is a perennial drought tolerant species found in sandy areas and also important as soil binder as well as food for cattle. The study on microsymbiont and root/stem bacterial endophytes in these genera will provide useful information on its adaptation to arid and poor soil texture. In view of the climate change and global food security, research and development activity on wild/native legumes should also be prioritized at national level to protect and conserve these important bio-resources of arid regions.

## REFERENCES

Abdel-Gahffar, A.S. (1989). Aspects of microbial activities and nitrogen fixation in Egyptian desert soil. *Arid. Soil. Res. Rehabi.* 3: 281-294.

Allen, O.N., & Allen, E.K. (1981). The Leguminosae: a source book of characteristics, uses and nodulation. University of Wisconsin Press, Madison, Wisconsin, and Macmillan Publisher Ltd., London.

Benata, H., Mohammed, O., Noureddine, B., Abdelbasset, B., Abdelmoumen, H., Muresu, R., Squartini, A., & El Idrissi, M.M. (2008). Diversity of bacteria that nodulates *Prosopis juliflora* in the eastern area of Morocco. *Syst. Appl. Microbiol.* 31(5): 378-386.

Benhizia, Y., Benhizia, H., Benguedovar, A., Muresu, R., Giacomini, A., & Squartini, A. (2004). Gamma Proteobacteria can nodulate legumes of the genus *Hedysarum*. *Syst. App. Microbiol.* 27: 462-468.

Bhandari, M.M. (1990). Flora of the Indian Desert. MPS Repros, 39 BGKT Extn. New Pali Road, Jodhpur.

Burbano, C.S., Hurek, T., & Reinhold-Hurek, B. (2008). Study of the microbial community structure and functional diazotrophic diversity in *Colophospermum mopane*. *8th European nitrogen fixation conference* Ghent, Belgium. pp. 152.

Dhir, R.P. (1977). Western Rajasthan soils, their characteristics and properties. *In desertification and its control,* ICAR. 102-115.

Endalkachew, W.M., Trygve, B., Peters, N.K., & Frostegard, A. (2004). Nodualtion status of native woody legumes and phenotypic characteristics of associated rhizobia in soils of Southern Ethiopia. *Biol. Fertil. Soils.* 40: 55-66.

Fagg, C.W., & Stewart, J.L. (1994). The value of *Acacia* and *Prosopis* in arid and semi arid environments. *J. Arid. Environ.* 27: 3-25.

Faria, S.M.de., Lima, H.C.de., Franco, A.A., Mucci, E.S.F., & Sprent, J.I. (1987). Nodualtion of legume tree from South-East Brazil. *Plant and Soil.* 99: 347-356.

Faria, S.M.de., & Lima, H.C.de. (1998). Additional studies of the nodulation status of legume species in Brazil. *Plant and Soil.* 200: 185-192.

Faria, S.M.de., H.C.de., Olivere, F.L. (1999). Nodulacao em epecie florestais especificidade hospedeira e implicacoes na sistematica de leguminosae. *In: Soil Fertility, Soil biology and Plant Nutrition Interrelationship,* (eds), by Siqueira, J.O. et al. Sociedade Brasileira de Ciencia do Solo, Unioversidade Federal da Lauras Brazil. Vicosa, Brazil. pp. 667-686.

Gehlot, H.S., Panwar, D., Tak, N., Tak, A., Sankhla, I.S., Poonar, N., Parihar, R., Shekhawat, N.S., Kumar, M., Tiwari, R., Ardley, J., **James, E.K.,** and Sprent, J.I. (2012) Nodulation of legumes from the Thar desert of India and molecular characterization of their rhizobia. *Plant and Soil* **357:** 227–243.

Graham, P.H., & Vance, C.P. (2003). Legumes: Importance and constraints to greater utilization. *Plant Physiol.* 131: 872-877.

Jha, P.K. (1988). Wild legume Rhizobium symbiosis: ecology and technology. Ph.D. thesis, University of Delhi, India.

Jha, P.K., Nair, S., Gopinathan, M.C., & Babu, C.R. (1995). Suitability of rhizobia inoculated wild legumes *Argyrolobium flaccidum, Astragalus gravelens, Indigofera gangatica,* and *Lespedeza stenocarpaq* in proving a vegetational cover in an unreclaimed limestone quarry. *Plant and Soil.* 177: 139-149.

Lewis, G.P., Schrire, B., Mackinder, B. & Lock, M. (2005). Legumes of the World. Royal Botanical Garden, Kew.

Naisbitt, T., James, E.K., & Sprent, J.I. (1992). The evolutionary significance of the genus *Chamaecrista* as determined by nodule structure. *New phytologist.* 122: 487-492.

Panwar, D., Gehlot, H.S., & Shekhawat, N.S. (2008). New nodulation report from Indian Thar desert. *The Bean Bag,* Royal Botanical Garden Kew.

Panwar, D. (2009). On genetic resources of arid-regions: Nodulating legumes and associated rhizobia. PhD thesis J.N. Vyas University, Jodphur.

Pokhriyal, T.C., Chaukiyal, S.P., & Naithani, H.B. (1993). *Ind For.* 119: 310.

Pokhriyal, T.C., Singh, H., Vijayrawat, K.C., Pokhriyal, A.K., & Kumar, R. (2003). Introduction of nitrogen fixing plants: A sustainable approach for the plantation forestry programs. *2nd International congress of plant physiology,* New Delhi.

Polhill, R.M. et al. (1981).Evolution and systematics of the Leguminosae. *In Advances in legume systematics* (R.M. Polhill & P.H. Raven, eds.). Royal Botanic Gardens, Kew. pp.1-26.

Polhill, R.M., & Raven P.H. (1981). Biogeography of the Leguminosae. *In*: *Advances in Legume Systematics Part 1,* (eds.), by Polhill, R.M., and Raven, P.H., Royal Botanical Garden, Kew. pp. 27-34.

Polhill, R.M. (1994). Classification of the leguminosae. *In: Phytochemical Dictionary 1: xxxxv-lvii,* (eds.), by Bisby, F.A., Buckingham, J., and Harborne, J.B., Chapman and Hall, London.

Ramakrishna, Y.S., Agnihotri, C.L., & Rao, R.S. (1992). Climatic resources of Thar Desert and adjoining arid areas of India. *In perspective of Thar and Karakum* (eds.), Amalkar, Abichandani, R.K., Anantha Ram, and Joshi, D.C., DST, Govt of India, New Delhi. pp. 10-20.

Ray Choudhury, S.P., Agarwal, R.K., Gupta, N.K., & Thomas, P.K. (1963). *Soils of India.* ICAR, New Delhi. pp. 495.

Rewari, R.B. (1979). All India coordinated pulse improvement project. IARI New Delhi, India.

Rewari, R.B., & Tilak, K.V.B.R. (1988). *Microbiology of pulses. In pulse crops* (eds.) Baldev, B., Ramanujam, S., and Jain, H.K., Oxford and IBH, New Delhi. 373-411.

Roy, B.B., Dhir, R.P., & Kolarkar, A.S. (1978). Soil of Rajasthan desert and their characteristics. *Proc. Winter School on Development of Rajasthan Desert,* CAZRI, Jodhpur.

Sandison, M.S., & Harris, P.J.C. (1991). *Prosopis cineraria*: a multipurpose tree for arid areas. *Henry Doubleday Research Association* (HDRA), Coventry, U.K.

Sanjappa, M. (1992). Legumes of India. Bisan Singh, and Mahendra Pal Singh, Dehradun.

Schimel, D.S. (2010). Dry lands in the earth system. *Science.* 327: 418-419.

Schrire, B.D., Lewis, G.P., & Lavin, M. (2005). *Biogeography of the Leguminosae.* In: G Lewis,

Shrire, B Mackinder, M Lock. Legumes of the World. Royal Botanic Gardens, Kew pp 21-54.

Shetty, B.V., & Singh, V. (1987). Flora of Rajasthan, Volume I. Published by Botanical Survey of India, Calcutta.

Simon, M.F. et al. (2011). The evolutionary history of Mimosa (Leguminosae): Toward a phylogeny of the sensitive plants. *Am. J. Bot.* 98:1201-1221.

Sprent, J.I. (2001). Nodulation in legumes. Royal Botanical Garden, Kew.

Sprent, J.I., & Gehlot, H.S. (2010). Nodulated legumes in arid and semi-arid environments: are they important? *Plant Ecology and Diversity*. 3: 211-219.

Subba Rao, N.S., Sen, A.N., & Gaur, Y.D. (1972). *Final Rep., PL 480 SchemeDivision of Microbiology*, IARI, New Delhi, India.

Subramaniam, B., & Babu, C.R. (1994). New nodulating legumes of potential agriculture and forestry value from subtropical Himalayan ecosystem. *Biological Agriculture and Horticulture*. 10: 297-302.

Sundra Rao, W.V.B., Sen, A.N., & Gaur, Y.D. (1969). Survey and isolation of root nodule bacteria in Indian soils. *Final Rep. Pl-480 Scheme, Division of Microbiology*. IARI, New Delhi, India.

Sylla, S.N., Samba, R.T., Neyra, M., Ndoye, I., Giraud, E., Willems, A., de Lajudei, P., & Dreyfus, B. (2002). Phenotypic and genotypic diversity of rhizobia nodulating *Pterocarpus erinaceus* and *P. lucens* in Senegal. *Syst. Appl. Microbiol.* 25 (4): 572-583.

Tarafdar, J.C., Yadav, R.S., Bareja, M., & Singh, G. (2006). Phosphorus fractionation under crops, trees, and grasses. *J. Indian Society Soil Sci.* 54: 38-44.

Tilak, K.V.B.R. (1993). *Bacterial fertilizers*. IARI, New Delhi, India. 4-33.

Yadav, R.S. et al. (2011). Soil biological properties under different tree based traditional agroforestry systems in a semi-arid region of Rajasthan, India. *Agroforestry systems*. 81(3): 195-202.

Zahran, H.H. (1991). Variation in growth pattern of three Egyptian strains of *Rhizobium leguminosarum* grown under sodium chloride and ammonium nitrate treatments. *Bull. Fac. Sci. Assiut. Univ.* 20: 161-169.

Zahran, H.H. (1999). *Rhizobium*-Legume symbiosis and nitrogen fixation under severe condition and in an arid climate. *Microbiology and Molecular Biology Reviews*. 63 (4): 968-989.

Zahran, H.H., (2001). Rhizobia from wild legumes: diversity, taxonomy, ecology, nitrogen fixation, and biotechnology. *Journal of Biotechnology*. 91: 143-153.

Zakhia, F., Jeder, H., Willems, A., Gilis, M., Dreyfus, B., & Lajudie, P. de. (2006). Diverse bacteria associated with root nodules of spontaneous legumes in Tunisia and first report for *nifH* like gene within the genera *Microbacterium* and *Starkeya*. *Microbial ecology*. 51: 375-393.

# 17

# Current Trends in Diagnosis, Prevention and Treatment of Cancer

**ANITHA KARTHA**[1], **R.K. KALE**[1,2], **AND RANA P. SINGH**[1,2,*]

[1]*School of Life Sciences, Central University of Gujarat, Gandhinagar, India*
[2]*Cancer Biology Laboratory, School of Life Sciences, Jawaharlal Nehru University, New Delhi, India*
**Correspondence Author: E-mail: ranaps@hotmail.com; ranasingh@cug.aca.in*

**ABSTRACT**

Cancer is a complex group of diseases characterized by uncontrolled cell growth and spread of cells to other parts of the body. Successful treatment of disease is dependent on the early detection, staging to determine the extent of the disease at the time of diagnosis, prompt treatment and adoption of appropriate prevention strategies. Following the success of screening for specific tumor markers, such as á-fetoprotein (AFP) and prostate-specific antigen (PSA), several new cancer biomarkers are being included in routine medical examinations. Moreover, screening of specific molecular targets will enable the development of personalized medicine. Current treatment options available are surgery, chemotherapy and radiation but have acute and long-term complications associated. Hence, a transition from emphasis on treatment of cancer to that of prevention is being implemented at present. Combining both treatment and prevention will be an effective strategy for controlling cancer. There also have been many exciting breakthroughs in the development of therapies that offer new ways of treating or preventing cancer. The major advances have been made in the treatment strategies including the targeted approaches, viz., immunology and vaccine development, antiproliferative mechanisms, photodynamic therapy, gene therapy, cell cycle and apoptosis etc. To date, the FDA has approved more than a dozen molecularly targeted agents against cancer, including trastuzumab, different aromatase inhibitors for breast cancer. A new strategy, cancer chemoprevention has emerged which mainly involves the use of phytochemicals which are part of our regular diets or non-dietary plants with less toxicity having equal or better effectiveness than chemically synthesized drugs. Targeting cancer stem cells with phytochemicals is yet another promising approach against cancer. Herein, we have discussed the advances made in the field of oncology which are of considerable importance.

**Keywords:** Cancer, Phytochemicals, Oncology, Prevention.

## INTRODUCTION

The development of cancer represents a complex series of events leading to the progressive evolution of normal cells into malignant cells which results in change in cell phenotype as well as genotype. Cancer is a class of diseases in which abnormal cells proliferate uncontrollably mostly to form solid tumor, invade other tissues and metastasize to various locations of the body through lymphatic or blood circulatory system. Tumorigenesis, i.e., the formation of tumors, is the result of interplay among genetic, epigenetic, biochemical and environmental factors. Some of the common environmental factors contributing to cancer include diet, lifestyle, infections with specific viruses and bacteria, radiation exposure and environmental carcinogenic pollutants (Anand et al., 2008), which cause abnormalities in the genetic or epigenetic materials of the cells (Kinzlerand Vogelstein, 2002). The regulation of cell growth and proliferation is maintained by various classes of genes, including proto-oncogenes and tumor suppressor genes, and certain mutations in these two classes of genes have been implicated in the onset of cancer.

The increasing insight into the causes and mechanisms of cancer, in previous two decades, has paved the way for better diagnosis and treatment of various types of cancers. Primarily, advances in diagnostic procedures have improved the ability to detect cancers at the early stages and also to diagnose them more accurately than was the case a few years back. In fact, some tests can identify precancerous tumors even before symptoms appear and hence can be used to arrest their growth and progression to malignancy. Two of the most commonly used screening tests in women are the Papanicolaou test to detect cervical cancer and mammography to detect breast cancer. Both the tests have not only aided in early detection but have also resulted in significant reduction in cancer-related mortality in the screened female population (Al -Hajj et al., 2003). Other cancers where early detection tests currently exist are colon, rectum, prostate, oral cavity, and skin cancers. Furthermore, conventional cancer treatments have improved remarkably and can cure many cases of cancer, and new therapeutic strategies offer great promise of being even more effective in combating the disease. The symptoms and complications of cancer vary greatly among different types of cancers.Diagnostic testing varies depending on the suspected type and stage of the cancer. Cancer is curable if it is diagnosed in the earliest stages including benign stages of the disease. Furthermore, the prognosis for cancer patients varies depending on many factors including the site and stage of advancement of the cancer, age of the patient, general health status and response to the intervention strategy. In the following sections, we have discussed the advances in diagnosis and prevention of cancer.

## DIAGNOSIS OF CANCER

By now, many reliable techniques have been developed to diagnose tumors. To detect tumor or cancer, several medical tests like blood tests, X-rays, computed tomography (CT) scans and endoscopy are commonly used. In some cases, it also includes a combination of imaging tests, such as magnetic resonance imaging (MRI), positron emission tomography (PET) scan and/or CT scan, which uses computerized technology to create detailed images of the body or an area of interest such as the organ hosting a tumor. An angiogram illuminates blood vasculature system that feeds blood to a solid tumor, and helps in deciding the course of treatment strategy. Once a cancer is diagnosed, it is staged by the pathologist. It provides an estimate of how advanced the cancer has become, including the size and metastasis.The TNM system is the most widely used staging system for cancer which is accepted by the International Union Against Cancer (IUAC) and the American Joint Committee on Cancer (AJCC). It is based on the extent of tumor (T), the extent of spread to the lymph nodes (N), and the presence of metastasis (M). It is done on the basis of information provided by imaging procedures, laboratory tests, pathology and surgical reports. It has been very helpful for clinicians to decide the course of treatment for the cancer patients.

In oncology, a biomarker refers to a substance or process that is indicative of the presence of cancer in the body. Generally, any biologically derived entity or processes which lead to cancer diagnosis (in prognosis, screening and risk assessment), at the stage of diagnosis or post-diagnosis, are potential biomarkers (Mishra and Verma, 2010). The age old example of cancer biomarkers are urinary Bence Jones protein and tumor specific antigen carcinoembryonic antigen (CEA) in colon carcinomas (Gold and Freedman, 1965). Prostate specific antigen (PSA) level in the circulation is a biomarker for prostate cancer.

## PREVENTION OF CANCER

Cancer prevention is defined as active measures adopted to decrease the incidence of cancer. It is classified into three levels, namely, primary, secondary and tertiary prevention. Primary prevention refers to the elimination of risk factors so that cancers do not develop. Preventive strategies like avoidance of exposure to UV solar radiation, cigarette smoking, junk food, etc are proven effective in controlling the development of cancer. Apart from these, some medical procedures like immunization against hepatitis B to prevent liver cancer, and against papillomavirus to prevent cervical cancer are also adopted. Secondary prevention is the measure adopted to prevent cancer morbidity and mortality. It is associated with screening and early detection. Screening targets at high risk individuals include those who have the strong hereditary background but are asymptomatic. This is necessary considering the fact that early detection and diagnosis leads to successful treatment and hence better survival rate. Secondary prevention measures include carcinogen exposure assessment, self-examination, utilization of diagnostic and detection technology like Pap-smear for early detection of cervical cancer, mammography for the early detection of breast cancer and in some cases, genetic testing. Tertiary prevention includes the measure adopted to limit the extend of death and diability and to provide rehabilitation to the cancer patients. It helps to assess the risk and monitor the recurrence of cancer in a secondary site through utilization of chemopreventive agents or application of suitable detection and diagnostic methods.

Cancer chemoprevention can be described as a strategy to reverse or suppress the process of carcinogenesis using chemical compounds (Sporn and Suh, 2002). The tremendous rise in the incidence of cancer together with the failure of conventional methods of treatment of advanced invasive disease emphasizes the need for new approaches of treatment. Extensive information is now available on the use of anti-estrogens, anti-inflammatory agents, antioxidants, and other diet derived agents as chemopreventive agents. Etiology of several types of cancers including breast, endometrium, prostate, testis, ovary, thyroid, and others can be attributed to sex steroid hormones (Henderson and Feigelson, 2000). Hence anti-estrogens, aromatase inhibitors and 5-alpha reductase inhibitors are most suitable candidates for endocrine-related cancer. Three chemopreventive agents namely tamoxifen (Fisher et al., 1998), raloxifene (Cummings et al., 1999), and fenretinide (Veronesi et al., 1999) are proved for their clinical efficacy of breast cancer chemoprevention. Inflammation has long been suspected to be linked with cancer. So anti-inflammatory drugs including aspirin, piroxicam and indomethacin are likely to provide chemoprotective effect (Kelloff et al., 1994; Vajnio, et al., 1997). Over expression of cyclooxygenases (COX-1 and COX-2), which are the key enzymes involved in the synthesis of inflammatory prostaglandins is documented in the development of epithelial tumors (Presscot and Fitzpatrick, 2000). Non-steroidal anti-inflammatory drugs (NSAIDS) are known to inhibit these enzymes and play beneficial role in regulating carcinogenesis.

Epidemiological studies have shown protective effect of diets rich in fruits and vegetables against cancer (Willett, 1994; Temple, 2000; Willett, 2002). Phytochemicals are plant derived chemicals and can be classified as carotenoids, phenolics, alkaloids, nitrogen containing compounds, and organosulfur compounds (Figure 1). Studies till date have demonstrated that these chemicals have complementary and overlapping mechanisms of action including antioxidant activity, regulation of gene expression such as proto-oncogenes and tumor suppressor genes in cell proliferation, cell differentiation, induction of cell cycle arrest and apoptosis, and stimulation of the immune system and regulation of epigenetic mechanisms (Sun et al., 2002; Chu et al., 2002; Dragsted et al., 1993; Waladkhani and Clemens, 1998).

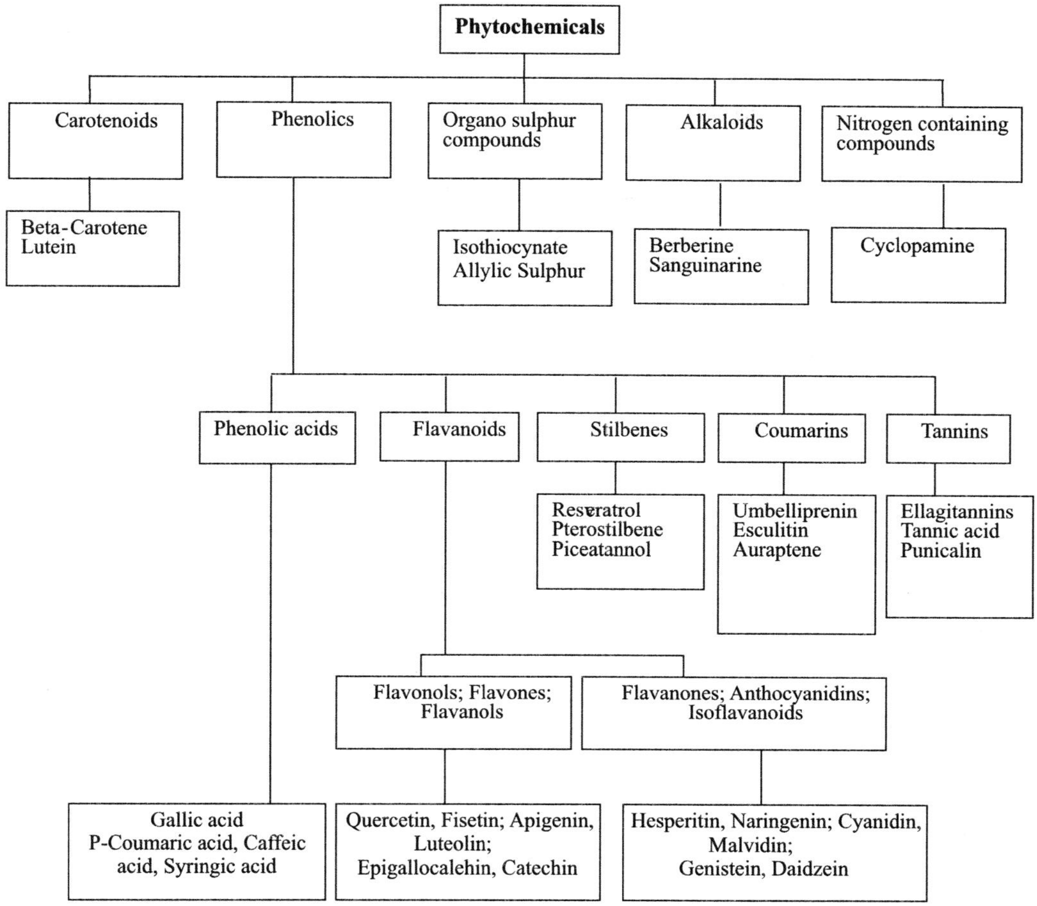


**Fig. 1.** Classification of dietary phytochemicals.

## CANCER MANAGEMENT

There are many management strategies as far as cancer is considered including surgery, chemotherapy, radiation therapy, immunotherapy including monoclonal antibody therapy and so on. The selection of a particular treatment depends upon the type of cancer, its location, the stage of the disease and also the patient's general health.

### Surgery of Cancer

All solid tumors or cancers can be cured if entirely removed by surgery, provided it has not been metastasized. If possible and appropriate, surgery is commonly performed before other forms of treatment. Surgery involves the removal of a wide surgical margin or a free margin. The width of the free margin depends on the type of the cancer and the method of removal (CCPDMA, Mohs surgery, POMA, etc.). Examples of surgical procedures for cancer include mastectomy for breast cancer, prostatectomy for prostate cancer, and lung cancer surgery for non-small cell lung cancer. The goal of the surgery can be either the removal of only the tumor, or the entire organ. Many other types of surgical methods for treating cancer and precancerous conditions exist, and investigators continue to research new ones. Some common types of cancer surgery are described below.

Cryosurgery uses very cold material, such as liquid nitrogen spray or a cold probe, to freeze and destroy cancer cells or cells that may become cancerous, such as irregular cells in cervical cancer. Electrosurgery involves killing cancer cells by applying high-frequency electrical currents. Laser surgery uses beams of high-intensity light to shrink cancer cells. It is accomplished by heat of the laser. In other cases, the laser is used to activate a previously administered chemical that cancer cells absorb. When stimulated by light, the chemical gets activated and in turn kills the cancer cells. Mohs' surgery involves careful removal of cancer layer by layer with a scalpel. It is useful for removing cancer from sensitive areas of the skin, such as near the eye and for assessing how deep a cancer has grown. All the abnormal cells are removed till the surrounding tissue shows no evidence of cancer.

In laparoscopic surgery a laparoscope is used to see inside the body without making large incisions. Instead, several small incisions are made and a tiny camera and surgical tools are inserted into the body. The advantage of smaller incisions includes faster recovery and reduced risk of complications. Laparoscopic surgery is used in cancer diagnosis, staging, treatment and symptom relief. Natural orifice surgery is currently being used as a way to operate organs in the abdomen without making incision in the skin. Instead, surgical tools are passed through a natural orifice, such as mouth, rectum or vagina. A small incision is made in the wall of the stomach and surgical tool is passed into the abdominal cavity in order to take a sample of organ affected, i.e., biopsy. Natural orifice surgery is in the early stages and is expected to reduce the risk of infection, pain and other complications of surgery. For many types of cancers, surgery is the first and foremost crucial option to provide long-term remission. Cancer surgery is continuously evolving with development of techniques and tools that aim at less invasive procedures.

## Radiotherapy of Cancer

Radiation therapy uses high-energy radiation to shrink tumors and kill cancer cells (Lawrence et al., 2008). Radiations used for the therapy includes X-rays, gamma-rays and charged particles. Radiation kills cancer cells by primarily damaging their DNA. Radiations can damage DNA directly or indirectly by generating free radicals within the cells through decomposition of water.Radiation therapy is of two types—external-beam radiation therapy, where the radiation may be delivered by a machine outside the body, and internal radiation therapy, where the source of radiation is kept inside the body of the patient. One of the most common types of external-beam radiation therapy is called 3-dimensional conformal radiation therapy (3D-CRT). 3D-CRT uses very sophisticated computer software and advanced treatment machines to deliver radiation to very precisely shaped target areas. Many other methods of external-beam radiation therapy are currently being tested and used in cancer treatment. These methods include intensity-modulated radiation therapy (IMRT), which increases the radiation dose to the precise areas and reduces radiation exposure to specific sensitive areas of surrounding normal tissue. Image-guided radiation therapy (IGRT), in which repeated imaging scans (CT, MRI or PET) are performed during the treatment. These imaging scans track the changes in a tumor's size and location achieved due to treatment, and thereby allow the position of the patient or the prescribed radiation dose to be adjusted accordingly.

Internal radiation therapy involves the treatment with radioactive material placed in the body near cancer cells, which is also known as brachytherapy. Brachytherapy causes less damage to normal tissues as they deliver higher doses of radiation to the cancers than external-beam radiation therapy (Patel and Arthur, 2006). There are two types of brachytherapy, namely permanent and temporary. Permanent brachytherapy involves the placement of source of radiation at the tumor site and sealing it there even after all of the radiation has been given off. This is not harmful for the patient as it is a low-dose-rate therapy. Temporary brachytherapy uses tubes (catheters) or other carriers to deliver the radiation sources, and both the carriers and the radiation sources are removed after the treatment. Temporary brachytherapy can be either low-dose-rate or high-dose-rate treatment.

Systemic radiation therapy refers to the use of radioactive isotopes/substances, such as radioactive iodine attached to an antibody which circulates in the blood to kill cancer cells. For the treatment of certain types of B-cell non-Hodgkin lymphoma (NHL), the drug ibritumomab tiuxetan (Zevalin®) has been approved by the Food and Drug Administration (FDA). It is targeted to a protein found on the surface of B lymphocytes as the antibody, part of this drug, recognizes it and binds to it.

Combined treatment modalities involving radiotherapy with chemotherapy is the widely adopted treatment regimen. However, a large number of tumors are poorly responsive or non-responsive to any kind of therapy. Increasing the doses of radiation or chemotherapeutic drugs is not a preferred method as it would not increase the response and might also exhibit resistance and increase systemic toxicity. So the ideal solution is to identify anticancer agents that trigger cell death preferentially in cancer cells as compared to the normal cells. Dietary phytochemicals are mostly devoid of toxicity when taken in limited doses, and are reported to possess both antioxidant properties which protect normal cells and pro-oxidant activity which contributes to their therapeutic actions. Several studies, for instance using polyphenols, like resveratrol, curcumin, have altered cell cycle progression and cytotoxic effects due to ionizing radiation (Zoberi et al., 2002; Chendil et al., 2004). Moreover, presence of phytochemical compound during irradiation amplifies their effects by mechanisms including toxic reactions of free radicals. Those compounds that are given after radiation show the effect by inhibiting repair of the radiation induced lethal and sub-lethal damage together with down-regulation of numerous pro-survival factors.

### Chemotherapy of Cancer

Chemotherapy refers to the use of drugs for the treatment of cancer. Chemotherapeutic agents act on cells, which grow and divide quickly, and arrest or slow down the progression of cancer. Most chemotherapeutic drugs work by impairing mitosis (cell division), effectively targeting fast-dividing cells to induce apoptosis. As these drugs cause damage to cells, they are termed cytotoxic. But the side effect of the therapy is that it will also harm healthy cells that divide quickly, like those that line the mouth, intestines and hair follicles. Chemotherapy can completely cure the cancer, or control it from metastasizing or reduce the cancer symptoms depending upon the type of cancer and the stage of diagnosis.

The chemotherapeutic drugs can be divided into alkylating agents, antimetabolites, anthracyclines, plant alkaloids, topoisomerase inhibitors, and other antitumor agents (Takimoto and Calvo, 2008). These drugs affect cell division or DNA synthesis. Alkylating agents alkylate many nucleophilic functional groups under the conditions in cells. Cisplatin, carboplatin, oxaliplatin are commonly used alkylating agents which impair cell function by forming covalent bonds with the amino, carboxyl, sulfhydryl, and phosphate groups in biologically important molecules (Takimoto and Calvo, 2008). Other agents are mechlorethamine, cyclophosphamide, chlorambucil, ifosfamide which work by chemically modifying DNA of the cell. Antimetabolites are compounds that inhibit the use of metabolite by the cell. The antimetabolites have toxic effects on cells, such as inhibition of cell growth and cell division, so these compounds are used for chemotherapy of cancer (Peters et al., 2000). The chemotherapeutic agents interfere with DNA synthesis and therefore cell division and in turn the growth of tumors. Anti-metabolites act like a purine (azathioprine, mercaptopurine) or a pyrimidine, which are the building blocks of DNA. They prevent the incorporation of purines into DNA during the S phase of the cell cycle, and stop the normal development and cell division (Takimoto and Calvo, 2008). Studies are in progress to identify or discover suitable chemotherapeutic agents with specific targeting ability and having lesser toxicity.

## ADVANCES IN CANCER BIOLOGY

The major advances made in cancer biology are the approaches in the treatment strategies adopted, namely immunology and vaccine development, antiproliferative approaches including targeting of angiogenesis and metastasis, photodynamic therapy, gene therapy, cell cycle and apoptosis. The relevant advancements in these approaches, helpful in cancer control and management, are discussed below.

### Cancer Immunotherapy

Immunotherapy involves the combination of antigen with a substance called an adjuvant. Advances in immunotherapy have resulted in development of particular adjuvant systems that can regulate and re-activate our immune system. It produces a heightened immune response as compared to the previous methodologies employed. The individual's own Natural Killer (NK) cells and Cytotoxic T-

Lymphocytes (CTL) are used in cell-based immunotherapy. This method of treatment has been in practice in Japan since 1990 based on the fact that these cells kill the cancer cells when they develop. This treatment is given together with other modes of treatment such as surgery, radiotherapy or chemotherapy, and is referred as Autologous Immune Enhancement Therapy (AIET) (Damodar et al., 2006; Sivaraman et al., 2008). The problem we face, however, is that immunoselection continually occurs during tumor development with the malignant tumors that eventually emerge in patients often being resistant to elimination by the immune system. Additionally, the challenges are to identify patients for which immunotherapy will be efficacious (Disis and Schiffman, 2001), and to explore means of reactivating the immune response against tumors.

### Gene Therapy for Cancer Control

The gene therapy involves the transfer of genes into cancer cells to make them more sensitive to chemotherapy, radiotherapy and other treatments. It also involves the transfer of genes into the normal cells to make them more resistance to toxic treatments. For instance, healthy hematopoietic stem cells are removed from the body, gene is inserted that makes these cells more resistant to the side effects of high doses of anticancer drugs, and then the cells are injected back into the patient.

Several gene therapy treatments are in late stage trials. TNF-alpha gene therapy involves the transfer of this gene exclusively to the cancer cells by intratumoral injections and a promoter that is activated by radiation therapy (Rasmussen et al., 2002). Another exciting gene therapy treatment agent is Rexin-G, the first injectable gene therapeutic agent approved by Food and Drug Administration for the treatment of pancreatic cancer (Gordonand Hall, 2005). INGN 201, an adenoviral vector containing p53 gene transfer, is currently in phase III testing for squamous cell carcinoma of the head and neck, and has completed phase I studies on prostate, ovarian, glioma and bladder cancers (Merritt et al., 2001; Swisher et al., 2003). Hence, gene therapy for cancer treatment holds great promise for increasing the effectiveness of currently available chemotherapeutic treatment regimens.

### Photodynamic Therapy of Cancer

Recent studies have shown that lasers, when targeted at lymphatic vessels, may suppress the spread of cancerous tumors. Lasers and drugs have already been used to treat tumors by directly targeting them. Herein, a light-sensitive drug is injected into tumors followed by irradiating an infrared laser on the area. This method is more appropriate for the tumors growing near the skin or the tissue which could be easily irradiated. The light induces the drug mainly to produce oxygen free radicals that destroy cancer cells. Photodynamic therapy (PDT) can also be useful in removing traces of malignant tissue after surgical removal of large tumors (Dolmans et al., 2003). So far many light sensitive chemicals have been identified, however, their non-selectivity for the cancer cells versus normal cells is still a bottleneck in the success of PDT.

## Anti-angiogenic Therapy of Cancer

Angiogenesis implies the formation of new blood vessels. Anti-angiogenic therapy of cancer involves the restoration of angiogenic balance and normalizing the blood vessels, and suppressing the new blood vessel formation to inhibit the growth and metastasis of vascular tumors.Several natural and synthetic angiogenesis inhibitors, known as anti-angiogenic agents, prevent or slow down the growth of cancer by blocking the formation of new blood capillaries. These agents also normalize tumor vasculature and tumor microenvironment. FDA has approved drugs with antiangiogenic activity as cancer therapies, and the first anti-vascular endothelial growth factor (VEGF) drug was Bevacizumab against colorectal cancer (Giantonio et al., 2007). Similarly, kinase inhibitor sunitinib was approved for the treatment of gastrointestinal stromal tumors (Motzer et al., 2007); sorafenib against renal cell carcinoma (Escudier et al., 2007); herceptin against breast cancer which down-regulates growth factors such as VEGF, TGF-alpha and Ang-1(Izumi et al., 2007).Chemopreventive agents, especially natural compounds, have also been shown to target and inhibit angiogenesis process and can be used conveniently as they are mostly non-toxic and could be part of our daily diet (Bhatand Singh, 2008). However, it is also necessary to target multiple pro-angiogenic factors and down-stream signaling pathways to accelerate tumor regression taking into consideration that the tumors may become resistant. Nevertheless, the antiangiogenic therapy has been quite successful in controlling the growth of tumors, as its main targets, the endothelial cells are less likely to acquire drug resistance.

## Targeting Cancer Stem Cells with Phytochemicals

In recent years, the discovery of cancer stem cells (CSCs) has changed the landscape of cancer research.CSCs were first observed in hematological malignancies (Lapidot et al., 1994), but have now been identified in solid tumors of breast (Al-Haj et al., 2003), prostate (Patrawala et al., 2006; Hurt et al., 2008), brain (Singh et al., 2004), colon (Ricci-Vitiani et al., 2007),and pancreas (Li et al., 2007; Hermann et al., 2007). They are known to have the power to generate or regenerate the whole tumor masses, thus the elimination of these cells is critical to the development of therapeutic interventions. CSCs are more resistant to existing therapies than the bulk of differentiated cells within the tumors. The identification of solid tumor stem cells provided researchers with a firm basis on which to re-evaluate cancer therapies to target and eliminate not only the bulk population of tumor cells, but also the rare but potent self-renewing cells that initiate and sustain cancers.The ability of phytochemicals to inhibit tumor formation both *in vitro* and *in vivo* is well documented (Ichikawa et al., 2007).An extensive amount of research is being conducted on the benefits of phytochemicals. More recently, the inhibitory effect of these compounds on a variety of cancers including leukemia, prostate, breast, colon, brain, melanoma, and pancreatic cancers (Christensen and LeBlanc, 1996; Lev-Ariet al., 2007; Karmakar et al., 2007) have been identified.

There are a number of clinical trials that have evaluated the natural products for their efficacy in preventing the disease which can be attributed to the antioxidant, antiproliferative, and pro-apoptotic effects of these compounds.For instance, epigallocatechin gallate (EGCG),an antioxidant and chemopreventive polyphenol from green tea has offered a great promise in chemoprevention trial on patients with high-grade prostate intraepithelial neoplasia (HG-PIN), the main premalignant lesion of prostate cancer (Bettuzzi et al., 2006). It reduced the conversion rate from HG-PIN to prostate cancer. Similarly, another clinical trial investigated the effects of vitamin D on patients with progressive, metastatic androgen-independent prostate cancer. Vitamin D and its analogs are anti-proliferative and anti-angiogenic in a wide range of cancers (Deeb et al., 2007). The study inferred that vitamin D in combination with the standard chemotherapy agent, docetaxel, led to an improvement in overall patient survival (Beer et al., 2007).

Genistein is a major isoflavonoid found in soy products and has been proposed to be responsible for lowering the rate of pancreatic cancer (Shah et al., 2007; Wang et al., 2009). *In vitro* studies demonstrated that genistein is able to induce apoptosis and inhibit the activation of the NF-êB pathway through the Notch signaling pathway in pancreatic cancer (Wang et al., 2006; Wang et al., 2009). It has been shown to act synergistically with vitamin D to inhibit the growth of prostate cells through the regulation of genes involved in stem cell self-renewal (Rao et al., 2002; Regenbrecht et al., 2008; Slusarz et al., 2010). It has been demonstrated that Notch pathway is highly expressed in pancreatic cancer cells with EMT phenotype consistent with the acquisition of the stem cell features (Shah et al., 2007; Wang et al., 2009; Sarkar et al., 2010). Therefore, genistein could cause reversal of EMT phenotype and eliminate the stem cells. These studies support the use of genistein for targeted elimination of pancreatic CSCs.

A significant number of studies describe the antiproliferative mechanism of retinoic acid (RA). It is the active form of vitamin A found in animals, whereas â-carotene is a precursor to the active form of vitamin A in plants (Krista et al., 2004). RAs play a critical role in embryogenesis and hematopoiesis (Marklund et al., 2004; Oren et al., 2003). It has been seen as a potential mediator of differentiation therapy and used in the treatment of acute promyelocytic leukemia (APL) (Clarke et al., 2004). Treatment with RA induces differentiation of promyelocytic leukemic cells to mature neutrophils (Drumea et al., 2008). A study with RA has shown the dramatically improved patient outcome, with complete remission rates of greater than 90% and long-term survival in about 75% of patients (Okuno et al., 2004).

In summary, targeting CSCs, especially by natural agents such as EGCG, genistein, sulforaphane curcumin and quercetin may be a novel strategy towards designing approaches for combination therapy using conventional therapeutics for achieving better treatment outcome for cancer patients.

**Table 1** Various tumor markers currently in use

| *Category of Tumor Marker* | *Tumor Marker* | *Cancer Type* | *Application Based on ASCO/NACB Recommendations* | *References* |
|---|---|---|---|---|
| Oncofetal antigens | Alpha fetoprotein | Germ cell hepatoma | • Diagnosis<br>• Differential diagnosis of NSGCT<br>• Staging<br>• Detecting recurrence<br>• Monitoring therapy | Abelev et al., 1963; Diamandis, 2002 |
| | Carcino embryonic antigen (CEA) | Colon | • Prognosis<br>• Detecting<br>• Recurrence<br>• Screening for hepatic metastasis | Ludwig and Weinstein, 2005; Diamandis, 2002 |
| Hormones | Calcitonin | Medullary thyroid carcinoma | • Diagnosis<br>• Monitoring therapy | Melvin et al., 1971; Sturgeon, 2002 |
| | Human chorionic gonadotropin-beta | Testicular | • Diagnosis<br>• Staging<br>• Detecting recurrence | Bagshawe et al., 1980; Diamandis, 2002 |
| Mucins and glycoproteins | CA15-3 | Breast | • Monitoring therapy | Kufe et al., 1984; Hilkens et al., 1984; Bast et al., 2001 |
| | CA125 | Ovarian | • Progonosis<br>• Detecting recurrence<br>• Monitoring therapy | Bast et al., 1981; Diamandis, 2002 |
| | CA19-9 | Pancreatic | • Monitoring therapy | Koprowski et al., 1979; Ludwig and Weinstein, 2005 |
| Receptors | ER and PgR | Breast | • Select patients for endocrine therapy | McGuire et al., 1977; Bast et al., 2001 |
| | HER2 | Breast | • Select patients for trastuzumab therapy | Yamamoto et al., 1986; Bast et al., 2001 |
| Enzymes | Lactate dehydrogenase Staging | Germ cell | • Diagnosis<br>• Detecting recurrence | Hill and Levi, 1954; Diamandis, 2002 |
| Specific proteins | Prostate specific antigen | Prostate | • Screening<br>• Diagnosis | Wang et al., 1979; Diamandis, 2002 |
| | Thyroglobulin | Thyroid | • Monitoring | Carayanniotis and Rao, 1977; Sturgeon, 2002 |

Abbreviations: ER, Estrogen Receptor; NACB, National academy of clinical biochemistry; NSGCT, Non-seminomatous germ cell tumor; PgR, progesterone receptor.

**Table 2** Proposed mechanisms by which dietary phytochemicals prevent cancer

| *Proposed Mechanisms* | *Proposed Molecular Targets* | *References* |
|---|---|---|
| *Antioxidant Activity* | *Exert Their Protective Action Either by Suppressing the Formation of Free Radicals or by Scavenging Free Radicals* | *Papas,1999* |
| Inhibition of cell proliferation through blocking G2-M phase progression | G2-M checkpoint of cell cycle progression, modulation in cyclin B1, cdc25 and chk1/2, as well as wee1, p53 and p21/cip1 | Knowles and Milner, 2001; Moragoda et al., 2001 |
| Induction of cell differentiation | Differentiation morphology (anti-progression) | Jang et al., 1996; Zi and Agarwal, 1999; Agarwal et al., 2000 |
| Inhibition of oncogene expression | *H-ras, c-myc, K-ras, src* | Avila et al., 1996; Ranelletti et al., 2000 |
| Induction of tumor suppressor gene expression | *p53, Rb, p21, p27* | Singh et al., 2002 |
| Induction of cell cycle arrest by blocking G1-S phase transition | Down-regulation of cyclin A, cyclin E, cyclin D1, CDK2/4/6 | Ahmed et al., 1997; Zi et al.,1998 |
| Induction of apoptosis | Bcl-2, Bcl-xL, IAP-1, IAP-2, XIAP, survivin, cFLIP, and TRAF-1,caspases activation-3,8 & 9, mitochondrial dysfunction,activation of JNK pathway | Tang, 2006 |
| Inhibition of signal transduction pathways | ErbB1, IGF1-R, MAPK/ERK1/2 and PI3K-AKT | Grandis et al.,1996; Zi et al., 1998; Sharma et al., 2001 |
| Enzyme induction and enhancing detoxification | Phase II enzymes<br>Glutathione peroxidase<br>Catalase<br>Superoxide dismutase | Sreerama et al., 1995; Hageman et al., 1997 |
| Enzyme inhibition | Phase I enzymes (block activation of procarcinogens)<br>Cyclooxygenase-2<br>Inducible nitric oxide synthase<br>Xanthine oxidase | Surh et al., 2001 |
| Enhancement of immune functions and surveillance | Enhanced NK cell activity, delayed-type hypersensitivity, cytokine production | Kandil et al., 1988; Bogden et al., 1994 |
| Antiangiogenesis | VEGF, PDGF | Lamy et al., 2002 |
| Inhibition of cell adhesion and invasion | MMP2,9 | Hwang and Lee, 2006 |
| Inhibition of nitrosation and nitration | Inhibit N-nitroso compound formation by destroying nitrosating agents | Knight and Forman, 1987; Bartsch and frank, 1996 |
| Prevention of DNA binding of transcriptionfactor | pRb or pRb-related protein-E2F | Tyagi et al., 2002 |
| Regulation of steroid hormone metabolism | Alter the potency of testosterone, estrogen, and their derivatives via oxidation and hydroxylation reactions | Aoyama et al., 1990 |

**Table 3** A brief account of current evidence regarding cancer chemoprevention in major target organs

| *Class of Agent* | *Chemical/ Generic/brand Name®* | *Target Cancer Site* | *Agents Activity* | *References* |
|---|---|---|---|---|
| Anti-Androgens | Flutamide/Eulexi® | Prostate | Finasteride blocks new testosterone production while flutamide blocks the androgen receptor | Tiammi et al., 2002 |
| Carotenoids | Lycopene | Prostate | Lycopene provides the natural red pigment in tomatoes. It is also found in smaller amounts in grapefruit, guava, watermelon, and papaya. | Van Breemen et al., 2008; Etminan et al., 2004 |
| Isothiocyanates | Phenethyl-isothiocyanate (PEITC) | Colon | PEITC occurs naturally in horseradish, vegetables such as watercress, turnip, cabbage, broccoli, cauliflower. Consuming PEITC may decrease cancer risk | Zhang, 2010 |
| NSAIDS | Celecoxib/Celeb® Sulindac/Clinoril® Aspirin | Colon, breast, prostate, bladder, esophagus and skin | Aspirin is the first pain-relieving drug developed in a group of pharmaceuticals called nonsteroidal anti-inflammatory drugs (NSAIDS). They works through cyclo-oxygenase (COX) inhibition. Celecoxib is COX -2 selective, Aspirin and sulindac are mixed COX-1 and COX-2 inhibitors. | Meyskens et al., 2008 |
| Polyphenols | Green Tea Catechins Grape seed polyphenols | Colon, lung, skin, prostate, basal cell carcinoma and osteosarcoma | Tea polyphenols are compounds in the tea leaves that contain natural plant antioxidants. Antioxidants prevent damage caused by free radicals to DNA and other molecules | Yang et al., 2001 |
| Resveratrol | Grape/wine constituent | Colon | The resveratrol content of wine is related to the length of time the grape skins are present during the fermentation process | Athar et al., 2007 |
| Retinoids | All-trans-retinoic acid, Tretinoin/ retinA®, 9-cis-retinoic acid, alitretinoin/Panre® 13-cis-retinoic acid Isotretinoin/ Acn® 4Hydroxyretinamide, (4-HPR) fenretinide Tararotene/Zorac® | Cervix, esophagus, larynx, lung, nasopharynx, oral cavity, and ovary | Vitamin A-like compounds | Chiesa et al., 1993; Sankaranarayanan and Mathew, 1996 |

**Table 3** Contd.

| *Class of Agent* | *Chemical/ Generic/brand Name®* | *Target Cancer Site* | *Agents Activity* | *References* |
|---|---|---|---|---|
| Selective Estrogen Receptor Modulators (SERMS) | Tamoxifen/ Nolvadex® Arzoxifene Acolbifene Raloxifene/Evist® | Breast | Arzoxifene, a selective estrogen receptor modulator (SERM), is an improved form of the drug raloxifene. Arzoxifene is more easily absorbed by the body than raloxifene, which is used to prevent osteoporosis in postmenopausal women, and it is being clinically tested against the drug tamoxifen for breast cancer prevention. | Vogel et al., 2010; Fisher et al., 2005 |
| Selenium compounds | Selenized yeast Selenomethionine Selenomethyl-Cystine | Bladder, breast, colon, rectum, lung, bladder and prostate | L-Selenome-thionine is a synthetic form of selenium, an essential nonmetallic trace element in grains, meat, yeast, and certain vegetables. | Clark et al., 1996 |
| Soy Isoflavones | Genistein, Daidzein and Glycitein | Breast and colon | Soy isoflavones are plant-derived substances known to interact with human estrogen receptors and commonly known as phytoestrogens. Soy isoflavones also produce non-hormonal effects. Isoflavones, such as genistein and daidzein, are found in soybeans and soy food products like tofu and textured vegetable protein. | Raffoul et al., 2007; Bielak-Zmijewska et al., 2000 |
| Sulfur containing antioxidants | Anethole trithione/Sialor® (Canadian) | Breast, lung, colon and bladder | They consists of sulforaphane found in vegetables like cabbage and brussels sprouts. | Fimognari et al., 2008 |
| Vitamin D analogs | Provitamin D Vitamin D2/ ergocalciferol Vitamin D3/ cholecalciferol | Skin, ovary, colon and breast | Vitamin D (D2 and D3) compounds are needed for healthy bone growth. The skin can produce Vitamin D3 from sunlight, and both compounds are also found naturally | Lamprecht and Lipkin, 2003; Giova-nnucci, 1998; Robsahm, et al., 2004; Corder et al., 1993; Schwartz, 2005 |

**Table 4** Clinically approved monoclonal antibodies in cancer immunotherapy

| *Cancer immunotherapy: monoclonal antibodies* | | | | | | |
|---|---|---|---|---|---|---|
| *Antibody* | *Brand name* | *Approval date* | *Type* | *Target* | *Approved treatments* | *References* |
| Alemtuzumab | Campath | 2007 | Humanized | CD52 | Chronic lymphocytic leukemia | Demko et al., 2008 |
| Bevacizumab | Avastin | 2004 | Humanized | Vascular endothelial growth factor | Colorectal cancer | Los et al., 2007 |
| Cetuximab | Erbitux | 2004 | Chimeric | Epidermal growth factor receptor | Colorectal cancer | Van et al., 2009 |
| Sunitinib | Sutent | 2006 | Humanized | CD33 | Kidney cancer | Hartmann and Kanz, 2008 |
| Ibritumomab tiuxetan | Zevalin | 2002 | Murine | CD20 | Non-Hodgkin lymphoma (with yttrium-90 or indium-111) | Milenic et al., 2004 |
| Ofatumumab | Arzerra | 2009 | Human | CD20 | Chronic lymphocytic leukemia | Zhang, 2009 |
| Panitmumab | Vectibix | 2006 | Human | Epidermal growth factor receptor | Colorectal cancer | Giusti, et al., 2007 |
| Ipilimumab | Yervoy® | 2011 | Human | Blocks the immunosupressive effects of CTLA-4 | Melanoma | Ralph and Mark, 2010 |

## CONCLUSION

The progression of cancer occurs by overcoming the barriers of defense that are softwared into all cells and tissues. The likelyhood of cancer incidence is increased by many factors that exert carcinogenic effects like diet and lifestyle, tobacco use, exposure to radiations, obesity, high fat or junk foods, which will directly or indirectly cause genetic and epigenetic alterations that create mutations and changes in gene expression pattern that are implicated in tumor growth and progression. Therefore, the adoption of healthier diet and lifestyle as well as risk identification can result in the reduction of cancer incidence and hence mortality. The most practical approach for reducing deaths from cancer is through early detection, as early stage cancers are curable than late stage cancers. Survivors of certain cancers are at greater risk of developing second primary cancers, so they should be aware of the proper nutrition and healthy practices as well as screening tests available for various types of cancers. Medical oncologists, the connecting link between cancer research and medical practice, should counsel patients and their families and emphasize early interventions with appropriate dietary and exercise patterns as well as risk identification and modification *via* genomic and proteomic tools available.

Success of medical oncology depends on the translation of basic scientific research to clinical application. Overall aim of basic cancer research is to account for and design a complete list of all the events that are required in converting a normal human cell into a malignant cell, which will enable the researcher to sketch a complete picture documented in detail that will identify the molecular targets for therapeutic intervention. However, even after considerable advancement in diagnosis and therapy of cancer in previous few decades, the cure or control of cancer remains a major challenge before us. Therefore, the current advances in chemopreventive approaches, if applied practically, are expected to reduce the burden of cancer control and management.

## ACKNOWLEDGEMENTS

The fellowship support from UGC to A. Kartha through CUG, Gandhinagar, India, is gratefully acknowledged.

## REFERENCES

Abelev, G.I., Perova, S.D., Khramkova, N.I., Postnikova, Z.A. and Irlin, I.S. (1963). Production of embryonal alpha-globulin by transplantable mouse hepatomas. *Transplantation*, **1:** 174-180.

Agarwal, C., Sharma, Y., Zhao, J. and Agarwal, R. (2000).A polyphenolic fraction from grape seed causes irreversible growth inhibition of breast carcinoma MDA-MB468 cells by inhibiting mitogen activated protein kinases activation and inducing G1 cell arrest and differentiation. *Clinical Cancer Research*, **6:** 2921-2930.

Ahmed, N., Feyes, D.K. and Nieminen, A.L. (1997). Green tea constituent epigallocatechin-3-gallate and induction of apoptosis and cell cycle arrest in human carcinoma cells. *J. National Cancer Institute*, **177:** 79-84.

Al-Hajj, M., Wicha, M.S., Ito-Hernandez, A., Morrison, S.J. and Clarke, M.F. (2003). Prospective identification of tumorigenic breast cancer cells. *Proceedings of National Academy of Sciences USA* **100:** 3983-3988.

Anand, P., Kunnumakara, A.B., Kuzhuvelil, C.S., Harikumar, B., Tharakan, S.T., Lai, O.S., Sung, B. and Agarwal, B.B. (2008). Cancer is a preventable disease that requires major lifestyle changes. *Pharmaceutical Research*, **25:** 2097-2212.

Aoyama, T., Korzekwa, K., Nagata, K., Gillette, J., Gelboin, H.V. and Gonzalez, F.J. (1990). Estradiol metabolism by complementary deoxyribonucleic acid-expressed human cytochrome P450s. *Endocrinology*, **126:** 1-6.

Athar, M., Back, J.H. and Tang, X. (2007). Resveratrol: a review of preclinical studies for human cancer prevention. *Toxicology and Applied Pharmacology* **224:** 274-283.

Avila, M.A., Cansado, J., Harter, K.W., Velasco, J.A. and Notario, V. (1996). *Advances in Experimental Medicine and Biology*, **401:** 101-110.

Bagshawe, K.D., Wass, M. and Searle, F. (1980) Markers in gynaecological cancer. *Archives of Gynecology*, **229:** 303-310.

Bartsch, H. and Frank, N. (1996). Blocking the endogenous formation of N-nitroso compounds and related carcinogens. *In:* Stewart, B.W., McGregor, D. and Kleihues, P. (Eds.). Principles of chemoprevention. Lyon, France. *International Agency for Research on Cancer,* **139:** 189-201.

Bast, R.C, Feeney, M. and Zazarus, H. (1981). Reactivity of a monoclonal antibody with human ovarian carcinoma. *J. Clinical Investigation,* **68:** 1331-1337.

Bast, R.C. Jr., Ravdin, P., Hayes, D.F., Bates, S., Fritsche, Jr. H., Jessup, J.M., Kemeny, N., Locker, G.Y., Mennel, R.G. and Somerfield, M.R. (2001). Update of recommendations for the use of tumor markers in breast and colorectal cancer: clinical practice guidelines of the American Society of Clinical Oncology. *J. Clinical Oncology,* **19:** 1865-1878.

Beer, T.M., Ryan, C.W. and Venner, P.M. (2007). Double-blinded randomized study of high-dose calcitriol plus docetaxel compared with placebo plus docetaxel in androgen-independent prostate cancer: A report from theASCENT Investigators. *J. Clinical Oncology,* **25:** 669-674.

Bettuzzi, S., Brausi, M., Rizzi, F., Castagnetti, G., Peracchia, G. and Corti, A. (2006). Chemoprevention of human prostate cancer by oral administrationof green tea catechins in volunteers with high-grade prostate intraepithelial neoplasia: A preliminary report from a one-year proof-of principle study. *Cancer Research,* **66:** 1234-1273.

Bhat, T.A. and Singh, R.P. (2008). Tumor angiogenesis—A potential target in cancer chemoprevention. *Food and Chemical Toxicology,* **46:** 1334-1378.

Bielak-Zmijewska, A., Koronkiewicz, M., Skierski, J., Piwocka, K., Radziszewska, E. and Sikora, E. (2000). Effect of curcumin on the apoptosis of rodent and human non-proliferating and proliferating lymphoid cells. *Nutrition and Cancer,* **38:** 131-138.

Bogden, J.D., Bendich, A. and Kemp, F.W. (1994). Daily micronutrient supplements enhance delayed hypersensitivity skin test responses in older people. *Americal J. Clinical Nutrition,* **60:** 437-447.

Carayanniotis, G. and Rao, V.P. (1977). Searching for pathogenic epitopes in thyroglobulin: parameters and caveats. *Immunology Today,* **18:** 83-88.

Chendil, D., Ranga, R.S., Meigooni, D., Satishkumar, S. and Ahmed, M.M. (2004). Curcumin confers radiosensitizing effect in prostate cancer cell line PC-3. *Oncogene,* **23:** 1599-1607.

Chiesa, F., Tradati, N. and Marazza, M. (1993). Fenretinide (4-HPR) in chemoprevention of oral leukoplakia. *J. Cell Biochemistry,* **17:** 255-285.

Christensen, J.G. and LeBlanc, G.A. (1996). Reversal of multidrug resistance *in vivo* by dietary administration of the phytochemical indole-3-carbinol. *Cancer Research,* **56:** 574-581.

Chu, Y.F., Sun, J., Wu, X. and Liu, R.H. (2002). Antioxidant and antiproliferative activities of vegetables. *J. Agricultural and Food Chemistry,* **50:** 6910-6916.

Clark, L.C., Combs, G.F. Jr. and Turnbull, B.W. (1996). Effects of selenium supplementation for cancer prevention in patients with carcinoma of the skin. A randomized controlled trial. Nutritional Prevention of Cancer Study Group. *J. American Medical Association,* **276:** 1957-1963.

Clarke, N., Germain, P., Altucci, L. and Gronemeyer, H. (2004). Retinoids: Potential in cancer prevention and therapy. *Expert Reviews in Molecular Medicine,* **6:** 1-23.

Corder, E.H., Guess, H.A., Hulka, B.S., Friedman, G.D. and Sadler, M. (1993) Vitamin D and prostate cancer: a prediagnostic study with stored sera. *Cancer Epidemiology, Biomarkers and Prevention,* **2:** 467-472.

Cummings, S.R., Eckert, S., Krueger, K.A., Grady, D., Powler, T.J., Cavely, J.A., Norton, L., Nickelsen, T., Bjarnason, N.H., Morrow M., Lippman., M.E., Black, D., Glusman, J.E., Costa, A. and Jordan, V.C. (1999). The effect of raloxifene on risk of breast cancer in post menopausal women: results from the MORE randomized trial. *J. American Medical Association*, **281:** 2189-2197.

Damodar, S., Terunuma, H. and Abraham, S. (2006). Autologous immue enhancement therapy (AIET) for a case of acute myeloid leukemia (AML) - our experience. *J. Stem Cells and Regenerative Medicine*, **1:** 40-41.

Deeb, K.K., Trump, D.L. and Johnson, C.S. (2007).Vitamin D signalling pathways in cancer: Potential for anticancer therapeutics. *Nature Reviews Cancer*, **7:** 684-700.

Demko, S., Summers, J., Keegan, P. and Pazdur, R. (2008). *The Oncologist*, **13:** 167-174.

Diamandis, E.P. (2002). Tumor markers: Past, present, and future. In Tumor markers: Physiology, pathobiology, technology, and clinical applications. (Ed.). Diamandis, E.P., Fritsche, E.P. H. Jr; Lilja, H., Chan, D. and Schwartz, M., Washington DC: AACC Press. 3-8.

Disis, M.L. and Schiffman K. (2001). Issues on clinical applications of cancer vaccines. *J. Immunotherapy*, **24:** 104-108.

Dolmans, D.E., Fukumura, D. and Jain, R.K. (2003). Photodynamic therapy for cancer. *Nature Reviews Cancer*, **3:** 380-386.

Dragsted, L.O., Strube, M. and Larsen, J.C. (1993). Cancer-protective factors in fruits and vegetables: biochemical and biological background. *Pharmacology and Toxicology*, **72:** 116-135.

Drumea, K., Yang, Z.F. and Rosmarin, A. (2008). Retinoic acid signaling in myelopoiesis. *Current Opinion in Hematology*, **15:** 37-41.

Escudier, B., Pluzanski, A. and Koralewski, P. (2007). Bevacizumab plus interferon alpha-2a for treatment of metabolic renal carcinoma: a randomized, double-blind phase-111 trial. *Lancet*, **37:** 2103-2111.

Etminan, M., Takkouche B. and Caamaño-Isorna F. (2004). The role of tomato products and lycopene in the prevention of prostate cancer: a meta-analysis of observational studies. *Cancer Epidemiology, Biomarkers & Prevention*, **13:** 340-346.

Fimognari, C., Lenzi, M., Cantelli-Forti, G. and Hrelia, P. (2008). Induction of differentiation in human promyelocytic cells by the isothiocyanate sulforaphane. *In-Vivo* **22:** 317-320.

Fischer, B., Constantine, J.P., Wicherham, D.L., Redmond, C.K., Kavanah, M., Crocin, W.M., Vogel, V., Robidoux, A., Dimitrov, N., Atkins, J., Daly, M., Weiand, S., Tan-Chiu, E., Ford, L and Wolmark, N. (1998). Tamoxifen for prevention of breast cancer: report of the national surgical adjuvant breast & bowel project P-1 study. *J. National Cancer Institute*, **90:** 1371-1388.

Fisher B., Costantino J.P., Wickerham, D.L., Cecchini R.S., Cronin, W.M., Robidoux, A., Bevers, T.B., Kavanah, M.T., Atkins, J.N. and Margolese, R.G. (2005). Tamoxifen for the prevention of breast cancer: current status of the National Surgical Adjuvant Breast and Bowel Project P-1 study. *J. National Cancer Institute*, **97:** 1652-1662.

Giantonio, B.J., Catalano, P.J., Meropol, N.J., O'Dwyer P. J., Mitchell, E. P., Alberts, S.R,. Schwartz, M.A. and Benson, A.B. (2007).Bevacizumab in combination with oxaliplatin, fluorouracil, and leucovorin (FOLFOX4) for previously treated metastatic colorectal cancer: results from the Eastern Cooperative Oncology Group Study E3200. *J. Clinical Oncology*, **25:** 1539-1544.

Giovannucci, E. (1998). Dietary influences of 1,25(OH)2 vitamin D in relation to prostate cancer: a hypothesis. *Cancer Causes Control*, **9:** 567-582.

Giusti, R.M., Shastri, K.A., Cohen, M.H., Keegan, P. and Pazdur, R. (2007). FDA drug approval summary: panitumumab (Vectibix). *The Oncologist*, **12:** 577-583.

Gold, P. and Freedman, S.O. (1965). Demonstration of tumor-specific antigens in human colonic carcinoma by immunological tolerance and absorption techniques. *J. Experimental Medicine*, **121:** 439-462.

Gordon, E.M. and Hall, F.L. (2005). Nanotechnology blooms, at last. *Oncology Reports*, **13:** 1003-1007.

Grandis, J.R., Zeng, Q. and Tweardy, D.J. (1996). Retinoic acid normalizes the increased gene transcription rate of TGF-alpha and EGFR in head and neck cancer cell lines. *Nature Medicine*, **2:** 237-276.

Hageman, G., Krul, C., Van Herwijnen, M., Schilderman, P. and Kleinjans J. (1997). Assessment of the anticarcinogenic potential of raw garlic in humans. *Cancer Letters*, **114:** 161-162.

Hartmann, J.T. and Kanz, L. (2008). Sunitinib and periodic hair depigmentation due to temporary c-KIT inhibition. *Archives of Dermatology*, **144:** 1525-1531.

Henderson, B.E. and Feigelson, H.S. (2000). Hormonal carcinogenesis. *Carcinogenesis*, **21:** 427-459.

Hermann, P.C., Huber, S.L., Herrler, T., Aicher, A., Ellwart, J. W., Guba, M., Bruns, C.J. and Heeschen, C. (2007). Distinct populations of cancer stem cells determine tumor growth and metastatic activity in human pancreatic cancer. *Cell Stem Cell*, **1:** 313-323.

Hill, B.R. and Levi, C. (1954). Elevation of a serum component in neoplastic disease. *Cancer Research*, **14:** 513-515.

Hilkens, J. (1984). Monoclonal antibodies against human milk-fat globule membranes detecting differentiation antigens of the mammary gland and its tumors. *International J. Cancer*, **34:** 197-206.

Hurt, E.M., Kawasaki, B.T., Klarmann, G.J., Thomas, S.B. and Farrar, W.L. (2008). CD44+CD24– prostate cells are early cancer progenitor/stem cells that provide a model for patients with poor prognosis. *British J. Cancer*, **98:** 756-765.

Hwang, E.S. and Lee, H.J. (2006). Allyl isothiocyanate and its N-acetylcysteine conjugate suppress metastasis via inhibition of invasion, migration, and matrix metalloproteinase-2/-9 activities in SK-Hep 1 human hepatoma cells. *Experimental Biology and Medicine*, **231:** 421-430.

Ichikawa, H., Nakamura, Y., Kashiwada, Y. and Aggarwal, B.B. (2007). Anticancer drugs designed by mother nature: Ancient drugs but modern targets. *Current Pharmaceutical Design*, **13:** 3400-3416.

Izumi, Y., Xu, L. and di Tamao, E. (2007). Tumor biology: herceptin acts as an anti-angiogenic- cocktail. *Nature*, **116:** 279-280.

Jang, M., Cai, L., Udiam, G.O., Sloving, K.O., Thomas C.F. and Beecher, C.W. (1997). Cancer chemopreventive activity of resveratrol, a natural product derived from grapes. *Science*, **275:** 218-220.

Kandil, O., Abdullah, T., Tabuni, A.M. and Elkadi, A. (1988). Potential role of *Allium sativum* in natural cytoxicity. *Archives of AIDS Research*, **1:** 230-231.

Karmakar, S., Banik, N.L. and Ray, S.K. (2007). Molecular mechanism of inositol hexaphosphate-mediated apoptosis in human malignant glioblastoma T98G cells. *Neurochemical Research*, **32:** 2094-2102.

Kelloff, G.J., Boone, C.W. and Steele, V.E. (1994). Mechanistic considerations in chemopreventive drug development. *J. Cell Biochemistry*, **20:** 1-24.

Kinzler, K.W. and Vogelstein, B. (2002). Colorectal tumors. *In:* Vogelstein, B., and Kinzler, K. W. (Eds.). The Genetic Basis of Human Cancer (2nd edition). McGraw-Hill, New York. 583-612.

Knight, T.M. and Forman, D. (1987). The availability of dietary nitrate for the endogenous nitrosation of Lproline. *In:* Bartsch, H., O'Neill, I.K. and Schulte-Hermann R, (Eds.). The relevance of N-nitroso compounds to human cancer: exposure and mechanisms. Lyon, France: *International Agency for Research on Cancer,* **84:** 518-523.

Knowles, L.M. and Milner, J.A. (2001). Possible mechanism by which allyl sulfides suppress neoplastic cell proliferation. *J. Nutrition,* **131:** 1061-1066.

Koprowski, H., Steplewski, Z. and Mitchell, K. (1979). Colorectal carcinoma antigens detected by hybridoma antibodies. *Somatic Cell Genetics,* **5:** 957-971.

Kristal, A.R., Platz, E. A., Parnes, H.L. and Albanes, D. (2004). Vitamin A, retinoids and carotenoids as chemopreventive agents for prostate cancer. *J. Urology,* **171:** 54-58.

Kufe, D., Inghirami, G., Abe, M., Hayes, D., Justi-Wheeler, H. and Schlom, J. (1984). Differential reactivity of a novel monoclonal antibody (DF3) with human malignant versus benign breast tumors. *Hybridoma,* **3:** 223-232.

Lamy, S., Gingras, D. and Beliveau, R. (2002). Green tea catechins inhibit vascular endothelial growth factor receptor phosphorylation. *Cancer Research,* **62:** 381-385.

Lamprecht, S.A. and Lipkin, M. (2003). Chemoprevention of colon cancer by calcium, vitamin D and folate: molecular mechanisms. *Nature Reviews Cancer,* **3:** 601-614.

Lapidot, T., Sirard, C. and Vormoor, J. (1994). A cell initiating human acute myeloid leukaemia after transplantation into SCID mice. *Nature,* **367:** 645-648.

Lawrence, T.S., Ten Haken, R.K. and Giaccia, A. (2008). Principles of Radiation Oncology. *In:* DeVita, V.T. Jr., Lawrence, T.S. and Rosenberg, S.A. (Eds.). *Cancer: Principles and Practice of Oncology,* 307-314.

Lev-Ari, S., Vexler, A., Starr, A., Shkenazy-Voghera, M., Greif, J., Aderka, D. and Ben-Yosef, R. (2007). Curcumin augments gemcitabine cytotoxic effect on pancreatic adenocarcinoma cell lines. *Cancer Investigation,* **25:** 411-418.

Li, C., Heidt, D.G., Dalerba, P., Burant, C.F., Zhang, L., Adsay, V., Wicha, M., Clarke, M.F. and Simeone, D.M. (2007). Identification of pancreatic cancer stem cells. *Cancer Research,* **67:** 1030-1037.

Los, M., Roodhart, J.M. and Voest, E.E. (2007). Target practice: lessons from phase III trials with bevacizumab and vatalanib in the treatment of advanced colorectal cancer. *The Oncologist,* **12:** 443-450.

Ludwig, J.A. and Weinstein, J.N. (2005). Biomarkers in cancer staging, prognosis and treatment selection. *Nature Reviews Cancer,* **5:** 845-856.

Marklund, M., Sjodal, M., Beehler, B.C., Jessell, T.M., Edlund, T. and Gunhaga, L. (2004). Retinoic acid signaling specifies intermediate character in the developing telencephalon. *Development,* **131:** 4323-4332.

McGuire, W.L., Horwitz, K.B. and Pearson, O.H. (1977). Current status of estrogen and progesterone receptors in breast cancer. *Cancer,* **39:** 2934-2947.

Melvin, K.E., Miller, H.H. and Tashjian, A.H. Jr. (1971). Early diagnosis of medullary carcinoma of the thyroid gland by means of calcitonin assay. *New England J. Medicine,* **285:** 1115-1120.

Merritt, J.A., Roth, J.A. and Logothetis, C.J. (2001). Clinical evaluation of adenoviral-mediated p53 gene transfer: review of INGN 201 studies. *Seminars in Oncology,* **28:** 105-114.

Meyskens, F.L Jr., McLaren, C.E. and Pelot, D. (2008). Difluoromethylornithine plus sulindac for the prevention of sporadic colorectal adenomas: a randomized placebo-controlled, double-blind trial. *Cancer Prevention Research,* **1:** 32-38.

Milenic, D.E., Brady, E.D. and Brechbiel, M.W. (2004). Antibody-targeted radiation cancer therapy. *Nature Reviews Drug Discovery*, **3:** 488-499.

Mishra, A. and Verma, M. (2010). Cancer biomarkers: are we ready for the prime time? *Cancers*, **2:** 190-208.

Moragoda, L., Jaszewski, R. and Majumdar, A.P. (2001). Curcumin induced modulation of cell cycle and apoptosis in gastric and colon cancer cells. *Anticancer Research*, **21:** 873-880.

Motzer, R.J., Huston, T.E. and Tomezak, P. (2007). Sunitinib versus interferon alpha in metastatic renal cell carcinoma. *New England J. Medicine*, **356:** 115-124.

Okuno, M., Kojima, S., Matsushima-Nishiwaki, R., Tsurumi, H., Muto, Y., Friedman, S.L. and Moriwaki, H. (2004). Retinoids in cancer chemoprevention. *Current Cancer Drug Targets*, **4:** 285-298.

Oren, T., Sher, J.A. and Evans, T. (2003). Hematopoiesis and retinoids: Development and disease. *Leukemia and Lymphoma*, **44:** 1881-1891.

Papas A. (1999). Vitamin E: Tocopherols and Tocotrienoids in Papas A.M. ed. Antioxidant Status, Diet, Nutrition and Health. Boca Raton, FL: CRC Press 189-210.

Patel, R.R. and Arthur, D.W. (2006). The emergence of advanced brachytherapy techniques for common malignancies. *Hematology/Oncology Clinics of North America*, **20:** 97-118.

Patrawala, L., Calhoun, T. and Schneider-Broussard, R. (2006). Highly purified CD44+ prostate cancer cells from xenograft human tumors are enriched in tumorigenic and metastatic progenitor cells. *Oncogene*, **25:** 1696–1708.

Peters, G.J., Van der Wilt, C.L, Van Moorsel, C.J., Kroep, J.R., Bergman, A.M. and Ackland, S.P. (2000). Basis for effective combination cancer chemotherapy with antimetabolites. *Pharmacology and Therapeutics*, **87:** 227-253.

Prescott, S.M. and Fitzpatrick, F.A. (2000). Cyclooxygenase-2 and carcinogenesis. *Biochimica et Biophysica Acta*, **1470:** 69-78.

Raffoul, J.J., Banerjee, S., Che, M., Knoll, Z.E., Doerge, D.R., Abrams, J., Kucuk, O., Sarkar, F.H. and Hillman, G.G. (2007). Soy isoflavones enhance radiotherapy in a metastatic prostate cancer model. *International J. Cancer*, **120:** 2491-2498.

Ralph, B. and Mark, S. (2010).Invasion of the Prostate Snatchers: No More Unnecessary Biopsies, Radical Treatment or Loss of Sexual Potency. Other Press, LLC. 227.

Ranelletti, F.O., Maggiano, N., Serra, F.G., Ricci, R., Larocca, L.M., Lanza, P., Scambia, G., Fattorossi, A., Capelli, A. and Piantelli, M. (2000). *International J. Cancer*, **85:** 438-445.

Ranieri, G., Paturno, R. and Ruggeri, E. (2008). Vascular endothelial growth factor (VEGF) as a target of bevacizumab in cancer: from the biology to the clinic. *Current Medicinal Chemistry*, **13:** 1845-1857.

Rao. A., Woodruff, R.D., Wade, W.N., Kute, T.E. and Cramer, S.D. (2002). Genistein and vitamin D synergistically inhibit human prostatic epithelial cell growth. *J. Nutrition*, **132:** 3191-3194.

Rasmussen, H., Rasmussen, C., Lempicki, M., Durham, R., Brough, D., King, C.R., Weichselbaum, R. and Ferade, T.N. (2002). Biologic: preclinical toxicology of a novel adenovector with a radiation-inducible promoter, carrying the human tumor necrosis factor alpha gene. *Cancer Gene Therapy*, **9:** 951-957.

Regenbrecht, C.R., Jung, M., Lehrach, H. and Adjaye, J. (2008). Themolecular basis of genistein-induced mitotic arrest and exit of self-renewal in embryonal carcinoma and primary cancer cell lines. *BMC Medical Genomics*, **1:** 49.

Ricci-Vitiani, L., Lombardi, D. G., Pilozzi, E., Biffoni, M., Todaro, M., Peschle, C. and De Maria, R. (2007). Identification and expansion of human colon-cancer-initiating cells. *Nature*, **445:** 111-115.

Robsahm, T.E., Tretli, S., Dahlback, A. and Moan, J. (2004). Vitamin D from sunlight may improve the prognosis of breast, colon and prostate cancer. *Cancer Causes Control*, **2:** 149-158.

Sankaranarayanan, R. and Mathew, B. (1996). Retinoids as cancer-preventive agents. *In:* Stewart, B.W., McGregor, D. and Kleihues, P. (Eds.). Principles of chemoprevention. *IARC Scientific Publications*, **139:** 47-59.

Sarkar, F.H., Li, Y., Wang, Z. and Kong, D. (2010).The role of nutraceuticals in the regulation of Wnt and Hedgehog signaling in cancer. *Cancer Metastasis Review*, **29:** 383-394.

Schwartz, G.G. (2005). Vitamin D and the epidemiology of prostate cancer. *Semin Dial.* **18:** 276-289.

Shah, A.N., Summy, J.M., Zhang, J., Park, S.I., Parikh, N.U. and Gallick, G.E. (2007). Development and characterization of gemcitabine-resistant pancreatic tumor cells. *Annals of Surgical Oncology*, **14:** 3629-3637.

Sharma, Y., Agarwal, C., Singh, A.K. and Agarwal, R. (2001). Inhibitory effect of silibinin on ligand binding to erbB1 and associated mitogenic signaling, growth and DNA synthesis in advanced human prostate carcinoma cells. *Molecular Carcinogenesis*, **30:** 224-236.

Singh, R.P., Dhanalakshmi, S. and Agarwal, R. (2002). Phytochemicals as cell cycle modulators—a less toxic approach in halting human cancers. *Cell Cycle*, **1:** 156-161.

Singh, S.K., Hawkins, C., Clarke, I.D., Squire, J.A., Bayani, J., Hide, T., Henkelman, R.M., Cusimano, M.D. and Dirks, P.B. (2004). Identification of human brain tumour initiating cells. *Nature*, **432:** 396-401.

Sivaraman, G., Pandian, A. and Abraham, S. (2008). Autologous immune enhancement therapy for advanced carcinoma of pancreas—A case report. *J. Stem Cells and Regenerative Medicine*, **4:** 13.

Slusarz, A., Shenouda, N.S., Sakla, M.S., Drenkhahn, S.K., Narula, A.S., MacDonald, R.S., Besch-Williford, C.L. and Lubahn, D.B. (2010).Common botanical compounds inhibit the hedgehog signaling pathway in prostate cancer. *Cancer Research*, **70:** 3382-3390.

Sporn, M.B. and Suh, N. (2002).Chemoprevention: an essential approach to controlling cancer. *Nature Reviews Cancer*, **2:** 537-543.

Sreerama, L., Hedge, M.W. and Sladek, N.E. (1995). Identification of a class3 aldehyde dehydrogenase in human saliva and increased levels of this enzyme, glutathione S-transferases, and DT-diaphorasin the saliva of subjects who continually ingest large quantities of coffee or broccoli. *Clincal Cancer Research*, **1:** 1153-1163.

Sturgeon, C. (2002). Practice guidelines for tumor marker use in the clinic. *Clinical Chemistry*, **48:** 1151-1159.

Sun, J., Chu, Y.F., Wu, X. and Liu, R.H. (2002). Antioxidant and antiproliferative activities of fruits. *J. Agricultural and Food Chemistry*, **50:** 7449-7454.

Surh, Y.J., Chun, K.S., Cha, H.H., Han, S.S., Keum, Y.S., Park, K.K. and Lee, S.S. (2001). Molecular mechanisms underlying chemopreventive activities of anti-inflammatory phytochemicals: down-regulation of COX-2 and iNOS through suppression of NF-kappa B activation. *Mutation Research*, **480:** 243-268.

Swisher, S.G., Roth, J.A., Komaki, R., Gu, J., Lee, J.J., Hicks, M., Ro, J., Y., Hong, W.K., Merritt, J.A., Ahrar, K., Atkinson, N.E., Correa, A.M., Dolormente, M., Dreiling, L., El-Naggar, A.K., Fossella, F., Francisco, R., Glisson, B., Grammer, S., Herbst, R., Huaringa, A., Kemp, B., Khuri, F.R., Kurie, J.M., Liao, Z., McDonnell, T.J., Morice, R., Morello, F., Munden, R., Papadimitrakopoulou, V., Pisters, K.M., Putnam, J.B. Jr., Sarabia, A.J., Shelton, T., Stevens, C., Shin, D.M., Smythe, W.R., Vaporciyan, A.A., Walsh, G.L. and

Yin, M. (2003). Induction of p53-regulated genes and tumor regression in lung cancer patients after intratumoral delivery of adenoviral p53 (INGN 201) and radiation therapy. *Clinical Cancer Research*, **9:** 93-101.

Takimoto, C.H. and Calvo, E. (2008). Principles of Oncologic Pharmacotherapy in Pazdur, R., Wagman, L.D., Camphausen, K.A., Hoskins, W.J. (Eds.). Cancer Management: A Multidisciplinary Approach. 11 ed.

Tang, W. (2006). Ganoderic acid T from *Ganoderma lucidum* mycelia induces mitochondria mediated apoptosis in lung cancer cells. *Life Sciences*, **80:** 205-211.

Temple, N.J. (2000). Antioxidants and disease: more questions than answers. *Nutrition Research*, **20:** 449-459.

Thomas, W.A. (2003). Immunotherapy: past, present and future. *Nature Medicine*, **9:** 269-277.

Tiammi, R.M., Lagiou, P. and Trichopoulos, D. (2002). Prospects for chemoprevention of cancer. *J. Internal Medicine*, **251:** 286-300.

Tyagi, A., Agarwal, C. and Agarwal, R. (2002). The cancer preventive flavonoid silibinin causes hypophosphorylationof Rb/p107 and Rb2/p130 via modulation of cell cycle regulators in human prostate carcinoma DU145 cells. *Cell Cycle*, **2:** 137-142.

Vajnio, H., Morgan, G. and Kleihues, P. (1997). An international evaluation of the cancer preventive potential of non-steroidal anti-inflammatory drugs. *Cancer Epidemiology, Biomarkers and Prevention*, **6:** 749-753.

Van, C.E., Köhne, C.H., Hitre, E., Zaluski, J., Chang Chien, C.R., Makhson, A., D'Haens G., Pintér, T., Lim, R., Bodoky, G., Roh, J.K., Folprecht, G., Ruff. P., Stroh, C., Tejpar, S., Schlichting, M., Nippgen, J. and Rougier, P. (2009). Cetuximab and chemotherapy as initial treatment for metastatic colorectal cancer. *England J. Medicine*, **360:** 1408-1424.

Van Breeman, R.B. and Pajkovic, N. (2008). Multitargeted therapy of cancer by lycopene. *Cancer Letters*, **269(2):** 339-351.

Veronasi, U., De Palo, G., Marubini, E., Costo, A., Formelli, F., Mariani, L., Decensi, A., Camerini, T., Del Turco, M.R., Di mianro, M.G., Muraca, M.G., Del vicchio, M., Pinto, C., D Auito, G., Boni, C., Campa, T., Magini, A., Miceli, R., Perloff, M., Malone, W.F. and Sporn, M.B. (1999). Randomised trial of fenerentide to prevent second breast malignancy in women with early breast cancer. *J. National Cancer Institute*, **91:** 1847-1856.

Vogel, V.G, Costantino, J.P. and Wickerham, D.L. (2010). Update of the National Surgical Adjuvant Breast and Bowel Project Study of Tamoxifen and Raloxifen (STAR) P-2 Trial: preventing breast cancer. *Cancer Prevention Research*, **3:** 696-706.

Waladkhani, A.R. and Clemens, M.R. (1998). Effect of dietary phytochemicals on cancer development. *International J. Molecular Medicine*, **1:** 747-753.

Wang, M.C. (1979). Purification of a human prostate specific antigen. *Investigative Urology*, **17:** 159-163.

Wang, Z., Zhang, Y., Banerjee, S., Li, Y. and Sarkar, F.H. (2006). Inhibition of nuclear factor activity by genistein is mediated *via*Notch-1 signaling pathway in pancreatic cancer cells. *International J. Cancer* **118:** 1930-1936.

Wang, Z., Li, Y., Kong, D., Banerjee, S., Ahmad, A., Azmi, A.S., Ali, S., Abbruzzese, J.L., Gallick, G.E. and Sarkar, F.H. (2009). Acquisition of epithelial–mesenchymal transition phenotype of gemcitabine resistant pancreatic cancer cells is linked with activation of thenotch signaling pathway. *Cancer Research*, **69:** 2400-2407.

Willett, W.C. (1994). Diet and health: what should we eat. *Science*, **254:** 532-537.

Willett, W.C. (2002). Balancing life-style and genomics research for disease prevention. *Science*, **296:** 695-698.

Yamamoto, T., Ikawa, S., Akiyama, T., Semba, K., Nomura, N., Miyajima, N., Saito, T., Toyoshima, K. (1986). Similarity of protein encoded by the human c-erb-B-2 gene to epidermal growth factor receptor. *Nature*, **319:** 230-234.

Yang, C., Landau, J., Huang, M. and Newmark, H. (2001). Inhibition of carcinogenesis by dietary polyphenolic compounds. *Annual Review of Nutrition*, **21:** 381-397.

Zhang, B. (2009). "Ofatumumab". m Abs **1(4):** 326-331.

Zhang, Y. ( 2010). Allyl isothiocyanate as a cancer chemopreventive phytochemical. *Molecular Nutrition and Food Research*, **54:** 127-135.

Zi, X., Grasso, A.W., Kung, H.J. and Agarwal, R.A. (1998). Flavonoid antioxidant, silymarin, inhibits activation of erbB1 signaling and induces cyclin-dependent kinase inhibitors, G1 arrest, and anticarcinogenic effects in human prostate carcinoma DU145 cells. *Cancer Research*, **58:** 1920-1928.

Zi, X. and Agarwal R. (1999). Silibinin decreases prostate-specific antigen with cell growth inhibition via G1 arrest, leading to differentiation of prostate carcinoma cells: implications for prostate cancer intervention. *Proceedings of National Academy of Sciences USA*, **96:** 7490-7494.

Zoberi, I., Bradbury, C.M. and Cury, H.A. (2002). Radiosensitizing and antiproliferative effects of resveratrol in two human cervical cell lines. *Cancer Letters*, **175:** 65-73.

# 18

# The Immune System

**Sukrat Sinha**

*Postdoctoral Fellow, Centre for Biotechnology, Nehru Science Complex, University of Allahabad, Allahabad-211002, U.P. India*

**ABSTRACT**

This chapter on immunology highlights different aspects of immunology in a more summarized manner. The chapter throws light on molecules to cells and organs, and finally to the immune system as an integrated system protecting the body from infection and helping to maintain the health of the body.

The chapter introduces the basic concepts and is essential for an understanding of immunology which continues with the antigen recognition molecules, i.e., antibodies, T-cell receptors and the molecules encoded by the major histocompatibility complex. It further deals with immune physiology, the role of the cells and organs of the immune system. It discusses the innate immune system and its connections to the adaptive immune system, immune techniques, historical aspects and future prospects.

**Keyword:** Immune System, Antibodies, Molecules, Immune Techniques.

## INTRODUCTION

The term "immune" derives from the Latin word "*immunitas*" meaning "freedom from". The immune system is a dynamic and versatile network of cells, tissues and organs which has two main functions of **recognition** and **response,** i.e., it recognizes **antigens** (foreign particles, organisms etc.) and then triggers pathways to destroy them. Immune system can be divided into **innate**, which is intrinsic and present before the prior contact with an antigen, and **adaptive**, which evolves itself on coming in contact with the antigen. Whenever any antigen enters into the body and if the immune system recognizes it as a foreign, it provokes an **effector response**. If on the other side the same antigen invades the body later on, the immune system induces a heightened **memory response**. Another property of a healthy immune system is to differentiate with the self and non-self, thereby preventing the body's own cells from attack during response. The immune system can be divided into two branches—the **Humoral** and **Cellular**. The humoral immune system relies on B lymphocytes to produce **plasma cells**, which thereby lead to the production of soluble antibodies that will bind the foreign antigens. The cellular immune system uses helper T lymphocytes (**Th**) and killer T lymphocytes (**CTLs**) to recognize and destroy the antigens. The primary immune response occurs on initial contact with a foreign antigen and results in production of immunoglobulin M (IgM) and then immunoglobulin G (IgG). If the same antigen is encountered again, immunological memory leads to a secondary immune response that produces a much more rapid and larger increase in specific IgG production. A large number of antibody-producing cells exist in an animal even before they encounter a foreign antigen, each cell producing only one specific antibody and displaying this on its cell surface. An antigen binds to cells that display antibodies with appropriate binding sites and causes proliferation of those cells to form clones of cells secreting the same antibody in high concentration. Antibodies bound to an invading microorganism activate the complement system via the classical pathway. This consists of a cascade of proteolytic reactions leading to the formation of membrane attack complexes on the plasma membrane of the microorganism that cause its lysis. Polysaccharides on the surface of infecting microorganisms can also activate complement directly in the absence of antibody via the alternative pathway.

Parts of the immune system which interact to provide overall protection for the animal:

- The **humoral immune response** (*humor* is an ancient term meaning fluid) relies on the production of soluble proteins called antibodies (or immunoglobulins) by **B lymphocytes**, so called because the cells mature in the bone marrow. B lymphocytes are often called simply **B cells** in abbreviated form.
- The **cellular immune response** is mediated by **T lymphocytes**, so called because their maturation from stem cells occurs in the thymus. In cellular immunity, it is the intact T lymphocytes themselves that are responsible for the recognition and killing of foreign invaders. These cells are the **cytotoxic T lymphocytes (CTL)**, also called **killer T cells**. Other T lymphocytes have another role; they provide essential help for B lymphocytes to produce antibodies and so are called **helper T cells**.

## PRIMARY AND SECONDARY IMMUNE RESPONSE

The presence of a foreign antigen stimulates the production of a specific antibody in the bloodstream which will recognize and bind tightly to it. The initial encounter of the naïve immunocompetent lymphocyte with an antigen produces primary immune response. The antibody molecules first produced after antigen injection are of the immunoglobulin M class and are called IgM molecules. However, about 10 days after antigen injection, the amount of IgM in the bloodstream (the **titer** of antibody) declines and there is a concurrent increase in another class of antibody called immunoglobulin G (IgG). In primary immune response, there is a lag period of 5 to 7 days before the antibody levels rise and peak is attained on $14^{th}$ day approximately, and then the antibody levels decline.

One of the most important features of the immune system is that, once an animal has encountered a particular pathogen, the system confers protection against future infection. This **immunological memory** means that if the same pathogen or antigen is encountered a second time, perhaps even decades after the previous occurrence, then the system reacts much faster and more dramatically to produce a large titer of specific IgG to counter the antigen. This is the **secondary immune response** and is mediated by long-lived memory T cells and memory B cells. In secondary immune response the lag is much shorter i.e., 1-2 days and antibody levels are consequently 100 to 1000 fold higher.

## CLONAL SELECTION

The **clonal selection theory** elucidates the mode of operation of the immune system by taking into account the following points:

- **Specificity:** Prior to the encounter of an antigen, each B cell develops a specific type of antibody molecule, which is anchored in the plasma membrane to facilitate its exposure to the foreign environment. Numerous B cells exist in the body against a number of antigens. A particular cell, which interacts with the antigen, gets stimulated to divide and forms an identical population of cells referred to as **clone**.
- **Memory:** Memory cells are also generated to counter the antigen if it invades the body in future. Hence, immunologic memory is also a consequence of clonal selection. During clonal selection, specific lymphocyte against a specific antigen is amplified. The lymphocytes commonly referred to as memory cells appear to have a longer life span than the naive lymphocytes from which they arise.

The antibody now produced does not remain anchored to plasma membrane, but is instead secreted.

## SELF-TOLERANCE

Cells that produce antibody, which reacts with normal body components and might create havoc in the organism, are killed early in fetal life so that the adult animal is normally unable to make antibodies against self. This inability to make antibodies against self is called **self-tolerance**.

## COMPLEMENT

Complement comprises a complex set of over 30 proteins found in plasma and on cell surfaces which, along with blood clotting, fibrinolysis and kinin formation,

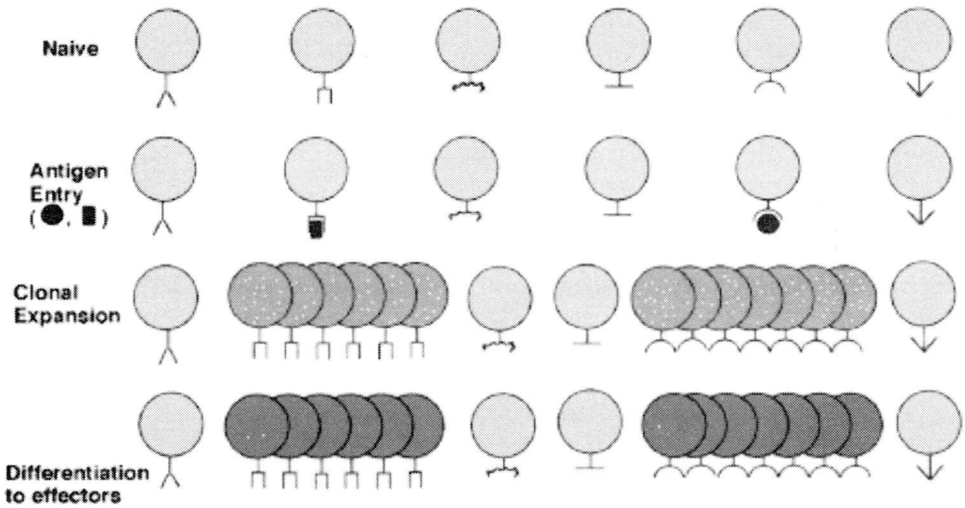


**Fig. 1.** The clonal selection theory of lymphocytes. Schematic diagram of clonal selection hypothesis illustrating the idea that each naïve lymphocyte has a different receptor specificity, each of which can bind a different antigenic determinant. When a pathogen is recognized by the cells, in this case by two different antigenic determinants, then the cells that bind to these determinants are selected to proliferate or undergo clonal expansion, and then differentiate into effector cells that either secrete antibody or mediate various effector mechanisms of cell-mediated immunity. From Abbas and Janeway *Cell* 100:129, 2000.

forms one of the triggered enzyme systems found in plasma. The complement system characteristically produces a fast, highly amplified response to a trigger stimulus mediated by a cascade phenomenon where the product of one reaction is the enzymic catalyst of the next. The activated or the split products of the cascade have a variety of defensive functions, and the complement proteins can therefore be regarded as a crucial part of the innate immune system. Some of the complement components are designated by the letter "C" followed by a number, which is related more to the chronology of its discovery than to its position in the reaction sequence. The most abundant and the most pivotal component is C3.

When the recognition function of the humoral immune system has been carried out by the production of specific antibodies and their binding to foreign antigens, destruction of the invading pathogen is the next step. One main defense pathway is the **complement system,** which is activated by the antibodies bound to the invading microorganism and eventually causes it to lyse by punching holes in its plasma membrane.

Immunoglobulin molecules bound to the surface of the microorganisms activate **C1**, the first component of the complement pathway. The activation occurs through the Fc portion of the bound antibody. Only bound antibody can activate complement; soluble antibody not bound to an antigen has no such effect.

Therefore, the early steps in complement activation consist of a **proteolytic cascade** in which more and more molecules are activated at each step.

Component C3 is the key component whose cleavage leads to the assembly of **membrane attack complexes** on the plasma membrane of the microorganisms, which create holes in the plasma membrane that lead to cell death. Various white blood cells also become activated during this process and phagocytose the pathogen. This pathway of complement activation, which starts when antibody has bound to antigen on the microbe surface, is called the **classical pathway**. An **alternative pathway** of activation also exists which is activated directly by polysaccharides in the cell wall of microorganisms even in the absence of antibody. The alternative pathway therefore defends the body against attack in the early stages before an immune response can occur and also augments the effects of the classical pathway of complement activation when the immune response has occurred.

## Antibodies

### *Light and heavy chains*

An antibody molecule is composed of two identical heavy chains and two identical light chains linked by disulphide bonds. Each heavy and light chain is composed of variable region and a constant region. Each light chain also called '**Bence-Jones**' protein is composed of about 200-220 amino acids where as each heavy chain is composed of about 330 ($\delta$, $\gamma$ and $\alpha$) to 440 ($\mu$ and $\varepsilon$) amino acids. Every molecule of an antibody acquires a '**Y**' shaped structure. The N-terminal ends of a light chain and its associated heavy chain coordinate to form an antigen binding site, so that an antibody molecule has two binding sites for antigen, hence called **bivalent**. Because of this, a single antibody molecule can bind two antigen molecules and so cross-link and precipitate antigens.

### *Antibody domains*

Each light chain consists of two repeating segments of about 110 amino acids that fold into two compact three-dimensional **domains**, one representing the variable region of the light chain and the other domain representing the constant region. Each heavy chain is also made up of repeating units about 110 amino acids long. Since each IgG heavy chain is about 440 amino acids long, it forms four domains, one domain for the variable region and three domains in the constant region. The similarity of amino acid sequence between the various domains suggests that they arose in evolution by gene duplication. **Fab and Fc** Papain, a protease, cuts the IgG molecule to release the two arms of the Y **fragments** shaped molecule, each of which has one antigen-binding site and is called an **Fab fragment** (Fragment antigen binding). Because Fab fragments have only one antigen-binding site (i.e. are **univalent**), they cannot cross-link antigens. The released stem of the Y-shaped molecule (consisting of the identical C-terminal parts of the two H chains) is named the **Fc fragment** (so called because it readily crystallizes). The Fc fragment carries

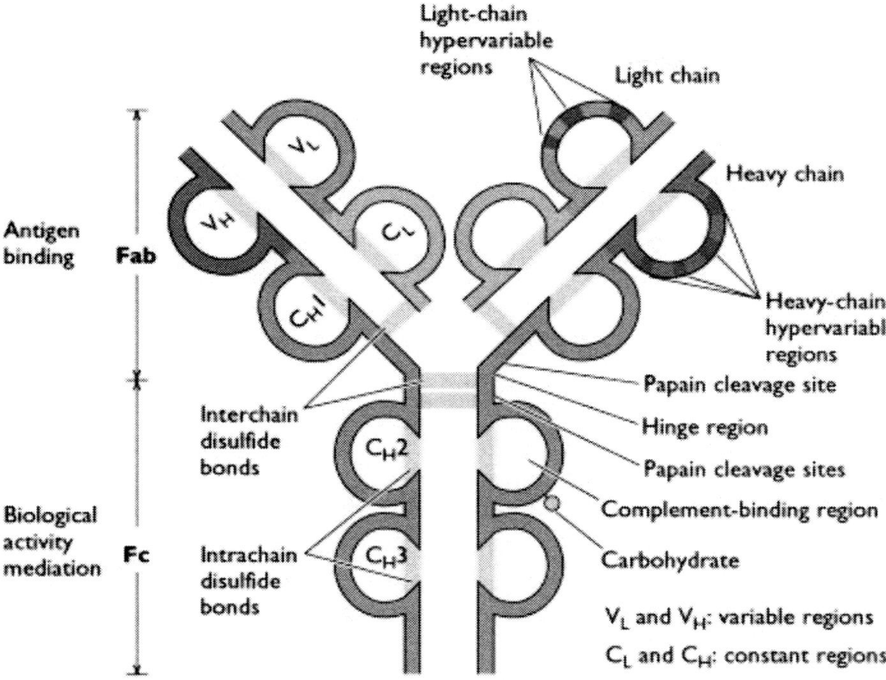


**Fig. 2.** Schematic diagram showing heavy and light chains of antibody molecules. Antibody molecule is composed of two heavy chains and two light chains. The molecule has two antigens binding sites each consisting of a heavy chain & light chain. Each heavy & light chain further includes a variable region (v) and a constant region (c) Antigen binding occurs at the variable region.

the effector sites that trigger the destruction of the antigen, for example triggering of the complement system and inducing phagocytosis of pathogens by other white blood cells. In contrast to papain, pepsin (another protease) cuts the IgG molecule to release the two arms of IgG still linked together and hence this fragment has two antigen-binding sites (i.e. is bivalent) and can still cross-link antigens. This is called the **F(ab′)$_2$ fragment**.

Humans have five different classes of antibody molecule which differ both in structure and in function. These are called immunoglobulin A (IgA), IgD, IgE, IgG and IgM and each has its own type of heavy chain; α, δ, ε, γ and μ, respectively.

Thus IgA molecules have two identical α heavy chains, IgD molecules have two identical ä heavy chains, etc. The human IgG class of antibodies is further divided into four IgG subclasses; IgG1, IgG2, IgG3 and IgG4, having $\gamma_1$, $\gamma_2$, $\gamma_3$ and $\gamma_4$ heavy chains respectively. The different heavy chains confer different properties and functions on each of the immunoglobulin classes:

- **IgM** has μ heavy chains and exists as a pentamer in combination with another polypeptide called the J chain, which is responsible for initiating the polymerization to form the pentameric structure. With its large number of antigen-binding sites, each IgM molecule binds very tightly to any pathogen that has multiple copies of the same antigen on its surface. The binding induces the Fc region to activate the complement pathway which eventually causes the death of the pathogen. IgM also activates macrophages to phagocytose pathogens. Not surprisingly given these functions, IgM is the first antibody produced when an animal responds to a new antigen.

- **IgG** is the main immunoglobulin in the bloodstream late in the primary immune response and particularly during the secondary immune response. Like IgM, it can activate complement and trigger macrophages, but is the only antibody that can pass through the placenta and so provide immunological protection for the fetus. It is also secreted into the mother's milk and is taken up from the gut of the newborn animal into the bloodstream, thus providing continuing protection after birth.
- **IgA** is the main class of antibody in secretions such as tears, saliva, and in secretions of the lungs and the intestine. It is the first line of immunological defense against infection at these sites.
- **IgE** occurs in tissues where, having bound the antigen, it stimulates mast cells to release a range of factors. Some of these in turn activate white blood cells (called eosinophils) to kill various types of parasite. However, the mast cells can also release biologically active amines, including histamine, which cause dilation and increased permeability of blood vessels and lead to the symptoms seen in allergic reactions such as hay fever and asthma. It is a villain responsible for leading to allergic reactions.
- **IgD** is found on the surface of mature B lymphocytes and in traces in various body fluids, but its exact function remains unclear.

Two different forms of light chains also exist. Antibody molecules in any of the antibody classes or sub-classes can have either two κ light chains or two λ light chains. Unlike the different heavy chains described above, no difference in biological function between κ and λ light chains is known.

## Antibodies

### *Polyclonal and monoclonal antibodies*

**Polyclonal antibodies** If an antigen is injected into an animal, a number of antibody-producing cells will bind that antigen, albeit with varying degrees of affinity, and so the antibody which appears in the bloodstream will have arisen from several clones of cells, that is it will be a **polyclonal antibody**. Different antibody molecules in a preparation of polyclonal antibody will bind to different parts of the macromolecular antigen and will do so with different binding affinities. The binding region recognized by any one antibody molecule is called an **epitope**. Most antibodies recognize particular surface structures in a protein rather than specific amino acid sequences (i.e. the epitopes are defined by the conformation of the protein antigen). A preparation of polyclonal antibodies will bind to many epitopes on the protein antigen.

### *Monoclonal antibodies*

If a single clone of antibody-producing cells could be isolated, then all of the antibody produced from that clone would be identical; all antibody molecules in such a **monoclonal antibody** preparation would bind to the same antigen epitope. The problem is that if an individual antibody-producing cell is isolated and grown in culture, its descendants have a limited lifespan that severely limits their use for the routine preparation of monoclonal antibodies. In 1975, Milstein and Kohler discovered how monoclonal antibodies of almost any desired antigen specificity can be produced indefinitely and in large quantities. Their method was to fuse a B lymphocyte producing antibody of the desired specificity with a cell derived from a cancerous lymphocyte tumor, called a myeloma cell, which is immortal. The cell fusion is called a **hybridoma**, which is both immortal and secretes the same specific antibody originally encoded by the B lymphocyte.

Monoclonal antibodies produced using these technologies are now common tools in research because of their very high specificity. For example, they can be used to locate particular molecules within cells or particular amino acid sequences within proteins. If they are first bound to an insoluble matrix, they are also extremely useful for binding to and hence purifying the particular molecule from crude cell extracts or fractions. They are also increasingly of use in medicine, both for diagnosis and as therapeutic tools, for example to inactivate bacterial toxins and to treat certain forms of cancer.

## ANTIBODY SYNTHESIS

### Somatic Recombination

In most animals, it is possible to distinguish **germ-line** cells from **somatic** cells. The germ-line cells are those that give rise to the male and female **gametes** (sperm and ova, respectively) whilst the somatic cells form the rest of the body structures of the individual animal. The importance of the germ-line cells is that it is these that ultimately give rise to the next generation. The human genome is thought to contain fewer than 105 genes yet a human can make at least 1015 different types of antibody in terms of antigen-binding specificity. Clearly the number of genes is far too small to account for most of this antibody diversity. Thus a **germ-line hypothesis**, whereby all antibodies are encoded by genes in germ-line cells, must be incorrect. In fact, there are no complete genes in germ-line cells that encode complete light chains or complete heavy chains. Instead the genes exist in separate coding sections and are assembled during B-lymphocyte maturation by a process called **somatic recombination.** This process of assembly takes place in every B lymphocyte. By assembling different fragments of DNA, completely new immunoglobulin genes can be created and hence this gives an enormous potential reservoir of antibody diversity.

### *Recombination of light chain genes*

The variable (V) region of a λ **light chain** is encoded by a separate DNA sequence from that encoding the constant (C) region. These two segments lie on the same chromosome but they are sited some way apart. During maturation of the B lymphocyte, the V and C region DNAs are moved next to each other and are joined together to create a functioning light chain gene. This somatic recombination usually occurs by deletion of the DNA between the germ-line V and C regions but can also occur in some cases by an inversion mechanism. In fact, the situation is rather more complex than this simple model implies. The germ-line **V gene segment** encodes only the first 95 amino acids of the variable region of the light chain polypeptide. The remaining few amino acids of the light chain variable region (residues 96–108) are encoded by a piece of DNA called the **J gene segment.** This J segment (for 'joining') must not be confused with the J polypeptide in IgM pentamers. In the germ-line, the J gene segment lies just upstream of the **C gene segment** and separated from it only by an intron. Furthermore, there are multiple V gene segments (about 300) and five J gene segments, one of them inactive. During B-lymphocyte differentiation, one of the 300 or so V regions becomes joined precisely to one of the J gene segments to create a light chain gene. This somatic recombination process is called **VJ joining**. V3 has been chosen for recombination and has joined to J2. Transcription now starts just upstream of the recombined V segment (just upstream of V3) and continues until the end of the C segment. The other J segment sequences are also transcribed, but these sequences are lost during subsequent RNA splicing that removes the intron upstream of the C segment. Thus the final mRNA contains only V3J2C sequences and encodes a corresponding light chain polypeptide. A very large number of different light chains can be made depending on which one of the 300 V segments is chosen and joined to which one of the J gene segments. The λ **light chain genes** also arise by somatic recombination during maturation of the B lymphocyte, but there are far fewer V and J gene segments than for κ chain genes. Most antibody molecules have κ light chains and not λ light chains. **Recombination of** Heavy chains are synthesized in an analogous manner but are encoded by four **heavy chain genes** gene segments, V, J, D and C. There are about 200–1000 VH segments, about 15 active DH segments (D for diversity, H for heavy) and four JH segments. Thus whereas the variable region of a light chain polypeptide is encoded by V and J segments, the variable region of a heavy chain is encoded by V, D and J segments. In the heavy chain gene system, there are also several C segments, one for each class of heavy chain; Cμ, Cδ, the various Cγ subclasses, Cε and Cα encoding μ, δ, γ, ε and α heavy chain constant regions respectively. During lymphocyte maturation, **two heavy chain gene rearrangements occur**. First a chosen DH segment joins a JH segment (**DJ joining**) and then the recombined DHJH joins a chosen VH segment (**VDJ joining**). For the assembly of both heavy chain and light chain genes, the ends of the various DNA segments to be joined can also undergo modification during the recombination process and this modifies existing codons at these junctions or even creates new codons, thus increasing antibody diversity still further. In addition, antibody genes exhibit a higher than normal rate of mutation.

**Class switching** When the complete heavy chain gene is transcribed, it generates a heavy chain for an IgM antibody since it is always the first C segment after the recombined VDJ that is transcribed. To switch to making a heavy chain for a different antibody class, say IgA, the lymphocyte DNA has to undergo yet another recombination event that moves Cá next to the recombined VDJ and deletes the intervening C segments. When this new gene is expressed, it will synthesize a $C\alpha$ heavy chain for an IgA antibody instead of the earlier $C\mu$ heavy chain for an IgM antibody. One crucial point about this **class switching** or **heavy chain switch** is that only the C region of the synthesized heavy chain changes; the variable region stays the same as before the switch. The specificity of the antibody is determined by the antigen binding site, which is formed by the variable regions of the heavy and light chains and not by the constant regions. Thus even when class switching occurs so that the lymphocyte now makes IgA instead of IgM, the specificity of the antibody for antigens stays the same.

## ANTIBODIES AS WORKING TOOLS

**Immunolocalization** The availability of an antibody (immunoglobulin) against a specific antigen offers the opportunity to use that antibody in a range of immunological methods. The site recognized by an antibody on the antigen is called the **antigenic determinant** or **epitope**. The high specificity of an antibody for its epitope allows it to be used as a reagent for determining the location of the antigen in a cell (**immunocytochemistry**), for example by coupling a fluorescent label to the antibody and then using fluorescence to localize its sites of binding by **immunofluorescence light microscopy**. Even higher resolution can be achieved using antibody to which electron-dense particles, such as ferritin or colloidal gold, have been coupled and then viewing the binding sites using electron microscopy. Indeed, **immuno-electron microscopy** can map the position of protein antigens within macromolecular complexes such as ribosomes.

**ELISA** Specific antibodies can also be used to quantify the amount of the corresponding antigen in a biological sample. Several types of immunological assays exist. An increasingly popular version is **enzyme-linked immunosorbent assay (ELISA)** which can readily detect and quantify less than a nanogram of a specific antigenic protein. In ELISA, the specific antibody is coupled to a solid support. A convenient format for ELISA is to use a plastic tray that has molded wells in it (a **microtiter tray**) where the antibody has been coupled to the plastic forming the wells. Samples to be assayed are added to the wells. If antigen is present that is recognized by the antibody, it becomes bound. The wells are then washed to remove unbound protein and incubated with a second antibody that recognizes the protein but at a different epitope than the first antibody. The second antibody is attached to an enzyme that can catalyze the conversion of a colorless or nonfluorescent substrate into a colored or fluorescent product. The intensity of the color or fluorescence produced for each sample is then measured to determine the amount of antigen present in each sample. Several machines are now commercially available that scan the wells of microtiter plates following ELISA and quantify the amount of antigen bound in each well.

### *Western blotting*

The **Westernblot** (alternatively, **protein immunoblot**) is a widely used analytical technique used to detect specific proteins in the given sample of tissue homogenate or extract. It uses gel electrophoresis to separate native or denatured proteins by the length of the polypeptide (denaturing conditions) or by the 3-D structure of the protein (native/non-denaturing conditions). The proteins are then transferred to a membrane (typically nitrocellulose or PVDF), where they are probed (detected) using antibodies specific to the target protein.

Western blotting can be used for detection of one or more antigens in a mixture. The sample is electrophoresed on an SDS–polyacrylamide gel that separates the proteins on the basis of size, resulting in a series of protein bands down the gel. Because the gel matrix does not let large proteins such as antibodies enter readily, the sample proteins must be first transferred to a more accessible medium. This process is called **blotting.** The gel is placed next to a nitrocellulose or nylon sheet and an electric field is applied so that proteins migrate from the gel to the sheet where they become bound. This particular form of blotting (i.e. blotting of proteins) is called **Western blotting** to distinguish it from blotting of DNA (Southern blotting) or RNA (Northern blotting). Next, the Western blot is incubated with a protein such as casein to bind to nonspecific protein-binding sites and hence prevent spurious binding of antibody molecules in subsequent steps. This step is said to 'block' nonspecific binding sites. The Western blot is then reacted with labeled antibody, unreacted antibody is washed away and those protein bands that have bound the antibody become visible and hence are identified. The method of visualization depends on how the antibody was labeled. If the labeling is by the incorporation of a radiolabel, then autoradiography is carried out to detect the radioactive protein bands. Alternatively, the antibody may be detected by incubating the sheet with a second antibody that recognizes the first antibody (e.g. if the first specific antibody was raised in rabbits, the second antibody could be a goat anti-rabbit antibody). The second antibody could be radiolabeled and its binding detected by autoradiography or it could be conjugated to an enzyme that generates a colored product as in ELISA. Western blotting can also be used to analyze specific antigens after two-dimensional gel electrophoresis which resolves proteins as spots, separated on the basis of both charge and size.

### *Immunoaffinity chromatography*

**Immunoaffinity chromatography** is one example of a range of different separation procedures generally called affinity chromatography which depend on high affinity interactions between two components. In immunoaffinity chromatography, a specific antibody to a protein antigen may be coupled to an inert matrix such as polysaccharide beads. These can be placed in a column and the cell sample loaded on. The protein antigen will bind to the matrix but other components will flow through the column. The protein antigen can then often be eluted in pure, or almost pure form; well over 1000-fold purification is routinely achieved in this single step. If the protein antigen is normally exposed on the plasma membrane of a desired cell type, these cells can be purified from a cell mixture by passing the mixture through the column. Only cells bearing the antigen on their surface will bind and can be eluted subsequently.

### *Apoptosis: cell suicide*

The immune system is the only organ to require massive cellular expansion within a period of hours in response to infection. This capacity brings with it some danger: activated cytotoxic cells, complete with toxic mediators and potent cytokines, must be controlled to avoid tissue damage by friendly fire. The removal of unwanted, activated cells is achieved by inducing them to commit suicide, a process termed apoptosis. Major advances in the understanding of the molecular events of apoptosis are beginning to offer new therapeutic options. The main molecule concerned with the form of apoptosis used to regulate death of activated lymphocytes (activation induced cell death) is Fas (CD95), widely expressed on T and B lymphocytes. Fas expression on these cells increases after encounter with antigen. Binding of Fas to its ligand leads to apoptosis of the cell expressing Fas through activation of an intracellular cascade of proteases. The expression of the Fas ligand is tightly regulated in the immune system, but it begins to be expressed on mature CD4 and CD8 T cells after activation. Thus, towards the end of an immune response the Fas ligand on activated T cells is able to bring about activation induced cell death in cells which have upregulated Fas. Understanding the mechanisms of activation induced cell death has had great bearing on some clinical problems. Firstly, a rare early childhood syndrome characterised by lymphoproliferation with lymphadenopathy, autoimmune haemolysis, thrombocytopenia, and hypergammaglobulinaemia has recently been attributed to mutations in the Fas gene. Activation induced cell death in these patients is severely defective, allowing lymphoproliferation to go unchecked.

Secondly, Fas expression has been noted on thyroid epithelial cells in patients with autoimmune thyroiditis. Thyrocyte damage in this condition is thought to be due to apoptosis induced by the interaction between activated T lymphocytes bearing the Fas ligand and thyrocytes bearing Fas. This is a new explanation for the pathogenesis of destructive autoimmune disease and may allow new intervention strategies which interfere with the Fas system to be devised. Finally, Fas may be important in transplantation. It has been known for many years that foreign grafts transplanted into certain sites (notably the eye and testis) are not rejected. The conventional explanation for this immune privilege has been a lack of lymphatics in these organs. However, the Fas ligand is permanently expressed in both the eye and the testis. Thus the immune privilege in these sites may be attributed to the killing of any lymphocytes expressing Fas in response to the allograft by the cells with the Fas ligand. This fascinating possibility is already being developed therapeutically to protect allografts. In experimental animals, islet cell allografts wrapped in myocytes genetically engineered to express the Fas ligand are not rejected, whereas those wrapped in unmanipulated myocytes are. Not all of the information about Fas and its ligand is good, however. Recent reports of cancer cells expressing the Fas ligand and inducing activation induced cell death in tumour specific cytotoxic T cells have lead to articles entitled "The tumour cell strikes back."

### *Apoptosis as a therapeutic tool in IBD?*

Defective apoptosis of mucosal cell populations seems to be a relevant pathogenetic mechanism in inflammatory bowel disease (IBD). It has been suggested that the induction of apoptosis in various effector cells may be a relevant therapeutic mechanism in IBD. Indeed, it was shown that different drugs used for treatment of IBD have the capacity to induce apoptosis in T cells or monocytes in vitro and in vivo. However, it remains unclear whether these observations are related to clinical efficacy of these agents. TNF-alpha is one of the most relevant proinflammatory mediators in IBD and anti-TNF treatment has been shown to be of particular benefit for patients with IBD. It could subsequently be shown that various anti-TNF-alpha agents, such as infliximab and adalimumab, can induce apoptosis in activated monocytes and lymphocytes in vitro and in vivo. This mechanism requires reverse signaling via transmembranous TNF, thereby eliciting a signal transduction cascade that results in programmed cell death. Although other mechanisms might also contribute to the clinical effect of anti-TNF-alpha, current data suggest that apoptosis is a relevant mechanism that is associated with clinical efficacy of anti-TNF agents. Induction of apoptosis in activated monocytes or T cells may be regarded as therapeutic tool not only for anti-TNF agents, but also for other drugs used in IBD. Future strategies should focus on identification of mechanisms that prevent apoptosis in the mucosa of patients with IBD and in targeting apoptotic pathways as a therapeutic strategy in IBD.

### Antigen recognition by T cells

T cell receptors for antigen do not bind soluble antigen like their antibody counterparts on B cells. Instead, an elaborate and complex process exists, whereby polypeptide antigens are internalised and processed by specialist antigen presenting cells, giving rise to short peptides. These in turn bind to molecules of the major histocompatibility complex and the combination is recognised by the T cell receptor. It is understandable, therefore, why in 1987 the immunology world was captivated by the first *x* ray crystallographic study of the structure of a molecule of the major histocompatibility complex. The crystal structure shows a long groove, perched on the backbone of the molecule, in which antigen derived peptides are embedded for recognition by the T cell receptor. Owing to difficulties in solubilising sufficient quantities of these receptors, the equivalent crystallographic view of a T cell receptor bound to peptide in complex with major histocompatibility complex has taken 10 years to be developed. The crystal structure shows that the T cell receptor sits diagonally over the groove in the molecule from the major histocompatibility complex. As a consequence, large portions of the receptor interact with the walls of the molecule of the major histocompatibility complex itself. Only a small region of the receptor interacts with the peptide and then only with a small part of the peptide antigen. With this view, it is more understandable why foreign molecules of the major histocompatibility complex encountered in an allograft look so different and thus induce such potent T cell responses. Functional studies on T cell recognition of antigen have also revealed some surprises. It was assumed for many years that T cells were activated in an all or none fashion, but by altering peptide antigens slightly it has become apparent that T cells have many grades of response, from activation to complete unresponsiveness (anergy).Thus T cells behave just like a pharmacological receptor, with agonists, partial agonists, and antagonists. This has led to the possibility of using slightly altered peptides as ligands to switch off unwanted T cell responses even damaging Th1 responses can be switched to protective Th2-like activity. This and other developments in antigen recognition are important for diseases and treatments in which immune effectors have a role—for example, autoimmunity and immunisation. It has now been shown unequivocally that a single T cell, reactive with a single self antigen, can also be stimulated by a peptide antigen from a virus. In this study the T cells came from a patient with multiple sclerosis and were directed against myelin components from the central nervous system, but they could also be switched on by peptides from common viruses such as herpes simplex and Epstein-Barr. It is easy to see why the role of viruses in autoimmune disease is one of the hottest topics in immunology.

### *Progress in cellular immunology leads to insights into T-cell effector functions: "two signal models" for lymphocyte activation*

The origins of cellular immunology date back to Metchnikoff, who discovered the important role that phagocytes play in immunity. Although he mistakenly identified the phagocyte, itself, as the cell type responsible for antibody production, it would take another 58 years before any further understanding of the cellular basis of immunity would exist. In 1942, Landsteiner and Chase demonstrated that cells, rather than antibodies, were recognized to contribute to specific immune phenomena. As there was no technology available to separate different types of circulating leukocytes, the nature of the immune-conferring cell remained elusive. It was not until the experiments of Gowans in 1959 that lymphocytes were shown to confer specific immunity. In many ways, Gowans' discovery of the central role of lymphocytes in immunity was analogous to the molecular genetic experiments of Avery, McLeod, and Lederberg, who demonstrated that DNA was the substance that conferred heredity.

The focus on the lymphocyte led to the discovery of the involvement of the thymus in cellular immunity and the distinction between T-cells, which provided "help" or cytotoxicity, and B-cells, which produced antibody (Jacques Miller and Graham Mitchell, 1961-1968). These workers also discovered the importance of T- and B-cell collaboration in the immune response. In 1968, Bretscher and Cohn proposed the first two-signal model of lymphocyte activation. Although the original model proved to be incorrect, the model has been refined over the years to take into account new experimental findings.

The two signal hypothesis basically states that, in addition to the delivery of antigen/MHC-mediated signals to the lymphocyte, there must be another signal provided by the MHC-bearing antigen-presenting cell (Fig. 3). The nature of this "second stimulus" was unknown until the the interaction of CD28 on the T-cell and B7 (CD80/CD86) on the antigen-presenting cell was necessary for optimal production of antibody. Later studies (1992) identified CD40 ligand (CD154) on activated T-cell as a critically important molecule that provides co-stimulation to B-cells; in humans. In the following year, it was demonstrated that mutations in the gene for CD40 ligand was responsible for a human disease, X-linked hyper-IgM immunodeficiency

However, even before the precise identity of co-stimulatory molecules was known, a theoretical framework for T-cell function was proposed by Burnet (1959) and later refined over the ensuing years. These models for T-cell function incorporated the "two signal" concept of lymphocyte activation, originally proposed by Bretscher and Cohn, into a model of immune recognition based on "self vs non-self." (Table 1).

**Table 1** Evolution of "two signal" model of lymphocyte activation

| *Basis of Immune Recognition* | *Cellular Determinants* | *Authors* |
|---|---|---|
| Self vs. Non-self | B-cell/Ag | Burnet, 1959 |
| | Self vs. Non-self | Signal one (B cell/Ag) |
| Signal two (T helper) | Bretscher and Cohn, 1970 | |
| | Self vs. Non-self | Signal one (T cell-Ag) |
| Signal two (APC) | Lafferty and Cunningman, 1975 | |
| | Non-infected self vs. Infected non-self | Signal one (T cell-Ag) |
| Signal two | | |
| (APCs activated by PRR*) | Janeway, 1989 | |
| Danger | Signal one (T cell-Ag) | Matzinger, 1994 |
| | Signal two (APCs activated by distressed cell) | |

*Pattern recognition receptor

How valid are these models? There is no doubt that the activation of T-cells requires multiple signals from antigen-presenting cells. Where the models diverge, and where the most controversy exists, is the very basis of immune recognition. What constitutes "non-self"? Because the immune response in any individual

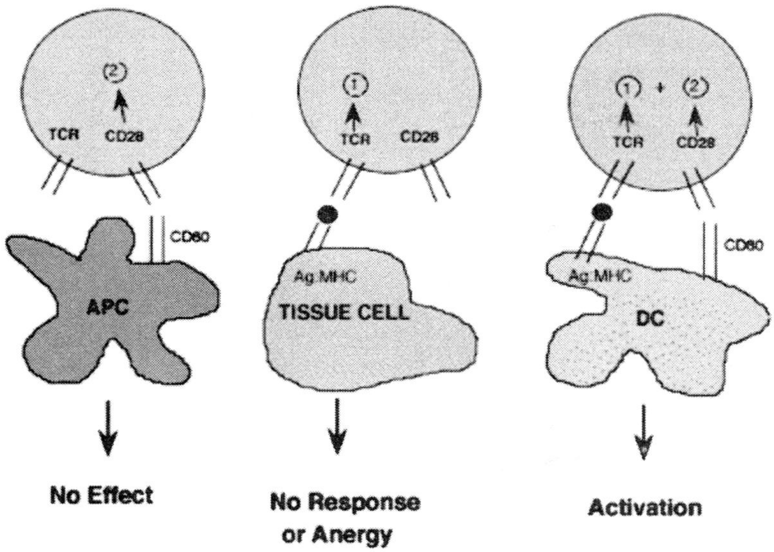


**Fig. 3.** The Two-signal theory of T-cell activation. When a T-cell encounters an antigen presenting cell (APC) that expresses costimulatory molecules but no foreign antigens, there is no apparent signal (left panel); when a T cell encounters a tissue cell or an APC that expresses antigens recognized by the TCR in the absence of costimulatory signals, the result is either no response or inactivation of the cell (anergy). However, when a T cell encounters an activated dendritic cell or other APC expressing both antigen and costimulatory molecules, then the T cell is activated to proliferate and undergo differentiation to an effector cell. Thus, T cells only become activated by their ligands when they are presented by a cell that expressescostimulatory molecules. From Abbas and Janeway *Cell* 100:129, 2000.

depends upon the selection of a small subset of lymphocytes out of a total of $10^5$ to $10^6$ possible clones, the very existence of these clones implies that they were positively selected by the immune system. But this occurs through an ongoing low-level presentation of self antigens, thus constituting a paradox: recognition of "non-self" depends on prior recognition of "self." How can this be resolved?

**Table 2** History of immunology time line

| | | |
|---|---|---|
| 1798 | Smallpox vaccination | Edward Jenner |
| 1876 | Validation of germ theory of disease by discovering bacterial basis of anthrax | Robert Koch |
| 1879 | Chicken cholera vaccine development | Louis Pasteur |
| 1890 | Discovery of Diphtheria"anti-toxin" in blood | Emil Von Behring, Shibasaburo Kitasato |
| 1882 | Isolation of the tubercule bacillus | Robert Koch |
| 1883 | Delayed type hypersensitivity | Robert Koch |
| 1884 | Phagocytosis; Cellular theory of immunity | Elie Metchnikoff |
| 1891 | Proposal that antibodies are responsible for immunity | Paul Ehrlich |
| 1891 | Passive immunity | Emil Roux |
| 1894 | Complement and antibody activity in bacteriolysis | Jules Bordet |
| 1900 | A, B, and O blood groupings | Karl Landsteiner |
| 1901 | Cutaneous allergic reaction | Maurice Arthus |

**Table 2** Contd.

| | | |
|---|---|---|
| 1903 | Opsonization by antibody | Almroth Wright & Stewart Douglas |
| 1907 | Discipline of immunochemistry founded | Svante Arrhenius |
| 1910 | Viral immunology theory | Peyton Rous |
| 1917 | Haptens dicovered | Karl Landsteiner |
| 1921 | Cutaneous reactions | Carl Prausnitz & Heinz Kustner |
| 1924 | Reticuloendothelial system | Ludwig Aschoff |
| 1939 | Discovery that antibodies are gamma globulins | Elvin Kabat |
| 1942 | Adjuvants | Jules Freund & Katherine McDermott |
| 1942 | Cellular transfer of sensitivity in guinea pigs (anaphylaxis) | Karl Landsteiner & Merill Chase |
| 1944 | Immunological hypothesis of allograft rejection | Peter Medawar |
| 1948 | Demonstration of antibody production in plasma B cells | Astrid Fagraeus |
| 1949 | Distinguishing self vs. non-self and its role in maintaining immunological unresponsiveness (tolerance)to self | Macfarlane Burnet & Frank Fenner |
| 1952 | Discovery of agammaglobulinemia (antibody immunodeficiency) | Ogden Bruton |
| 1953 | Immunological tolerance hypothesis | Rupert Billingham, Leslie Brent, Peter Medawar & Milan Hasek |
| 1955-59 | Clonal selection theory | Niels Jerne, David Talmage, & Macfarlane Burnet |
| 1957 | Discovery of Interferon | Alick Isaacs & Jean Lindemann |
| 1958 | Identification of first autoantibody and first recognition of autoimmune diseases | Henry Kunkel |
| 1959-62 | Elucidation of antibody structure | Rodney Porter & Gerard Edelman |
| 1959 | Lymphocytes as the cellular units of clonal selection; Discovery of lymphoid circulation | James Gowans |
| 1961 | Discovery of thymus involvement in cellular immunity | Jaques Miller |
| 1968 | Recognition of B- and T-cells in immunodeficiencies | Robert Good |
| 1968 | Distinction of bone marrow- and thymus-derived lymphocyte populations; Discovery or T- and B-cell collaboration | Jacques Miller & Graham Mitchell |
| 1965 | Proof that B-cells demonstrate allelic exclusion | Benvenuto Pernis |
| 1968-70 | Elaboration of two-signal model of lymphocyte activation | Peter Bretscher & M. Cohn |
| 1970 | Discovery of membrane immunoglobulins | Benvenuto Pernis & Martin Raff |
| 1971 | Recognition of hypervariable regions in Ig chains | Elvin Kabat |
| 1972 | Elucidation of the major histocompatibility complex | Baruj Benacerraf, Jean Dausset, & George Snell |
| 1974 | Discovery of MHC restriction | Rolf Zinkernagel & Peter Doherty |
| 1974 | HLA-B27 predisposes to an autoimmune disease | Derek Brewerton |
| 1975 | Monoclonal antibodies used in genetic analysis | Georges Kohler & Cesar Milstein |
| 1976 | First demonstration of cross-priming | Michael Bevan |
| 1978 | Direct evidence for somatic rearrangement in immunoglobulin genes | Susumu Tonegawa |
| 1978 | Recognition that dendritic cells are distinctive and highly potent antigen-presenting cells | Ralph Steinman |
| 1979 | Discovery of leukocyte adhesion molecules and their role in lymphocyte trafficking | Eugene Butcher |
| 1984-87 | Identification of genes for the T cell antigen receptor | Mark Davis, Leroy Hood, Stephen Hedrick, & Gerry Siu |
| 1987 | Crystal structure of MHC-peptide solved | Pam Bjorkman, Jack Strominger, & Don Wiley |
| 1989 | Emerging field of innate immunity. Infectious non-self model of immune recognition ("stranger hypothesis") | Charlie Janeway |

**Table 2** Contd.

| | | |
|---|---|---|
| 1994 | Danger hypothesis of immune responsiveness | Polly Matzinger |
| 1986 | Discovery of T helper subsets | Tim Mossmann & Bob Coffman |
| 1989 | Discovery of first chemokines | Edward Leonard, Teizo Yoshimura, & Marco Baggiolini |
| 1991 | Discovery of the first co-stimulatory pathway (CD28/B7) for T-cell activation | Kevin Urdahl & Mark Jenkins |
| 1992 | Cloning of CD40 ligand and recognition of its role in T-cell -dependent B-cell activation | Armitage, Spriggs, Lederman, Chess, Noelle, Aruffo, et al. |
| 1996-97 | Discovery of the role of Toll and Toll-like receptors in immunity | Ruslan Medzhitov, Charlie Janeway, & Jules Hoffmann |
| 2001 | Discovery of FOXP3—the gene directing regulatory T cell development | Hori S, Nomura T, Sakaguchi S |
| 2005 | Development of human papillomavirus vaccine | Ian Frazer |

## THE FUTURE OF IMMUNOLOGY?

*"Predictions are hard to make, especially ones about the future."*

— Peter Medawar

It is a pure speculation (but fun) to gauge the direction that immunology is headed in the next 20-30 years, since immunology continues to fascinate and frustrate researchers the world over. Further advances in the cell biology of the immune system will no doubt occur, which will lead to novel vaccines for infectious and non-infectious diseases, such as cancer and diseases of aging. New receptor — or cytokine—modifying therapeutics will be developed based on insights obtained from experimental immunology. The application of the human genome project to diseased populations will identify new drug targets, and high-throughput screens and combinatorial chemistry will accelerate the pace of drug discovery. Gene and protein microarray techniques and proteomics will reveal new components of immunity that will expand our knowledge of how the immune system works. Perhaps more importantly, we will begin to appreciate the fundamental role that the immune system plays in nearly all human diseases, and exploit this knowledge to alter the natural history of these diseases. Eventually, what has evolved as a precise and powerful tool created by nature to ensure the continued survival of the species could perhaps be manipulated to yield a better quality of life.

## REFERENCES


A Concise History of Immunology- Steven Greenberg

Akira, S., Takeda, K. and Kaisho, T. (2001). Toll-like receptors: Critical proteins linking innate and acquired immunity. *Nature Immunol,* **2:** 675.

Apoptosis as a therapeutic tool in IBD? Ann NY *Acad. Sci.* 2006 Aug; **1072:** 62-77

Burnet, F.M. (1959). *The Clonal Selection Theory of Acquired Immunity.* Cambridge University Press, Cambridge.

Cohen, S.G. and Samter, M. (1992). *Excerpts from Classics in Allergy.* Symposia Foundation, Carlsbad, California.

Desour, L. (1922). *Pasteur and His Work* (translated by A.F. and B.H. Wedd). T. Fisher Unwin Ltd., London. Fritig, B., Heitz, T. and Legrand, M. 1998. Antimicrobial proteins in induced plant defense. *Curr. Opin. Immunol.* **10:** 12.

James, B.D. and Hooper, N.M. Instant Notes in Biochemistry

Kimbrell, D.A. and Beutler, B. (2001). The evolution and genetics of innate immunity. *Nature Rev. Genet.* **2:** 256.

Kindt, T.J. and Capra, J.D. (1984). *The Antibody Enigma.* Plenum Press, New York.

Landsteiner, K. (1947). *The Specificity of Serologic Reactions.* Harvard University Press, Cambridge, Massachusetts.

Lawson, P.R. and Reid, K.B. (2000). The roles of surfactant proteins A and D in innate immunity. *Immunologic Reviews* **173:** 66.

Medawar, P.B. (1958). *The Immunology of Transplantation. The Harvey Lectures 1956–1957.* Academic Press, New York.

Medzhitov, R. and Janeway, C.A. (2000). Innate immunity. *N. Eng. J. Med.* **343:** 338.

Metchnikoff, E. (1905). *Immunity in the Infectious Diseases.* MacMillan, New York.

Milestones in Immunology.

Otvos, L. (2000). Antibacterial peptides isolated from insects. *J. Peptide Sci.* **6:** 497. Paul, W. (Ed.) 1999. *Fundamental Immunology,* 4th ed. Lippincott- Raven, Philadelphia.

Recent Advances in Immunology. BMJ. (1998). 316. 600-603. Mike Kemeny and Mark Peakman

Roitt, I.M. and Delves, P.J. (Eds.). (1998). *An Encyclopedia of Immunology,* 2nd ed., vols. 1-4. Academic Press, London.

Timeline in Immunology-Wikipedia

# 19

# Sleep Benefits in Health and Cognition

VIBHA MADAN AND SUSHIL K. JHA

*School of Life Sciences, Jawaharlal Nehru University, New Delhi-110067, India*

**ABSTRACT**

Two distinct sleep stages, rapid eye movement (REM) and non-REM (NREM) sleep, have evolved in birds and mammals possibly from the stem reptiles, but the purpose of its evolution remains unclear. Several reports have suggested that sleep plays an important role in the proper functioning of various physiological systems, such as the immune system, energy expenditure and metabolism, homeostatic balance of cellular milieu, etc. Additionally, sleep helps in the maturation of neuronal circuitries during early life. Apart from this, sleep facilitates the encoding of learnt information and forms memory in the respective brain areas throughout life (in children, middle and old age humans). Thus, it seems that sleep plays a notable role in our life and its chronic/acute loss induces several different kinds of disorders such as diabetes, hypertension, etc. We have attempted to review the detrimental effects in the event of its acute and/or chronic loss and the role of sleep in maintaining good health and cognition.

**Keywords:** Diabetes, Hypertension, Immune Function, Learning and Memory, Sleep Deprivation, Obesity.

## INTRODUCTION

Sleep is a natural state of rest characterized by a reduction in voluntary body movement and sensory perceptions from the surroundings (Zeplin et al., 2005). Phylogenetically, more evolved animals, for example, birds and mammals, demonstrate two major distinct sleep states: non-rapid eye movement (NREM) sleep and rapid eye movement (REM) sleep. The sleep states in humans progress in five stages (stage I – V), while it comprises two to three stages in the rat, cat, dog, ferret and other mammals (Zeplin et al., 2005; Jha et al., 2006). On the other hand, lower vertebrates such as fishes, amphibians and reptiles show a single sleep state (analogous to mammalian NREM sleep). REM sleep has not been noted in these animals till date (Nicolau et al., 2000), suggesting that REM sleep has evolved in birds and mammals from the stem reptiles from which the birds and mammals originated. Interestingly, REM sleep is also absent in some aquatic mammals exhibiting convergent evolution such as the dolphin, whale, manatee etc., suggesting that the natural forces have facilitated the evolution of sleep states only in more advanced terrestrial animals, birds and mammals.

Since nature has purposely favored the evolution of complex sleep states in some of the highly evolved terrestrial animals, anyone would be quite inquisitive to know what purpose it serves to our brain and body. Numerous theories have been proposed over the years regarding the functions of sleep, for example, sleep boosts up the immune functions, has a role in brain maturation and helps in energy conservation etc. (Rechtschaffen, 1998; Graves et al., 2001). According to the ontogenic hypothesis of REM sleep, the brain activity occurring during neonatal REM sleep seems to be particularly important to the developing organism (Marks et al., 1995). Studies investigating the effects of REM sleep loss during early life have shown that it can result in behavioral problems later in life because REM sleep loss decreases total brain mass (Mirmiran et al., 1983; Mirmiran et al., 1983) and also causes an abnormal amount of neuronal cell death (Morrissey et al., 2004). Further, it has also been observed that sleep helps encode the information of several different types of memory in the brain (Graves et al., 2001; Stickgold et al., 2001; Jha et al., 2005; Chowdhury et al., 2011). For example, disturbed sleep or no sleep after successful learning of a task causes dearth in performance in rats (Smith and Rose, 1996; Chowdhury et al., 2011) as well as in humans (Stickgold et al., 2000; Stickgold et al., 2001). It has also been noted that REM sleep and/or its associative component such as pontine P-wave density (PGO wave) are augmented following successful learning in the rat. It is believed that neurons in the brain replay the memory information (obtained during initial training period) during different sleep stages for their consolidation and long-term retention (Skaggs and McNaughton, 1996). All these studies thus suggest that the evolution of sleep has some purposeful meaning and its loss induces serious detrimental effects on our health and cognitive functions. In this review, we summarize the current concepts indicating the consequences of sleep loss along with the obligatory functional benefits of sleep in maintaining good health and cognition.

## SLEEP DISRUPTION IMPAIRS IMMUNE FUNCTION

Chronic sleep disruption may cause health impairments. It has been found that short-term sleep deprivation in humans (up to 64 hours) produced marked cognitive impairments (Horne, 1998) with minor abnormal physiological signs, for example, total sleep deprivation caused homeostatic imbalance of iron concentration in blood plasma, augmented drug-induced normocytic anemia etc. (Naitoh et al., 1990). However, prolonged sleep deprivation (up to eight days and so) produced several abnormal neurological and physiological signs (Kollar et al., 1969). It has been reported that very prolonged sleep deprivation (for nine or more days) ended in collapse and hospitalization of the subjects (Luby et al., 1960). In animal studies, it has also been observed that with short period of sleep deprivation, there are no gross detectable disturbances in intermediary metabolism, clinical chemistry or hematological indexes that provide substantial clues to the intervention of sleep deprivation effects. However, it has been reported that prolonged sleep deprivation in rats causes an unexplained hypercatabolic state, secondary malnutrition symptoms and mortality (Everson, 1993). Further, blood culture examination revealed invasion by opportunistic microbes to which there was no febrile response. These results suggest that the life-threatening condition caused by prolonged sleep deprivation is a breakdown of host defense against indigenous and pathogenic microorganisms (Everson, 1993). Nevertheless, the cause of death after prolonged sleep deprivation is not clearly known, but some studies suggest that the cause of death due to sleep deprivation may be attributed to bloodstream infection by some of the opportunistic facultative anaerobes generally ascribed to gut origin, such as *Pseudomonas aeruginosa, Klebsiella pneumoniae, Staphylococcus aureus, Streptococcus agalactiae* and *Corynebacterium jejeikum* (Everson, 1993). These organisms do not cause primary bacteremia and threaten life unless the host is immune compromised (Relman and Falkow, 1995). Interestingly, it has been found that sleep deprivation caused incursion of pathogenic bacteria to normal body tissues during the first few days of sleep deprivation. Throughout the course of sleep deprivation, abnormal control of bacteria was indicated by bacterial overgrowth in the intestine, polymicrobial infection of the mesenteric lymph nodes and increased invasion of extraintestinal body tissues. This chronic internal septic state is associated with progressive development of a deep negative energy balance and is known to culminate in lethal bacterial invasion of the bloodstream. The findings indicate that the processes underlying bacterial disease begin early after the onset of prolonged sleep deprivation and that infection of normally sterile tissues precedes overt signs of morbidity (Everson and Toth, 2000). All these studies suggest that immune system diffuses into a compromised state after sleep deprivation leading to a condition which makes a subject susceptible to infection and ultimately sick. Hence, sleep seems to act like a regulated immune system enforcer and might help to keep the pathogenic bacteria under guard.

## SLEEP DISRUPTION AND PREVALENCE OF OBESITY AND DIABETES

Several epidemiologic studies have demonstrated a close link between chronic sleep loss and changes in basal metabolic rate, which are ultimately involved in inducing obesity and diabetes [for review see (Knutson and Van Cauter, 2008)]. The energy status in the body is regulated by the hypothalamic area of the brain through a hormone called "Leptin", which is released by the adipocytes (Ahima et al., 2000). Circulating leptin levels in humans show a rapid decline or increase in response to acute caloric shortage or surplus, respectively (Ahima et al., 2000). It has been observed that total sleep deprivation induces changes in circulating leptin level (Morselli et al., 2010). Since, leptin acts as a feedback signal from adipose tissue to the hypothalamus for long-term regulation of energy balance, hence the alteration of its release with sleep loss may affect the body mass significantly (Mullington et al., 2003). Another peptide, "Ghrelin" produced predominantly by the stomach, is also involved in energy balance regulation. Leptin induces satiety by activating hypothalamic neurons associated with food intake (Morselli et al., 2010), whereas ghrelin stimulates appetite (Havel, 2001). The reciprocal interaction between these two hormones thus helps maintain a balance of our meal consumption. The studies suggest that the decrease in the level of leptin or increase in the level of ghrelin causes obesity and also alters glucose tolerance (Morselli et al., 2010). Interestingly, it has been noticed that one night sleep deprivation induces subjective hunger (Schmid et al., 2007). Thus, it seems that sleep loss down-regulates the satiety hormone leptin and concomitantly up-regulates the appetite-stimulating hormone ghrelin, which increases hunger and food intake causing inturn obesity.

The detrimental effects of sleep loss on glucose metabolism have been investigated by several researchers. Kuhn et al. using an oral glucose tolerance test (OGTT) had compared glucose tolerance in the same subject (total 28 young healthy human volunteers) before and after a prolonged period of total sleep deprivation. They observed an altered glucose tolerance during sleep deprived condition compared to the normal sleep-waking pattern (Kuhn et al., 1969). In another similar study, it has been noticed that the activity of enzymes associated with energy metabolism such as glycolysis and Kreb's cycle in the skeletal muscle decreased after sleep loss suggesting that sleep loss induces prediabetic type of muscle metabolism (Vondra et al., 1981). Interestingly, they have also noticed an increase in fasting blood glucose levels at the end of the sleep deprivation period (Vondra et al., 1981). Further, studies demonstrate that fasting insulin (a hormone that regulates the blood glucose level and is a marker of insulin response to OGTT) increased after sleep deprivation, suggesting that sleep deprivation induced insulin resistance (VanHelder et al., 1993). These evidences imply that regular sleep loss alters several enzymatic and hormonal activities associated with a constellation of metabolic and endocrine consequences that may contribute or predispose us to an increased risk of obesity and/or diabetes.

## SLEEP DISRUPTION AS A RISK FACTOR FOR HYPERTENSION AND CARDIOVASCULAR DISEASE

Sleep loss may induce the causal factors associated with obesity and/or diabetes, which may further lead to the prevalence of hypertension and/or coronary heart disease. But several direct studies have also analyzed the longitudinal impact of short-term sleep to determine whether or not short sleep duration increased the incidence of hypertension in short sleepers (Nagai et al., 2010). It was observed that the subjects having less than 5 hours sleep per night manifested an increased risk of hypertension in the middle aged group, but not the old aged group (Gangwisch et al., 2006). Further, Irwin and Ziegler in 2005 performed sleep deprivation in normal and alcoholic subjects. They observed that heart rate, blood pressure and sympathetic catecholamine levels were similar in normal and alcoholics (after alcohol abstinence), but partial sleep deprivation during night induced greater increase in heart rate in alcohol-dependent men compared to normal subjects (Irwin and Ziegler, 2005). Further, they also found that partial late night sleep disturbances in normal healthy young volunteers significantly increased nocturnal level of circulating norepinephrine and epinephrine compared to their undisturbed sleep periods (Irwin et al., 1999). These two circulating hormones increase the vasomotor tone of the blood vessels and increase blood pressure. Also, these two hormones/ neurotransmitters directly increase the heart rate by accelerating the heart pumping mechanism through S-A node of the heart. Hence, it seems that short sleep duration increased the sympathetic nervous system activity that, in turn, increases blood pressure (Nagai et al., 2010). It has been found that a person who works more than 67 hours or more has the highest prevalence of coronary heart disease and acute myocardinal infarction (Buell and Breslow, 1960; Netterstrom et al., 1999). These studies clearly suggest that chronic sleep disturbances induce sympathetic nervous system and also increase circulating norepinephrine and epinephrine level. The activation of these induces hypertension and is also associated with several cardiovascular diseases.

The prevalence of hypertension is greater in patients with sleep breathing disorder such as obstructive sleep apnea (OSA) (Wolk et al., 2003). In OSA patients, there is a recurrent episode of cessation of respiratory airflow due to upper airway inspiratory collapse during sleep. This causes intermittent blockage of air passage through trachea and ultimately decreases the oxygen saturation during sleep. With the inspiratory blockage, sleep gets disturbed and patients suffering from OSA experience significant sleep loss. Several cross-sectional and longitudinal studies have observed that OSA patients manifest greater chance of prevalence of hypertension [for review see (Wolk et al., 2003)]. The association between sleep-disordered breathing and hypertension has been observed in men and women across all ages (older and younger subjects) and ethnic groups (in normal and overweight individuals) (Nieto et al., 2000). In fact, OSA causes acute nocturnal surges in blood pressure in response to hypoxia mediated stimulation of sympathetic nervous system, which may sometimes reach high levels such as 240/120 mm Hg (Somers et al., 1995). Interestingly, it has been observed that effective treatment of OSA with continuous positive airway pressure (CPAP) helps decrease blood pressure and sleep disturbances (Becker et al., 2003). All these studies present compelling evidence supporting close association of hypertension with chronic sleep loss. The underlying cause atleast in part may be ascribed to notable sympathetic overactivity and also to humoral, neuroendocrine and metabolic abnormalities.

## SLEEP HELPS CONSERVE METABOLIC ENERGY AND DETOXIFY CELLULAR COMPARTMENTS

One of the functions of sleep has been proposed to conserve metabolic energy, otherwise one would need more energy intake per day in absence of sleep to perform our day-to-day activities. Recently, Jung et al., have quantified energy expenditure during sleep, sleep deprivation and during recovery sleep in humans (Jung et al., 2011). They have observed that energy expenditure was higher during sleep deprivation compared to the corresponding sleep period. Further, they found that energy expenditure was lower during an 8-hour recovery sleep after sleep deprivation compared to the corresponding normal sleep period (Jung et al., 2011). Their finding suggests that metabolic cost was quite high after missing one night sleep (Jung et al., 2011). The intriguing question is how does sleep contribute to save metabolic energy for our brain and body? The possible explanation for the decreased energy consumption during sleep may be the net result of suppression of metabolically costly physiological processes such as respiration, heart rate, gut motility and muscular activity (Kumar et al., 1990; Trinder et al., 1992; Shinar et al., 2006). Also, it has been noted that along with the augmented energy expenditure, food intake increases during sleep deprivation, because the levels of satiety hormone "leptin" decrease (Mullington et al., 2003) and the levels of hunger hormone "ghrelin" increase (Havel, 2001; Schmid et al., 2008). Thus, sleep loss contributes to increased food intake and energy expenditure by activating cellular metabolic processes.

Several reports suggest that sleep deprivation also causes cellular accumulation of several toxic biomolecules due to increased metabolic activity. Cells primarily derive energy from tightly controlled biochemical oxidation of substrates such as carbohydrates, fats etc. by increasing cellular metabolic activity. This augmented metabolic strategy to obtain energy in the form of ATP is extremely efficient, but this process does lead to the formation of reactive oxygen species and other free radicals as byproducts, which causes oxidative damage to biomolecules such as DNA and proteins (Spitz et al., 2004). Additionally, reactive oxygen species and free radicals also cause apoptosis (programmed cell death) and/or necrosis (Spitz et al., 2004). A variety of antioxidative enzymes such as superoxide dismutase (SOD) and glutathione peroxidase (GPx) help regulate the level of ROS. With prolonged sleep deprivation in the rats, it was observed that the activities of SOD and GPx significantly decreased in the hippocampus and brainstem, suggesting an alteration in the metabolism of ROS resulting in oxidative stress (Ramanathan et al., 2002). Sleep deprivation not only causes increased production of ROS, but also changes the neuronal membrane properties causing neuronal death (Mallick et al., 1995; Pal et al., 2005; Biswas et al., 2006). Hence, sleep possibly down-regulates the production of toxic substances such as ROS and free radicals by decreasing cellular metabolic activity and helps guard cellular damage.

## THE ROLE OF SLEEP IN BRAIN MATURATION

The brain sensory neuronal circuitries remain in naïve form at the time of birth and post natal maturation of sensory pathways are dependent on stimulus-driven neuronal activity (Hubel and Wiesel, 1970) and sleep (Madan and Jha, 2008). Altered visual experience during the critical period after birth in kitten permanently impaired the physiologic response of the brain to visual stimulation. During the time of development, synaptic connectivity in the cortex exhibits a high level of plasticity as synapses are formed and retracted, a process strongly driven by sensory activity (Hubel and Wiesel, 1970). We have investigated that sleep potentiates the processes for the formation of new connections as well as the retraction from the altered pathways in the visual circuitry (Jha et al., 2005; Aton et al., 2009). As little as 6 hours of sleep is sufficient enough to enhance the circuit re-modulation induced by monocular deprivation in the visual cortex, and this process is blocked when animals are prevented from sleeping (Frank et al., 2001) or when postsynaptic activity in the visual cortex is reversibly silenced during sleep (Jha et al., 2005). We further demonstrated that sleep potentiates the synaptic strengthening through the activation of glutamatergic NMDAR receptor and protein-kinase activity, the key components of sleep-dependent remodeling of synaptic connections (Aton et al., 2009). These studies suggest that sleep is one of the factors that helps mature naïve circuitries to an adult form. Since, mammalian sleep in not a homogenous state, hence it may be asked if all sleep states have similar functional properties. We have observed that both NREM and REM sleep states play a similar role for synaptic remodeling in the visual circuitry (Aton et al., 2009), however, REM sleep alone can play some differential beneficial role in brain maturation.

A consistent evidence about the role of REM sleep in brain maturation has been obtained from REM sleep deprivation studies (Marks et al., 1995). It has been found that if vision in one eye is blocked for two weeks in kittens born with normal vision and they are simultaneously devoid of REM sleep during this period, it induces anatomically and functionally deleterious effects on the lateral geniculate nucleus (LGN) in the thalamus (Marks et al., 1995). Further, it has been demonstrated that the elimination of ponto-geniculo-occipital wave, a phasic activity during REM sleep, induced similar results in the LGN (Shaffery et al., 1999), and REM sleep deprivation delayed the development of synaptic plasticity in the LGN (Hogan et al., 2001) and retarded the maturational reduction of long-term potentiation (LTP) in visual cortex of immature rats (Shaffery et al., 2002). These results emphasize that REM sleep is also an important part of development of brain circuitries after birth. The formed circuitries once mellowed stay for lifetime, and sleep during the developing phase helps mature sensory circuitries. Sleep not only helps strengthen circuitry during early life, but also helps potentiate synapses during rest of the life to maintain ultimate functionality of the neurons (Madan and Jha, 2008).

## THE ROLE OF SLEEP IN MEMORY CONSOLIDATION

Several reports have offered compelling evidence supporting the role of sleep in consolidation of some forms of memories (Stickgold, 2005; Aly and Moscovitch, 2010). Recall of face-location associations augmented significantly in subjects following a retention interval containing sleep than having an equally long period of waking (Talamini et al., 2008), suggesting the beneficial effects of sleep in preserving these memories. Similarly, humans trained for the visual texture discrimination task (Gais et al., 2000), a motor sequence task (Walker et al., 2003) and a motor adaptation task (Hill et al., 2008) showed better performance on post-training days, if they experienced undisturbed sleep (in the night of the training days) as compared to subjects kept awake for a similar period of time (Stickgold, 2005). Further, deprivation of selective sleep stage, for example, rapid eye movement (REM) sleep, caused impairment of spatial learning task (eight-box task) (Bjorness et al., 2005) and inability to recognize spatial cues to locate the hidden-platform but not the visible-platform of the water maze (Smith and Rose, 1996). These studies suggest the propitious nature of sleep in facilitating the encoding of information of some memories.

Several other studies have investigated the role of sleep in memory consolidation by depriving the subjects/animals for sleep and have noted down the detrimental effects on memory processes after sleep deprivation. For example, total sleep deprivation for 6 hours was performed on rats and it was observed that memory consolidation of contextual fear conditioning was impaired but consolidation processes of cued fear-conditioned memory remained unscathed (Graves et al., 2003). Similarly, REM sleep deprivation retards the extinction memory of cued fear-conditioning but not of contextual fear-conditioning (Silvestri, 2005). We have recently observed that sleep deprivation for 6 hours impaired the memory encoding of trace conditioned memory (Chowdhury et al., 2011). Thus, it seems that the consolidation of some specific memories needs sleep for information encoding, while the consolidation process of some memory types is sleep-independent.

The memory consolidation may ensue over a period of hours to days after training (Abel and Lattal, 2001). Several reports suggest that consolidation of some memories, for example, fear conditioned memory, can be disrupted by blocking the signaling machinery such as protein kinase A (PKA) and protein synthesis during the initial critical period (at 0-4 hours) after training. Further, learning could not be impaired if such pathways are blocked after the critical period ends (Bourtchouladze et al., 1998). It has been reported that sleep deprivation soon after training (0–5/0-6 hours) impairs the consolidation of recognition and fear-conditioned memory, but if sleep deprivation is performed 5–10/7-12 hours after training, it did not induce any impairment (Graves et al., 2003; Palchykova et al., 2006). Similarly, we have observed in a canonical form of *in vivo* cortical plasticity triggered by monocular deprivation that sleep soon after induction of ocular dominance plasticity (6 hour period) potentiates the synaptic changes, while its deprivation during the same period did not change it appreciably (Frank et al., 2001; Jha et al., 2005). Hence, sleep during the initial 6 hour period after training seems to be indispensable for encoding the information of some form of memories. These studies offer persuasive evidences for sleep function in memory consolidation.

## CONCLUSION

Sleep performs an indispensable role in our body and chronic sleep loss causes several deficits. Sleep has evolved in birds and mammals with some distinctive features and has become obligatory for survival. Few important functions of sleep have been discussed in this article and future studies are awaited to get a robust answer for "why do we sleep". Impending studies may bring out more information to solve the mystery of sleep. It is known that sleep is a key component for proper functionality of several physiological functions, but it raises few intriguing questions: (a) whether similar or analogous form of sleep exists or had ever existed in lower animals such as reptiles, amphibians etc.; (b) if it exists, does it have similar usual properties as observed in mammals and birds; (c) if similar form does not exist, then what is/are the stand-in or quasi stages in lower animals having similar functional distinctiveness. Future studies will bring out answers of several such questions, but the current available information undoubtedly suggests that the sleeping brain performs ably in birds and mammals to modulate several imperative functions.

## ACKNOWLEDGEMENTS

Funding to SKJ from CSIR, DBT, UGC resource networking, DST-PURSE is highly acknowledged.

## REFERENCES

Abel, T. and Lattal, K.M. (2001). Molecular mechanisms of memory acquisition, consolidation and retrieval. *Current Opinion in Neurobiology*, **11:** 180-187.

Ahima, R.S., Saper, C.B., Flier, J.S. and Elmquist, J.K. (2000). Leptin regulation of neuroendocrine systems. *Front Neuroendocrinol*, **21:** 263-307.

Aly, M. and Moscovitch, M. (2010). The effects of sleep on episodic memory in older and younger adults. *Memory*, 1-8.

Aton, S.J., Seibt, J., Dumoulin, M., Jha, S.K., Steinmetz, N., Coleman, T., Naidoo, N. and Frank, M.G. (2009). Mechanisms of sleep-dependent consolidation of cortical plasticity. *Neuron*, **61:** 454-466.

Becker, H.F., Jerrentrup, A., Ploch, T., Grote, L., Penzel, T., Sullivan, C.E. and Peter, J.H. (2003). Effect of nasal continuous positive airway pressure treatment on blood pressure in patients with obstructive sleep apnea. *Circulation*, **107:** 68-73.

Biswas, S., Mishra, P. and Mallick, B.N. (2006). Increased apoptosis in rat brain after rapid eye movement sleep loss. *Neuroscience*, **142:** 315-331.

Bjorness, T.E., Riley, B.T., Tysor, M.K. and Poe, G.R. (2005). REM restriction persistently alters strategy used to solve a spatial task. *Learning and Memory*, **12:** 352-359.

Bourtchouladze, R., Abel, T., Berman, N., Gordon, R., Lapidus, K. and Kandel, E.R. (1998). Different training procedures recruit either one or two critical periods for contextual memory consolidation, each of which requires protein synthesis and PKA. *Learning and Memory*, **5:** 365-374.

Buell, P. and Breslow, L. (1960). Mortality from coronary heart disease in California men who work long hours. *J. Chronic. Dis.* **11:** 615-626.

Chowdhury, A., Chandra, R. and Jha, S.K. (2011). Total sleep deprivation impairs the encoding of trace-conditioned memory in the rat. *Neurobiol Learn Mem.* **95:** 355-360.

Datta, S. (2000). Avoidance task training potentiates phasic pontine-wave density in the rat: A mechanism for sleep-dependent plasticity. *J. Neurosci*, **20:** 8607-8613.

Everson, C.A. (1993). Sustained sleep deprivation impairs host defense. *Am. J. Physiol.* **265:** R1148-1154.

Everson, C.A. and Toth, L.A. (2000). Systemic bacterial invasion induced by sleep deprivation. *Am. J. Physiol. Regul. Integr. Comp. Physiol.* **278:** R905-916.

Frank, M.G., Issa, N.P. and Stryker, M.P. (2001). Sleep enhances plasticity in the developing visual cortex. *Neuron*, **30:** 275-287.

Gais, S., Plihal, W., Wagner, U. and Born, J. (2000). Early sleep triggers memory for early visual discrimination skills. *Nature Neuroscience*, **3:** 1335-1339.

Gangwisch, J.E., Heymsfield, S.B., Boden-Albala, B., Buijs, R.M., Kreier, F., Pickering, T.G., Rundle, A.G., Zammit, G.K. and Malaspina, D. (2006). Short sleep duration as a risk factor for hypertension: analyses of the first National Health and Nutrition Examination Survey. *Hypertension*, **47:** 833-839.

Graves, L., Pack, A. and Abel, T. (2001). Sleep and memory: a molecular perspective. *Trends Neurosci*, **24:** 237-243.

Graves, L.A., Heller, E.A., Pack, A.I. and Abel, T. (2003). Sleep deprivation selectively impairs memory consolidation for contextual fear conditioning. *Learning and Memory*, **10:** 168-176.

Havel, P.J. (2001). Peripheral signals conveying metabolic information to the brain: short-term and long-term regulation of food intake and energy homeostasis. *Exp. Biol. Med. (Maywood)*, **226:** 963-977.

Hill, S., Tononi, G. and Ghilardi, M.F. (2008). Sleep improves the variability of motor performance. *Brain Research Bulletin*, **76:** 605-611.

Hogan, D., Roffwarg, H.P. and Shaffery, J.P. (2001). The effects of 1 week of REM sleep deprivation on parvalbumin and calbindin immunoreactive neurons in central visual pathways of kittens. *J. Sleep Res.* **10:** 285-296.

Horne, J.A. (1998). Why we sleep: The Functions of sleep in humans and other mammals. *In:* New York: Oxford. pp. 13-103.

Hubel, D.H. and Wiesel, T.N. (1970). The period of susceptibility to the physiological effects of unilateral eye closure in kittens. *J. Physiol.* **206:** 419-436.

Irwin, M., Thompson, J., Miller, C., Gillin, J.C. and Ziegler, M. (1999). Effects of sleep and sleep deprivation on catecholamine and interleukin-2 levels in humans: clinical implications. *J. Clin. Endocrinol. Metab.* **84:** 1979-1985.

Irwin, M.R. and Ziegler, M. (2005). Sleep deprivation potentiates activation of cardiovascular and catecholamine responses in abstinent alcoholics. *Hypertension*, **45:** 252-257.

Jha, S.K., Coleman, T. and Frank, M.G. (2006). Sleep and sleep regulation in the ferret (Mustela putorius furo). *Behav Brain Res.* **172:** 106-113.

Jha, S.K., Jones, B.E., Coleman, T., Steinmetz, N., Law, C.T., Griffin, G., Hawk, J., Dabbish, N., Kalatsky, V.A. and Frank, M.G. (2005). Sleep-dependent plasticity requires cortical activity. *J. Neuroscience* **25:** 9266-9274.

Jung, C.M., Melanson, E.L., Frydendall, E.J., Perreault, L., Eckel, R.H. and Wright, K.P. (2011). Energy expenditure during sleep, sleep deprivation and sleep following sleep deprivation in adult humans. *J. Physiol.* **589:** 235-244.

Knutson, K.L. and Van Cauter, E. (2008). Associations between sleep loss and increased risk of obesity and diabetes. *Ann. N.Y. Acad. Sci.* **1129:** 287-304.

Kollar, E.J., Pasnau, R.O., Rubin, R.T., Naitoh, P., Slater, G.G. and Kales, A. (1969). Psychological, psychophysiological, and biochemical correlates of prolonged sleep deprivation. *Am. J. Psychiatry,* **126:** 488-497.

Kuhn, E., Brodan, V., Brodanova, M. and Rysanek, K. (1969). Metabolic reflection of sleep deprivation. *Act Nerv Super* (*Praha*) **11:** 165-174.

Kumar, D., Idzikowski, C., Wingate, D.L., Soffer, E.E., Thompson, P. and Siderfin, C. (1990). Relationship between enteric migrating motor complex and the sleep cycle. *Am. J. Physiol.* **259:** G983-990.

Luby, E.D., Frohman, C.E., Grisell, J.L., Lenzo, J.E. and Gottlieb, J.S. (1960). Sleep deprivation: effects on behavior, thinking, motor performance, and biological energy transfer systems. *Psychosom. Med.* **22:** 182-192.

Madan, V. and Jha, S.K. (2008). What is sleep and why is it needed? *International J. Life Sciences and Technology* **1:** 9-23.

Mallick, B.N., Thakkar, M. and Gangabhagirathi, R. (1995). Rapid eye movement sleep deprivation decreases membrane fluidity in the rat brain. *Neurosci. Res.* **22:** 117-122.

Marks, G.A., Shaffery, J.P., Oksenberg, A., Speciale, S.G. and Roffwarg, H.P. (1995). A functional role for REM sleep in brain maturation. *Behav Brain Res.* **69:** 1-11.

Mirmiran, M., Scholtens, J., van de Poll, N.E., Uylings, H.B., van der Gugten, J. and Boer, G.J. (1983). Effects of experimental suppression of active (REM) sleep during early development upon adult brain and behavior in the rat. *Brain Res.* **283:** 277-286.

Mirmiran, M., Uylings, H.B. and Corner, M.A. (1983). Pharmacological suppression of REM sleep prior to weaning counteracts the effectiveness of subsequent environmental enrichment on cortical growth in rats. *Brain Res.* **283:** 102-105.

Morrissey, M.J., Duntley, S.P., Anch, A.M. and Nonneman, R. (2004). Active sleep and its role in the prevention of apoptosis in the developing brain. *Med. Hypotheses,* **62:** 876-879.

Morselli, L., Leproult, R., Balbo, M. and Spiegel, K. (2010). Role of sleep duration in the regulation of glucose metabolism and appetite. *Best Pract. Res. Clin. Endocrinol. Metab.* **24:** 687-702.

Mullington, J.M., Chan, J.L., Van Dongen, H.P., Szuba, M.P., Samaras, J., Price, N.J., Meier-Ewert, H.K., Dinges, D.F. and Mantzoros, C.S. (2003). Sleep loss reduces diurnal rhythm amplitude of leptin in healthy men. *J. Neuroendocrinol,* **15:** 851-854.

Nagai, M., Hoshide, S. and Kario, K. (2010). Sleep duration as a risk factor for cardiovascular disease—a review of the recent literature. *Curr. Cardiol. Rev.* **6:** 54-61.

Naitoh, P., Kelly, T.L. and Englund, C. (1990). Health effects of sleep deprivation. *Occup. Med.* **5:** 209-237.

Netterstrom, B., Nielsen, F.E., Kristensen, T.S., Bach, E. and Moller, L. (1999). Relation between job strain and myocardial infarction: a case control study. *Occup. Enviorn. Med.* **56:** 339-342.

Nicolau, M.C., Akaarir, M., Gamundi,A., Gonzalez, J. and Rial, R.V. (2000). Why we sleep: the evolutionary pathway to the mammalian sleep. *Prog. Neurobiol.* **62:** 379-406.

Nieto, F.J., Young, T.B., Lind, B.K., Shahar, E., Samet, J.M., Redline, S., D'Agostino ,R.B., Newman, A.B., Lebowitz,M.D. and Pickering, T.G. (2000). Association of sleep-disordered breathing, sleep apnea, and hypertension in a large community-based study. Sleep Heart Health Study. *Jama* **283:** 1829-1836.

Pal, D., Madan, V. and Mallick, B.N. (2005). Neural mechanism of rapid eye movement sleep generation: Cessation of locus coeruleus neurons is a necessity. *Sheng Li Xue Bao* **57:** 401-413.

Palchykova, S., Winsky-Sommerer, R., Meerlo, P., Durr, R. and Tobler, I. (2006). Sleep deprivation impairs object recognition in mice. *Neurobiology of Learning and Memory* **85:** 263-271.

Ramanathan, L., Gulyani, S., Nienhuis, R. and Siegel, J.M. (2002). Sleep deprivation decreases superoxide dismutase activity in rat hippocampus and brainstem. *Neuroreport* **13:** 1387-1390.

Rechtschaffen, A. (1998). Current perspectives on the function of sleep. *Perspect Biol. Med.* **41:** 359-390.

Relman, D.A. and Falkow, S. (1995). A molecular perspective of microbial pathogenicity. *In:* Mandell, G.L., Bennett, J.E., Dolin, R., Mandell, G.L., Bennett, J.E., Dolin, R.S. Principles and Practice of Infectious Disease. New York: Churchill Livingstone. pp. 19-25.

Schmid, S.M., Hallschmid, M., Jauch-Chara, K., Bandorf, N., Born, J. and Schultes, B. (2007). Sleep loss alters basal metabolic hormone secretion and modulates the dynamic counterregulatory response to hypoglycemia. *J. Clin. Endocrinol. Metab.* **92:** 3044-3051.

Schmid, S.M., Hallschmid, M., Jauch-Chara, K., Born, J. and Schultes, B. (2008). A single night of sleep deprivation increases ghrelin levels and feelings of hunger in normal-weight healthy men. *J. Sleep Res.* **17:** 331-334.

Shaffery, J.P., Roffwarg, H.P., Speciale, S.G. and Marks, G.A. (1999). Ponto-geniculo-occipital-wave suppression amplifies lateral geniculate nucleus cell-size changes in monocularly deprived kittens. *Brain Res. Dev. Brain Res.* **114:** 109-119.

Shaffery, J.P., Sinton, C.M., Bissette, G., Roffwarg, H.P. and Marks, G.A. (2002). Rapid eye movement sleep deprivation modifies expression of long-term potentiation in visual cortex of immature rats. *Neuroscience,* **110:** 431-443.

Shinar, Z., Akselrod, S., Dagan, Y. and Baharav, A. (2006). Autonomic changes during wake-sleep transition: a heart rate variability based approach. *Auton Neurosci* **130:**17-27.

Silvestri, A.J. (2005). REM sleep deprivation affects extinction of cued but not contextual fear conditioning. *Physiology & Behavior* **84:** 343-349.

Skaggs, W.E. and McNaughton, B.L. (1996). Replay of neuronal firing sequences in rat hippocampus during sleep following spatial experience. *Science* **271:** 1870-1873.

Smith, C. and Rose, G.M. (1996). Evidence for a paradoxical sleep window for place learning in the Morris water maze. *Physiol. Behav* **59:** 93-97.

Somers, V.K., Dyken, M.E., Clary, M.P. and Abboud, F.M. (1995). Sympathetic neural mechanisms in obstructive sleep apnea. *J. Clin. Invest,* **96:** 1897-1904.

Spitz, D.R., Azzam, E.I., Li, J.J. and Gius, D. (2004). Metabolic oxidation/reduction reactions and cellular responses to ionizing radiation: a unifying concept in stress response biology. *Cancer Metastasis Rev.* **23:** 311-322.

Stickgold, R. (2005). Sleep-dependent memory consolidation. *Nature,* **437:** 1272-1278.

Stickgold R, Hobson JA, Fosse R and Fosse M. (2001). Sleep, learning, and dreams: off-line memory reprocessing. *Science,* **294:** 1052-1057.

Stickgold, R., Whidbee, D., Schirmer, B., Patel, V. and Hobson, J.A. (2000). Visual discrimination task improvement: A multi-step process occurring during sleep. *J. Cogn. Neurosci.* **12:** 246-254.

Talamini, L.M., Nieuwenhuis, I.L., Takashima, A. and Jensen, O. (2008). Sleep directly following learning benefits consolidation of spatial associative memory. *Learning and Memory* **15:** 233-237.

Trinder, J., Whitworth, F., Kay, A. and Wilkin, P. (1992). Respiratory instability during sleep onset. *J. Appl. Physiol.* **73:** 2462-2469.

VanHelder, T., Symons, J.D. and Radomski, M.W. (1993). Effects of sleep deprivation and exercise on glucose tolerance. *Aviat Space Environ Med* **64:** 487-492.

Vondra, K., Brodan, V., Bass, A., Kuhn, E., Teisinger, J., Andel, M. and Veselkova, A. (1981). Effects of sleep deprivation on the activity of selected metabolic enzymes in skeletal muscle. *Eur. J. Appl. Physiol. Occup. Physiol.* **47:** 41-46.

Walker, M.P., Brakefield, T., Seidman, J., Morgan, A., Hobson, J.A. and Stickgold, R. (2003). Sleep and the time course of motor skill learning. *Learning and Memory* **10:** 275-284.

Wolk, R., Shamsuzzaman, A.S. and Somers, V.K. (2003). Obesity, sleep apnea, and hypertension. *Hypertension* **42:** 1067-1074.

Zeplin, H., Siegel, J.M. and Tobler, I. (2005). Mammalian sleep. In: Dement WC, editor. Dement WCs. Principles and practice of sleep medicine. Philadelphia: WB Saunders Company. pp. 91-100.

# 20

# Climate Change: India's Perceptions, Policies and Action Plans for Sustainable Development

M.H. FULEKAR, BHAWANA PATHAK AND JYOTI FULEKAR

*School of Environment and Sustainable Development, Central University of Gujarat Gandhinagar-382030, Gujarat*

**ABSTRACT**

Climate change is one of the most important global environmental challenges of sustainable development. The impact of climate variability and change, associated with socio-economic development will affect the ability of countries to achieve sustainable development goals. The global response to climate change, notably the United Nations Framework Conversation on climate change and Kyoto protocol, seek to implement this objective through voluntary greenhouse gases (GHG) emission reduction targets. Climate change impacts, policies contributing to climate mitigation, renewable energy generation for reducing emission of greenhouse gasses, including India's National Action Plans and its implementation as well as international participation addressing climate change have been highlighted. The interlinks of climate change and sustainable development have been cited in the chapter.

**Keywords:** Climate Change, Sustainable Development, Greenhouse Gases, Kyoto Protocol, Renewable Energy.

## INTRODUCTION

Climate change is one of the most important global environmental challenges, with implications for food production, water supply, health, energy, etc. Addressing climate change requires a good scientific understanding as well as coordinated action at national and global level. Historically, the responsibility for greenhouse gas emissions' increase lies largely with the industrialized world, though the developing countries are likely to be the source of an increasing proportion of future emissions. The projected climate change under various scenarios is likely to have implications in food production, water supply, coastal settlements, forest ecosystems, health, energy security, etc. The adaptive capacity of communities likely to be impacted by climate change is low in developing countries. The efforts made by the UNFCCC and the Kyoto Protocol provisions are clearly inadequate to address the climate change challenge. The most effective way to address climate change is to adopt a sustainable development pathway by shifting to environmentally sustainable technologies and promotion of energy efficiency, renewable energy, forest conservation, reforestation, water conservation, etc. The issue of highest importance in developing countries is reducing the vulnerability of their natural and socio-economic systems to the projected climate change. India and other developing countries will face the challenge of promoting mitigation and adaptation strategies, bearing the cost of such an effort, and its implications for economic development.

The climate change issue is a part of the larger challenge of sustainable development. As a result, climate policies can be more effective when consistently embedded within broader strategies designed to make national and regional development paths more sustainable. The impact of climate variability and change, climate policy responses, and associated socio-economic development will affect the ability of countries to achieve sustainable development goals. The pursuit of these goals will in turn affect the opportunities for, and success of, climate policies. In particular, the socio-economic and technological characteristics of different development paths will strongly affect emissions, the rate and magnitude of climate change, climate change impacts, the capability to adapt, and the capacity to mitigate.

The UN Conference on Environment and Development (UNCED) in 1992 at Rio de Janeiro led to FCCC (Framework Convention on Climate Change), which laid the framework for the eventual stabilization of greenhouse gases in the atmosphere, recognized the common but differentiated responsibilities and respective capabilities, and social and economic conditions. The Convention came into force in 1994. Subsequently, the 1997 Kyoto protocol, which came into force in 2005, reasserted the importance of stabilizing greenhouse gas concentrations in the atmosphere and adhering to sustainable development principles. The Protocol laid out guidelines and rules regarding the extent to which a participating industrialized country should reduce its emissions of six greenhouse gases—carbon dioxide, methane, nitrous oxide, chlorofluorocarbon, hydrofluorocarbons and perfluorocarbons. It requires industrialized countries to reduce their greenhouse gas emissions by a weighted average of 5.2%, based on the 1990 greenhouse gas emissions. The reduction is to be achieved by the end of the five-year period, 2008 to 2012. The Kyoto Protocol does not require the developing countries to reduce their greenhouse gas emissions.

The issue of combating global climate change inherently addresses the ability of economics, primarily in developing nations, to reduce emission of greenhouse gases (GHG) in industrial production. The global response to climate change, notably the United Nations Framework Convention on Climate Change (UNFCCC) and Kyoto Protocol, seek to implement this objective through voluntary GHG emission reduction targets. Compliance is often assessed through GHG inventories which measure carbon intensities of production. However, as on date, developing countries do not have binding GHG emission reduction targets under the Kyoto Protocol.

For example, a developed country faced a binding GHG emission reduction target can seek to consume 'polluting' products (products and processes with high carbon intensities) by importing the same from developing countries. As a result, the developing country has to increase its production of the product without worrying about binding climate compliance. It is worth noting that foundries, ship breaking and other polluting industries have shifted to the developing countries from the US and EU shores. A large part of China's industrial production seeks to meet demands of developed nations. According to the Human Development Report 2004, countries facing rising GHG emission levels between 1990-04 include China (103 percent), Korea (roughly 93 percent) and India (97 percent). On the contrary, emissions from Germany and UK fell roughly by 17 and 10 percent over the same period. In the manner, international trade spurs shift in emissions from developed to developing nations. While positive fallout is higher economic growth, carbon intensity is certainly on the rise.

Climate change affects almost every sector of the economy—agriculture, forestry, power, manufacturing industries and even defense, shipping and marine resources. Some policies that have significant overlap with climate change include India's Integrated Energy Policy.

## CLIMATE CHANGE-IMPACTS

According to the IPCC, the likely range of global average surface warming over the period of this century vary from 0.3°C to 6.4°C, depending on the model used for simulation. The corresponding average rise in sea levels ranges from 0.18m to 0.59m, excluding the impact of dynamic changes in the ice flow. In Southern Asia, the mean annual increase in temperature by the end of the century is projected to be around 3.8°C in the Tibetan plateau and 3.3°C in South Asia and 2.5°C in South East Asia. There is still an ongoing debate over the pace at which the temperature is expected to rise over the coming century, potential mechanisms by which the change will affect the region are clear. These are the changes in subcontinent's river systems that flow from the Tibetan plateau to the Indian Ocean, and rising sea levels and their impact on river-deltas and low-lying islands. In addition, a third mechanism pertinent to this study is extreme weather—cyclones, droughts, floods etc., that do not exclusively result from global warming, but are both vitiated by it and complicate our response to the disasters it causes.

## Projected Impacts of Climate Changes

Let us discuss the projected impacts of climate changes in detail.

### *(a) Glacial recession*

The glaciers on the Tibetan plateau are the source of Asia's biggest rivers, including the Brahmaputra, the Indus, the Sutlej and several of the northern tributaries of the Ganges that irrigate the subcontinent. Geopolitically, the source of most of these rivers, except the main Ganges, lies in China. The melting of the Himalayan glaciers as a result of the rise in the earth's temperature will first increase the drainage through the major river systems into the ocean, followed by reduction in their volumes, once the glaciers begin to disappear. It is projected that some of the mightiest Himalayan Rivers might end up as seasonal, monsoon-fed rivers like those in southern India.

### *(b) Rising sea levels*

The rise in global sea levels due to the melting of polar ice caps and glaciers around the world is expected to result in the submergence of low lying areas including river deltas, coastlines and small islands. This places highly populated regional cities like Karachi, Dhaka, Mumbai, Kochi and Mangalore at risk. The entire country of Maldives could disappear under the Indian Ocean by the middle of the century. In addition, the coastline could advance inland across several heavily populated parts of Bangladesh, Sri Lanka, Myanmar and Pakistan (as indeed, several parts of India).

### *(c) Extreme weather*

In addition, climate change risks worsening the impact of natural disasters like cyclones, floods and droughts that affect the subcontinent. This could take several forms: increasing the intensity of cyclones or floods, changing the geographical area where these occur, occurring at increased frequencies and, in general, complicating adaptation efforts. For instance, a cyclone of record strength striking a river delta that is itself deluged as a result of glacial recession would transform the disaster into a much more complicated one.

## Climate Change Impacts: Indian Scenario

Let us study the impact of climate change in Indian scenario in detail.

### *(a) Water resources*

The hydrological cycle is likely to be altered and the severity of droughts and intensity of floods in various parts of India are likely to increase. Further, a general reduction in the quantity of available run-off is predicted.

### *(b) Agriculture*

Simulations using dynamic crop models indicate a decrease in yield of crops as temperature increases in different parts of India. However, this is offset by an increase in $CO_2$ at moderate rise in temperature and at higher warming. Negative impact on crop productivity is projected due to reduced crop durations.

### *(c) Forests*

Climate impact assessments using BIOME-3 model and climate projections for the year 2085 show that 77% and 68% of the forested grids in India are likely to experience shift in forest types. Indications show a shift towards wet forest types in the northeastern region and dry forest types in the northwestern region in the absence of human influence. Increasing atmospheric $CO_2$ concentration and climate warming could also result in a doubling of net primary productivity.

### *(d) Coastal zone*

Simulation models show an increase in frequencies of tropical cyclones in the Bay of Bengal; particularly intense events are projected during the post-monsoon period. Sea level rise is projected to displace populations in coastal zones, increase flood in low-lying coastal areas, loss of crop yields from inundation and salinization.

### *(e) Human health*

Malaria is likely to persist in many states and new regions may become malaria-prone and the duration of the malaria transmission windows is likely to widen in northern and western states and shorten in southern states.

### *(f) Desertification*

Globally, about 1900 Mha of land are affected by land degradation, of which 500 Mha each are in Africa and the Asia-Pacific and 300 Mha in Latin America. Climate change leading to warming and water stress could further exacerbate land degradation, leading to desertification. The United Nations Convention to Combat Desertification (UNCCD) aims to address the problem of land degradation, which is linked to climate change.

It is important to note that the climate-sensitive sectors (forests, agriculture, coastal zones) and the natural resources (groundwater, soil, biodiversity, etc.) are already under stress due to socio-economic pressures. Climate change is likely to exacerbate the degradation of resources and socio-economic pressures. Thus, countries such as India with a large population dependent on climate-sensitive sectors and low adaptive capacity have to develop and implement adaptation strategies.

## Economic Losses due to Climate Change—Case Study

Rakesh Kumar et. al; (2008), National Environmental and Engineering Research Institute has performed economic analysis to understand climate impacts with a view to plan out adaptability strategies for Mumbai, as a case study, for the period up to 2050. These cost considerations took into account the most conservative figures with regard to the period of impact, population affected, loss of earnings and material damage (Table 1).

**Table 1** Estimated economic losses due to the impact of climate change in Mumbai

| *Type of Impact* | *Type of Costs and Period of Impact* | *Cost in Rupees (crores)* |
|---|---|---|
| Dislocation due to extreme events of flooding of low-lying areas-every five years till 2050* | Cumulative costs over the period 2005–2050 | 407.6 |
| Material damage to low-lying areas due to extreme events – every five years till 2050* | Cumulative costs over the period 2005–2050 | 6413 |
| Mortality costs due to extreme events of flooding – every five years till 2050* | Cumulative costs over the period 2005–2050 | 3050 |
| Disability-adjusted life years (DALYs) lost due to diseases like malaria, diarrhoea and leptospirosis | Cumulative costs over the period 2005–2050 | 3153 |
| Building-foundation damages for the period till 2050 due to sea-level rise | Single-cost estimate for the year 2050 | 15 01 725 |
| Tourism loss: less number of tourists visiting Mumbai | Single-cost estimate for the year 2050, as compared with the base year 2005 | 19 63 500 |

Indians are concerned about climate change because our country is particularly vulnerable to its impacts. With a coastline of 7,600 kilometers, for instance, we have to be worried about sea level rise. Some parts of the country, such as the Sundarbans across the Hooghly delta and the low-lying coastal area of Kutch on the western side, are particularly vulnerable to sea level rise, because even with a small increase in sea level, large parts of these locations would be threatened with major damage and destruction and with complete inundation, resulting from storm surges and cyclonic activity. In the Sundarbans, in particular, some islands have already disappeared and others are under similar threat.

The impacts of climate change on India would be diverse and serious. There is already an evidence in some parts of the country of changes in precipitation patterns. While some parts of India show a perceptible decline in rainfall and there is reduced snow in the Himalayas, a major concern also arises from projected increases in the frequency and intensity of extreme precipitation events. These are not only likely to pose a major danger to those who would be affected directly, but they could also affect the livelihoods of hundreds of millions of small farmers, who are entirely dependent on rain-fed agriculture. India is also vulnerable to the increase in frequency, intensity, and duration of floods, droughts, and heat waves. Human health will be affected by climate change, not only on account of these occurrences but also as a result of increased vector-borne diseases. Another area of deep concern to Indian society stems from the impacts of climate change on agriculture. There is already growing evidence, on the basis of ongoing research, that some crop yields are declining on account of climate change.

## CLIMATE CHANGE IMPACTS AND PRIORITIES: KEY SECTORS

Since India does not have to adhere to voluntary commitment on climate change, it is a net gainer at least till the first commitment period (2008-12). However, the fact that developed countries can consume products without having compensate for the damages caused by GHG emission, simply by relocating industries to the developing countries. It raises the question of whether carbon taxes or any other penal measures should be imposed on the importing country, or a separate set of regulations should be introduced to tackle GHG emissions arising out of international trade.

### *(a) Agriculture*

The agricultural sector is a priority view of climate change and our overall economic development. Shifts in temperature and pressure zones may have strong repercussions on the overall productivity of Indian agriculture. Temperature shifts lead to shift in the vegetation patterns. This means farmers have to change either their crops or cropping patterns. These uncertainties are compounded by limitations of knowledge on weather and agro markets. As a net exporter of agricultural products, any drop in productivity adversely affects India's exports and the balance of trade. Worse, it affects livelihoods of several agro-based communities across the country.

### *(b) Aviation*

While admitting that it is a significant source of employment and growth, air travel's share in GHG emission is steadily increasing. Countries facing voluntary GHG emission reduction targets are likely to impose carbon taxes on airlines, as some already have. For commercial reasons, airline companies will aim to transfer the financial liability on passengers leading to an escalation in ticket prices. Airlines in countries but sharing the same airspace do not face the same taxes will thus enjoy a price advantage.

### *(c) Manufacturing Industries*

Climate change impacts on manufacturing industries are felt solely on the nations in terms of their binding commitments under the Kyoto Protocol. It is in this sector that widespread and long-term impacts of leakage of GHG emissions due to international trade are evident. Experts estimate a total of almost 5.3 giga tonnes of $CO_2$, resembling 20 per cent of the total global emissions arising out of trade among the nations. Hence, international trade, is a major cause of anthropogenic climate change, whereby the concern nations should be able to successfully import goods that will not have high intensity of carbon production. Some argue on the practicability of carbon tax on consumption especially in the case of products that are highly carbon intensive in production. In the absence of quantified emission reduction targets for countries, a consumer tax may be a workable solution. In the long run, however, specific 'carbon taxation' both on production and consumption can be an option.

Combating climate change is a global responsibility and factors like deforestation, rampant in India and Asia, are as much responsible for global warming as industrial activities. Further, it is well known that a large part of India's population does not contribute to climate change due to poverty and low access to means that can generate GHG emissions. Finally, India, with a large and diverse agricultural and allied services sector should seriously consider a strategy for adaptation to climate change.

## POLICIES CONTRIBUTING TO CLIMATE MITIGATION

The country under the Kyoto Protocol, India, is not subject to binding emission targets. India has unequivocally rejected taking on quantitative restrictions under any subsequent international collective agreement, but has committed to prevent its per capita emissions from ever exceeding those of developed countries through a range of mitigation and adaptation techniques adopted. The Prime Minister's Council on Climate Change (PMCCC) was constituted in June 2007 to develop India's first ever national action plan on climate change. The policy was published on 30th June 2008, and the salient features of this policy are highlighted in Table 2.

**Table 2** India's Government policy on climate change

| | | |
|---|---|---|
| 1. | Mitigation Policy | • It discusses climate change mitigation & adaptation options for the nation<br>• Mitigation options—energy conservation, renewable energy, CDM measures etc<br>• Adaptation measures—crop improvement, drought proofing, health care, risk financing etc which take over 2 percent of the GDP |
| 2. | Environmental Policy | • Absolves India of any 'significant responsibility' on issue of climate change<br>• Express concerns on trans-boundary pollution issues endorses polluter pays rationale for distorting trade and investment |
| 3. | Energy Policy | • Focuses on sale of GHG emission reductions & carbon credits |
| 4. | Various Plan Documents | • Concentrate on economic opportunities of CDM projects<br>• Supports India's position against voluntary GHG emission reduction targets<br>• No effort of addressing the trade and climate change linkages |

## ENERGY OVERVIEW AND POLICIES

India is the world's fourth largest economy and fifth largest GHG, accounting for about 5% of global emissions. India's emissions increased 65% between 1990 and

2005 and are projected to grow another 70% by 2020 by other measures. India's emissions are low compared to those of other major economies. India accounts for only 2% of cumulative energy-related emissions since 1850. On a per capita basis, India's emissions are 70% below the world average and 93% below those of the United States.

India remains home to the world's largest number of poor people, with nearly 35% living on less than a dollar a day. Its economy is growing rapidly, however, with GDP rising about 8% a year over the past five years. As the economy has grown, emissions intensity (GHGs per unit of GDP) has declined significantly. India's GHG intensity is currently 20% lower than the world average (and 15% and 40% lower than the United States' and China's, respectively). Factors contributing to the decline in energy intensity include improved energy efficiency, increased use of renewable and nuclear power, expanded public transport, and energy pricing reform. With rapid economic growth, rising income, and greater availability of goods and services, energy demand rose 68% between 1990 and 2005, about 3.5% annually. The Government projects energy demand growth of 5.2% a year for the next 25 years, driven by annual GDP growth rates of 8-10%.

Coal accounts for 39% of the total primary energy demand, followed by biomass and waste (29%), oil (25%) and natural gas (5%). The high proportion of biomass and waste reflects the fact that some 500 million people have no access to electricity or other modern energy services. Coal is projected to remain the primary energy source, with demand growing nearly three-fold by 2030.

India has a number of policies that, while not driven by climate concerns, contribute to climate mitigation by reducing or avoiding GHG emissions. Specific estimates of the emission impacts of the policies described below are in most cases not available. However, a recent analysis by The Energy and Resources Institute (TERI) concluded that in the absence of a number of energy policies that are currently being implemented, $CO_2$ emissions would be nearly 20% higher compared to business as usual scenarios in both 2021 and 2031.

Many of these policies are contained in the Five Year Plans developed by the Planning Commission to guide economic policy in India (the 11th Five Year Plan covers 2007-2012). Other policies are found in the Integrated Energy Policy approved by the Planning Commission in 2006 with the broad objective of meeting energy demand "at the least cost in a technically efficient, economically viable and environmentally sustainable manner."

## Energy Policies

### *Renewable energy*

Currently, modern renewable energy constitutes 4% of the total installed capacity of the power generating sector. Between 2002 and 2007, 6800 mewgawatts (MW) of renewable power capacity was added, about 3000 MW more than the 10th Five Year Plan target. The 11th Five Year Plan sets a target of increasing the installed capacity to 23,500 MW by 2012, or more than 10% of total installed capacity, with wind comprising 72% and biomass and hydro power about 14% each.

The Electricity Act (2003) encourages the development of renewable energy by mandating that State Electricity Regulatory Commissions (SERCs) allow connectivity and sale of electricity to any interested person and permit off-grid systems for rural areas. The National Tariff Policy (2006) stipulates that SERCs must purchase a minimum percentage of power from renewable sources, with the specific shares to be determined by each SERC individually. The states of Himachal Pradesh and Tamil Nadu have the highest quotas—20% by 2010 and 10% by 2009, respectively.

Under the Rural Electrification Policy (2006), electrification of all villages must be completed by 2012. Of the 80,000 villages that have no access to electricity, 18,000 villages are in remote areas that must be electrified through the use of renewable energy. Currently, about 3000 villages have been electrified, primarily through solar systems.

- **Wind power:** Wind power comprises over 65% of renewable capacity, ranking India fourth in terms of wind power generation worldwide. The Ministry of New and Renewable Energy estimates the overall potential for wind power at 45,000 MW, with only about 6270 MW currently developed.

  In recent years, the policy framework has been strengthened to reduce upfront costs to investors. Long-term low-interest loans are being provided by the Indian Renewable Energy Development Agency. Cumulative loan approvals amounted to $1.9 billion at the end of 2006. The Central Government allows 80% accelerated depreciation for the first year; concessions on import duties, sales tax and excise duties; and a 10-year income tax exemption for profits from wind generation. Subsidies are also provided for demonstration projects in states where commercial activity has not begun.
- **Solar power:** Solar thermal projects receive financial assistance in the form of capital subsidies, sales incentives, and reimbursement of fees. To encourage foreign investment in solar photovoltaic technology, the government allows an automatic approval procedure for up to 74% of foreign direct investment in joint venture projects. Up to 100% foreign direct investment is permitted if approved by the Foreign Investment Promotion Board. Various subsidies and loans are also available for manufacturers and users of solar power. Raw materials and photovoltaic components are exempt from excise duties and benefit from concessional import duties. In New Delhi, the use of solar water heating systems in certain categories of buildings has been made mandatory. In 2006, a rebate scheme was introduced in the domestic sector to encourage the use of these systems. The government recently introduced a demonstration program to support large grid interactive solar projects. The program sets a feed-in tariff of 12 rupees (about 25 cents) per kilowatt-hour (kWh) for solar photovoltaic power, and Rs.10/kWh (~22c/kWh) for solar thermal power generation through 2009 for qualifying projects.
- **Other renewables:** Biomass projects for power generation receive fiscal incentives including subsidies, income tax, holidays, excise duty, sales tax exemptions, and accelerated depreciation. Currently, the CDM also attracts developers to build biomass projects. Hydropower contributes 33,642 MW (or

26%) of electricity generated in India. The 11th Five Year Plan calls for an additional capacity of 15,585 MW by 2012.

Hydro Development Plan targets 50,000 MW of new capacity by 2025-26. Small hydropower projects (up to 25 MW) are eligible for incentives such as concessional customs duties and income tax exemptions for 10 years.

- **Biofuels:** The Ministry of Petroleum and Natural Gas is implementing a mandatory program for the introduction of ethanol-blended gasoline (5% gasohol) nationwide by April 2008. However, due to fluctuations in the supply of ethanol, the program is currently running behind schedule. The Biodiesel Price Policy (2005) fixed the initial purchase price of biodiesel at Rs. 25/liter (~60c/liter). The Government is formulating a national policy on biofuels to introduce financial incentives, develop R&D for production and commercialization of ethanol and jatropha, and establish a national biofuel development board.
- **Forestry:** In 2005, the forest and tree cover in India was 24%.17. The $11^{th}$ Five Year Plan proposes an increase in the forest and tree cover of 1% a year through 2012. In 2007, the Prime Minister announced the Green India program to reforest 6 million hectares of degraded forest lands.

## INDIA'S NATIONAL ACTION—PLANS

Policy States: ***"Our people have a right to economic and social development and to discard the ignominy of widespread poverty. For this we need rapid economic growth. But I also believe that ecologically sustainable developments need not be in contradiction to achieving our growth objectives. In fact, we must have a broader perspective on development. It must include the quality of life, not merely the quantitative accretion of goods and services. Our people want higher standards of living, but they also want clean water to drink, fresh air to breathe and a green earth to walk on."***

In June 2008, Prime Minister Singh released India's first National Action Plan on Climate Change (NAPCC), outlining existing and future policies and programs addressing climate mitigation and adaptation. The plan identifies eight core National Missions:

- **National Solar Mission:** The NAPCC aims to promote the development and use of solar energy for power generation and other uses with the ultimate objective of making solar competitive with fossil-based energy options.
- **National Mission for Enhanced Energy Efficiency:** Current initiatives are expected to yield savings of 10,000 MW by 2012.
- **National Mission on Sustainable Habitat:** To promote energy efficiency as a core component of urban planning.
- **National Water Mission:** With water scarcity projected to worsen as a result of climate change, the plan sets a goal of 20% improvement in water use efficiency through pricing and other measures.
- **National Mission for Sustaining the Himalayan Ecosystem:** The plan aims to conserve biodiversity, forest cover, and other ecological values in the Himalayan region, where glaciers, which are a major source of India's water supply, are projected to recede as a result of global warming.

- **National Mission for a "Green India":** Goals include the afforestation of 6 million hectares of degraded forest lands and expanding forest cover from 23% to 33% of India's territory.
- **National Mission for Sustainable Agriculture:** The plan aims to support climate adaptation in agriculture through the development of climate-resilient crops, expansion of weather insurance mechanisms, and agricultural practices.
- **National Mission on Strategic Knowledge for Climate Change:** To gain a better understanding of climate science, impacts and challenges, the plan envisions a new Climate Science Research Fund, improved climate modeling, and increased international collaboration. It also encourages private sector initiatives to develop adaptation and mitigation technologies through venture capital funds.

The National Mission - plans will be running through 2017 and directs ministries to submit detailed implementation plans to the PMCCC by December 2008.

## Implementation

India's response to the challenge of climate change can perhaps be best described by referring to the NAPCC, which actually consists of eight separate missions involving both mitigation and adaptation measures. As far as mitigation is concerned, the Solar Energy Mission, which has set a goal of 20,000 megawatts of solar capacity being installed by 2020, is clearly the most ambitious and progressive plan that could have been drawn up in this field. The impacts of climate change would, of course, seriously affect agriculture and availability of water, and the NAPCC will target adequate adaptation measures in both these areas.

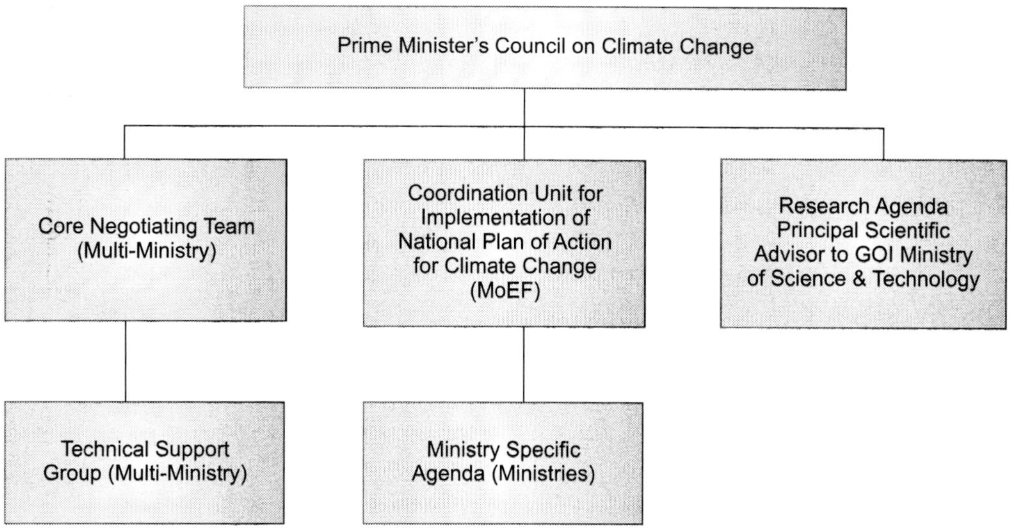


**Fig. 1.** Structure of institutional arrangement of NAPCC.

According to the NAPCC, the 8 National Missions are to be institutionalized by "respective ministries, and will be organized through inter-sectoral groups including, in addition to related Ministries, Ministry of Finance and the Planning Commission, experts from industry, academia and civil society.

Each Mission has been given the task of evolving specific objectives for the remaining duration of the $11^{th}$ Plan and the 12th Plan period (through to 2017). These objectives were to be stated in comprehensive documents along with strategies, plans of action, timelines and monitoring and evaluation criteria, which were to be submitted to the (PMCCC) by December 2008. The Council is to periodically review the progress of these Missions and each Mission is to report its performance publically every year.

The implementation strategy is to be supported by increasing public awareness through media and communication strategy, civil society involvement, capacity building, curricula reform and awards.

## INTERNATIONAL PARTICIPATION

India is a country which is and will continue to be severely impacted by climate change precisely at a time when it is confronted with huge development imperatives. It is expected that the Copenhagen outcome not only provides with the space we require for accelerated social and economic development, in order to eradicate widespread poverty, but also create a global regime which is supportive of our national endeavours for ecologically sustainable development. India has already declared that even as it pursues its social and economic development objectives, it will not allow its per capita GHG emissions to exceed the average per capita emissions of the developed countries. India is resisting the setting of a specific emission reduction target on a global basis for 2050, even though there is enough scientific evidence to show that to keep global warming within 2°C increase (the maximum permissible to avoid possible catastrophic consequences of climate change), this is the minimum reduction required. This stand prevents global action on climate change.

There are proposals which would require developing countries, except the LDCs, to commit to at least 20% to 30% reduction in their GHG emissions from business as usual, while developed countries commit themselves to absolute reductions in their emissions, upto 2020. The cost of such deviation which cannot be met by domestic resources could be posed for international financial and technological support.

India has made written submissions to the United Nations Framework Convention on Climate Change (UNFCCC) on each of the following issues being considered in the negotiations, as a constructive contribution to negotiations. These are as follows:

(i) Submission on long term co-operative action
(ii) Submission on enhancing action on adaptation
(iii) Financing architecture for meeting financial commitments under the UNFCCC

(iv) Submission on technology transfer mechanism
(v) Submission on mitigation actions of developing countries
(vi) Submission on measurement, reporting and verification
(vii) Submission on reduced deforestation in developing countries (REDD), Sustainable Forest Management (SFM), and Afforestation and Reforestation (A&R), Under the Bali Action Plan (BAP)
(viii) Submission on Nationally Appropriate Actions of Developing countries
(ix) Submission on Financing Flows

More submissions will be made as the negotiations proceed.

India has also joined together with its G 77 + China partners to make several constructive contributions to the ongoing multilateral negotiations. While India has announced a NAPCC, it is resisting proposals for the actions specified under the Plan, to constitute commitments by India in a global Climate Change agreement.

It may be difficult for India to accept emission reduction targets on a national basis. These targets can be set, taking into account the different levels of economic development of different countries. This will also address the apprehension of developed countries that assumption of strict emission standards by them would render their industries uncompetitive relative to those in major developing countries.

Overall, a strategic relationship between the United States and India to deal with the challenge of climate change would have benefits not only for the two countries themselves but for the world as well by providing a model for similar arrangements between other developed and developing countries. India is also trying to promote collaborative ventures with the member nations of the South Asian Association for Regional Cooperation, since they face similar challenges, as well as with the European Union (E.U.), which has a major program for funding technological developments involving organizations based in the E.U. and those in "third countries," such as India.

India is a party to both the UN Framework Convention on Climate Change and the Kyoto Protocol. Developing country like India has no binding emission limits under the Protocol. However, India is an active participant in the Clean Development Mechanism (CDM) established by the Protocol. The CDM grants marketable emission credits for verified reductions in developing countries. Developed countries buying these credits can apply them toward their Kyoto targets. India has more than 345 registered CDM projects, more than any other country and about a third of all projects globally (In terms of the overall volume of CDM reductions, China ranks first with 51% followed by India at 14%). The largest project categories are biomass and wind power. Most projects in India are undertaken on a unilateral basis— developed independently by local stakeholders without the direct involvement of the country.

## UNDP IN INDIA: ADDRESSING CLIMATE CHANGE

United Nations Devlopment Programme (UNDP) in India is committed to the cause of protecting the global environment and has been engaged in climate relevant activities for over a decade. The overall objective has been to support the Government of India in meeting its national commitments under the UNFCCC and the Kyoto Protocol, and building national capacity for sustainable use and protection of natural resources. These initiatives are funded through resources from UNDP, GEF (Global Environmental Facility) and resources from the Montreal Protocol (the Montreal Protocol is an international treaty designed to protect the ozone layer by phasing out the production substances believed to be responsible for ozone depletion) to achieve environmental goals in line with the national priorities.

The climate change activities of UNDP India include:

- Technical assistance and capacity building of all stakeholders to address climate change issues and meet commitments under the UNFCCC and the Kyoto Protocol
- Mitigation of GHG with a focus on the energy sector
- Developing adaptation strategies to cope with the impacts of climate change

These initiatives have resulted in enhanced capacities through availability of reliable baseline information (on institutional and technical capacities); strengthened institutional capacity for climate protection and development of GEF and CDM projects (CDM is an arrangement under the Kyoto Protocol allowing industrialized countries with a greenhouse gas reduction commitment to invest in projects that reduce emissions in developing countries), advancement of energy efficiency and cleaner energy technologies and services. In addition to these specific initiatives that directly deal with the global environmental problem, UNDP is including climate change responses in all its programme areas, such as poverty, governance and disaster management.

**Table 3** UNDP in India: Climate change protection

| | |
|---|---|
| **1. Technical assistance and capacity building** | • Technical assistance and capacity building include a project on 'Selected options for stabilizing GHG emissions for sustainable development' which is designed to identify specific investment projects and assess related institutional and policy options for stabilising GHG emissions. UNDP also facilitated the 'Asian Least-Cost Greenhouse Gas Abatement Study (ALGAS) project, the Initial National Communication to the UNFCCC (NATCOM) and the National Capacity Needs Self Assessment (NCSA). As a follow up to the Initial NATCOM, UNDP India is now supporting the implementation of a project to prepare the Second National Communication to the UNFCCC. |
| | • In addition, UNDP India is providing institutional support to the Government of India through the Ministry of Environment and Forests (MoEF) to access environmental financing by strengthening the GEF Cell in the Ministry; and building capacities of state governments and agencies to develop CDM projects. |

**Table 3** Contd.

| | |
|---|---|
| **2. Mitigating climate change** | • UNDP India recognizes the urgency to mitigate GHG emissions through sustainable energy interventions, which should enable growth for all. The organization facilitates activities for GHG reduction by improving energy efficiency in different economic sectors.<br>• Promotion of cleaner and renewable energy fuels is carried out by technology demonstrations. This is also to demonstrate their financial viability. These projects, apart from reducing the GHG emissions also enhance access to energy in rural areas. |
| **3. Promoting energy efficiency** | • UNDP promotes cleaner energy across different sectors of the economy with particular focus on the industrial, commercial and transport sectors. Initiatives have been taken up in energy intensive small and medium enterprises (SMEs). In the steel re-rolling mills, entrepreneurs have received assistance to enhance energy conservation by the use of technologies that emit less carbon.<br>• Similar initiatives in the tea processing units will help reduce the biomass being used as fuel, and in the brick kilns to reduce coal consumption. Within the transport sector, UNDP India is looking at pilot projects in nine cities with a view to improving urban transportation. Another project in the pipeline addresses the energy efficiency of Indian Railways.<br>• In addition, UNDP India has initiated a project to support promotion of energy efficiency standards to help reduce the GHG emissions. It is also partnering with civil society groups to encourage green ratings of buildings. To bring energy and environment consciousness in industry, environmental impact assessments are being promoted. Other programmes that are planned will look at advanced technology in electric motors, green cremation systems, and energy efficiency in new commercial buildings. |
| **4. Cleaner energy fuels** | • UNDP India promotes resource assessments and commercialization of cleaner and renewable energy fuels and technologies – with particular focus on livelihood generation. The technologies include the capturing of coal bed methane (CBM), transforming waste to energy. Key to the success of many of the renewable initiatives has been the involvement of the communities so that the activities can be taken forward. These initiatives were supported through a appropriate technologies (e.g. biogas-based cooking, biomass-based power generation), through capacity building as well as support in financing.<br>• UNDP India supported the first large initiative on small hydro in India. This initiative helped pilot implementation of 20 projects, which led to the development of a database for small hydro plants, and helped rural communities to gain access to energy for enterprises. The project enabled the government to look at the tariff policy for renewable energy projects for |

**Table 3** Contd.

| | |
|---|---|
| | the first time. In recent years, programmes have been supported to convert waste to energy through bio-methanation. The gas generated was used for heating and electricity. The substrates used were solid (e.g. vegetable and slaughter-house wastes, municipal waste, and cow dung), liquid (e.g. industrial effluents) and some hazardous waste such as leather shavings.<br>• An ongoing project 'Biomass Energy for Rural India (BERI)' is addressing the removal of barriers to promote biomass as clean technology. A 'Power Purchase Agreement' was signed between the Gram Panchayat (local governance body) and the electricity distribution company owned by the government for the first time in India.<br>• Collaboration between public, private and gram panchayat is underway to establish a power plant with equity share. In this regard, the Renewable Energy for Rural Livelihoods (RERL) project is an important project. These initiatives align with the Government's national schemes of Remote Rural Energy and Village Energy Security Programmes. The lessons learnt at the policy level would be utilized to replicate the models in the 18,000 un-electrified villages in India.<br>• The Coal Bed Methane project, in the eastern part of India, demonstrates that the large scale production of methane gas in coal mines is technically and economically viable. The methane captured from these mines will be used for power and as fuel for cars. |
| 5. **Other initiatives** | • Under the Small Grants Programme of GEF-UNDP, approximately 60 climate change mitigation projects have been supported. These were designed to remove the cultural, institutional, technical and economic barriers, and promote dissemination of accessible and sustainable measures (technology, policy and practises) at the grass root level.<br>• UNDP is one of the implementing agencies for the Montreal Protocol in India. UNDP has assisted the Government of India in the phasing out of chlorofluorocarbons (CFCs) in the consumption sector. Approximately, 400 metric tons of CFCs in the foam, commercial refrigeration, servicing and carbon tetrachloride (CTC) sectors have been phased out so far. A pipeline of Persistent Organic Pollutants projects (Persistent organic pollutants (POPs) are organic compounds that are resistant to environmental degradation) is being developed with the MoEF. |
| 6. **Adaptation to climate change** | • UNDP India recognizes the importance of integrating effective adaptation strategies into planning at national and local levels. This helps in assessing climatic vulnerabilities, identification of gender sensitive adaptive strategies, expansion of livelihood options of vulnerable communities and building resilience to adverse impacts of climate change. |

**Table 3** Contd.

| | |
|---|---|
| | • UNDP India plans to develop risk transfer options for better response to the climatic impacts, as part of the UNDP India's Disaster Recovery and Management programme. |
| **7. UNDP Country Programme from 2008 to 2012** | • In the new UNDP Country Programme cycle, UNDP India will continue to support the Government's efforts to meet commitments under multilateral environmental agreements through a two-pronged approach. This involves leveraging of additional environmental finance as well as fostering strategic partnerships on the ground that seek to safeguard environmental resources.<br>• UNDP will also support national efforts at strengthening the capacities of governments and communities, to effectively cope with and adapt to climate change. Strategic initiatives and partnerships with a focus on the poor are being developed for greenhouse gases and national environmental benefits. These initiatives build upon ongoing efforts and involve the private sector. Other areas are designing interventions focusing on the poor so that technical solutions reach them and policies are adapted to include their needs. An additional initiative looks at environmental governance and the capacity development of state and district level officials, and elected representatives in the planning, implementation and monitoring of climate change measures. |

## CLIMATE CHANGE AND SUSTAINABLE DEVELOPMENT

Sustainable development has become a part of all climate change policy discussions at the global level, particularly due to adoption of Agenda 21 and the various Conventions resulting from the UNCED-1992. The generally accepted and used definition as given by the Brundtland Commission is "development that meets the needs of the present without compromising the ability of future generations to meet their own needs". Sustainable development has become an integrating concept embracing economic, social and environmental issues. Sustainable development does not preclude the use of exhaustible natural resources, but requires that any use be appropriately offset. This concept is not acceptable to many developing countries since it seems to disregard their aspirations for growth and development. Further, sustainable development cannot be achieved without significant economic growth in the developing countries.

Three critical components in promoting sustainable development are economic growth, social equity and environmental sustainability. The question often asked is, should the current economic growth (GNP, employment, etc.) be sacrificed for long-term environmental conservation? Policy makers in developing countries often perceive a tradeoff between economic growth and environmental sustainability. However, there is a growing evidence to show that environmental conservation for sustainability of natural resources is not a luxury but a necessity when considering long-term economic growth and development, particularly in the least developed countries. The decline and degradation of natural resources such as land, soil, forests, biodiversity and groundwater, resulting from current unsustainable use patterns are likely to be aggravated due to climate change in the next 25 to 50 years. Africa, South Asia and some regions of Latin America are already experiencing severe land degradation and freshwater scarcity problems.

There are many ways to pursue sustainable development strategies that contribute to mitigation of climate change. A few examples are presented below:

**(a) Energy-Efficient Technology:** Adoption of cost-effective energy-efficient technologies in electricity generation, transmission distribution, and end-use can reduce costs and local pollution in addition to reduction of greenhouse gas emissions.

**(b) Renewable-Energy:** Shift to renewables, some of which are already cost effective, can enhance sustainable energy supply, and reduce local pollution and greenhouse gas emissions.

**(c) Forest–Protection:** Adoption of forest conservation, reforestation, afforestation and sustainable forest management practices can contribute to conservation of biodiversity, watershed protection, rural employment generation, increased incomes to forest dwellers and carbon sink enhancement.

**(d) Transportation-Efficiency:** Efficient, fast and reliable public transport systems such as metro railways can reduce urban congestion, local pollution and greenhouse gas emissions.

**(e) Development–Activities Control:** Adoption of participatory approach to forest management, rural energy, irrigation water management and rural development, in general, can promote sustained development activities and ensure long-term greenhouse gas emission reduction or carbon sink enhancement.

**(f) Energy-Economic Policy:** Rational energy pricing based on long-run-marginal cost principle can level the playing field for renewables, increase the spread of energy-efficient and renewable energy technologies, and the economic viability of utility companies, ultimately leading to greenhouse gas emission reduction.

## CONCLUSION

Climate change is one of the most important global environmental challenges, with implications for food production, water supply, health care and energy generation. Climate change affects almost every sector of the economy—agriculture, forestry, power, manufacturing industries and even defense, shipping and marine resources. The UNFCCC and the Kyoto Protocol provisions are addresses on the climate change challenges and mitigating strategies. The Protocol laid out guidelines and rules regarding the extent to which a participating industrialized country should reduce its emissions of six greenhouse gases—carbon dioxide, methane, nitrous oxide, chlorofluorocarbon, hydrofluorocarbons and perfluorocarbons. UNDP in India addresses on climate change are focused. The India's Policies on Climate change and National Action Plan for mitigating the adverse impacts are highlighted in the paper. The effective strategies for mitigating the impact of climate change are to adopt a sustainable development pathway by shifting to environmentally sustainable technologies and promotion of energy efficiency, renewable energy, forest conservation, reforestation, water conservation, etc.

## REFERENCES


As of 31st December 2006. Ministry of Power, 2007. "Report of the Working Group on Power for Eleventh Plan (2007-2012)" Volume II. Government of India

Bolin, B. and Sukumar, R., Global perspective. In *Land use, Land use Change and Forestry* (Eds. Watson, R.T., Noble, I.R. and Bolin, B.), Intergovernmental Panel on Climate Change, Cambridge University Press, Cambridge, 2000.

Bruce, J.P., Lee, H. and Hates, E.F. (1996). *Climate Change 1995: Economicand Social Dimensions of Climate Change*, Contribution of Working Group III to the Second Assessment Report of Intergovernmental Panel on Climate Change (IPCC), Cambridge University Press, Cambridge.

Chandler, W., Schaffer, R., Dadi, Z., Shukla, P.R., Tudela, F., Davidson, O. and Alpan-Atamar, S. (2000). Climate change mitigation in developing countries, Report, The Pew Center on Global Climate Change, Washington DC, October, 2002.Nakicenovic, N. et al., Special report on emissions scenarios. Intergovernmental Panel on Climate Change, Cambridge University

Climate Change (2001). Synthesis Report, Intergovernmental Panelon Climate Change, Geneva, Switzerland, 2001.

Climate Change (2001). The Scientific Basis, Summary for Policy Makers and Technical Summary of the Working Group I Report, Intergovernmental Panel on Climate Change, Geneva, Switzerland, 2001.

*Climate Change 2001: Impacts, Adaptation, and Vulnerability*, Summary for Policy Makers and Technical Summary of the Working Group II Report, Intergovernmental Panel on Climate Change(2001), Geneva, Switzerland, IPCC.

*Climate Change 2001: Mitigation*, Summary for Policy Makers and Technical Summary of the Working Group III Report, Intergovernmental Panel on Climate Change (IPCC), Geneva, Switzerland, 2001.

Data as of 14th July 2008. Data from Capacity Development for the CDM, available at http://www.cd4cdm.org

Data as of 14th July 2008. Data from UNFCCC CDM Statistics, available http://cdm.unfccc.int/Statistics/Registration/AmountOfReductRegisteredProjPieChart.html

Data from IEA, (2007). "$CO_2$ Emissions from Fossil Fuel Combustion 1971-2005"; IEA, 2007. "World Energy Outlook 2007: China and India Insights" and USEPA, 2006. "Global Anthropogenic Non-$CO_2$ Greenhouse Gas Emissions: 1990-2020". All emissions figures are for total greenhouse gas emissions as estimated by the International Energy Agency ($CO_2$ from fossil fuels), the US EPA (other greenhouse gases) and the World Energy Outlook (projections), except for GHG emissions by sector which rely on these same sources but are estimated by Ecofys, 2006 in this report: http://www.ecofys.com/com/publications/documents/Ecofys_Factors_underpinning_future_actionOct2006.pdf. Charts on page 1 use 2005 data.

Economic policy in India is implemented through five-year plans which are developed, executed and monitored by the Planning Commission. The five-year plans are focused on strong GDP growth rate, though recently other targets for economic performance, for instance education and health services, have been included. The most recent 10th

Five Year Plan (2002-2007) aimed at a GDP growth rate of 8%. The 11th Five Year Plan (2007-2012) is currently being developed and the working group papers are available. The fact sheet uses both the 10th Five Year Plan and the 11th Five Year Plan working group papers as reference.

Federation of Hotel and restaurants Association of India [Internet]. New Delhi (India): FHRAI.http://www.fhrai.com. – accessed 20 March 2008.11. Fung, W.Y., Lam, K.S., Hung, W.T., Pang, S.W. and Lee, Y.L. Impact of urban temperature on energy consumption of Hong Kong. *Energy*. 2006 Nov; **31(14):** 2623-2637.

Frontline: India's National Magazine [Internet].Chennai (India): The Hindu, c2008. http://www.hinduonnet.com/fline/ - accessed 19March 2008.

Galitsky, C., Worrell, E., Radspieler, A., Healy, P., Zechiel, S. and Fetzer Vineyards (2005). *BEST Winery Guidebook: Benchmarking and Energy and Water Savings Tool for the Wine Industry*, LBNL 3184.

Garg, A., Ghosh, D. and Shukla, P.R. (2003). Energy sector policies and mitigation of GHG emissions from India. In *Climate Change Economics and Policy: Indian Perspectives* (Ed. Toman, M.), Resources for the Future Publication, Washington DC.

Glisters International [Internet]. Mumbai (India): Glisters Internationals. http://www.glisters.com/rates.html - accessed 10March 2008.

*Global Environmental Outlook*, United Nations Environment Program (UNEP), Oxford University Press, 1999.

Goldemberg, J., Squitieri, R., Stiglitz, J., Amano, A., Shaoxiong, X. and Saha, R., (1996). Introduction: scope of the assessment. In *Climate Change 1995: Economic and Social Dimensions of Climate Change*, Contribution of Working Group III to the Second Assessment Report of the Intergovernmental Panel on Climate Change, Cambridge University Press, Cambridge.

Government of India, (2008). "National Action Plan on Climate Change". Prime Minister's Council on Climate Change.

Government of India, (2008). "National Action Plan on Climate Change". Prime Minister's Council on Climate Change.

Government of India, (2008). "National Action Plan on Climate Change". Prime Minister's Council on Climate Change.

*Human Development Report: 2003*, United Nations Development Programme (UNDP), Oxford University Press, New York.

IGES, (2008). "CDM Country Fact Sheet: India". Institute of Global Environmental Strategies

*India's Initial National Communications to the United Nations Framework Convention on Climate Change*, Ministry of Environment and Forests, New Delhi, 2004.

*Indian Vision 2020*, SP Gupta Committee Report. Planning Commission, New Delhi, 2002.

*International Energy Outlook 2005*, US Department of Energy, Energy Information Administration of Government of United States of America, 2005.

Jacob James, (2005). The science, politics and economics of global climate change: Implications for the carbon sink projects, *Curr. Sci.*, **89(3):** 464-474.

Jepma, C.J. and Munasinghe, M., *Climate Change Policy; Facts, Issues and Analyses*, Cambridge University Press, Cambridge, UK, 1998.

Kshirsagar, N.A., Shinde, R.R. and Mehta, S. (2006). Floods in Mumbai: impact of public health service by hospital staff and medical students. *J. Postgrad. Med.* **52(4):** 312-314.

Kumar, K.R., Sahai, A.K., Kumar, K.K., Patwardhan, S.K., Mishra, P.K., Revadekar, J.V., Kamala, K. and Pant, G.B. (2006). High-resolution climate change scenarios for India for the 21st century. *Current Science*, **90(3):** 334-345.http://www.ias.ac.in/currsci/feb102006/334.pdf -accessed 20 March 2008.

Loske, R., *Scope of the Report: Setting the Stage: Climate Change and Sustainable Development,* Third Assessment Report of Working Group III to Intergovernmental Panel on Climate Change, 1996.

Mills, Insurance in a climate of change. *Science,* 2005, **309:** 1040-1044.

Ministry of Environment and Forests, (2005). "Forest Survey of India". Government of India

Ministry of New and Renewable Energy, (2007). "Annual Report 2006-07". Government of India.

Ministry of New and Renewable Energy, (2007). "Annual Report 2006-07". Government of India

Nair, R., Shukla, P.R., Kapshe, M., Garg, A. and Rana, A., (2003). Analysis of long-term energy and carbon emission scenarios for India. In *Mitigation and Adaptation Strategies for Global Change,* Kluwer Academic Publishers, Dordrecht, vol. 8, pp. 53-69.

National Climatic Data Center (NCDC) [Internet].Asheville (North Carolina): U.S. Department of Commerce.http://www.ncdc.noaa.gov/oa/ncdc.html -accessed on 20 March 2008.

*Our Common Future,* Report of the World Commission on Environment and Development (WCED), Oxford University Press, New York, 1987.

Pielke, R.A.J. (1998). Rethinking the role of adaptation in climate policy. *Global Environ. Change,* **8.2:** 159-170.

Planning Commission, (2006). "Integrated Energy Policy Report of the Expert Committee". Government of India.

Press, Cambridge.

Prüss-Üstün, Annette; Campbell-Lendrum, Diarmid; Corvalán, Carlos; Woodward, Alistair. *Assessing the environmental burden of disease at national and local levels: introduction and methods.* Environmental Burden of Disease Series, No1.Geneva: World Health Organization, 2003.http://whqlibdoc.who.int/publications/2003/9241546204.pdf - accessed on 5 March 2008.

Ravindranath, N.H. and Sathaye, J., *Climate Change and Developing Countries,* Kluwer Academic Publishers, Dordrecht, Netherlands, 2002.

Ravindranath, N.H., Joshi, N.V., Sukumar, R. and Saxena, A. (2005). Impact of climate change on forests in India. *Curr. Sci.,* **90:** 354-361.

Rupa Kumar, K. et al., (2005). High-resolution climate change scenarios for India for the 21st century. *Curr. Sci.,* **90:** 334-345.

Shukla, P.R., Rana, A., Garg, A., Kapshe, M. and Nair R. (2004). *Climate Change Assessment for India: Applications of Asia Pacific Integrated Model (AIM),* Universities Press, Hyderabad.

Shukla, P.R., Sharma, S.K., Ravindranath, N.H., Bhattacharya, S. and Garg, A. (Eds.). (2003). *Climate Change and India: Vulnerability Assessment and Adaptation,* Universities Press, Hyderabad.

Solar PV Technology Programs apply to Solar Street Lighting Systems, Solar Lanterns, Solar Home Lighting Systems/Solar Home Systems, Stand-alone PV Power Plants, Solar PV Water Pumping systems and other applications of PV Technology including new applications.

Tata Energy Research Institute [Internet]. New Delhi (India): TERI, c2008.http://www.teriin.org/teri_news.php?num=11 –accessed on 20 March 2008).

*Tenth Plan Document,* Planning Commission, New Delhi, 2002.

The compounded annual growth rate for the world in the same time period was 1.3%. Data from IEA, 2007. "World Energy Outlook 2007: China and India Insights". International Energy Agency

This category includes industries, hospitals and nursing homes, government buildings, educational institutions and corporate and residential buildings, having an area of 500 sq m. and above.

United Nations Framework Convention on Climate Change 1992 (UNFCCC), Geneva, Switzerland, United Nations Environment Program Information Unit on Climate Change, 1992.

Unpublished analysis prepared for Ministry of Environment and Forests, 2005

World Climate .com [Internet]. Isle of Man: Buttle and Tuttle, c1996-2008.http://www.worldclimate.com/ - accessed 19 March 2008.

World Health Organization. National burden of disease studies: a practical guide. 2nd edition. Geneva: WHO, 2001.

*World Resources 2000–2001, People and Ecosystems: The Fraying Web of Life*, World Resources Institute, Oxford University Press, New York, 2001.

Worrell, E. and Price, L. (2005). An integrated benchmarking and energysavings tool for the iron and steel industry. *Int. J. Green Energy,* in press.

# Index